STRUCTURE AND DYNAMICS OF MEMBRANOUS INTERFACES

STRUCTURE AND DYNAMICS OF MEMBRANOUS INTERFACES

Edited by

Kaushik Nag
Memorial University
St. Jhon's, Newfoundland and Labrador
Canada

A John Wiley & Sons, Inc., Publication

Copyright © 2008 by John Wiley & Sons, Inc. All rights reserved.

Published by John Wiley & Sons, Inc., Hoboken, New Jersey
Published simultaneously in Canada

No part of this publication may be reproduced, stored in a retrieval system, or transmitted in any form or by any means, electronic, mechanical, photocopying, recording, scanning, or otherwise, except as permitted under Section 107 or 108 of the 1976 United States Copyright Act, without either the prior written permission of the Publisher, or authorization through payment of the appropriate per-copy fee to the Copyright Clearance Center, Inc., 222 Rosewood Drive, Danvers, MA 01923, (978) 750-8400, fax (978) 750-4470, or on the web at www.copyright.com. Requests to the Publisher for permission should be addressed to the Permissions Department, John Wiley & Sons, Inc., 111 River Street, Hoboken, NJ 07030, (201) 748-6011, fax (201) 748-6008, or online at http://www.wiley.com/go/permission.

Limit of Liability/Disclaimer of Warranty: While the publisher and author have used their best efforts in preparing this book, they make no representations or warranties with respect to the accuracy or completeness of the contents of this book and specifically disclaim any implied warranties of merchantability or fitness for a particular purpose. No warranty may be created or extended by sales representatives or written sales materials. The advice and strategies contained herein may not be suitable for your situation. You should consult with a professional where appropriate. Neither the publisher nor author shall be liable for any loss of profit or any other commercial damages, including but not limited to special, incidental, consequential, or other damages.

For general information on our other products and services or for technical support, please contact our Customer Care Department within the United States at (800) 762-2974, outside the United States at (317) 572-3993 or fax (317) 572-4002.

Wiley also publishes its books in a variety of electronic formats. Some content that appears in print may not be available in electronic formats. For more information about Wiley products, visit our web site at www.wiley.com.

Library of Congress Cataloging-in-Publication Data:

Nag, Kaushik.
  Structure and dynamics of membranous interfaces / Kaushik Nag.
     p. cm.
  Includes index.
  ISBN 978-0-470-11631-9 (cloth)
  1. Interfaces (Physical sciences) 2. Membranes (Technology) I. Title.
  QC173.4.I57N35 2009
  571.6'4–dc22
                                            2008009440

Printed in the United States of America

10 9 8 7 6 5 4 3 2 1

CONTENTS

INTRODUCTION   vii

PART I   MEMBRANE STRUCTURE   1

1. The Membrane Interface as a Structured Compartment and a Substrate for Enzyme Action   3
   Ole G. Mouritsen, Luis A. Bagatolli, and Adam C. Simonsen

2. ToF-SIMS Imaging of Lipid/Protein Model Systems   19
   Michael Seifert, Mohammed Saleem, Daniel Breitenstein, Hans-Joachim Galla, and Michaela C. Meyera

3. Flexibility and Structure of Fluid Bilayer Interfaces   45
   Michael Rappolt and Georg Pabst

4. X-Ray Diffraction Studies of Lung Surfactant Membrane Structures   83
   Marcus Larsson

5. Neutron and X-Ray Scattering from Isotropic and Aligned Membranes   107
   J. Katsaras, J. Pencer, M.-P. Nieh, T. Abraham, N. Kučerka, and Thad A. Harroun

PART II   DYNAMICS AND MOLECULAR EVENTS AT MEMBRANE INTERFACES   135

6. Interaction of Plasma Proteins with Phospholipids at Interfaces   137
   Chia-Lin Yin, D. Dorcas, Anna Dudek, and Chien-Hsiang Chang

7. Monitoring of Membrane-Associated Protein Binding and of Enzyme Activity in Monolayers at the Air–Water Interface by Infrared Spectroscopy   165
   Sylvain Bussières, Julie Boucher, Philippe Desmeules, Michel Grandbois, Bernard Desbat, and Christian Salesse

8. **Chirality and Dipolar Interactions of Membrane Mimetic Amphiphilic Molecules** 191
   Nilashis Nandi, K. Thirumoorthy, and Dieter Vollhardt

9. **Organic and Inorganic Osmolytes at Lipid Membrane Interfaces** 227
   Peter Westh and Günther H. Peters

10. **Protein Lipid Interactions from a Molecular Dynamics Simulation Point of View** 267
    Christian Kandt, Edit Màtyus, and D. Peter Tieleman

PART III COMPLEX MEMBRANOUS SYSTEMS 283

11. **Molecular Analysis of Bacterial Membranous Systems** 285
    Salim Sioud, Nicolas Joly, Patrick Martin, and Joseph Banoub

12. **Thermodynamics of the Nervous Impulse** 317
    Thomas Heimburg and Andrew D. Jackson

13. **Relationships Between Surface Viscosity, Monolayer Phase Behavior, and the Stability of Lung Surfactant Monolayers** 341
    Joseph A. Zasadzinski, Coralie Alonso, Junqi Ding, Frank Bringezu, Heidi Warriner, Tim Alig, Siegfried Steltenkamp, and Alan J. Waring

14. **A Cursory Glance at the Phyiscochemical Properties of Oppositely Charged Surfactants in Solution and at the Air–Water Interface** 385
    Amiya Kumar Panda and Kaushik Nag

15. **Phase Transitions, Cholesterol and Raft Structures in Films and Bilayers of a Natural Membranous System** 417
    Kaushik Nag, Mauricia Fritzen-Garcia, Ravi Devraj, Ashley Hillier, and Doyle Rose

Index 441

INTRODUCTION

The initial idea for this book occurred over two decades ago, when this editor was a graduate student in an obscure university located in an obscure city in eastern Canada. While attending various lectures on membrane-related topics and Biophysical Society of America meetings, and having a mentor who was an expert in some fields of membranous system, I was fascinated by membranous system research in the late 1980s. This was mainly due to a graduate project on the development of an instrument to observe domains in monolayer films undergoing phase transitions. Over the last century, membrane research had developed classically from the seminal findings of Langmuir, of Gorter and Grendall, and of Singer and Nicholson's fluid mosaic model; then in the early 1990s, proteins and genes had taken center stage in biomolecular research. Lipids had taken a back seat despite the fact that the chemical diversity of lipids in membranes was unknown. Even with the development of enormously sophisticated biophysical technology, only a few membrane protein structures are known to date, and we know very little about membrane structures. I was also quite intrigued with the use of protein analysis and molecular biology techniques in researching the "elusive" proteins such as flippases (which are supposed to maintain lipid asymmetry) when we knew so little about the structure-dynamics aspect of one (cholesterol) or more membrane components or details of their physicochemical behavior. Later, I came to understand that a major hurdle in studying membranes is their physical property of being liquid crystalline material, and thus general biochemical approaches to study such systems were limited. The available tools for studying the physicochemical behavior of membranes, their dynamics, and atomic structures were only accessible to elite "atom smashing scientists" and others doing research in diverse nonbiological fields such as chemical physics or condensed matter. Observing the behavior of various amphipathic molecules using physicochemical methods also required an integrated theoretical framework, which could suggest the molecular structure and dynamic properties of membranous systems such as surfactants, colloids, thin films, liquid crystals, and soft condensed matter. Some of these properties of amphipathic molecules, although complex, had some similarity to the physical properties of "soap." This integrated theoretical framework of membranes, surfactants, liquid crystals, and self-assembling systems being "soft materials" was described by Professor P. DeGennes in the early 1990s. To quote this master of analogies (and a 1992 Nobel laureate): "it is perhaps amusing to note there is similarity in thought between those who study string theories and those of description of soaps." This framework brought about a revolution in our theoretical understanding of soft-material and liquid crystal membranous systems, as well as in defining the polymorphic behavior of such materials.

Although there are a number of excellent review books in the field of cell membrane research as well as specialized books and monographs on surfactants and liquid crystalline materials, the idea that some of these diverse systems can be observed from the viewpoint of "membranous systems" had not yet been developed to date. There are a number of "membranous systems" that are diverse in composition and function compared to cell bilayers. Some are extracellular secretions such as lung surfactant; others, such as lipoproteins, skin barrier and prokaryotic double membranes, and colloidal surfactants, have similarities to the self-assembly and polymorphic behavior of membrane components. For example, lung surfactant, a highly evolutionary conserved lipid–protein secretion from type II lung cells, helps in normal respiration for all air-breathing vertebrates. The internal structures found in the postsecretory lung surfactant suggest a fascinating array of lipid polymorphic structures such as bilayer vesicles, planar multilayers, multilamellate vesicles, and tubular hexagonal type I phases. Others, such as lipoprotein particles, which carry cholesterol and triglycerides in plasma, are membranous in the sense that lipid–protein spherical monolayers hold these structures intact. Other systems, such as the neural axons, have peculiar coiled membranous sheaths and conduct electrical impulses via ion exchange. Bacterial membranes consist of some ubiquitous glycolipids that form double bilayers, remnants of structures of mitochondrial membranes. It has been proposed that mitochondria developed from bacterial insertion in primordial cells. Some of the lipid-associated proteins in these systems, although diverse in function, have similar types of lipid–protein organizations, membrane orientations, structure–function domain organization, physical properties, and lipid dynamics as those observed in studies of membrane models. I believe that research on such diverse membranous systems has greatly enhanced our understanding of not only cellular membranes, but also surfactant, colloids, and other soft materials. Most research on membranes conducted using state-of-the-art physicochemical techniques is also regularly utilized in studies of the physicochemical behavior of soft and hard materials. This book was originally planned to incorporate as many diverse membranous systems as possible, but focusing on recent developments in the field of cellular membrane research. Certain fields such as lipoproteins have not been studied in detail as general membranous systems; also, some authors balked at contributing to a book based heavily on biophysics and structure–function studies of membranous interfaces. The word "interface" as used in the title of this book means mainly the lipid–protein bilayer and film interface with a polar medium, as well as the "interface" between various membranous systems.

Part I of this book is focused on direct experimental studies on the polymorphic structures of models and some natural membranous systems. This area is quite extensive since there has been an explosion of recent research based on the discovery of real structure–function domains or lipid rafts in cell membranes. Over the last three decades, there has been speculation on membrane domains; however, they were finally demonstrated to exist and were imaged in the plasma membranes of cells (see details in Chapter 15). These structure-function rafts were formed by phase-segregated association of sphingomyelin–lipid–cholesterol into nanoscale domains on which membrane proteins reside during function. The chapters in Part I discuss the supramolecular arrangement of lipids and protein in membranes. Although structural domains of specific lipids were first observed in monolayer films using fluorescence techniques, their demonstration and imaging in native and model bilayer membranes was of critical importance to understanding lipid–protein interactions in membranous systems. Thus Chapter 1 discusses the structured domains in model membranes, as well as some functional implications of such structures toward enzyme function. One of the coauthors of this chapter (Dr. Bagatolli) was the first to apply a two-photon bilayer imaging technique for structural observation of domains in freestanding bilayers using a unique giant unilamellar vesicle system. Since I was involved in observing such structures in monolayer films over a decade earlier I was relieved to find such structures were actually found in bilayers as well. The domains (rafts) in free-standing bilayers can be imaged using the fluorescence methods employed for monolayer films; however, due to the extremely complex composition of most natural membranous systems, these "domain-raft" structures are difficult to analyze compositionally. In Chapter 2, a relatively new technique of time-of-flight secondary ion mass spectrometry (ToF-SIMS), which is used to chemically map the domains, is discussed in detail. The method uses sputtering of the surface or interface of a membranous system with an inert ion beam, and thus generating secondary ion fragments or sometimes the complete molecule. Each single secondary ion from a lipid or protein or their fragments is then processed through a mass spectral analyzer and mapped. This allows for compositional analysis of membrane domains and recently has been utilized to understand nanoscale organization of lipids and proteins in rafts. These structures, formed in membranes of different compositions, may allow for specific functional flexibility and are discussed in Chapter 3, where approaches from microscale supramolecular organization to those at the molecular scale are covered. Chapter 4 gives a description of the molecular and atomic scale approach in studying noncellular lung surfactant (LS) membranous structures using X-ray diffraction. The LS system also shows some specific lipid–protein polymorphism and unique phase segregation of lipids, such as hexagonal type I phase structures called tubular myelin as imaged directly at the lung air–water interface and studied using computer simulations. Using cryotransmission electron microscopy and X-ray scattering, the molecular arrangement of lipids of tubular nonlamellar structures can be deciphered and modeled; the role of cholesterol in such systems is discussed in Chapter 4. Although X-ray diffraction has been applied over the last four decades in membrane structural studies, the use of neutron diffraction in such studies in relationship to the atomic scale models developed by X-ray techniques is discussed in detail in Chapter 5. I have purposely allowed the authors of Chapter 5 to include some images of their neutron diffraction (Chalk River, Ontario, Canada) grand scale facility and meta-complex instrumental details ("atom smashing") involved in studying such mundane materials as membrane lipid organization, to suggest at what level researchers are using subatomic particles to comprehend membrane structures.

About 2500 years ago an Eastern ascetic on his enlightened view of the world stated that all "forms are emptiness, and emptiness is nothing other than form." The statement can be justified in the sense that all structures (form) observed in any system are dynamic and eventually dissolve into other forms and thereby are only transitory. I feel this gives a view of the dynamics with which all membrane structures are "formed," evolving into other superstructures as well as finally dissolving into some others or into a general membrane environment. As shown by the "snapshots" of membrane domain structures in Part I, these domains are created by dynamic conditions that exist in the bilayer due to the fast and slow motions of the membrane constituents. The question has always arisen as to what the "equilibrium structures" of domains actually are. At what time frames do the membranous structures such as domains exist for them to function, and how do we study these dynamic events? Recent studies have utilized quantum dot imprinting and single molecule techniques to understand events at different time scales. In Part II, from understanding how soluble plasma proteins may interact with the membranous interface in models of lung disease, as well as how cytoplasmic proteins may interact with the lipid interface (Chapter 6), or how specific enzyme activity leads to single molecular events (Chapter 7), we can get some idea about how membranous structures are related to dynamics at least at the supramolecular level. However, since infrared spectroscopy has allowed us to measure hydrocarbon and amide vibrations that occur at very short bond vibrational time scales, the dynamics of other average motions in lipid and protein molecules in model membranes can also be measured (Chapter 7). In some cases, the chirality of the lipids and amino acids in proteins may be an important factor in membrane organization; domain formation and their dipolar orientation at an interface helps in further understanding how such structurally organized interfaces can be modeled (Chapter 8). One of the most important developments in membrane function has been a model of how various ions (and specifically water) cross the membrane hydrophobic barrier. As the recent discovery of aquaporins (water pores) has demonstrated, a basic understanding of membranes as a "semipermeable" barrier cannot be complete without understanding how such water and ion (pH) balance is maintained across membranes (Chapter 9). All biological energetics or the final formation of the high energy phosphate compounds (adenosine triphosphate) occur at an inner mitochondrial membrane interface via the rotation and dynamics of simple proton pumps, which makes understanding the revised ion barrier dynamics of membranes important even after four decades. Finally, dynamic events and molecular motions at the membranous interfaces are probably best studied using computer simulations, since this method allows for a more atomistic model organization of lipids and proteins, while incorporating the various force fields and degrees of freedom of the molecules at very small time scales, which cannot be deciphered by using a single experimental method

(Chapter 10). Although molecular dynamics simulations are models created by specialized programs, because of the speed with which molecular motions occur at a membranous interface— from single vibrations of bonds in a lipid to lateral and spinning motions of specific or groups of molecules— these highly dynamic events can only be understood via computation (Chapter 10). The snapshots of such dynamic events are thus imaged and modeled using high performance computers and slowed down to our optical ten frames per second understanding of membrane dynamics. With parallel processors, supercomputers, fuzzy logic, and quantum mechanics in MD simulations field calculations, eventually a relatively dynamic view of the membranous system may soon be emerging, which may allow for future evaluation of the specific functional domains.

Part III of this book deals mainly with many complex membranous systems, such as bacterial and neural membranes as well as lung surfactant and other colloidal systems. There are several lipid–protein systems of diverse functionality, which are similar in structure and dynamic behavior to those observed in the cellular bilayers or their models. Some of the earlier studies of plasma membrane models have contributed immensely to our understanding of diverse noncellular membranous systems. Perhaps one of the first models of lung surfactant function was developed from using Langmuir films of extracted (lavage or washed) materials from mammalian lungs. Surfactant is packaged and secreted by type II pneumocyte cells and supposedly forms a highly surface active "mono-molecular layer" or film at the air–water interface. These films reduce the surface tension of the lung interface during respiration and prevent lung collapse. During the late 1950s, when the material was first extracted, Langmuir films were the only tool available to study thin films of LS and the films' low surface tension reducing ability (see details in Chapter 15 for discovery).

One of the major developments in analyzing the complexity and diversity of lipids in membranous systems was the study of lipidomics using mass spectrometry. The method has been refined to a point where whole membranes can be literally "extracted" using organic solvents and the lipidome can be analyzed, where each single lipid species is analyzed to a mass resolution of less than one unit. The bacterial membranes are unusual in the sense that certain species contain a double bilayer as well as a large number of glycolipids and some unusual structured lipids with four to six fatty acyl chains (Chapter 11). Previous analyses of these complex systems were possible by complicated acylation–deacylation reactions, esterification, and fragmentation using gas chromatography. With the development of tandem and "soft" ionization mass spectrometers, the complete lipidomes are mapped; in addition, single species are detected at mass resolutions where the isotropic distribution of carbon-13 in a single species can be discriminated (Chapter 11). The mass spectral method has also been modified to actually map or image the localization of specific lipids or proteins in domains in model and natural membranes as discussed in Chapter 2, thus helping in chemically mapping membrane structures.

The other complex membranous systems dealt with in Part III are on nerve signal conducting pathways (Chapter 12), lung surfactant (Chapters 13 and 15), and general inorganic and organic detergents, which form some of the most peculiar nonbilayer phase structures, due to their general "surfactant-like" nature (Chapter 14). The nerve ("ous"!) systems are dealt with from the viewpoint of thermodynamics, since the phase states of lipids as well as some of the structures involved with these phases are critical for the functioning of the myelin-covered cellular membranous extensions of axons and dendrons.

As editor, I must profess a heavy handedness in dealing with a number of chapters on lung surfactant, since this complex system has recently been rediscovered by membrane physical chemists, physicists, engineers, and nanotechnologists. In Chapter 15, we have tried to speculate on how normal cell membrane research as well as its discoveries regarding phase states, lipid-raft structures, and lipid–protein interactions can easily be compared to a completely functionally different system present in all air-breathing species lungs. After all, fascination with lung surfactant research was planted in my mind by cell membrane studies performed over a century ago. I have also been more fascinated by recent developments in the fields of soft materials, nanotechnology, biophysics, surface and interface science, and many other still emerging fields over the last decade, as pertaining to membranous systems.

My humble hope is that this book will contribute to the reader's imagination in diverse fields and in the broader overview of dynamic yet structured soft materials in nature. If that happens, my work on this volume will not have been in vain. If not, then I humbly apologize for the many mistakes, lack of knowledge, and naivety with which I may have pursued this project from a simple fascination with a system that perhaps played a critical role in the evolution of life on our planet.

I would like to thank a number of people with whom I have had the opportunity to meet and discuss membranes over the last two decades. My initial interest in the area developed about three decades ago, when I was an undergraduate student at Presidency College, in Kolkata, India. Dr. Haripada Chattopadhyay, in a dark, mosquito-infested chronobiological physiology laboratory, described to his students the methods to analyze bilateral respiratory rhythms of humans, and ignited my initial curiosity of things wonderful but yet unknown and their role in the evolution of life. Incidentally, this room was one floor above the laboratory where the Nobel Laureate C. V. Raman had worked on the development of the first Raman spectroscope, as well close to an office space used by Dr. Satyen Bose, who had developed the ideas on Higgs–Boson condensates. I was also fascinated by a lecture on cosmology by Dr. A. K. Roychowdhury of the same institution, which led to my pursuing degrees in physics and human physiology at this institution. Later, in Canada, working in a biochemistry laboratory under Dr. Kevin Keough for a second master's degree and later a doctorate, my scientific imagination was allowed to run wild in trying to develop an instrument to visualize Langmuir films via fluorescence. In fact, this laboratory was the first in the world to synthesize "real" cell membrane phospholipids that had chain melting transitions well below 0 °C and develop theories on interdigitated phases in membranes as well as study them using calorimetry. I spent about seven years during this phase looking at structures, domains, and black "strings and galactic spirals" (Chapter 15) formed in Langmuir films for every possible compositional mixture of lung surfactant. However, to my utter dismay, we discovered that the natural porcine lung surfactant does not behave like any of its single- or multi-component models; instead, it acts as an alloy. My biochemist's interest in the physics of soft materials, of membranous systems, of fractal dimension, of quantum jitters, of chaos, and of superstrings was highly encouraged by Dr. David Pink (St. Xavier University, Antigonish, Nova Scotia), who unfortunately could not contribute a chapter due to physical reasons. I feel immense gratitude toward him and basically this book is indirectly a product of his initial encouragement. (Thank you, David; live long and prosper!). Recently, I developed doubts (more questions?) regarding the real functional role (and not the "laboratory assigned" one) of LS in the lungs as only a "surface-active" agent. Perhaps this material is a unique extracellular membranous system with some yet unknown functions, some components and structures having similarities to the cell membranes. I am not sure anymore after two decades of research.

I am indebted to the Canadian Institute for Health Research (CIHR) for a New Investigator Award and operating grants from the CIHR, National Scientific and Engineering Research Council (NSERC) of Canada, Janeway Children's Health Advisory Committee, as well as major infrastructure funding for the Atomic Force and Raman Microscopes from the Canadian Foundation for Innovation. I would also like to thank Ms. Anita Lekhwani of John Wiley & Sons for her perseverance, encouragement, and patience from the beginning of the project, as well as Ms. Rebecca Amos for help with final publication.

KAUSHIK NAG

St. John's, Newfoundland and Labrador, Canada
April 2008

PART I
MEMBRANE STRUCTURE

CHAPTER 1

The Membrane Interface as a Structured Compartment and a Substrate for Enzyme Action

OLE G. MOURITSEN, LUIS A. BAGATOLLI, and ADAM C. SIMONSEN

MEMPHYS—Center for Biomembrane Physics, Department of Physics and Chemistry, University of Southern Denmark, Odense Denmark

CONTENTS

1.1	Introduction	3
1.2	Models of Lipid Membranes	5
1.3	Lateral Structure of Lipid Bilayers	6
1.4	Enzymology of Secretory Phospholipase A_2 (s-PLA$_2$)	9
1.5	Imaging of s-PLA$_2$ Action on Lipid Bilayers	11
Abbreviations		15
Acknowledgments		15
References		16

1.1 INTRODUCTION

Our picture of the biological membrane has changed considerably over the last few decades mainly due to the advancements in instrumentation that allow us to image membranes with an increasing resolution in space and time. Whereas the picture of the membrane since the introduction of the celebrated Singer–Nicolson fluid-mosaic model [1] has always been one imparting the membrane assembly with considerable dynamics and disorder, the current picture is more refined, describing the membrane as a structured bimolecular and fluid flexible sheet with a certain degree of local lateral organization [2–4] in terms of differentiated lipid domains, in some cases called rafts [5]. Although the lipid-bilayer component

---

Structure and Dynamics of Membranous Interfaces, edited by Kaushik Nag
Copyright © 2008 John Wiley & Sons, Inc.

of the membrane is only about 5 nm thick, it is furthermore associated with a distinct transmembrane structure that is described by the so-called lateral pressure profile [6], which displays variations of local stresses corresponding to hundreds of atmospheres across the 5-nm thick bilayer. It is this highly dynamic and still structured and stressful environment that the proteins and enzymes associated with the membranes have to come to terms with in order to carry out their function.

This insight has led to an increasing understanding of the importance of lipids and lipid structure for cell function in general and for protein and enzyme function in particular. By adding to this picture that certain lipids are now also known to act as signaling molecules for a large range of biochemical processes, it becomes clear why lipids and the emerging fields of lipidology and lipidomics have now moved center stage, and the importance of lipids for life sciences is considered to be similar to that of genes and proteins.

This has led to a revival of the study of lipid–protein interactions and of the mutual influence of lipids and proteins on each other. Questions have arisen not only as to how proteins influence the lipid matrix but also as to how the lipids influence protein structure and function. Specific questions cover the insertion and folding of integral membrane proteins, the oligomerization of the protein segments in the plane of the membrane, the structural stability of membrane proteins, the anchoring of proteins in membranes and at membrane surfaces, and the requirement of specific lipids and a certain lipid structure for optimal protein functionality [4, 7].

However, the study of lipids and lipid membranes is complex because of the subtle elements of order that characterize lipid assemblies. Whereas the properties of genes and proteins are described in terms of well-defined molecular structure, the properties of lipids are characterized by terms like variability, diversity, plasticity, adaptability, fluidity, and complexity. The particular role played by lipids is most often determined by their collective properties—that is, properties that cannot be associated with the individual lipid molecule but are consequences of their interactions and cooperative behavior. Examples of such properties are membrane curvature and curvature stress, transbilayer pressure profile, acyl-chain order parameter, packing density, diffusional motion, phase state, and small-scale lateral organization in space and time characterized by a coherence length or equivalently by an average lipid-domain size. Properties like these require quantitative characterization by use of powerful biophysical techniques.

In this present chapter we review some of the results that have been obtained in our laboratories with regard to characterization of lateral order in model bilayer membranes as well as biological membranes, with particular focus on membrane domains in the submicron regime. The results are based mainly on fluorescence microscopy and two-photon laser scanning microscopy as well as atomic force microscopy (AFM). We then show by a specific example how a small enzyme, secretory phospholipase A_2 (s-PLA_2) becomes activated at membrane interfaces in a way that is controlled by the lateral structure of the lipid-bilayer substrate.

This serves as a clear-cut example of how certain collective properties of a lipid assembly are marshalling the function of a protein, hence highlighting the importance of lipids for membrane function. It furthermore shows how novel instrumentation and imaging techniques open up a new window to allow for quantitative description of biochemical processes.

1.2 MODELS OF LIPID MEMBRANES

The generic properties of membranes are conveniently studied by means of well-defined model systems of varying complexity. Here we shall be concerned with rather simple models, specifically unilamellar or multilamellar vesicles (liposomes) as well as solid-supported single bilayers or double bilayers as illustrated in Fig. 1.1. Dispersions of unilamellar or multilamellar vesicles lend themselves to be studied by bulk techniques, such as calorimetry and spectroscopy, whereas individual giant unilamellar vesicles (GUVs, 20 μm mean diameter) conveniently can be investigated by fluorescence microscopy-related techniques [8, 9] or micromechanical techniques. Single- or double-supported bilayers on planar solid surfaces in water are ideally suited to be imaged by fluorescence microscopy or atomic force microscopy techniques. Use of a combination of the various techniques provides a rather complete picture of lipid-bilayer lateral structure, thermodynamics, thermomechanics, and molecular organization. Needless to say, these models can be composed of different types of lipids, for example, one-component systems, mixtures, or natural lipid extracts from cells, possibly reconstituted with proteins or subjected to enzymes that act on the bilayers. Hence the models reflect some properties of real membranes whereas other properties (e.g., asymmetry) are not represented.

Small unilamellar vesicles are formed by extrusion techniques, whereas GUVs are readily formed by electroformation techniques originally developed by Angelova et al. [10]. Multilamellar dispersions are prepared by simple hydration of dry lipids in buffer.

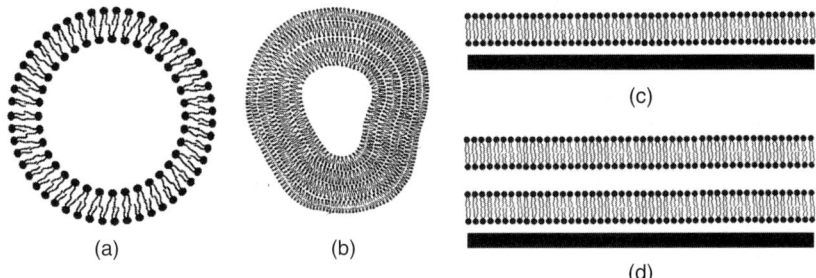

Figure 1.1 Models of membranes in the form of (a) a unilamellar vesicle (b) a multilamellar vesicle (c) a lipid bilayer on a solid support and (d) a double-supported lipid bilayer.

Single-supported bilayers can be formed by a variety of techniques, including vesicle fusion [11], Langmuir–Blodgett deposition [12] Langmuir–Schaefer deposition [13], self-assembly from solution [14], or hydration of thin, spin-coated lipid films [15]. Double-supported bilayers are more delicate to prepare but can result from vesicle explosion followed by appropriate washing procedures [16]. However, this procedure is not robust for all system compositions and we have found that double-supported membranes can be prepared more reliably by the spin-coating procedure mentioned above. A distinct feature of this method is that dehydrated multiple bilayers are formed during the spin-coating process. The subsequent hydration step merely serves to remove excess bilayers from the surface.

A common substrate surface for supported membranes is muscovite mica, which is hydrophilic and atomically planar, a property that is desirable for AFM measurements. Clean mica surfaces are easily prepared by cleaving the crystal, but it is only semi-transparent and therefore only suitable for high-resolution optical microscopy if the crystal is thin (<100 μm). Glass has ideal optical properties and is suitable as a membrane support if appropriately cleaned. However, most glass is nonplanar with an rms roughness of typically 2–3 nm and this roughness is partially superimposed on AFM measurements and may also interfere with lateral domain formation.

1.3 LATERAL STRUCTURE OF LIPID BILAYERS

The distribution of the fluorescent probes in GUVs and planar membranes formed by the same lipid material allows identifying and comparing lipid domains of the same nature and morphology in both systems. Moreover, AFM imaging of the same systems provides additional information on domain formation, which is free of possible interference due to the fluorescent probes. An example of data obtained by these techniques for lipid bilayers composed of POPC and brain ceramide is shown in Fig. 1.2.

Whereas AFM (Fig. 1.2a) furnishes a topological image from which the nature of the domains and phases can only be inferred, and ordinary fluorescence microscopy using probes like DiIC18 fluorescence microscopy (Fig. 1.2b, c) only provides for contrast between different domains, fluorescence microscopy based on for example, LAURDAN generalized polarization (GP) (Fig. 1.2d) allows one to establish structural properties of the coexisting phases, in this case coexisting solid (gel) and fluid domains [17]. Evidence for the solid nature of the domains comes from the LAURDAN emission properties that in turn is reflected in the GP function [9] and is further supported by the observation of the domains being shaped as crystallites with broken domain boundaries.

A more complex case is illustrated in Fig. 1.3 in the case of ternary mixtures of DOPC, DPPC, and cholesterol. In this case cholesterol serves to destabilize solid (gel) phases and introduce liquid-ordered domains whose liquid character is manifested by circular domain interfaces controlled by the line tension of the liquid domains.

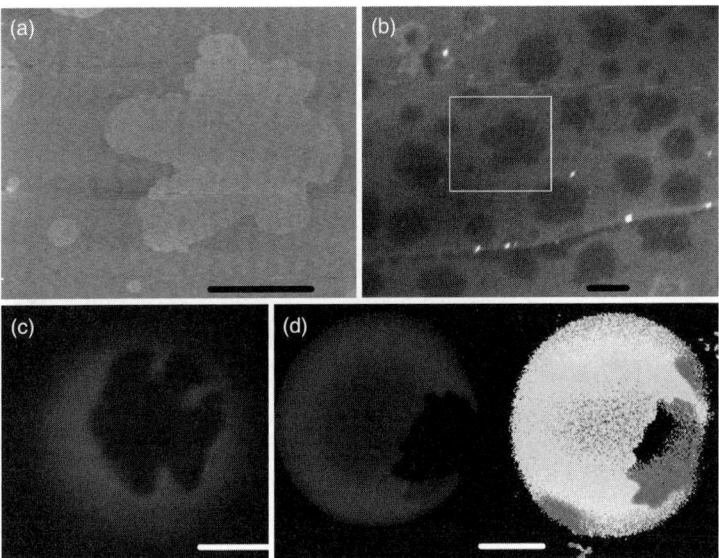

Figure 1.2 (a) Atomic force microscopy (AFM) image and (b) fluorescence microscopy image of solid-supported bilayers composed of a DiIC18-labeled POPC/brain ceramide 5:1 mol mixture. Giant unilamellar vesicles composed of (c) DiIC18-labeled and (d) LAURDAN-labeled POPC/brain ceramide 5:1 mol mixtures. The scale bars are 5 μm. (Adapted from Ref. 17 where the full color representation can be seen.)

The AFM image in Fig. 1.3a has been obtained on a system of a double-supported bilayer. The round domains are clearly seen in the top bilayer. Domains in the bottom bilayer can barely be seen through the top layer. Such double-supported bilayers can be useful to minimize the effect of the support, which, however, also to some extent can be decoupled from the support by using a 150-mM NaCl solution that tends to screen the electrostatic interaction between the lipid bilayers and the mica support. We have found that double-supported membranes are the class of planar model membranes that most closely mimic the behavior found in GUVs. The dark domains of Fig. 1.3a are in the liquid-ordered state and the bright background phase is in the liquid-disordered state, as established by independent AFM measurements. Fluid domains in such double-supported membranes have round shapes that closely resemble the domains found in GUVs (Fig. 1.3b). The fluid domains are highly mobile and they coalesce upon collision with each other. The domains in the double-supported bilayers and the free bilayers in GUVs resemble to some extent domains found in monolayers on air–water interfaces whose details, however, depend on the lateral pressure applied and the fact that they are bounded by the low-dielectric air space.

Finally, we turn to lipid bilayers with a more complex composition related to real biological membranes. Figure 1.4 shows images of the lateral structure

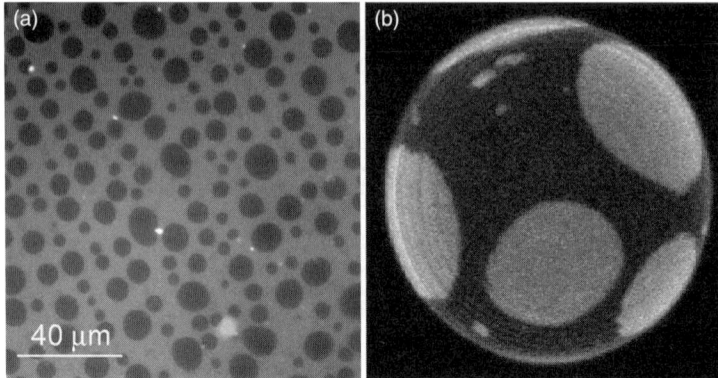

Figure 1.3 Fluorescence microscopy image of a solid-supported double bilayer formed by hydration of (a) a spin-coated lipid film and (b) a fluorescence microscopy image of a GUV both composed of a DOPC/DPPC/Chol of 2:2:1 mixture at $T = 20\ ^\circ$C. The probes DiIC18 and Bodipy-PC were used in the fluorescent experiments. (Adapted from Refs. 18 and 19.)

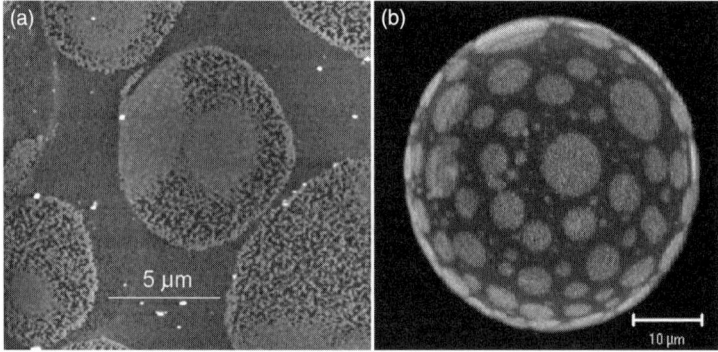

Figure 1.4 Atomic force microscopy image of (a) a solid-supported bilayer and (b) fluorescence microscopy image of a GUV both composed of native pulmonary surfactant membranes. The circular domains correspond to the liquid-disordered phase, and the fine structure seen in the AFM image (a) is caused by the pulmonary surfactant proteins B and C, which are integral membrane proteins. The probes DiIC18 and Bodipy-PC were used in the fluorescent experiments. (Adapted from Ref. 19.)

of bilayers formed by the natural extract of pig lung surfactants that consist of a complex many-component lipid mixture in addition to integral membrane proteins, the SP-B and SP-C pulmonary surfactant proteins [19]. The major lipid component is DPPC. In addition, the bilayers contain about 20 mol % cholesterol (with respect to phospholipids) whose effect has been shown to introduce a fluid–fluid coexistence pattern with round domains at physiological temperatures.

Additionally, in this particular membranous system, extraction of cholesterol but not extraction of the membrane proteins affects the observed membrane lateral pattern. Specifically, extraction of cholesterol generates a pattern that can be linked with the presence of gel/fluid-like phase coexistence observed in model lipid mixtures. The lateral structure observed in the native pulmonary surfactant at physiological temperatures has been related to the functional properties, for example, the spreading capabilities at air–water interfaces [19]. This observation suggests that pulmonary surfactant could be one of the first membranous systems reported where the coexistence of specialized membrane domains may constitute a structural basis for its function.

1.4 ENZYMOLOGY OF SECRETORY PHOSPHOLIPASE A_2 (s-PLA$_2$)

s-PLA$_2$ is a large class of low molecular weight, water-soluble enzymes typically with molar masses below 20 kD. They are found, for example, in venoms, gastric fluids, tear fluid, and inflamed and cancerous tissues. Some act on zwitterionic lipid substrates, whereas others require anionic lipids for being activated. However, they share a common characteristic in that they are interfacially active; that is, they require the lipid substrate to be presented in an organized form, such as a lipid monolayer or bilayer. s-PLA$_2$ carries out its function by catalyzing the hydrolytic cleavage of the acyl ester bond in the *sn*-2 position, thereby producing lysolipids and a free fatty acid.

The s-PLA$_2$ from snake venom has a peculiar enzymology that is quite different from Michaelis–Menten kinetics. Rather than being maximally active in the initial state, these enzymes display zero turnover initially and only after a so-called lag time, do they exhibit a rapid burst in activation. This phenomenon is referred to as lag-burst behavior and is illustrated in Fig. 1.5a [20].

Numerous studies have found that the activity of s-PLA$_2$ is strongly dependent on the structure (or so-called quality) of the lipid substrate and the temperature. In particular, it has been suggested that defects and fluctuations in the layer are nucleation sites for activating the enzyme. Based on rather general considerations, one would expect that there are at least three characteristic contributions to τ as illustrated in Fig. 1.5b: (1) a decrease in τ upon increasing temperature due to Arrhenius activation of the enzyme; (2) an increase in τ upon increasing temperature due to progressive denaturation of the enzyme; and (3) an anomalous minimum at the phase transition of the lipids reflecting the fluctuations in bilayer density in the neighborhood of the lipid-chain melting transition. Depending on the relative positioning and strengths of the three contributions, two possible scenarios for $\tau(T)$ as shown in Fig. 1.5c arise. Obviously, the location of the phase transition is important for determining whether there are one or two minima in the lag-time behavior.

Figure 1.6 illustrates experimental data for cases with either two minima or a single minimum. A single minimum is found for those lipid bilayers where the phase transition temperature is very low and outside the range of study,

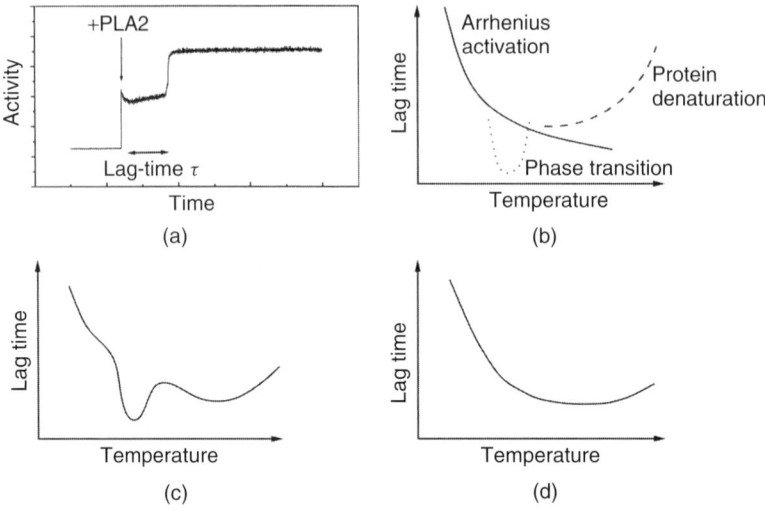

Figure 1.5 Enzymology of s-PLA$_2$. (a) Lag-burst behavior. After a lag period of duration τ (the lag time), there is a sudden burst in enzymatic activity here monitored by the intrinsic Trp-fluorescence of the enzyme. (b) There are three contributions to the lag time: Arrhenius activation that tends to enhance the enzyme activity, protein denaturation that diminishes the activity, and an anomalous variation near the phase transition. The three contributions add up to two different scenarios as illustrated in (c) with two local minima and in (d) with a single minimum in the lag time. (Adapted from Ref. 21.)

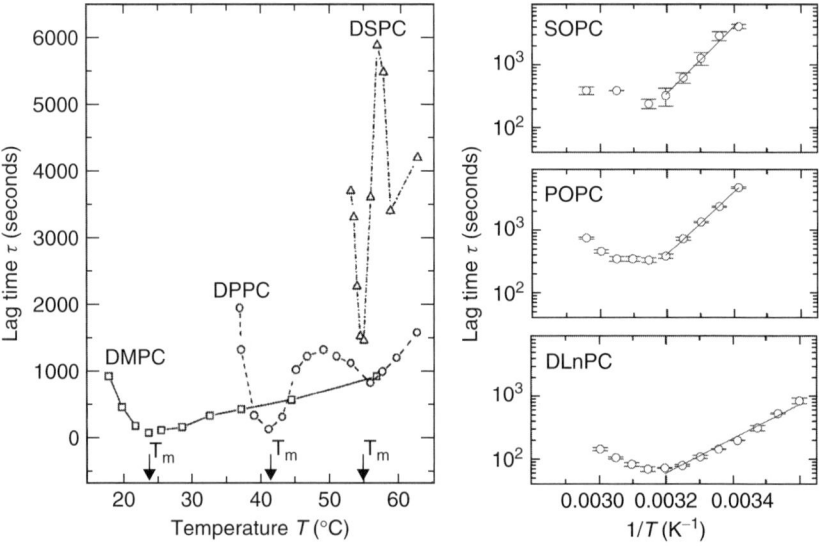

Figure 1.6 Lag time τ, as a function of temperature for the activation of s-PLA$_2$ for a series of saturated lipids (to the left) and a series of unsaturated lipids (to the right). (Adapted from Refs. 21 and 22.)

for example, e.g. for unsaturated lipids [21]. It is noteworthy that the Arrhenius activated branch for the unsaturated lipid bilayers is associated with an activation energy that correlates with the bending modulus of the bilayers; the softer the bilayer the lower the activation energy.

The anomalous thermal variation of the lag time displayed for the three different saturated lipid bilayers in Fig. 1.6—that is, a short chain lipid (DMPC), an intermediate chain lipid (DPPC), and a long chain lipid (DSPC)—shows that the activity tracks the phase transition with a minimum in $\tau(T)$ at the respective transition temperature. The minimum is deeper and broader, the shorter the lipid chains. This corroborates the viewpoint that the stronger the lipid-bilayer fluctuations, the more active the enzyme.

The activation characteristics are furthermore very dependent on the type of s-PLA$_2$. For example, human s-PLA$_2$ type IIA does not exhibit a clear lag-burst behavior and furthermore requires anionic lipids for activation [23]. It has been found that there is a lower threshold in terms of minimum charge density on the bilayer surface for activation of the enzyme. This lower threshold, however, can be obtained by local domain formation of the charged lipid species, even in cases where the global charge density is lower than the threshold (e.g., by bringing the bilayer into a phase coexistence region).

1.5 IMAGING OF s-PLA$_2$ ACTION ON LIPID BILAYERS

The effect of the s-PLA$_2$ on lipid bilayers can readily be investigated by the imaging techniques described earlier on either supported bilayers [24–29] using AFM or fluorescence microscopy, or on GUVs using light [30, 31] or fluorescence microscopy by means of appropriate fluorescence probes [32]. This work supplements the classical work by Ringsdorf and collaborators, who imaged the enzymatic action by fluorescence microscopy in a monolayer assay [33].

In Fig. 1.7 is shown time-resolved fluorescence microscopy images of the lateral structure of a fluid POPC solid-supported lipid bilayer in water subject to s-PLA$_2$ action. The initially smooth, defect-free bilayer with a few large holes (dark) is subjected to s-PLA$_2$, a lag phase follows with very little action, followed by a burst of activity that appears to be nucleated by the preexisting large holes (bottom left), defect lines, and newly formed smaller holes. Eventually, the whole bilayer is digested with some unidentified remains of the products and fluorescent probes at the surface of the support [27].

As shown in Fig. 1.7, the enzyme kinetics of such time-resolved images can be analyzed quantitatively [27, 28] by measuring from the images the areas of the different regions of the bilayer during degradation. The figure shows the total membrane area and its rate of change. Four distinct regimes can be discerned with a clear lag phase and a burst region. After a region of constant activity, there is a linear decrease of area, corresponding to a simple rate equation, $dA/dt \sim -\sqrt{A}$; that is, the area change scales as the perimeter of the domains, which is consistent with the enzyme being activated predominantly at the interfaces that present

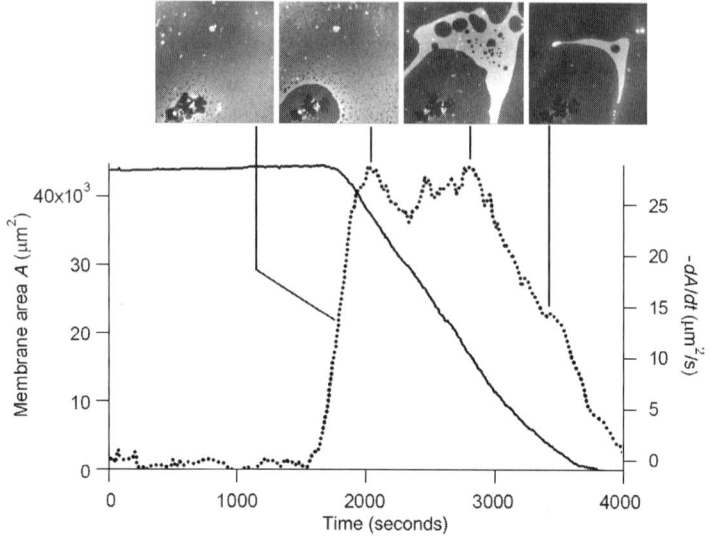

Figure 1.7 Time evolution of the intact bilayer area (solid line) is shown together with the rate of area change (dotted line) of a fluid POPC lipid bilayer subject to the action of s-PLA$_2$. The kinetics is characterized by four distinct regimes: a lag phase, a burst region, a region of constant activity, and a linear region where the area diminishes proportional to the perimeter of the bilayer domains. Typical 120-μm × 120-μm images of the lateral structure in the different regions are also shown. (Adapted from Ref. 27.)

themselves as defect lines to the enzyme. Further details of the enzymatic action can be revealed by AFM imaging techniques [24, 25, 34]. Such investigations have also revealed a lag phase and shown that the enzyme is activated at defect lines, such as the rim of holes or at induced defects caused by the accumulation of hydrolysis products.

A particular kind of well-defined nanoscale defect structure is formed in certain lipid bilayers in their so-called ripple phase (cf. Fig. 1.8) that for PC lipids persists over a range of temperatures just below the main phase transition. These ripples can be imaged by atomic force microscopy provided that the bilayer is sufficiently decoupled from the influence of the solid support (e.g., by sitting on top of another bilayer) [26]. In Fig. 1.8 is shown an AFM image of a binary 1:1 DMPC–DSPC bilayer in the ripple phase. The ripples have a periodicity around 26–30 nm and an amplitude around 2–3 nm.

Upon exposure to the enzyme, the ripples are progressively broadened and their amplitude is increased until a critical point, where the bilayer collapses into a flat state. It appears likely that the lipids most prone to attack are those that reside at the top of the ripples.

Turning now to imaging of s-PLA$_2$ action on GUVs, Figs. 1.9 and 1.10 illustrate the gross morphological changes that the vesicles suffer upon action of the enzyme [32]. Taking advantage of the LAURDAN probe's ability to provide

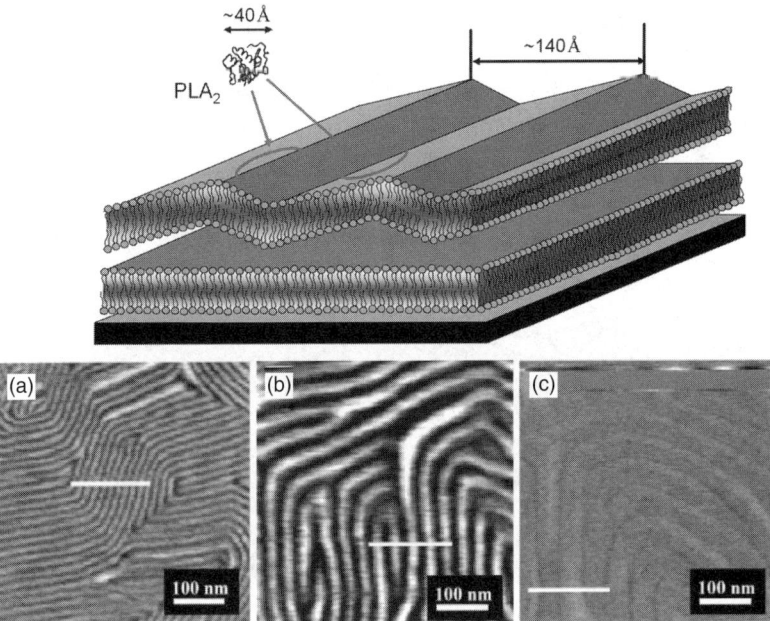

Figure 1.8 Schematic illustration of a double-supported lipid bilayer exhibiting a rippled structure in the top bilayer, which is only weakly coupled to the support. An approaching s-PLA$_2$ molecule is shown to scale. Frames (a)–(c) show time-resolved AFM images of the action of s-PLA$_2$ on solid 1:1 DMPC–DSPC double bilayers at room temperature where the lipids are in the ripple phase. Images are shown for times (a) 0, (b) 77, and (c) 84 min after adding the enzyme. (Adapted from Ref. 26.)

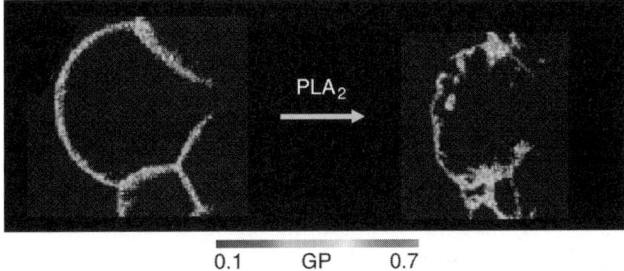

Figure 1.9 Time-dependent changes in the LAURDAN GP of DMPC GUVs upon addition of s-PLA$_2$ (*C. Atrox*) at $T = 26\ °C$. The detailed appearance of the GP images in the figure shows formation of small solid domains (orange dot accumulations) suggesting that these regions correspond to product-rich domains formed upon hydrolysis. Presumably, the fatty acid and lysolipid that remain in the bilayer form the basis for these small, high-GP regions. A high GP is indicative of low water penetration that in turn indicates high lipid order or domain formation. (Adapted from Ref. 32 where the correct color representation can be found.)

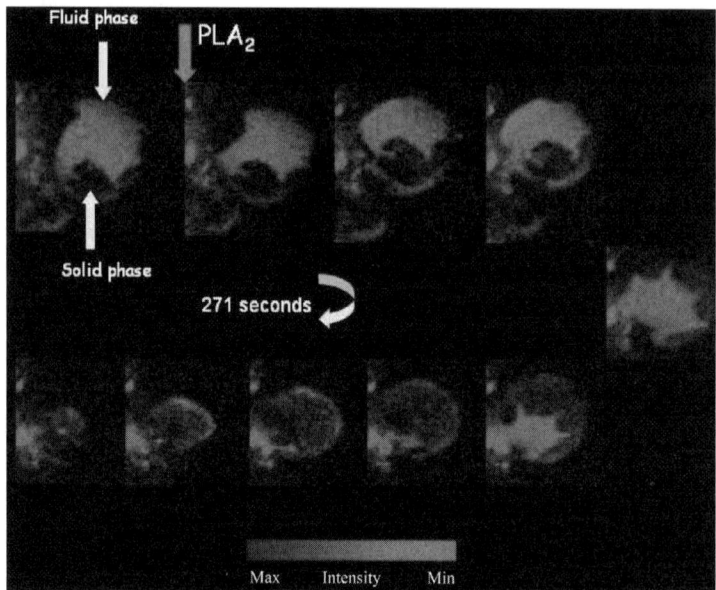

Figure 1.10 Series of two-photon excitation images of 1:1 DMPE–DMPC GUVs at $T = 44\ °C$ before and after the addition of a *C. atrox* s-PLA$_2$. The fluorescent probe is rhodamine-DPPE. The light and the dark areas in the vesicles correspond to the fluid and gel phases, respectively (grey-scale color representation is according to the scale on the bottom of the figure). (Adapted from Ref. 32.)

information on the structural properties of the lipid domains, the images in Fig. 1.9 indicate that solid domains are formed during the enzymatic action on a single-component GUV. These domains are likely to be enriched in hydrolysis products, in particular, the less water-soluble fatty acids that tend to order the lipid chains.

The effects of s-PLA$_2$ action on binary lipid mixtures in the form of GUVs are illustrated in Fig. 1.10. At the particular temperature chosen, the bilayer is in a gel/fluid-phase coexistence region and the images show than the fluid phase is more prone to hydrolysis than the solid, gel-phase domains. As observed in the figure the fluid-phase domains are hydrolyzed faster than the solid-phase domains. This observation was made also for GUVs composed of DLPC–DAPC lipid binary mixture. Consistent with these results, a preferential binding of *Crotalus atrox* s-PLA$_2$ to the fluid-phase domains was also observed in both mixtures.

It is likely that the range of observations of activation of s-PLA$_2$ described earlier can be rationalized in terms of a common and simple underlying molecular mechanism. It has been proposed [35] that such a mechanism could be protrusion of lipid molecules out of the lipid-bilayer surface as illustrated in Fig. 1.11. Molecular dynamics calculations have shown [21] that such protrusions appear on the time scale of tens of picoseconds with a coherence length of about 0.5 nm.

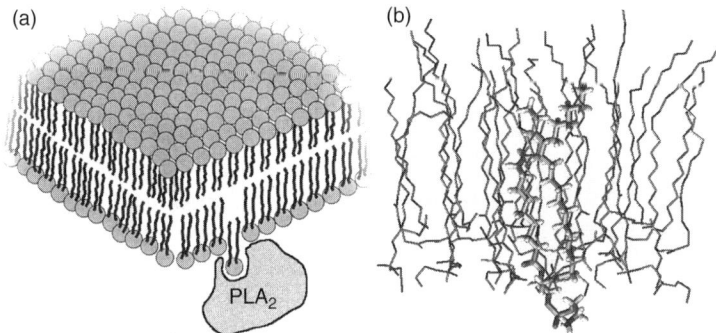

Figure 1.11 (a) Schematic illustration of an s-PLA$_2$ molecule bound to a lipid bilayer. The proposed mechanism of a lipid protrusion as a requirement for activating the enzyme to catalyze hydrolysis is shown. (b) A detailed protrusion event of a lipid molecule as observed in a molecular dynamics simulation. (Adapted from Ref. 21.)

Such protrusions would facilitate the entering of a lipid molecule into the active hydrophobic pocket of the enzyme, which is required to initiate hydrolysis. This model is supported by the various findings of phenomena that tend to activate the enzyme: for example, strong fluctuations, softening of lipid bilayers, and defect formation are all expected to enhance protrusion modes.

ABBREVIATIONS

AFM, atomic force microscopy; Bodipy-PC, 2-(4,4-difluoro-5,7-dimethyl-4-bora-3a,4a-diaza-s-indacene-3-pentanoyl)-1-hexadecanoyl-*sn*-glycero-3-phosphatidylcholine; Chol, cholesterol; DiIC18, 1,1-dioctadecyltetramethyl indotricarbocyanine iodide; DAPC, diarachioyl phosphatidylcholine; DLnPC, dilinoleoylphosphatidylcholine; DLPC, dilaureoyl phosphatidylcholine; DMPC, dimyristoyl phosphatidylcholine; DMPE, dimyristoyl phosphatidylethanolamine; DOPC, dioleoyl phosphatidylcholine; DPPC, dipalmitoyl phosphatidylcholine; DPPE, dipalmitoyl phosphatidylethanolamine; DSPC, distearoyl phosphatidylcholine; GP, generalized polarization; GUV, giant unilamellar vesicle; LAURDAN: 6-dodecanoil-2-dimethylaminonaphtalene; POPC, 1-palmitoyl,2-oleoyl phosphatidylcholine; SOPC, stearoyl-oleoyl phosphatidylcholine; s-PLA$_2$, secretory phospholipase A$_2$.

ACKNOWLEDGMENTS

MEMPHYS-Center for Biomembrane Physics is supported by the Danish National Research Foundation. Chad Leidy has kindly provided us with the illustration of the ripple phase in Fig. 1.8.

REFERENCES

1. S. Singer and G. L. Nicolson (1972). The fluid mosaic model of cell membranes. *Science* **172**: 720–730.
2. O. G. Mouritsen, (2005). *Life—As a Matter of Fat. The Emerging Science of Lipidomics*. Springer-Verlag, Heidelberg.
3. P. L. Yeagle, (Ed.) (2005). *The Structure of Biological Membranes*, 2nd edition. CRC Press, Boca Raton, FL.
4. L. K. Tamm (Ed.) (2005). *Protein–Lipid Interactions—From Membrane Domains to Cellular Networks*. Wiley-VCH, Weinheim.
5. K. Jacobson, O. G. Mouritsen, and G. W. Anderson (2006). Lipid rafts: at a cross road between cell biology and physics. *Nature Cell Biol.* **9**: 7–14.
6. R. S. Cantor, (1999). Lipid composition and the lateral pressure profile in bilayers. *Biophys. J.* **76**, 2625–2639.
7. M. Ø. Jensen and O. G. Mouritsen (2004). Lipids do influence protein function—the hydrophobic matching hypothesis revisited. *Biochim. Biophys. Acta* **1666**: 205–226.
8. S. L. Veatch and S. L. Keller (2005). Seeing spots: complex phase behavior in simple membranes. *Biochim. Biophys. Acta Mol. Cell Res.* **1746**: 172–185.
9. L. A. Bagatolli (2006). To see or not to see: lateral organization of biological membranes and fluorescence microscopy. *Biochim. Biophys Acta* **1758**: 1541–1556.
10. M. I. Angelova, S. Soleau, P. Meleard, J. F. Faucon, and P. Bothorel (1992). Preparation of giant vesicles by external AC fields. Kinetics and application. *Prog. Colloid Polym. Sci.* **89**: 127–131.
11. A. A. Brian and H. M. McConnell (1984). Allogeneic stimulation of cytotoxic T cells by supported planar membranes. *Proc. Natl. Acad. Sci. USA* **81**: 6159–6163.
12. K. B. Blodgett (1935). Films built by depositing successive monomolecular layers on a solid surface. *J. Am. Chem. Soc.* **57**: 1007–1022.
13. I. Langmuir and V. J. Schaefer (1938). Activities of urease and pepsin monolayers. *J. Am. Chem. Soc.* **60**: 1351–1360.
14. B. A. Cornell, V. L. B. Braach-Maksvytis L. G. King, P. D. O. Osman, B. Raguse L. Wieczorek, and R. J. Pace (1997). A biosensor that uses ion-channel switches. *Nature* **387**: 580–583.
15. A. C. Simonsen and L. A. Bagatolli (2004). Structure of spin-coated lipid films and domain formation in supported membranes formed by hydration. *Langmuir* **20**: 9720–9728.
16. C. Leidy, T. Kaasgaard, J. H. Crowe, O. G. Mouritsen and K. Jørgensen (2002). Ripples and the formation of anisotropic lipid domains: imaging two-component supported double bilayers by atomic force microscopy. *Biophys. J.* **83**: 2625–2663.
17. M. Fidorra, L. Duelund, C. Leidy, A. C. Simonsen, and L. A. Bagatolli (2006). Absence of fluid-ordered/fluid-disordered phase coexistence in ceramide/POPC mixtures containing cholesterol. *Biophys. J.* **90**: 4437–4451.
18. E. J. Morris, M. H. Jensen, and A. C. Simonsen (2006). Domain coarsening in single and double supported ternary membranes with coexisting fluid phases. *Langmuir* **23**: 8135–8141.
19. J. Bernardino de la Serna, J., Perez-Gil, A. C. Simonsen, and L. A. Bagatolli (2004). Cholesterol rules: direct observation of the coexistence of two fluid phases in native

pulmonary surfactant membranes at physiological temperatures. *J. Biol. Chem.* **279**: 40715–40722.
20. O. G. Mouritsen, T. L. Andresen, A. Halperin, P. L. Hansen, A. F. Jakobsen, U. B. Jensen, M. Ø. Jensen, K. Jørgensen, T. Kaasgaard, C. Leidy, A. C. Simonsen, G. H. Peters, and M. Weiss (2006). Activation of interfacial enzymes at membrane surfaces. *J. Phys. Condens. Matter* **18**: S1293–S1304.
21. P. Høyrup, T. H. Callisen, M. Ø. Jensen, A. Halperin, and O. G. Mouritsen (2004). Lipid protrusions, membrane softness, and enzymatic activity. *Chem. Phys. Phys. Chem.* **6**: 1608–1615.
22. T. Hønger, K. Jørgensen, R. L. Biltonen, and O. G. Mouritsen (1996). Systematic relationship between phospholipase A_2 activity and dynamic lipid bilayer microheterogeneity. *Biochemistry* **28**: 9003–9006.
23. C. Leidy, L. Linderoth, T. L. Andresen, O. G. Mouritsen, K. Jørgensen, and G. H. Peters (2006). Domain-induced activation of human phospholipase A_2 type IIA: local versus global composition. *Biophys. J.* **90**: 3165–3175.
24. L. K. Nielsen, J. Risbo, T. H. Callisen, and T. Børnholm (1999). Lag-burst kinetics in phospholipase A_2 hydrolysis of DPPC bilayers visualized by atomic force microscopy. *Biochim. Biophys. Acta* **1420**: 266–271
25. M. Grandbois, H. Clausen-Schaumann, and H. Gaub (1998). Atomic force microscopy imaging of phospholipid bilayer degradation by phospholipase A_2. *Biophys. J.* **74**: 2398–2404.
26. C. Leidy, O. G. Mouritsen, K. Jørgensen, and G. H. Peters (2004). Evolution of a rippled membrane during phospholipase A_2 hydrolysis studied by time-resolved AFM. *Biophys. J.* **87**: 408–418.
27. U. Bernchou Jensen and A. C. Simonsen (2005). Shape relaxations in a fluid supported membrane during hydrolysis by phospholipase A_2. *Biochim. Biophys. Acta* **1715**, 1–5.
28. A. C. Simonsen, U. Bernchou Jensen, and P. L. Hansen (2006). Hydrolysis of fluid supported membrane islands by phospholipase A_2: time-lapse imaging and kinetic analysis. *J. Colloid Interface Sci.* **301**: 107–115.
29. P. Moraille and A. Badia (2005). Enzymatic lithography of phospholipid bilayer films by stereoselective hydrolysis. *J. Am. Chem. Soc.* **127**: 6546–6547.
30. J. Y. Lehtonen P. K. J. Kinnunen (1995). Phospholipase A_2 as a mechanosensor. *Biophys. J.* **68**: 1888–1894.
31. R. Wick, M. I. Angelova, P. Walde, and P. L. Luisi (1996). Microinjection into giant vesicles and light microscopy investigation of enzyme-mediated transformations. *Chem. Biol.* **3**: 105–111.
32. S. Sanchez, L. A. Bagatolli, E. Gratton, and T. Hazlett (2002). A two-photon view of an enzyme at work: *C. Atrox* PLA_2 interaction with single-lipid and mixed-lipid giant unilamellar vesicles. *Biophys. J.* **82**: 2232–2243.
33. D. Grainger, W. A. Reichert, H. Ringsdorf, and C. Salesse (1989). An enzyme caught in action: direct imaging of hydrolytic function of phospholipase A_2 in phosphatidylcholine monolayers. *FEBS Lett.* **252**: 73–82.
34. T. Kaasgaard, C. Leidy, J. H. Ipsen, O. G. Mouritsen, and K. Jørgensen (2001). *In situ* atomic force microscope imaging of supported bilayers. *Single Mol.* **2**: 105–108.
35. A. Halperin and O. G. Mouritsen (2005). Role of lipid protrusions in the function of interfacial enzymes. *Eur. J. Biophys. Biophys. Lett.* **34**: 967–971.

CHAPTER 2

ToF-SIMS Imaging of Lipid/Protein Model Systems

MICHAEL SEIFERT and MOHAMMED SALEEM

Institute of Biochemistry, Westfälische Wilhelms-Universität, Münster, Germany

DANIEL BREITENSTEIN

Tascon GmbH, Münster, Germany

HANS-JOACHIM GALLA and MICHAELA C. MEYER

Institute of Biochemistry, Westfälische Wilhelms-Universität, Münster, Germany

CONTENTS

2.1	Introduction		20
2.2	Principle and Setup of ToF-SIMS		20
	2.2.1	Primary Ion Sources	21
	2.2.2	Modulation of Primary Ion Beams: Pulsing, Mass Separation, and Spatial Focusing Process	22
	2.2.3	Impact of Primary Ions on the Probe Surface: Collision Cascade and Sputtering Process	24
	2.2.4	Lateral Resolution in ToF-SIMS Imaging	25
	2.2.5	Separation and Detection of Secondary Fragment Ions by the ToF Analyzer	27
2.3	Application of ToF-SIMS: Analysis of Biomaterial Surfaces		28
	2.3.1	Sample Preparation	29
	2.3.2	Chemical Identification of Biomolecules	29
	2.3.3	Imaging of Lipid/Protein Assemblies	34
	2.3.4	ToF-SIMS in Biomembrane Research	36
2.4	Conclusion and Outlook		39
References			39

Structure and Dynamics of Membranous Interfaces, edited by Kaushik Nag
Copyright © 2008 John Wiley & Sons, Inc.

2.1 INTRODUCTION

Currently, one of the most promising areas in nanotechnology science lays in the preparation and characterization of molecular surfaces and layers. Besides numerous applications in physical or chemical engineering like semiconductors, flat panel (LC/TFT) displays, or polymer brushes, these surfaces gain more and more importance in biomaterials science [1]. Within this research field the assembly of surface-supported lipid membranes or monolayers, including various proteins with essential functions, stands in focus of progressive research [2]. One crucial requirement for an accurate insight into molecular interactions and mechanisms is the topographic, structural, and chemical characterization in a submicrometer scale of these surfaces.

A powerful tool for high resolution surface, interface, and thin-film analysis is the time-of-flight secondary ion mass spectrometry (ToF-SIMS). In principle, SIMS is a destructive technique since bombarding a biomaterial probe with high energetic primary cations leads to the desorption of secondary ions from the surface [3, 4]. However, using very low primary ion dose densities guarantees a quasi nondestructive surface analysis. Emitted fragment ions are detected by a time-of-flight analyzer offering a high mass resolution, high mass range, and precise mass registration. It is even possible to simultaneously obtain the lateral distribution of the different surface components, so-called chemical maps or mass resolved images, by rastering the fine-focused ion beam over the surface [5].

Today, ToF-SIMS is a versatile and well-established analytical technique for material characterization. A wide range of materials can be characterized; this is the reason why ToF-SIMS is applied in a variety of fields such as microelectronics, materials science, nanotechnology, and even life sciences. In this chapter, we outline ToF-SIMS fundamentals and instrumentation, as well its manifold applications and perspectives, especially in the field of lung surfactant and biomembrane research.

2.2 PRINCIPLE AND SETUP OF TOF-SIMS

One of the grand advantages of the ToF-SIMS technique is the consolidation of chemical characterization and lateral imaging of heterogeneous sample surfaces. From techniques like scanning force microscopy (SFM) or scanning tunneling microscopy (STM), topographic conditions of the surface can be gained [6]; devices like X-ray photoelectron spectroscopy (XPS) or Auger electron spectroscopy (AES) are convenient to receive information about the elemental composition of the probe, even in lateral resolution [7]. However, for chemical imaging, not only in terms of elemental analysis but in the more detailed mode of molecular imaging, another technique has to be applied. For this purpose, laterally resolved ToF-SIMS is accounted to be the almost exclusive alternative. In principle, the spatial distribution of all chemical species that constitute the regarded surface can be obtained. In a limited mass range even intact and

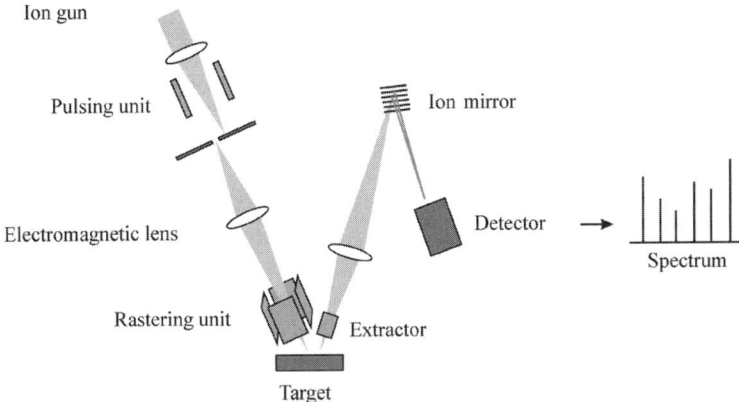

Figure 2.1 Schematic illustration of a general ToF-SIMS setup.

complete molecules (routinely in a mass range up to $m/z = 1000$, in special cases up to 10,000) of the regarded sample surface can be identified and detected [3, 8].

Consequently, such a persuasive and progressive method demands a highly complex device and underlies ongoing improvement and extension. Here, a current overview about the different components and the experimental setup of a ToF-SIMS is depicted and all measuring steps are declared in a convenient order.

Figure 2.1 shows a general setup of a ToF-SIMS used for the analysis of various biomaterial surfaces. In principle, electrically focused, high energetic primary ions are shot onto the sample surface and initiate a collision cascade, which leads to the emission of charged molecules and fragments as well as ions representing the chemical composition of the probe. In the ToF analyzer, an electric field accounts for an acceleration, separation, and detection of these ions in respect to their mass/charge (m/z) ratio.

2.2.1 Primary Ion Sources

One essential component of the ToF-SIMS instrument is the primary ion source. Within the four available types of ion sources (liquid metal ion guns (LMIGs), electron impact (EI) sources, surface ionization sources, and duoplasmatron sources), currently two kinds of sources are commonly used for bioanalysis: electron impact sources on the one hand and LMIGs on the other hand. Here a brief introduction is given, whereas a detailed overview about principles, properties, and applications of such primary ion sources shall not be presented here and can be found elsewhere [9, 10]. Within electron impact sources, accelerated electrons (produced by thermionic emission via heating an electrically charged metal filament surface) collide with the gaseous or vaporized source material and deliver primary ions with energies in the kilo electron volt (keV) range. Noble gases (argon), oxygen, or SF_6 typically serve as source material. Especially, polyatomic primary ions like SF_5^+ received attention due to several advantages like rather high secondary ion yields for the analysis of organic materials. Fullerene ion beams (C_{60}^+, C_{60}^{2+}, C_{60}^{3+}) generated from vaporized buckminsterfullerenes offer excellent secondary ion yields, especially for high molecular weight fragments, but feature rather low lateral resolution in the minor micrometer range.

The other important class of primary ion sources, which were also used in our studies (see Sections 2.3.2 and 2.3.3), are the liquid metal ion guns (LMIGs). A liquid metal flows from a reservoir to a small needle tip, where a high voltage (in the kilovolt range) is applied. Consequently, the liquid metal establishes a Taylor cone and above a certain threshold voltage, a jet of metal cations erupts from the cone similar to the phenomenon of electrospray ionization (ESI) [11]. The ions are accelerated due to their mass/charge (m/z) ratio. Until a few years ago, gallium with its melting point (mp) of 30 °C (and thus predestined as a liquid metal source at near room temperature) was commonly used in LMIGs [3]. In recent years other metals like gold (mp = 1064 °C) and bismuth (mp = 271 °C) gained more attention [4]. These metals have to be warmed up above their melting points. To enable a reasonable handling gold—but not bismuth—has to be disposed in eutectic alloys (e.g., gold–silicon (mp = 370 °C) or gold–germanium (mp = 356 °C)) for decreasing the melting point. The main difference between gold or bismuth sources and gallium sources is the generation of not only monoatomic, singly charged primary ions (Ga^+) but also cluster ions like Bi_3^+, Bi_3^{2+} or Au_3^+ [4]. These cluster formations have some great advantages regarding the generation of secondary fragment ions from the analyzed sample and shall be discussed later, where these mechanisms will be described.

2.2.2 Modulation of Primary Ion Beams: Pulsing, Mass Separation, and Spatial Focusing Process

After generation of primary ions, the ion beam has to be pulsed, mass selected, and spatially focused in order to obtain evaluable mass spectra of secondary ions [11]. Apparently, an unpulsed beam would lead to a continuous strike-out of fragment ions from the sample surface. No reasonable separation of secondary ions according to their m/z ratio and their "time of flight" in the ToF analyzer would then take place. In general, mass resolution (R) indicates the ability to distinguish a mass signal (m) and a neighboring mass signal ($m + \Delta m$). It is defined as $R = m/\Delta m$. Nevertheless, in ToF techniques the full width at half-maximum (FWHM) of a signal is used to define Δm. Furthermore, mass and charge discrepancies exist within the primary ion cluster, which lead to varying accelerations and velocities of these ions. As a result, lateral resolution of the spectrometer is reduced if the ion beam was not spatially focused with a convenient electrical lens system.

Special devices are needed to achieve the described requirements for a primary ion beam. One possibility to pulse and specifically select one single species of ions or ion clusters is the application of double blanking plates [4]. In principle, the ion beam passes through two electrical deflector fields (double blanking

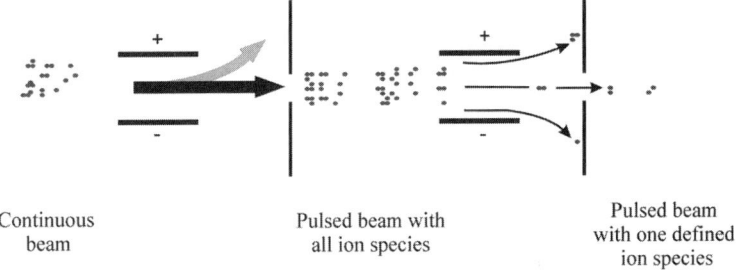

Figure 2.2 Schematic design of blanking plates.

plates), which are operated alternately (Fig. 2.2). First, a deflection plate is electronically pulsed across a small aperture, which leads to primary ion pulses of 10–100 ns. Then the second plate is operated after a controlled time delay letting pass only the primary ion species with a defined m/z ratio. After pulsing and specific mass selection, the ion beam is focused by application of an electrical lens system in order to obtain a laterally defined bombarding area on the sample surface.

The just described operating mode of the ToF-SIMS spectrometer is termed unbunched mode or "burst alignment mode" [12]. It is characterized by a high lateral resolution (down to <100 nm) and is therefore applied for imaging of surfaces.

One of the critical parameters to obtain a high mass resolution is the length of the primary ion pulse. A short pulse results in a short-term collision cascade and a well-defined starting time for the secondary ions. In principle, by using a short unbunched primary ion pulse, a high mass resolution is achievable. Nevertheless, the limited number of primary ions in each pulse will necessitate numerous repetitions of analysis cycles in order to obtain acceptable secondary, ion intensities. This would result in unacceptable long measuring times. Therefore, for spectra acquisition, most frequently a "bunched mode" is used, reducing the unbunched pulse below 1 ns. Such a bunching device consists of two back-to-back plates each with an aperture on the beam axis. When the ion pulse passes the space between the plates, a voltage is applied from the rear plate and the ions are accelerated. The rear ions of the beam are exposed to the accelerating electrical field for a slightly longer period of time and reach velocities somewhat larger than those of the ions in the front of the beam. This bunching process can be optimized in a way that all primary ions of different speed hit the probe at the same time. By application of buncher devices, the pulse duration can be decreased from several tens of nanoseconds to shorter than 1 ns by concurrent maintenance of convenient primary ion doses [4], which denotes possible mass resolution ($m/\Delta m$) of >10.000 in the bunched mode [13]. However, lateral resolution is low (only 2–5 μm) because of chromatic aberration of the electrostatic lens system, resulting from the different speed of the primary ions [12]. This constitutes primary ions with different velocities not being brought to focus at the same point behind the lens system.

Both operating modes of the spectrometer can contribute to ToF-SIMS analysis with their respective strengths. A combined use of either technique leads to a very precise and convenient analysis of probe surfaces. For example, the creation of databases for fragment ions and m/z ratios of varying substrates is accomplished in the bunched mode with high accuracy. By comparison with such databases, the use of the laterally higher resolved burst alignment mode can be applied to gain the exact spatial distribution of different chemical components of a surface probe.

2.2.3 Impact of Primary Ions on the Probe Surface: Collision Cascade and Sputtering Process

The modulated primary ion pulse hits the intrinsic sample surface and leads to a striking out of mainly uncharged fragment molecules and secondary ions, which are representative for the probe material. Nevertheless, a small fraction of 10^{-1} to 10^{-6} of the emitted fragments is either positively or negatively charged. Most of the produced secondary ions are singly charged, whereby the charge sign depends on the electron configuration of the particles [1, 5]. The limited number of available charged particles necessitates an effective and accurate detection.

When hitting the sample surface, a primary ion transfers its kinetic energy to the molecules and atoms of the probe (Fig. 2.3). A so-called collision cascade starts, whereby the deposited energy puts the molecules and atoms into motion, a process that also includes bond breaking. The energy transfer occurs either directly from the primary ion to the probe or indirectly by collision of already affected and moving sample molecules and atoms with other molecules and atoms. The collision cascade spreads without preferential direction isotropically from the original impact position of the primary ion. The cascade reaches

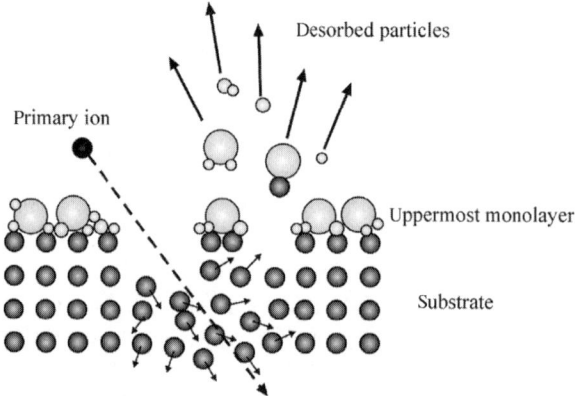

Figure 2.3 Scheme of collision cascade.

a certain lateral (cross section) and perpendicular (depth) dispersion. By moving away from the impact position (either lateral or perpendicular), the density of transferred collision energy becomes less and finally ceases. While proximate to the impact position mainly atomic fragments and ions are generated, along the collision cascade the decreased energy density leads to the emission of larger molecular fragments and ions or even complete molecules of the regarded surface. A certain ratio of these particles, which is statistically determined, acquires an impulse in the direction of the sample surface and concurrently features enough energy to emit from the sample surface [4, 5, 11].

One has to be conscious of the fact that the impact of primary ions with the resultant desorption of fragment molecules and ions representing the analyzed surface leads to ablation of material from the probe. An overvalued primary ion dose leads to a significant alteration of the surface. The intrinsic chemical nature of the sample will be destroyed and the quantities of detected ions will not represent the originally prepared probe surface. Image contrast and resolution of specific ion distribution can be altered or decreased by prolonged primary ion bombardment. The prevention of such probe damage can be ensured by limiting the primary ion dose to an amount that influences only about 1% of the surface structure. In this "static" SIMS regime, the alteration of the probe is expected to be neglectable and results in surface images that are representative for the original probe composition [14, 15].

Additionally, the detected fragment ions refer almost exclusively to the near surface region, which leads to an information depth of only one or a few monolayers [16]. In monomolecular films, the amount of molecules matches the range of 10^{14} particles/cm^2 [5]. The demanded primary ion dose density for "static" SIMS is on the order of 10^{11}–10^{13} ions/cm^2 [11], depending on different factors influencing the secondary ion yield such as ionization probability of the respective surface molecules or the primary ion efficiency.

Exceeding the "static" SIMS primary ion dose leads to the "dynamic" SIMS mode, whereby a systematic removal of surface material takes place. By application of "dynamic" SIMS, depth profiles of the respective probes can be gained. However, the drastically increased primary ion dose leads to the formation of mainly atomic particles and ions and represents an elemental analysis rather than an analysis of molecular surface components [3].

2.2.4 Lateral Resolution in ToF-SIMS Imaging

The already mentioned imaging process, performed with the burst alignment mode, offers an intrinsically advanced lateral resolution at the expense of a decreased nominal mass resolution compared to the bunched mode. The reason for this is the elimination of the buncher device and thus the avoidance of a high chromatic aberration in the following lens system resulting from the broad spread of primary ion velocities. But even by omitting the buncher, the primary ion pulse underlies a certain energy bandwidth. As a result, a certain ion pulse collision area cannot be undercut. Other device parameters, like quantity of emission current of the primary ion source or beam pulsing for accurate mass determination, influence and finally decrease the achievable lateral resolution of the instrument. These characteristics will not be described in detail here. Explicit information can be found elsewhere [3, 17, 18]. In the (virtual) case of a physically infinite focused primary ion beam, the lateral resolution of one single primary ion would be determined by the expansion of the above described collision cascade. This extent lays in the range of 2–10 nm, depending on primary ion properties (mass, energy, mono- or polyatomic character) as well as sample properties and chemical composition, respectively [3, 14, 15]. By taking account of the mentioned device parameters, one has to be aware that an achievable lateral resolution is in the range of 50–80 nm, which corresponds to the achievable focus of the ion beam and is not on the order of the collision cascade dimension anymore, but is still highly resolved [3].

Nevertheless, in organic samples the achievable lateral resolution often does not depend on the achievable focus spot size. It depends rather on the number of ions that can be generated from a given area. Therefore the term "useful lateral resolution" is used. Technically spoken, it is the side length of a square where a predefined number (most often four is chosen) of ions from one secondary ion species can be detected. The useful lateral resolution for most organic samples and analytes is in the range of 350 nm [19].

The mathematical definition necessitates the introduction of some crucial parameters based on primary ion and sample characteristics. The secondary ion emission yield (Y) is calculated as the number of detected secondary ions (N_D) by the number of primary ions hitting the sample surface (N_P) during mass spectrum achievement:

$$Y = \frac{N_D}{N_P} \quad (2.1)$$

where N_D and N_P are the number of detected secondary ions and the number of primary ions, respectively.

Furthermore, the term "disappearance cross section" (σ) has to be introduced. As a result of primary ion impact, the surface coverage of a certain analyte is reduced by desorption or fragmentation. Under prolonged bombardment, the signal intensity of a certain signal therefore decays exponentially following the equation

$$N_D = N_{D,\,t=0} \exp(-\sigma_D \cdot PIDD) \quad (2.2)$$

where $PIDD$ is the primary ion dose density (i.e., the number of primary ions per analyzed area). The disappearance (or damage) cross section (σ) is given by the slope of the exponential decay and corresponds to the mean area from which no further secondary ion (N_D) can be generated after primary ion impact.

Naturally, Y and σ are determined by different factors of primary ion and probe attributes. A detailed overview about the mathematical background is given elsewhere [3, 4, 19]. The termed parameters describe the essentially important "useful lateral resolution" (Δl) by the relation

$$\Delta l = \left(\frac{N_D \sigma}{Y}\right)^{1/2} \quad (2.3)$$

As already mentioned, gold or bismuth cluster primary ions feature some advantages compared to monoatomic ion sources like Ga^+ or Ar^+. Consequently, recent ToF-SIMS studies are often accomplished using these cluster ions sources. For example, analysis of lipid/protein compositions for modeling lung surfactant systems were conducted using Bi_3^+ cluster ion sources as is described in more detail in a later section. One important attribute for estimating primary ion quality is the secondary ion efficiency E, which is given as

$$E = \frac{Y}{\sigma} \quad (2.4)$$

Comparing the monoatomic Bi^+ with Bi_3^+ cluster, the polyatomic ion offers about five times higher secondary ion efficiency for organic analytes than the monoatomic projectile [19]. Naturally, the cluster exhibits a greater disappearance cross section σ than a single Bi^+ ion, but this effect is outperformed by the intrinsically higher secondary ion emission yield Y. In addition, the disappearance cross section σ of the Bi_3^+ cluster (4.14×10^{-13} cm^2) is obviously smaller than that of three single and separated Bi^+ ions ($3 \cdot 2.75 \times 10^{-13}$ cm^2) [19].

Whereas it has to be pointed out that the desorption and ionization process of secondary particles is not completely understood, a common hypothesis for the explanation of this effect is the near surface deposition of the impact energy: The polyatomic projectile is expected to break apart leading to a multiple and simultaneous bombardment of a surface area by the primary ion fragments having a lower kinetic energy each than a respective monoatomic projectile.

2.2.5 Separation and Detection of Secondary Fragment Ions by the ToF Analyzer

Separation with respect to their m/z ratio and detection of ionized species is performed by a time-of-flight analyzer. First, the ions are accelerated in the accelerating section by a homogeneous electric field [11]. Ions with a given charge acquire the same nominal kinetic energy,

$$E_{kin} = zUe \quad (2.5)$$

where z is the net charge of the secondary ion, e is the elementary charge, and U represents the applied voltage of the electric field. The kinetic energy E_{kin} of an ion and its velocity v are linked by

$$E_{kin} = \tfrac{1}{2}mv^2 \quad (2.6)$$

where the velocity of the ions is given as

$$v = \left(\frac{2E_{\text{kin}}}{m}\right)^{1/2} = \left(\frac{2zUe}{m}\right)^{1/2} \quad (2.7)$$

with m as the mass of the regarded fragment ion. The ions exhibit different velocities due to their mass and charge, represented as their m/z ratio:

$$\frac{m}{z} = \frac{2Ue}{v^2} \quad (2.8)$$

When entering the drift section, the ions fly a given distance (L) with their constant velocity, thus reaching the detector at differing times (t) of flight:

$$v = L/t \quad (2.9)$$

Linking Eqs. (2.8) and (2.9) and keeping in mind that U and L are constant, the proportionality of flight time (t) and mass to charge ratio (m/z) gets conspicuous:

$$\frac{m}{z} = \frac{2Uet^2}{L^2} \quad (2.10)$$

According to Eq. (2.10), highly charged and light fragments attain the detector after a shorter time of flight than the bigger ones.

Taking account of the above described sputtering process, one should remember that the collision cascade leads to a varying quantity of energy transfer to the probe material and thus to a certain velocity spread of desorbed fragment ions before reaching the accelerating unit. This discrepancy is still existent after leaving the acceleration unit and denotes a velocity bandwidth for one single kind of secondary ion when entering the drift section. This velocity discrepancy would worsen the mass resolution. It can be minimized, for example, by application of a so called ion mirror unit, where the flight direction of the ions is inverted by application of an additional electric field. Secondary ions with higher velocities penetrate this field further than the slower ones and thus feature a slightly longer invert curvature. Accurate modulating and adjustment of the ion mirror field leads to accumulation of one kind of secondary ions, which guarantees an optimal, isochronal detection for optimal mass resolution [11].

2.3 APPLICATION OF TOF-SIMS: ANALYSIS OF BIOMATERIAL SURFACES

When we began applying ToF-SIMS in the 1990s to analyze the distribution of lung surfactant components in Langmuir–Blodgett (LB) films, mainly organic molecules or polymers on solid supports had been studied with the combination of these techniques [17, 18]. ToF-SIMS had also successfully been applied for the detection, identification, and characterization of biomolecules such as peptides, vitamins, and nucleotides way back in 1978 by Benninghoven and Sichtermann [20]. Later, the same group applied this technique for structural characterization of the mutant apolipoprotein A-I and its variants [21, 22].

The ToF-SIMS technique had then already advanced to such an efficient tool for the characterization of a variety of materials immobilized to metal substrates that an application to lung surfactant model systems was enticing. Especially when imaging of surfaces with high lateral resolution became possible, we aimed at chemically resolving the domain structures in lipid/protein monolayers. From fluorescence microscopy experiments in combination with film balance measurements, we already had a good picture concerning the phase behavior of our lung surfactant model systems [23]. However, due to the limitations of microscopic techniques, we were not able to gain any information on the exact localization of individual molecules within specific phases. The rapid development of mass spectrometric imaging techniques with high lateral resolution promised to overcome this impediment and to provide the information we were seeking. In this section we want to give an overview of the evaluation of ToF-SIMS results and data interpretation using surfactant model systems as an example. Additionally, analytical capabilities for studying membrane systems are illustrated and applications in the biomembrane research field are discussed.

2.3.1 Sample Preparation

Lipids and also proteins can easily be analyzed in the bulk state or as dried films on a solid support to obtain ToF-SIMS mass spectra. For imaging of biomaterial surfaces, however, uniform thin films have to be prepared. LB technology is especially suitable since it enables the preparation of well-defined and highly ordered films of only one or more layers on a solid support. Lipids or lipid/protein mixtures are spread from organic solvent onto an aqueous subphase of a film balance and then transferred to the substrate. In order to obtain optimal ionization and to avoid charging of the surface due to bombardment of the surface with charged primary ions, LB films are ideally prepared on gold-coated glass [24, 25] or mica sheets [15, 26]. Morphology of the monolayers covering the subphase is generally conserved when transfer ratios are close to 1 [24, 27]. A major advantage of this technique is that LB films transferred on gold or mica can be investigated by surface analytical tools other than ToF-SIMS (e.g., SFM or SEM). The respective images obtained by these methods aid in verifying and complementing the results obtained from film balance and fluorescence microscopy measurements [27, 28]. The combination of different techniques is therefore necessary to provide reliable information on phase behavior of lipid/protein mixtures and lateral distribution of the different components.

2.3.2 Chemical Identification of Biomolecules

In order to characterize the fragmentation pattern of individual molecules and to identify typical secondary ions characteristic for the analyte, the so-called

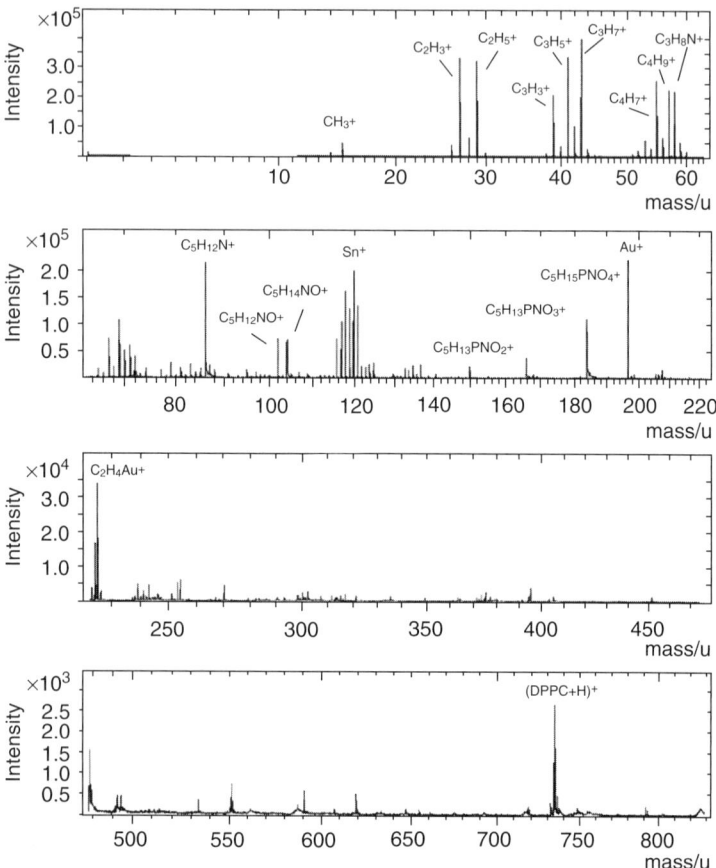

Figure 2.4 ToF-SIMS spectrum of positive SI from a DPPC monolayer.

fingerprint ions, basic studies on one-component systems have to be performed. An example for a ToF-SIMS spectrum of positive SI of DPPC is presented in Fig. 2.4. To achieve high mass resolution, the primary ion pulse was bunched. The fragments with m/z ratios of 102, 104, 125, 150, 166, 184, and 224 result from cleavages in the headgroup region (Fig. 2.5) and can therefore be used to identify this lipid. The fragmentation observed for DPPC corresponds well to the one obtained by other mass spectrometric methods [29, 30]. In the molecular ion region, DPPC is detected both as protonated and dehydrogenated quasimolecular ions (DPPC-H)$^+$ and (DPPC + H)$^+$ with nominal masses of 732.5 and 734.5 u/e, respectively. Similar molecular ion peaks resulting from DPPC have also been identified with MALDI [31]. Interestingly, at higher ion masses, nonspecific SI are found belonging to Au$^+$ (M197) and hydrocarbon fragments associated to gold (e.g., C$_2$H$_4$Au$^+$ with $m/z = 225$ u/e). They result from the solid support on which the LB film was prepared.

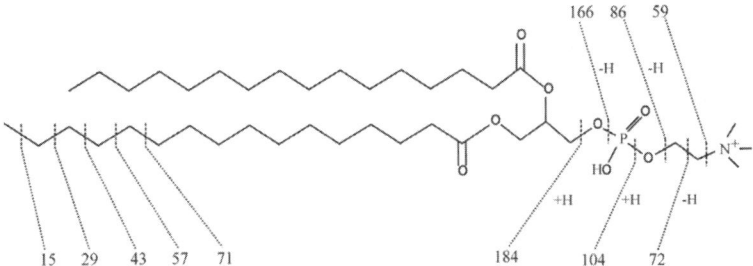

Figure 2.5 Fragmentation of DPPC. Typical cleavage sites for the formation of secondary ions are indicated.

Most of the fragments in the mass range below 100 u/e are nonspecific hydrocarbon SI and nitrogen-containing SI, which can be attributed to the fragmentation of the choline headgroup. Unfortunately, interference of some of these secondary ions occurs. For instance, $C_4H_8N^+$ (70.065 u/e) and $C_5H_{10}^+$ (70.078 u/e) both have a nominal mass of 70 u/e. Nevertheless, at the achieved mass resolutions ($R = 8000 = m/\Delta m$ in this case of M70) these fragments can be differentiated, as signals with a mass difference of $\Delta m = 0.009$ can be distinguished. Therefore $C_5H_{10}^+$ (70.078 u/e) and $^{13}CC_4H_9^+$ (70.074 u/e) could not be separated in the ToF analyzer. Also, taking into account the isotope contribution of carbon, where the portion of ^{13}C amounts to 1.1% per C atom and increases with the number of carbon atoms, at least 3% of the intensity of M70 is due to $^{13}CC_4H_9^+$.

However, even at nominal mass resolution the signal at a m/z ratio of 70 u/e could be assigned to $C_4H_8N^+$ which in this pure DPPC sample could only be derived from the DPPC headgroup. Because $C_5H_{10}^+$ has an unpaired electron, it contributes only 7% to the intensity of M70. This can be described by the nitrogen rule, which denotes that only nitrogen-containing SI with paired electrons and even nominal masses yield intense peaks [32]. Hydrocarbon ions, however, generally possess odd masses or odd numbers of hydrogen atoms to yield high SI intensities. As a result, only about 10% can be attributed to "nonspecific" SI.

All the fragments typical for a certain lipid species are listed for a better survey in Table 2.1. Hydrocarbon SI are not included since they are nonspecific and therefore not relevant for identification of monolayer components. However, some of the hydrocarbon SI such as M15, M29, M43, M57, and M71 are shown in Fig. 4.4.

Naturally, mass spectra of lipids acquired from ToF-SIMS experiments contain an immense number of peaks, which result from the acyl side chains. In the case of DPPC and DPPG, no discrimination of these lipids is possible from just studying these hydrocarbon fragments. Unfortunately, negatively charged lipids do not yield prominent positive fingerprint SI. Only some negative fragments together with the negatively charged quasimolecular ions are detectable, which makes exact chemical localization of anionic lipids difficult especially since the SI yields of intact species are comparatively low when standard instrumentation

TABLE 2.1 Secondary Ions Obtained from Components of Surfactant Model Systems

Component	m/z	Ion
DPPC	102	$C_5H_{12}NO^+$
	104	$C_5H_{14}NO^+$
	125	$C_2H_6PO_4^+$
	150	$C_5H_{13}NPO_2^+$
	166	$C_5H_{13}NPO_3^+$
	184	$C_5H_{15}NPO_4^+$
	224	$C_8H_{19}NPO_4^+$
	734	$(DPPC-H)^+$
	735	$(DPPC + H)^+$
DPPG	173	$C_3H_{10}PO_6^+$
	767	$(DPPG + 2\ Na-H)^+$
d62DPPG	34	$C_2D_5^+$
	50	$C_3D_7^+$
	62	$C_4D_7^+$
	66	$C_4D_9^+$
SP-B	28	CH_2N^+
	30	CH_4N^+
	44	$C_2H_6N^+$
	70	$C_4H_8N^+$
	110	$C_5H_8N_3^+/C_7H_{12}N^+$
	112	$C_5H_{10}N_3^+/C_7H_{14}N^+$
	120	$C_8H_{10}N^+$
SP-C	18	NH_4^+
	30	CH_4N^+
	44	$C_2H_6N^+$
	110	$C_5H_8N_3^+$
	4024	$(SP-C + H)^+$

is used [33]. Sometimes complexes with counterions such as sodium in the case of DPPG, namely $(DPPG + 2\ Na-H)^+$, could be used for identification of this lipid [25]. In the case of dipalmitoylphosphatidylserine (DPPS), calcium ions or calcium phosphates were the only secondary ions representative for this lipid [27].

In order to facilitate the analysis of mixtures containing negatively charged phospholipids, especially when it comes to monitoring the chemical distribution pattern within the monolayer, it is reasonable to use chemically modified lipid components. It is important that these lipid derivatives possess a phase and mixing behavior similar to its analog. Deuterated lipid components such as d62-DPPG, which possesses deuterated fatty acids, have been successfully applied in studies of surfactant model systems containing SP-B [28, 34]. The results obtained from these experiments are discussed in more detail later. Here we would like to focus on the SI obtained for this lipid, which will be used later for the purpose of identification (Table 2.1). As expected, a different fragmentation pattern is observable. The secondary ions obtained for the deuterated acyl chains showed an increased m/z ratio. The deuterocarbon species $C_2D_5^+$, $C_3D_7^+$, $C_4D_7^+$, and $C_4D_9^+$ yielded the fragments M34, M50, M62, and M66. As a comparison, the corresponding hydrogenated species had m/z ratios of 29 u/e, 43 u/e, 55 u/e, and 57 u/e, respectively.

Proteins also usually fragmentize in ToF-SIMS experiments. Typically, amino acid fragments with the composition M-CO$_2$H are generated in this fragmentation process. Hydrophobic proteins such as the surfactant proteins SP-B and SP-C have also been successfully analyzed by ToF-SIMS. In the case of SP-C, characteristic fragments are $m/z = 30$ u/e (Gly), M44 (Ala), and M110 (His) (Table 2.1). Being a relatively small protein, the quasimolecular ion (SP-C + H)$^+$ at $m/z = 4024$ u/e can be detected in spin-coated preparations of SP-C but not in mixed LB layers with lipids [24]. Due to the presence of nitrogen, SP-C specific fragments are easily identified. Except for M18, most fragments correspond to iminium ions of the amino acids glycine, alanine, proline, valine, leucine/isoleucine, and histidine after a cleavage of the COOH group [24]. For a better overview, a list comprising the secondary ions resulting from different amino acids for identification of proteins is given in Table 2.2. Fragments of basic amino acids such as lysine and arginine corresponding to M101 (Lys-45)$^+$ and M129 (Arg-45)$^+$, respectively, cannot be found. Some of the fragments interfere with the ones obtained from DPPC (e.g., $C_4H_{10}N^+$ with $m/z = 72$ u/e and $C_5H_{12}N^+$ with $m/z = 86$ u/e); others, however, are only specific for SP-C (e.g., M18, M30, and M110). Even though M18 and M30 also appear in DPPC spectra, they are called SP-C specific because of their significantly higher SI yields.

SP-B, too, shows a fragmentation pattern that is dominated by SI originating from amino acids after cleavage of COOH (Table 2.1). Apart from the fragments M30, M44, M70, M72, and M110, which were also characteristic for SP-C, phenylalanine (M120) was detected.

As can be seen from this short overview of mass spectroscopic analysis of biomolecules, lipids and proteins can easily be identified by characteristic secondary ions. However, interference of signals can disturb the exact compositional analysis of more complex mixtures such as native membranes. In such a case, it is necessary to use isotopically labeled compounds, as we did in the case of surfactant model systems, in order to obtain information on lateral distribution of individual molecules in a lipid/protein environment. A similar approach was followed by Marxer et al. [8], who studied isotopically labeled proteins and lipids with ToF-SIMS. A more detailed description of further ToF-SIMS applications in the biomembrane field is given later when imaging examples of biomaterial surfaces are discussed.

In summary, the methods used for evaluation of ToF-SIMS signals are based on similar methods used with conventional mass spectrometry. By now, several handbooks containing reference spectra and a comprehensive collection of static

TABLE 2.2 Characteristic Secondary Ions of Amino Acids for Mass Spectrometric Analysis of Proteins

Mass	Ion	Amino Acid
30	CH_4N^+	Glycine (Gly, G)
44	$C_2H_6N^+$	Alanine (Ala, A)
60	$C_2H_6NO^+$	Serine (Ser, S)
70	$C_4H_8N^+$	Proline (Pro, P)
72	$C_4H_{10}N^+$	Valine (Val, V)
74	$C_3H_8NO^+$	Threonine (Thr, T)
76	$C_2H_6NS^+$	Cysteine (Cys, C)
86	$C_5H_{12}N^+$	Leucine (Leu, L)
86.1	$C_5H_{12}N^+$	Isoleucine (Ile, I)
87	$C_3H_7N_2O^+$	Aspargine (Asn, N)
88	$C_3H_6NO_2^+$	Aspartate (Asp, D)
101	$C_4H_9N_2O^+$	Glutamine (Gln, Q)
101.1	$C_5H_{13}N_2^+$	Lysine (Lys, K)
102	$C_4H_8NO_2^+$	Glutamate (Glu, E)
104	$C_4H_{10}NS^+$	Methionine (Met, M)
110	$C_5H_8N_3^+$	Histidine (His, H)
120	$C_8H_{10}N^+$	Phenylalanine (Phe, F)
129	$C_5H_{13}N_4^+$	Arginine (Arg, R)
136	$C_8H_{10}NO^+$	Tyrosine (Tyr, Y)
159	$C_{10}H_{11}N_2^+$	Tryptophan (Trp, W)

SIMS data have been published [35, 36]. Also, an electronic spectral library was produced by Surface Spectra Ltd., which contains more than 1000 materials and over 1900 spectra [37]. Such tools facilitate the evaluation of ToF-SIMS spectra and are extremely helpful when new substances are analyzed for the first time.

2.3.3 Imaging of Lipid/Protein Assemblies

ToF-SIMS images of biomaterial surfaces are obtained by rastering a fine-focused ion beam over a defined area. Mass spectra are acquired pixel by pixel and are then digitally processed. The intensities for a given secondary ion are pictured in dependence of their localization. The resulting chemical maps semiquantitatively monitor the distribution of the analyzed fragments on the surface with high lateral resolution. The comparison of different SI images offers the unique possibility to gain a reliable picture of the two-dimensional localization of individual biomolecules. Some examples of ToF-SIMS applications in surfactant research are presented here in order to give an impression of the potential of this technique for imaging lipid and protein distributions with high lateral resolution.

As mentioned in the previous section, lipids are chemically very similar in their acyl chain moieties. They can therefore be identified with ToF-SIMS by only a few unique headgroup fragments or the respective molecular ions.

In order to study surfactant model systems containing DPPC and DPPG, it is reasonable to use deuterated lipids, which have a different fragmentation pattern and furnish superior information about the organization of the studied film. A recently published study reports the application of the deuterated DPPG analog d62DPPG to differentiate from the emitted DPPC fragments [28, 34] and to elucidate the chemical composition of domain structures observed earlier in the presence of surfactant proteins by fluorescence microscopy [23, 38–42], scanning force microscopy (SFM) [43–45], and scanning near-field optical microscopy (SNOM) [46]. The molecular distribution of surfactant model systems at collapse pressure was especially of interest since protrusion formation occurs under these conditions and chemical composition of these three-dimensional (3D) structures had so far been only indirectly characterized. Exemplary ToF-SIMS images of the SP-B-containing surfactant model systems studied previously [28, 34] are presented in Fig. 2.6. Mass-resolved images at nominal mass resolution of DPPC/d62DPPG/SP-B (4:1:0.2 mol %) monolayers were obtained in burst alignment mode (focus 300 nm) with a Bi_3^+ primary ion beam. The negatively charged lipid d62DPPG, recognizable by the secondary ion $C_3D_7^+$ with a m/z value of 50 u/e, showed an inhomogeneous distribution pattern. Distinct domains with high SI intensity appeared bright in a dark network with low secondary ion yields. The fragments emitting from DPPC (M125) and SP-B (M70) were found to possess highest signal intensities in the network surrounding the d62DPPG domains. It was therefore concluded from the obtained distribution pattern that a demixing of the monolayer components had occurred. The surfactant protein SP-B colocalizes with the DPPC, while the second lipid component, namely DPPG, phase separates from the DPPC/SP-B

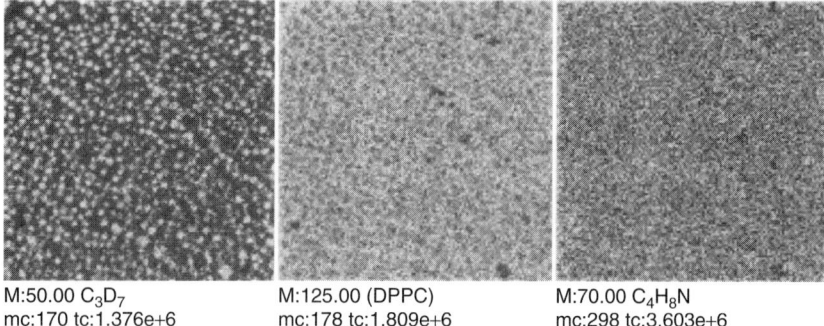

M:50.00 C_3D_7
mc:170 tc:1.376e+6

M:125.00 (DPPC)
mc:178 tc:1.809e+6

M:70.00 C_4H_8N
mc:298 tc:3.603e+6

Figure 2.6 ToF-SIMS images of positively charged secondary ions of the system DPPC/d62DPPG/SP-B (4:1:0.2 mol %) after LB transfer at 50 mN/m. The fragment ions with a mass/charge ratio of 50 (d62DPPG specific), 125 (DPPC specific), and 70 (SP-B specific) are shown. Ion intensities are represented by a grey scale, with dark grey representing low intensities and bright grey representing high intensities. For each image, white is set to the intensity in the brightest pixel (normalization). This intensity is referenced after "mc." The legend "tc" refers to the total number of detected secondary ions in the respective image.

containing network by forming discreet domains. This can be observed in an even more distinct manner in the color coated images depicted in Seifert et al. [28].

The detailed study of SP-B-containing surfactant model system performed by Breitenstein et al. [34] also provides additional information on the distribution of negatively charged secondary ions, which is necessary to obtain a complete picture of the lateral organization of all available fragments. Additionally, the secondary ion distribution at different surface pressures was verified and different preparations of solid-supported monolayers (Langmuir–Blodgett vs. Langmuir–Schaefer) were compared. Although these results, which were obtained on an aqueous subphase, did not show the expected colocalization of SP-B and DPPG that was suggested by Pérez-Gil et al. [47] from electron spin resonance experiments performed in a buffered system (pH 7.0), specific interactions between these two species cannot be ruled out. They are assumed to result from electrostatic interactions between the positively charged amino acid residues and the anionic headgroups of DPPG [48, 49]. It was argued that pH differences of the subphase used could account for the demixing of DPPG and SP-B. For this reason, a systematic study at pH 5.5 and pH 7.0 was performed combining film balance experiments, fluorescence microscopy, scanning force microscopy, as well as ToF-SIMS [28]. The comparison of DPPC/DPPG/SP-B monolayers at different pH conditions revealed that even at pH 7.0 SP-B remained localized in the DPPC-rich phase. It was suggested in this study that an extended hydrogen-bond network is formed in the headgroup region of DPPG, which leads to bridging of adjacent lipid molecules. Such a network would inevitably reduce the solubility of SP-B in the DPPG-rich phase. As a result, the protein would be excluded from the tightly packed DPPG domains and assemble in the surrounding DPPC matrix. Despite these results, SP-B/DPPG interactions still might exist. Under the experimental conditions chosen so far, however, they are most likely to be overcome by the intermolecular interactions existing in the extended hydrogen-bond network. Lipid systems containing unsaturated PGs should not be able to form such a network. It is possible that SP-B/PG interactions occur in such a model system under conditions favoring mixing of lipids (e.g., Ca^{2+}-free buffer, neutral pH, fluid lipid systems).

2.3.4 ToF-SIMS in Biomembrane Research

Biological membranes are highly complex mixtures consisting of a multitude of different lipids and proteins. Their lateral organization is assumed to be pivotal for their physiological function, as is discussed in numerous publications dealing with the raft theory [50–54]. Lipid rafts were originally defined as membrane fragments insoluble to treatment with cold nonionic detergent [55]. It was a matter of controversy for a long time if these rafts were only an artifact of membrane preparation conditions. However, recent studies provide convincing evidence that lipid/protein associates do exist in cellular membranes even though their lateral organization may be more complex than first suggested by raft theories [56, 57].

Lipid rafts are assumed to consist of lipid/protein microdomains enriched in cholesterol, sphingolipids, and a large amount of signaling molecules such as protein kinases and phosphatases, heterotrimeric G proteins, and small GTP binding proteins [58]. Cholesterol-rich domains are therefore thought to play an essential role in membrane traffic and signaling processes. In order to understand the role of cholesterol for the formation of lipid rafts in cell membranes, the mixing properties of lipid mixtures containing cholesterol were studied. They were found to be characterized by the formation of two fluid phases, which are conventionally denoted as liquid-ordered (*lo*) and liquid-disordered (*ld*) phases [59]. A major drawback of the techniques mainly used for the investigation of raft formation and composition is that they require the addition of fluorescent labels. When lipid monolayers at the air–water interface of a film balance are studied with fluorescence microscopy, the existence of different phases due to the exclusion of the fluorescent dye from the highly ordered domains is visualized. The exact chemical composition of the film, however, remains unclear. Additionally, an effect of the fluorophor on the phase behavior of the studied system cannot be ruled out. The great advantage of ToF-SIMS is that it allows a label-free analysis of model membranes in order to study raft formation, as has been described by McQuaw et al. [60]. Different mixtures of POPC, 18:0 sphingomyelin (SM) and cholesterol (CH) were studied with ToF-SIMS. Apart from the protonated molecular ions $[M + H]^+$, characteristic fragments for all lipids were found within each tested model system. The results of this study indicate that in a mixture of POPC/SM/CH (30:47:23), the sterol forms separate domains from which POPC is excluded. Sphingomyelin was found to mainly colocalize with cholesterol. In contrast to this, all the binary mixtures (e.g., POPC/CH (4:3), POPC/SM (2:3), and SM/CH (2:1)) were found to form homogeneous monolayers. These results indicate that with ToF-SIMS it was possible to characterize the mixing behavior of raft-forming lipid systems and to illustrate the complexity of lipid interactions.

Other examples of nonraft lipid mixtures studied with ToF-SIMS intended to visualize the lateral organization of membrane compounds [27, 61, 62] or even to quantify the degree of phase separation [62]. One study dealt with DPPC/DPPS monolayers that were analyzed with ToF-SIMS to determine if the negatively charged DPPS, which is present in the inner leaflet of cellular membranes, forms separate domains in the presence of calcium ions. This domain formation was assumed to be physiologically relevant for the binding of peripheral membrane proteins such as the annexins, which play an important role in membrane traffic and signal transduction pathways [27]. Demixing of these lipids was found to indeed be induced by calcium ions, to be reversible upon addition of EGTA, and to present a binding platform to different annexins as was evidenced with SFM [63, 64]. Another study focused on mixtures of DPPE with cholesterol [62]. Phosphatidylethanolamines are known to be the most abundant lipids in the inner leaflet of the plasma membrane. It was investigated if cholesterol domains are formed when mixed with DPPE and if such domains could exist in the inner leaflet of cellular membranes. Indeed, cholesterol-rich phases where observed at cholesterol contents below 50 mol%, whereas DPPE was homogeneously distributed over the whole area of the film at all concentrations tested. Phase segregation therefore was not complete. Interestingly, these condensed DPPE/CH phases grew steadily when cholesterol content increased and connected at the percolation threshold value until at 50 mol% a homogeneous phase was formed.

These attempts to gain molecular insight into spatial organization of membranes stress the potential of ToF-SIMS for this research field. However, so far one phenomenon has yet to be characterized in more detail to improve interpretation of ToF-SIMS data: the so-called matrix effect. It describes the phenomenon that the ionization efficiency of secondary ions is not necessarily associated with the actual local surface concentration of the respective components. As a result, not only chemical but also physical differences within a monolayer determine the yields of ion fragments, which complicates data interpretation as well as quantification [11, 18, 24, 34]. For example, the counting rate of fragment ions in multilayer structures is decreased by the factor 1.5–5 compared to monolayer structures. In general, the matrix effect accounts for the fact that ToF-SIMS cannot be considered an absolutely quantitative method. Yet the matrix effect can be used for structural analysis of surface films and thus offers interesting information about the regarded system.

An interesting example for secondary ion contrasts observed due to the matrix effect is described in Bourdos et al. [25]. DPPC monolayers transferred in the region of coexistence of the liquid-expanded (*le*) and liquid-condensed (*lc*) phase displayed significant contrast due to different ion yields of the respective phases. Generally, SI yields obtained from the *le* phase were higher, which led to the observed, as we call it, positive contrast in the ToF-SIMS images. As was shown in this study, signal intensities generally decreased with increasing surface density, which was especially the case for the hydrocarbon ions. Some of the secondary ions, however, showed higher SI yields in the *lc* phase (e.g., the fragment from the choline headgroup with $m/z = 104$) and therefore showed a negative contrast. Electrostatic interactions in the headgroup region of the closer packed DPPC molecules were assumed to account for this inverse effect. The matrix effect can obviously lead to contrasts due to physical differences in membrane systems.

Structural effects also influence ionization of membrane constituents, as was reported most recently [65]. It was shown that signal intensities of molecular ions of POPC were highest when lipid bilayers were prepared on a solid support. LB monolayers and disordered POPC films, however, evidenced only low SI yields. Other fragments such as phosphocholine were not involved. It was proposed that molecular ions yielded higher signal intensities due to weaker interactions in the upper leaflet of the bilayers. These results suggest that the molecule peak could be used as a marker signal for membrane structure and especially as an indicator for the presence of lipid bilayers.

The molecular environment of a lipid also affects signal intensities, as was reported for the phosphocholine fragment M184 of DPPC [66]. Changes in substrate material and film composition were more effective in altering SI yields than differences in packing density. Addition of cholesterol, for example, increased the phosphocholine signal of DPPC, on the one hand, but masked the headgroup fragment signal of DPPE, on the other hand. This hiding or enhancing of phospholipid headgroup peaks by membrane constituents such as cholesterol has to be considered when cellular systems are investigated with ToF-SIMS.

2.4 CONCLUSION AND OUTLOOK

The technical advances made in the past twenty years led to ToF-SIMS being a frequently used and highly efficient tool for the characterization of biomaterial surfaces. It offers the unique possibility to analyze the composition of surface layers with better spatial resolution than optical microscopy. Chemical images of biological samples are obtained, which provide detailed information on the location of individual biomolecules and, in the field of biomembrane research, the composition of different phases. ToF-SIMS therefore presents a key for understanding how biomolecules assemble and which parameters influence their localization in biological assemblies. However, improvements of sensitivity, resolution, and quantification of mass spectrometric images would extend the versatility and applicability of this technique in the biological field. Furthermore, the future perspectives of ToF-SIMS in biomembrane research strongly depend on understanding and controlling the matrix effect.

REFERENCES

1. A. M. Belu, D. J. Graham, and D. G. Castner (2003). Time-of-flight secondary ion mass spectrometry: techniques and applications for the characterization of biomaterial surfaces. *Biomaterials* **24**: 3635–3653.
2. A. Janshoff and C. Steinem (2006). Transport across artificial membranes—an analytical perspective. *Anal. Bioanal. Chem.* **385**: 433–451.
3. B. Hagenhoff (2000). High resolution surface analysis by TOF-SIMS. *Mikrochim. Acta* **132**: 259–271.
4. A. Brunelle, D. Touboul, and O. Laprevote (2005). Biological tissue imaging with time-of-flight secondary ion mass spectrometry and cluster ion sources. *J. Mass Spectrom.* **40**: 985–999.
5. A. Benninghoven (1994). Chemical analysis of inorganic and organic surfaces and thin films by static time-of-flight secondary ion mass spectrometry (ToF-SIMS). *Angew. Chem., Int. Ed. Engl.* **33**: 1023–1043.
6. Z. Reich, R. Kapon, R. Nevo, Y. Pilpel, S. Zmora, and Y. Scolnik (2001). Scanning force microscopy in the applied biological sciences. *Biotechnol. Adv.* **19**: 451–485.
7. H. J. Mathieu (2001). Bioengineered material surfaces for medical applications. *Surf. Interface Anal.* **32**: 3–9.

8. C. G. Marxer, M. L. Kraft, P. K. Weber, I. D. Hutcheon, and S. G. Boxer (2005). Supported membrane composition analysis by secondary ion mass spectrometry with high lateral resolution. *Biophys. J.* **88**: 2965–2975.
9. D. Weibel, S. Wong, N. Lockyer, P. Blenkinsopp, R. Hill, and J. C. Vickerman (2003). A C60 primary ion beam system for time of flight secondary ion mass spectrometry: its development and secondary ion yield characteristics. *Anal. Chem.* **75**: 1754–1764.
10. F. Kötter and A. Benninghoven (1998). Secondary ion emission from polymer surfaces under Ar^+, Xe^+ and SF_5^+ ion bombardment. *Appl. Surf. Sci.* **133**: 47–57.
11. D. Breitenstein (2008). The use of time-of-flight secondary ion mass spectrometry in biochemistry. *J. Biomacromol. Mass Spectrom.* (to be published).
12. R. N. Sodhi (2004). Time-of-flight secondary ion mass spectrometry (ToF-SIMS): versatility in chemical and imaging surface analysis. *Analyst* **129**: 483–487.
13. H. Nygren, K. Borner, B. Hagenhoff, P. Malmberg, and J. E. Mansson (2005). Localization of cholesterol, phosphocholine and galactosylceramide in rat cerebellar cortex with imaging ToF-SIMS equipped with a bismuth cluster ion source. *Biochim. Biophys. Acta* **1737**: 102–110.
14. J. Cheng and N. Winograd (2005). Depth profiling of peptide films with ToF-SIMS and a C-60 probe. *Anal. Chem.* **77**: 3651–3659.
15. M. C. Biesinger, P. Y. Paepegaey, N. S. McIntyre, R. R. Harbottle, and N. O. Petersen (2002). Principal component analysis of ToF-SIMS images of organic monolayers. *Anal. Chem.* **74**: 5711–5716.
16. R. Van Ham, L. Van Vaeck, A. Adriaens, F. Adams, B. Hodgesc, and G. Groenewold (2002). Inorganic speciation in static SIMS: a comparative study between monatomic and polyatomic primary ions. *J. Anal. Atom. Spectrom.* **17**: 753–758.
17. S. D. Hanton (2001). Mass spectrometry of polymers and polymer surfaces. *Chem. Rev.* **101**: 527–569.
18. I. S. Gilmore and M. P. Seah (1996). Static SIMS: a study of damage using polymers. *Surf. Interface Anal.* **24**: 746–762.
19. D. Touboul, F. Kollmer, E. Niehuis, A. Brunelle, and O. Laprévotec (2005). Improvement of biological time-of-flight-secondary ion mass spectrometry imaging with a bismuth cluster ion aource. *J. Am. Soc. Mass Spectrom.* **16**: 1608–1618.
20. A. Benninghoven and W. Sichtermann (1978). Detection, identification and structural investigation of biologically important compounds by secondary ion mass spectrometry. *Anal. Chem.* **50**: 1180–1184.
21. H. U. Jabs, G. Assmann, D. Greifendorf, and A. Benninghoven (1986). High performance liquid chromatography and time-of-flight secondary ion mass spectrometry: a new dimension in structural analysis of apolipoproteins. *J. Lipid Res.* **27**: 613–621.
22. A.v. Eckardstein, H. Funke, M. Walter, K. Altland, A. Benninghoven, and G. Assmann (1990). Structural analysis of human apolipoprotein A-I variants. Amino acid substitutions are nonrandomly distributed throughout the apolipoprotein A-I primary structure. *J. Biol. Chem.* **265**: 8610–8617.
23. A. von Nahmen, A. Post, H. J. Galla, and M. Sieber (1997). The phase behavior of lipid monolayers containing pulmonary surfactant protein C studied by fluorescence light microscopy. *Eur. Biophys. J.* **26**: 359–369.
24. N. Bourdos, F. Kollmer, A. Benninghoven, M. Ross, M. Sieber, and H. J. Galla (2000). Analysis of lung surfactant model systems with time-of-flight secondary ion mass spectrometry. *Biophys. J.* **79**: 357–369.

25. N. Bourdos, F. Kollmer, A. Benninghoven, M. Sieber, and H. J. Galla (2000). Imaging of domain structures in a one-component lipid monolayer by time-of-flight secondary ion mass spectrometry. *Langmuir* **16**: 1481–1484.
26. M. C. Biesinger, D. J. Miller, R. R. Harbottle, F. Possmayer, N. S. McIntyre, and N. O. Petersen (2006). Imaging lipid distributions in model monolayers by ToF-SIMS with selectively deuterated components and principal component analysis. *Appl. Surf. Sci.* **252**: 6957–6065.
27. M. Ross, C. Steinem, H.-J. Galla, and A. Janshoff (2001). Visualization of chemical and physical properties of calcium-induced domains in DPPC/DPPS Langmuir–Blodgett layers. *Langmuir* **17**: 2437–2445.
28. M. Seifert, D. Breitenstein, U. Klenz, M. Meyer, and H. J. Galla (2007). Solubility vs. electrostatics: what determines the lipid/protein interaction in the lung surfactant. *Biophys. J.* **93**: 1192–1203.
29. E. Ayanoglu, A. Wegmann, O. Pilet, G. D. Marbury, J. R. Hass, and C. Djerassi (1984). Mass spectrometry of phospholipids. Some applications of desorption chemical ionization and fast atom bombardment. *J. Am. Chem. Soc.* **106**: 5246–5251.
30. T. Matsuhara and A. Hayashi (1991). FAB/Mass spectrometry of lipids. *Prog. Lipid Res.* **30**: 301–322.
31. D. J. Harvey (1995). Matrix-assisted laser desorption/ionization mass spectrometry of phospholipids. *J. Mass Spectrom.* **30**: 1333–1346.
32. F. W. McLafferty and F. Turecek (1993). *Interpretation of Mass Spectra*. University Science Books, Mill Valley, CA.
33. A. Benninghoven, F. G. Rüdenauer, and H. W. Werner (1987). *Secondary Ion Mass Spectrometry (SIMS)*. Wiley, Hoboken, NJ.
34. D. Breitenstein, J. J. Batenburg, B. Hagenhoff, and H.-J. Galla (2006). Lipid specificity of surfactant protein B studied by time-of-flight secondary ion mass spectrometry. *Biophys. J.* **91**: 1347–1356.
35. D. Briggs, A. Brown, and J. C. Vickermann (1989). *Handbook of Static Secondary Ion Mass Spectrometry (SIMS)*. Wiley, Chichester, UK.
36. J. G. Newmann, B. A. Carlson, R. S. Michael, J. F. Moulder, and A. H. Teresa (1991). *Static SIMS Handbook of Polymer Analysis*. Perkin-Elmer Corp.
37. Surface Spectra Ltd., www.surfacespectra.com.
38. B. M. Discher, K. M. Maloney, W. R. Schief, Jr., D. W. Grainger, V. Vogel, and S. B. Hall (1996). Lateral phase separation in interfacial films of pulmonary surfactant. *Biophys. J.* **71**: 2583–2590.
39. A. D. Horowitz, B. Elledge, J. A. Whitsett, and J. E. Baatz (1992). Effects of lung surfactant proteolipid SP-C on the organization of model membrane lipids: a fluorescence study. *Biochim. Biophys. Acta* **1107**: 44–54.
40. J. Perez-Gil, K. Nag, S. Taneva, and K. M. Keough (1992). Pulmonary surfactant protein SP-C causes packing rearrangements of dipalmitoylphosphatidylcholine in spread monolayers. *Biophys. J.* **63**: 197–204.
41. K. Nag, J. Perez-Gil, A. Cruz, and K. M. Keough (1996). Fluorescently labeled pulmonary surfactant protein C in spread phospholipid monolayers. *Biophys. J.* **71**: 246–256.
42. K. Nag, S. G. Taneva, J. Perez-Gil, A. Cruz, and K. M. Keough (1997). Combinations of fluorescently labeled pulmonary surfactant proteins SP-B and SP-C in phospholipid films. *Biophys. J.* **72**: 2638–2650.

43. M. Amrein, A. von Nahmen, and M. Sieber (1997). A scanning force- and fluorescence light microscopy study of the structure and function of a model pulmonary surfactant. *Eur. Biophys. J.* **26**: 349–357.
44. I. Panaiotov, T. Ivanova, J. Proust, F. Boury, B. Denizot, K. Keough, and S. Taneva (1996). Effect of hydrophobic protein SP-C on structure and dilatational properties of the model monolayers of pulmonary surfactant. *Colloids Surf. B* **6**: 243–260.
45. A. von Nahmen, M. Schenk, M. Sieber, and M. Amrein (1997). The structure of a model pulmonary surfactant as revealed by scanning force microscopy. *Biophys. J.* **72**: 463–469.
46. A. Kramer, A. Wintergalen, M. Sieber, H. J. Galla, M. Amrein, and R. Guckenberger (2000). Distribution of the surfactant-associated protein C within a lung surfactant model film investigated by near-field optical microscopy. *Biophys. J.* **78**: 458–465.
47. J. Perez-Gil, C. Casals, and D. Marsh (1995). Interactions of hydrophobic lung surfactant proteins SP-B and SP-C with dipalmitoylphosphatidylcholine and dipalmitoylphosphatidylglycerol bilayers studied by electron spin resonance spectroscopy. *Biochemistry* **34**: 3964–3971.
48. S. Hawgood, M. Derrick, and F. Poulain (1998). Structure and properties of surfactant protein B. *Biochim. Biophys. Acta* **1408**: 150–160.
49. J. Perez-Gil (2001). Lipid–protein interactions of hydrophobic proteins SP-B and SP-C in lung surfactant assembly and dynamics. *Pediatr. Pathol. Mol. Med.* **20**: 445–469.
50. D. A. Brown and E. London (2000). Structure and function of sphingolipid- and cholesterol-rich membrane rafts. *J. Biol. Chem.* **275**: 17221–17224.
51. K. Simons and E. Ikonen (1997). Functional rafts in cell membranes. *Nature* **387**: 569–572.
52. M. Edidin (2003). The state of lipid rafts: from model membranes to cells. *Annu. Rev. Biophys. Biomol. Struct.* **32**: 257–283.
53. S. Mayor and M. Rao (2004). Rafts: scale-dependent, active lipid organization at the cell surface. *Traffic* **5**: 231–240.
54. K. Jacobson, O. G. Mouritsen, and R. G. Anderson (2007). Lipid rafts: at a crossroad between cell biology and physics. *Nat. Cell Biol.* **9**: 7–14.
55. D. A. Brown and J. K. Rose (1992). Sorting of GPI-anchored proteins to glycolipid-enriched membrane subdomains during transport to the apical cell surface. *Cell* **68**: 533–544.
56. J. F. Hancock (2006). Lipid rafts: contentious only from simplistic standpoints. *Nat. Rev. Mol. Cell Biol.* **7**: 456–462.
57. S. Munro (2003). Lipid rafts: elusive or illusive? *Cell* **115**: 377–388.
58. L. J. Foster, C. L. De Hoog, and M. Mann (2003). Unbiased quantitative proteomics of lipid rafts reveals high specificity for signaling factors. *Proc. Natl. Acad. Sci. USA* **100**: 5813–5818.
59. J. H. Ipsen, G. Karlstrom, O. G. Mouritsen, H. Wennerstrom, and M. J. Zuckermann (1987). Phase equilibria in the phosphatidylcholine–cholesterol system. *Biochim. Biophys. Acta* **905**: 162–172.
60. C. M. McQuaw, L. Zheng, A. G. Ewing, and N. Winograd (2006). Localization of sphingomyelin in cholesterol domains by imaging mass spectrometry. *Langmuir* **23**: 5645–5650.

61. C. M. McQuaw, A. G. Sostarecz, L. Zheng, A. G. Ewing, and N. Winograd (2005). Lateral heterogeneity of dipalmitoylphosphatidylethanolamine–cholesterol Langmuir–Blodgett films investigated with imaging time-of-flight secondary ion mass spectrometry and atomic force microscopy. *Langmuir* **21**: 807–813.
62. C. M. McQuaw, L. Zheng, A. G. Ewing, N. Winograd (2007). Localization of sphingomyelin in cholesterol domains by imaging mass spectrometry. *Langmuir* **23**: 5645–5650.
63. A. Janshoff, M. Ross, V. Gerke, and C. Steinem (2002). Visualization of annexin I binding to calcium-induced phosphatidylserine domains. *ChemBioChem* **2**: 587–590.
64. M. Menke, M. Ross, V. Gerke, and C. Steinem (2004). The molecular arrangement of membrane-bound annexin A2-S100A10 tetramer as revealed by scanning force microscopy. *ChemBioChem* **5**: 1003–1006.
65. C. Prinz, F. Hook, J. Malm, and P. Sjovall (2007). Structural effects in the analysis of supported lipid bilayers by time-of-flight secondary ion mass spectrometry. *Langmuir* **23**: 8035–8041.
66. A. G. Sostarecz, D. M. Cannon, Jr., C. M. McQuaw, S. Sun, A. G. Ewing, and N. Winograd (2004). Influence of molecular environment on the analysis of phospholipids by time-of-flight secondary ion mass spectrometry. *Langmuir* **20**: 4926–4932.

CHAPTER 3

Flexibility and Structure of Fluid Bilayer Interfaces

MICHAEL RAPPOLT and GEORG PABST

Institute of Biophysics and Nanosystems Research, Austrian Academy of Sciences, Graz, Austria

CONTENTS

3.1	Introduction	45
3.2	Mechanical Methods for Membrane Elasticity Determination	49
	3.2.1 Micropipette Aspiration Method	49
	3.2.2 Shape Fluctuation Analysis	50
	3.2.3 Differential Confocal Microscopy	51
	3.2.4 Atomic Force Microscopy	52
3.3	Lipid Bilayer Structure and Elasticity	53
	3.3.1 A "Look" at Vesicles with Small-Angle X-Ray Diffraction	54
	3.3.2 Solid-Supported Bilayers Studied by Surface X-Ray Diffraction	60
	3.3.3 Monolayer Versus Bilayer Bending	62
3.4	Temperature and Composition Behavior of Fluid Bilayers	64
	3.4.1 Temperature Dependence of Bending Rigidity	64
	3.4.2 Membrane Fluctuation and Chain Length Dependence	66
	3.4.3 Role of Lipid Species for Bilayer Rigidity and Shape	68
3.5	Conclusion and Outlook	71
Appendix: List of Methods and Material Parameters		72
Acknowledgments		73
References		73

3.1 INTRODUCTION

Interfaces are ubiquitous in our daily lives, although we usually associate computer or semiconductor technology with this term. There are far more

Structure and Dynamics of Membranous Interfaces, edited by Kaushik Nag
Copyright © 2008 John Wiley & Sons, Inc.

important interfaces that we usually are not aware of and that occur in our own bodies or more generally in all living cells. Any living being—even the simplest bacterium—is permanently exposed to the natural tendency of events to drive a system from an ordered into a disordered state, formulated in the second law of thermodynamics. Thus life calls for powerful strategies and tools to maintain order and reduce losses. One of these tools is the membrane interface.

Eukaryotic membranes are, in general, a compound of lipids and proteins, which is centered between an associated cytoskeleton and the glycocalix [1, 2]. Usually lipids make up the structural matrix for membrane proteins and act as a highly selective barrier, whereas proteins are believed to be the "machines" that regulate any process on the cellular level [3, 4]. However, there is growing awareness that there is more to lipids than just making up a passive layer. In contrast, they actually play an active and crucial role in regulating cell function, including active and passive transport, or endo- and exocytosis [5–7]. Clearly, every function or purpose demands appropriate membrane characteristics, which enters the "homeostatic principle" [8], suggesting that the lipid composition of membranes is regulated in order to maintain a stable structure in response to various environmental conditions and particular needs. Thus one often observes three-dimensional (3D) saddle-like shaped membrane configurations, for instance, in the smooth endoplasmic reticulum, where a high interface to volume ratio is required for the various lipid synthesis processes [9], whereas it makes sense for the outer plasma membrane to be nearly planar and highly resistive to external strain, because one of its duties is to protect the inside of the cell from being exposed to the cell's exterior [10].

Accepting this active role of lipids, it also becomes understandable, at least to some extent, why so many lipids can be found in nature. Because membranes are always composed of more than one lipid species, the bilayer has to accommodate the different lipid properties that may lead to the formation of domains that are enriched in some particular lipid component. This latter aspect has triggered recently immense research efforts, as such lipid domains—also called rafts—are thought to be important for several cellular functions [11, 12].

Another case, where membrane composition is crucial, concerns the influence of membrane curvature on protein regulated processes in and across biointerfaces [13, 14]. The "balanced-spring" model as expressed by Laggner and Lohner [15] addresses this issue. Interactions between membrane active compounds, such as peptides, and lipids can lead to a modification of the spontaneous bilayer curvature through the release of the stored "frustration" energy [16] that is due to intrinsic packing properties given by lipid–lipid interactions. Although protein inclusion may lead also to other effects (e.g., pore formation, micellization) [14], this model provides a good starting point to look at the coupling between the lipids and membrane proteins.

Besides basic research, there is also a growing interest in nanotechnology for the development of smart membrane interfaces being applied in biosensors, nanoreactors, or submicron drug vehicles [17–20]. For example, nanostructured drug delivery systems such as cubosomes are promising candidates for drug administration [21].

The fluid lipid bilayer is the common element of nearly all biomembrane interfaces and arises by self-assembly in aqueous solutions due to the amphiphilic properties of lipids (hydrophobic effect) [22]. The fluid lamellar bilayer—also referred to as the L_α phase [23]—is classified in soft matter physics as a smectic-A type lyotropic liquid crystal [24, 25]. The lipid chains—although they are undergoing constant conformational changes (*trans/gauche* isomerizations) [26]—are on average aligned perpendicular to the membrane plane. While the lipid molecules are free to diffuse in the lateral direction, and hence provide membrane fluidity, bilayer stability is provided by the hydrophobic effect, which acts to minimize the polar/apolar interface area of each monolayer and counterbalances the lateral chain and headgroup pressure, respectively [27]. Chain pressure depends on the number of activated *trans* to *gauche* conformations in the chains while the headgroup repulsion is mainly understood to be caused by steric, hydrational, and electrostatic interactions [28, 29]. (See Fig. 3.1.)

Thus even if one strips all proteins and associated macromolecular networks of a natural membrane, the remaining naked lipid bilayer is still a complex material to characterize [6, 30]. Its material properties are closely related to the possible bilayer deformations and can be determined by many different means (for review

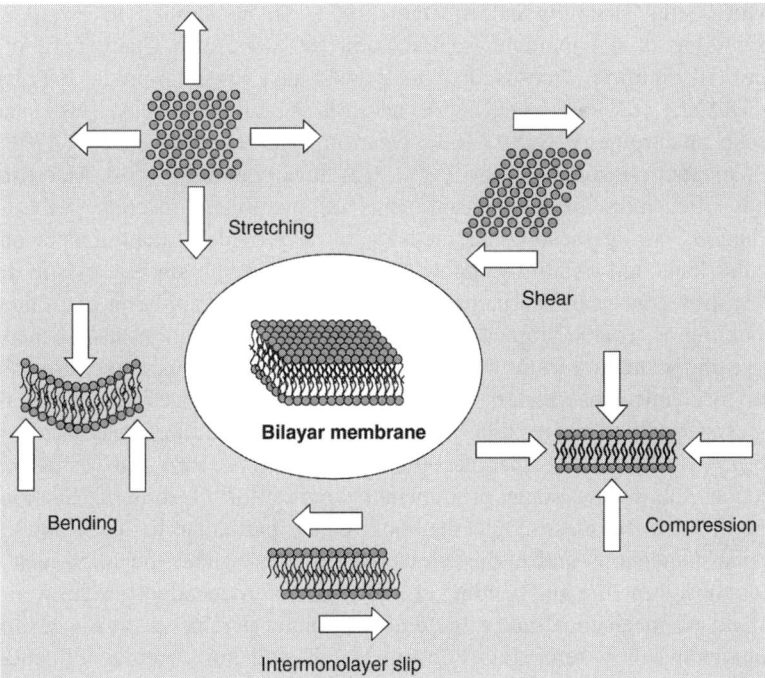

Figure 3.1 Schematic overview of possible modes of bilayer membrane deformation.

see the monograph of Hianik and Passechnik [31]). Shear and intermonolayer friction do not play an essential role for the membrane elasticity in its fluid state and usually are neglected. They only need to be considered when dealing with erythrocytes or membranes with polymerized lipids [1, 32]. In this case, membrane viscosity can be obtained from fluorescence photobleaching recovery (FPR) [33], tether formation studies [34], electrocompression measurements [35], or optical dynamometry [36]. A discussion on the consequences and effects of monolayer coupling on the elasticity of membranes can be found in Fischer [37]. In general, the mechanical properties of fluid bilayers are characterized by the moduli of bending, lateral dilatation, and volume compressibility. The volume compressibility can be deduced from combining the results of density [38] and ultrasound velocity [39–41] measurements. Methods to determine the area compression modulus K_A are micropipette aspiration (MPA) [42, 43] (compare Section 3.2.1), reflection interference contrast microscopy (RICM) [44, 45], dynamic light scattering [46, 47], optical tweezers studies [48], and combined osmotic pressure and small-angle X-ray scattering experiments [49, 50].

In this chapter we concentrate on the classical bending deformation of lipid bilayers. In the past thirty years a vast number of methods have been developed to measure this property, including the above mentioned "macroscopic methods" MPA [42, 43], tether formation [34, 51], RICM [44, 45], and optical dynamometry studies [36] as well as vesicle fluctuation analysis (VFA) [52–56], differential confocal microscopy (DCM) [57], and atomic force microscopy (AFM) [58, 59]. Differential scanning calorimetry (DSC) can be applied to estimate K_C in the regime of the main phase transition [60] (see also Chapter 12 in this volume). In addition, "mesoscopic methods" such as small-angle X-ray scattering (SAXS) [61–67], small-angle neutron scattering (SANS) [68], neutron spin-echo spectrometry (NSE) [69], electron spin resonance (ESR) [70], and nuclear magnetic resonance (NMR) [71, 72] have been developed. Macroscopic methods offer more direct measurements on membrane response to external perturbations, while mesoscopic methods often provide structural data on the molecular level and obtain mechanical properties through smectic elastic theory [73–75]. In recent years, an increasing number of works applying plain numerical modeling or "coarse-grained" molecular dynamic simulations established also a deeper understanding in the field [76, 77]. To give a first example, typical values for mechanical parameters together with the latest structural results of the egg-phosphatidylcholine (egg-PC) bilayer are summarized in Table 3.1.

We briefly discuss the macroscopic methods MPA, VFA, DCM, and AFM in Section 3.2, giving some prominent examples for illustration. Section 3.3 concentrates on the mesoscopic methods with a particular focus on SAXS on liposomal dispersions and highly aligned lipid films. Here, the main idea is to point out how structure and bending elasticity are interdependent and how one can get a hold on mechanical and structural membrane parameters at once. Finally, we demonstrate how temperature, chain length, and lipid species influence the flexibility and structure of bilayers (Section 3.4). We conclude with a summary and outlook in Section 3.5.

TABLE 3.1 Bilayer Properties of Egg-PC at Room Temperature[a]

Structural properties[b]		Structural properties[b]	
M_w (g/mol)	768.5	K_B (N/m^2)	$(1-3) \times 10^9$ [52]
v_L (mL/g)	0.988 (30 °C) [65]	K_A (N/m)	0.167 [50]
V_L (Å^3)	1261 (30 °C) [65]	E (Pa)	2×10^6 [98]
d_{HH} (Å)	35.4 (30 °C) [65]	K_C (J)	$(4-5) \times 10^{-20}$ [55]
A (Å^2)	69.4 (30 °C) [65]	μ (N/m)	~ 0

[a]Temperature different from room temperature given in parentheses.
[b]The molecular weight (M_w), lipid-specific volume (v_L), lipid volume (V_L), headgroup-to-headgroup distance of the bilayer (d_{HH}), lateral area/lipid (A), bulk compression modulus (K_B), area compression modulus (K_A), Young's modulus (E), and shear modulus (μ) are given.

3.2 MECHANICAL METHODS FOR MEMBRANE ELASTICITY DETERMINATION

3.2.1 Micropipette Aspiration Method

The micropipette aspiration technique consents to determine both the area compressibility modulus K_A as well as the bending modulus K_C. An example of such an experiment is given in Fig. 3.2a,b [78], in which a giant vesicle of diacyl PC is manipulated by the use of a micropipette. The video micrographs show how the vesicle is gradually pressurized by applying an aspiration pressure. This leads to a progressive increase in the projection length L inside the pipette. In the low-pressure regime, the response to aspiration pressure exhibits a logarithmic relationship between aspiration pressure p and L (entropic tension regime),

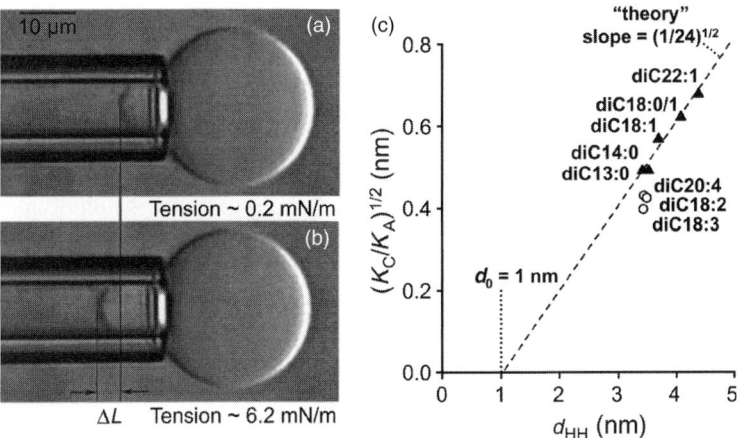

Figure 3.2 Aspiration micropipette experiments carried out on fluid PC bilayers [78]. Video micrographs show the vesicle at (a) low tension and (b) high tension. (c) The validity of the "brush" model is demonstrated for various PC/water systems (for details see text). (From Ref. 78 with permission.)

whereas at higher aspiration pressures, the relationship is linear [78, 79]. Usually K_C values are extracted at low pressures and K_A values at higher pressures. In the high-pressure regime ($\Delta p > 1000$ Pa) the membrane surface expands in response to the in-plane tension, and analysis of the linear Hookean elastic response yields an estimate of K_A (so-called apparent K_A). However, residual thermal undulations skew the high-tension regime to some small extent such that the direct elastic stretch moduli K_A, obtained after correction for smoothing of thermal undulations, have slightly higher values than the apparent ones [78].

Classical elasticity theory on thin materials shows that the bending modulus should scale with the square of its thickness [80]. If one treats each monolayer as a thin structureless sheet with a neglectable shear modulus ($\mu = 0$) and with zero intermonolayer slip, then the bending rigidity can be found to be

$$K_C = K_A (d_{HH} - d_0)^2 / 24 \tag{3.1}$$

in which K_A is the area compression modulus, d_{HH} is the headgroup to headgroup distance of the bilayer, and d_0 represents the nondeformable part of the bilayer (i.e., the headgroup size, d_H). In Fig. 3.2c the validity of this "brush" model (see Appendix in Rawicz et al. [78]) is impressively demonstrated for the L_α phase of PCs. In Section 3.4.2 we revisit this model and demonstrate how this simple relationship can be exploited together with structural X-ray data.

3.2.2 Shape Fluctuation Analysis

One of the few disadvantages of the micropipette aspiration or other related techniques that perturb the membrane with external forces [42, 81] is that large deformations may lead to a change of the lipid distribution in multicomponent bilayers [82]. A convenient nonintrusive method especially for in situ measurements is given by VFA. The membrane flickering phenomenon can be understood as an interplay between thermally induced bending excitations on one side and curvature elasticity on the other [83, 84]. Although the three-dimensional shape fluctuation of quasispherical vesicles was understood quite early [54], progress in the application of VFA has been rather slow. Enhanced sample preparation methods for giant vesicles [85], improved theoretical analysis [55, 86–88], and especially recent technical developments, which helped to increase the signal-to-noise ratio, have promoted this technique in the last few years [86, 89].

Figure 3.3a shows four different examples for vesicle flicker simulations [86]. In this special case the possible effect of gravity on the bending undulations of a quasispherical fluid lipid vesicles was investigated (the parameter g_0 scales with the ratio between the acceleration of gravity g and the bending elasticity modulus K_C). A scientific case for the application of VFA is given in Fig. 3.3b [90]. It is well known that cholesterol significantly increases the area expansion modulus, K_A [91], the lysis tension [92], and the bending rigidity modulus, K_C [93, 94]. In this study the stiffening effects of ergosterol (found in fungi and protozoan membranes) and lanosterol (plays a role in some prokaryotic membranes) have

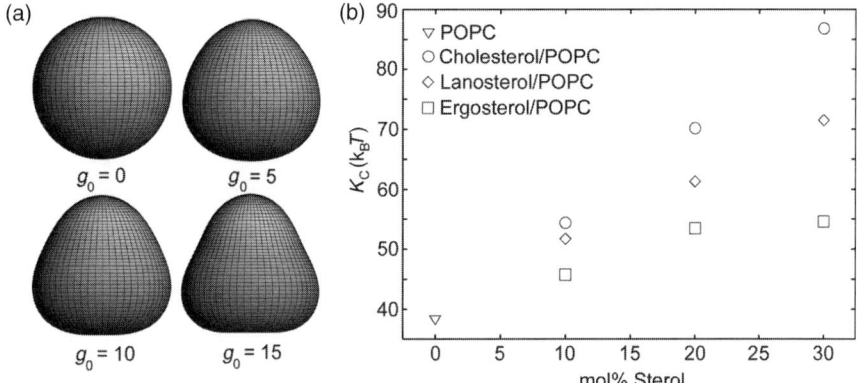

Figure 3.3 Vesicle fluctuation analysis for different POPC/sterol GUVs. (a) Different shape simulations under the influence of gravitation are displayed. (b) The influence of cholesterol, lanosterol, and ergosterol on the bending modulus of POPC is shown. (Panel (a) from Ref. 86 and panel (b) from Ref. 90 with permission.)

been compared to the effects imposed by cholesterol, which is the most abundant sterol in eukaryotic biointerfaces. The results show that cholesterol has the largest effect on the rigidity of fluid PC bilayers.

3.2.3 Differential Confocal Microscopy

In the last paragraph we have introduced the elegant noninvasive method of spectral shape analysis. Its application however, is, restricted to fluid bilayers because bending fluctuations become negligible below the chain melting transition. A promising novel technique combines optically induced vesicle deformations with differential confocal microscopy (DCM) [57]. The working principle is explained in Fig. 3.4a. Giant unilamellar vesicles (GUVs) are kept in a special chamber that reduces any perturbation from possible water flow. Vesicle deformations are induced with a green laser and are recorded with DCM. The bending modulus K_C is easily deduced from the work applied to change an initially spherical vesicle into an oblate vesicle [95].

Figure 3.4b demonstrates the advantage of the method in a case study on dipalmitoyl phosphatidylcholine (DPPC). The bending modulus can be determined within three different phases with 8.5% reliability (below the pretransition temperature T_p in the lamellar gelphase, thereafter in the ripple gel phase, and above the main transition T_m in the fluid lamellar phase). Other methods such as the optical dynamometry [36] work only in the gel-phase regime, where the shear viscosity is nonzero, and VFA can only be applied to the fluid phase as mentioned earlier [96] (cf. Fig. 3.4b, □). In Section 3.4.1 we come back to this topic, discussing in greater detail the coupling of structural and mechanical properties of the PC bilayer in the vicinity of the main transition.

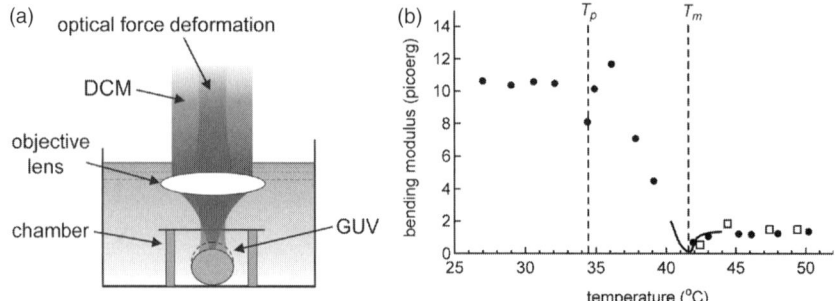

Figure 3.4 Examination of the pre- and main phase transition of DPPC by differential confocal microscopy (DCM) [57]. (a) Schematic of the working principle of the all-optical DCM setup. (b) The bending modulus of DPPC is displayed from 25 to 50 °C. The experimental data of Lee et al. [57] (•) is compared to data of Fernandez-Puente et al. [133] (□) and to an estimation from Heimburg [60] (solid line) (Panel (a) has been supplied by Chau-Hwang Lee and panel (b) is from Ref. 57 with permission.)

3.2.4 Atomic Force Microscopy

Atomic force microscopy (AFM) is usually applied to obtain surface topographical images with spatial resolution close to 1 Å by scanning a sharp tip fixed to a cantilever at a close distance to the surface [97]. However, recently AFM imaging and force measurements were carried out simultaneously on absorbed but integral unilamellar vesicles to provide shape and mechanical properties at once [58, 59]. Ng and colleagues investigated by means of force-versus-distance curves egg-PC vesicles with sizes of about 50 nm [98] (the detection limit is just above 10^{-12} N). In the AFM tip advancement and compression several characteristic breaks can be recorded. As schematically illustrated in Fig. 3.5, the first break occurs when the tip ruptures the freestanding top bilayer and the second when the lower adsorbed membrane patch is penetrated. The Young's modulus E of small unilamellar vesicles (SUVs) is determined within the Hertzian contact model [99] and the bending modulus K_C can be estimated from the relation [100]

$$K_C = E d_{HH}^3 / 12(1 - \nu^2) \tag{3.2}$$

where d_{HH} is the headgroup to headgroup distance (bilayer thickness) and ν is the Poisson ratio, which for vesicles is normally set to 0.5. We note, however, that the formula is strictly correct only for isotropic material [100]. For pure egg-PC, Liang et al. [98] found $E = 1.97 \times 10^6$ Pa and $K_C = 2.7 \times 10^{-20}$ J.

In Table 3.2 an overview of the different methods—including the typical vesicle sizes and shapes—is given with data concerning the bending modulus of egg-PC. The spread of results is clearly higher than one would expect from the accuracy of methods themselves [56]. We return to discuss this point in our conclusion.

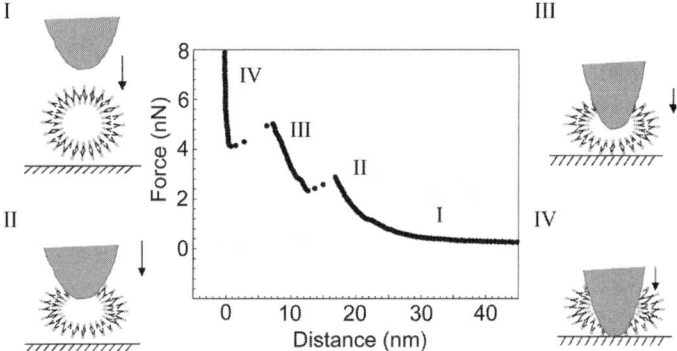

Figure 3.5 Scheme of the tip effect on the vesicle during a typical force-versus-distance measurement on an egg-PC/cholesterol vesicle. Region I is noncontact, region II illustrates the elastic deformation of the vesicle under tip compression, region III corresponds to further tip compression after the penetration of the top bilayer, and region IV shows the tip on the top of the mica substrate. (Adapted with permission from Ref. 98.)

TABLE 3.2 Comparison of the Bending Modulus K_C of Egg-PC Obtained by Different Methods

Method	Vesicle Shape (size)	K_C ($\times 10^{-20}$ J)	References
Vesicle fluctuation analysis	Tubular vesicles ($L = 11-34$ μm)	23	52
Vesicle fluctuation analysis	Cylindrical vesicles ($L > 10$ μm)	10–20	53
Vesicle fluctuation analysis	Quasispherical vesicles (>10 μm)	4–5	55
Entropic tension, E-field deformation	Spherical vesicles (>20 μm)	2.5	81
Entropic tension, E-field deformation	Spherical vesicles (15–70 μm)	4.5; 6.6	85
Entropic tension, AMP	Spherical vesicle (~ 20 μm)	5.8	67[a]
AFM force curve	Spherical vesicles (<60 nm)	2.1	98
SAXS and osmotic pressure	Multilamellar vesicles (~ 1 μm)	5.5	65

[a] Personnel information of Evan Evans given to John Nagle.

3.3 LIPID BILAYER STRUCTURE AND ELASTICITY

The last section demonstrated how bilayer elasticity can be measured directly through "force and effect" type experiments. In this section, we review how the elastic behavior of membranes can be deduced indirectly from scattering experiments. The number one goal of X-ray or neutron diffraction studies is to determine the sample structure and in particular that of the lipid bilayer in the present case. It follows from a thorough treatment of the lattice properties that the bilayer elasticity contributes to the scattered intensity, especially in the vicinity of Bragg reflections. This is in particular the case for static diffraction experiments, where X-rays are superior to neutrons because of their higher intensities that allow large reduction in the properties of the incident beam with respect to wavelength spread $\Delta\lambda/\lambda$, size, and divergence. In the case of dynamic scattering experiments, however, neutrons may be the optimum probes in the future. In the following we mainly concentrate on static X-ray scattering experiments and only mention briefly contributions from dynamic studies. We also give only a very rough overview of the involved theory. For more details, see the recent reviews by Rappolt et al. [101] and Pabst [62]. The full physics is covered in the article by de Jeu et al. [25].

3.3.1 A "Look" at Vesicles with Small-Angle X-Ray Diffraction

When X-rays impinge on a lipid vesicle they are scattered by the modulation of the electron density ρ. While there is little variation of ρ in-plane, ρ exhibits pronounced changes perpendicular to the lipid–water interface. Hence we are able to obtain information regarding the bilayer structure in the z-direction (Fig. 3.6). Below the chain melting transition, especially in the gel phases also, the hydrocarbon chains are organized on two-dimensional (2D) lattices. This yields reflections in the so-called wide-angle regime, which is not considered in the present chapter.

Figure 3.6 shows the principle of a small-angle X-ray diffraction (SAXD) experiment on a dispersion of multilamellar vesicles (MLVs). MLVs are simply prepared by hydrating a dry lipid system in an aqueous solution by standard protocols that depend on the type of lipid studied. The lipid dispersion is then filled into a thermostated glass capillary that is transparent to X-rays. Monochromated and collimated X-rays, frequently of wavelength $\lambda = 1.54$ Å, using either CuK$_\alpha$ radiation from laboratory generators or energies of 8 keV at synchrotron beamlines, are transmitted through the sample and the scattered intensity is recorded in the range of scattering values, typically $2\theta = 0.09$ degrees (corresponding to distances of 1000 Å to 10 Å within the sample) using specialized small-angle cameras (e.g., see Refs. 102–104). Usually, the scattered intensity is plotted as a function of the scattering vector $q = (4\pi \sin\theta)/\lambda$, which has the advantage of being a wavelength-independent variable (2θ is not!). Because of the random orientation of the lipid bilayers within the dispersion, intensity is scattered isotropically into concentric circles. The profile of such a pattern shows equally spaced Bragg peaks, if the sample exhibits a lamellar structure. The presence of peaks simply expresses the fact that there is an average separation between bilayers, meaning that we know how far and in which direction we need to walk in order to meet the next bilayer. In more scientific terms, the system is then described to exhibit positional correlations.

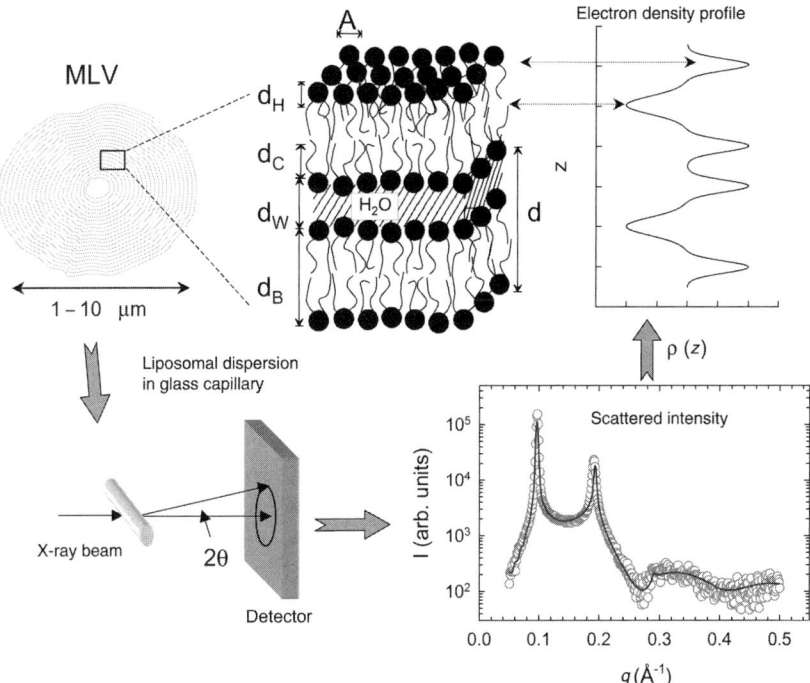

Figure 3.6 Principle of a SAXD experiment. The liposomal dispersion of MLVs within a glass capillary is exposed to X-rays and the scattered intensity is recorded as a function of the scattering vector. The analysis of the diffraction pattern then yields structural information through the electron density profile.

However, because of disorder the number of Bragg peaks is usually low for fully hydrated lipid samples. It is therefore desirable to model the global diffraction pattern in the full q-range in order to retrieve structural information (for details of the method see the recent review by Pabst [62]). In order to do so, the scattered intensity

$$I(q) = \frac{S(q)|F(q)|^2}{q^2} \qquad (3.3)$$

is set to be given as a product of two independent functions—the structure factor $S(q)$ and the absolute square of the form factor $F(q)$. The structure factor describes the lattice and its disorder in Fourier space and the form factor accounts for the modulation of the electron density at the lattice points described by $S(q)$.

There are two theories for disorder in lamellar systems that are of practical importance: the paracrystalline theory (PT) [105, 106] and the Caillé theory [73] in its modified form (MCT) [74]. The first considers stacking imperfections with a Gaussian distribution of the mean separation between perfectly flat layers. MCT, in turn, takes into account the energetic description of the lamellar system and derives the impact of bending fluctuations on the lamellar order. Both considerations show that the long-range order is destroyed, which is an effect of the low dimensionality of the system. Still, we observe Bragg peaks, but the shapes of the peaks differ markedly from that of ideal δ-functions [107] and the system is said to exhibit *quasi*-long-range order. In the case of MCT and for MLVs, the peak intensity of the structure factor decays as (see Fig. 3.7a, inset)

$$S(q - q_h) \propto (q - q_h)^{-1+\eta h^2} \qquad (3.4)$$

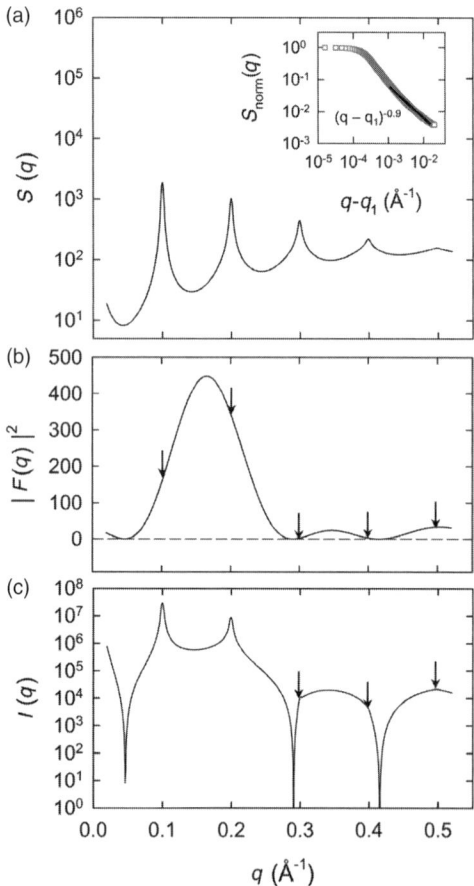

Figure 3.7 Theoretical scattering from MLVs. (a) The structure factor according to MCT. (b) The absolute square of the form factor. Both are combined by Eq. (3.3) to yield the scattered intensity shown in (c). The inset to panel (a) shows the normalized structure factor of the first-order Bragg peak on a double logarithmic scale, demonstrating the power law decay with $\eta = 0.1$ (see Eq. (3.4)). Arrows in panels (b) and (c) indicate the positions of the Bragg peaks.

where h refers to the lamellar reflection order and

$$\eta = \frac{\pi k_B T}{2d^2 \sqrt{K_C B}} \quad (3.5)$$

is the fluctuation or Caillé parameter (k_B is the Boltzmann constant). The fluctuation parameter is a function of temperature T and lamellar repeat distance d (Fig. 3.6). More importantly, however, it depends on the bending rigidity K_C of a single bilayer and the bulk modulus of interaction B. The B modulus is derived from a harmonic description of the bilayer interaction potential; that is, it is approximated to have a parabolic shape [24]. The full structure factor according to MCT is shown in Fig. 3.7a.

Hence determination of the peak shape opens an experimental window on the bending rigidity. Naturally this requires high instrumental resolution, which is given usually only at synchrotron beamlines. There is only one caveat. The fluctuation parameter is a function of the product of K_C and B, both of which are unknown. This means that η increases also if B drops, which is the case when bilayers are more separated (i.e., there is less confinement for fluctuation modes with large wavelengths). Hence in order to disentangle K_C and B, independent measurements for B are required.

This can be done by determining the interactions between adjacent bilayers through an osmotic pressure technique [66, 108, 109]. In brief, various osmotic pressures Π, applied upon addition of large neutral polymers to the aqueous phase, are equated to the pressures of van der Waals interactions, hydration, and undulation repulsion. This yields the equation of state, from which the bending rigidity can be determined by a fit. The B modulus is then calculated as

$$B = \left(\frac{k_B T}{2\pi}\right)^2 \frac{A_{fl}^2}{K_C} \exp(-2d_W/\lambda_{fl}) \quad (3.6)$$

where A_{fl} is the amplitude of the fluctuation pressure, λ_{fl} its decay length, and d_W the bilayer separation (Fig. 3.6). A_{fl}, λ_{fl}, and d_W are determined from fitting the osmotic pressure data. Results might be affected slightly by approximations for the van der Waals pressure, or by the simple summation over direct (van der Waals, hydration, electrostatic) and entropic (undulation repulsion) contributions, which strictly speaking is not valid [110, 111].

Alternatively, B can also be estimated from

$$B \approx -\left(\frac{\partial \Pi}{\partial d}\right)_T \quad (3.7)$$

for $\Pi \to 0$ and works likewise for d_W instead of d. This way, no additional assumption on K_C has to be made. However, it implies that fluctuations do not contribute to the total balance of forces. Henceforth Eq. (3.7) is strictly only applicable if one can neglect bending fluctuations. Nevertheless, Eq. (3.7) is a tractable route that can be applied to get insight on the behavior of the elastic parameters as a function of, for example, temperature, as shown in Section 3.4.1.

Let us return now to the original scattering problem and show how structural information can be obtained. Again, we go through the concepts only briefly. For details see Rappolt et al. [101] and Pabst [62]. The form factor appearing in Eq. (3.3) describes the modulation of the electron density profile and is obtained by Fourier transforming $\rho(z)$. It is then convenient to apply a model description of the electron density profile that catches the main features of $\rho(z)$ and involves at the same time a minimum number of parameters. This is realized in the simple Gaussian model [62, 112, 113] that uses a summation of two Gaussians of width σ_H at the positions $\pm z_H$ to account for the electron dense headgroup region and a third Gaussian of negative amplitude and width σ_C at $z = 0$ that describes the electron sparse hydrocarbon chain region. The Fourier transform of this function can be solved analytically and is shown in Fig. 3.7b. Figure 3.7c shows how the modulation of $F(q)$ affects the observed intensity according to Eq. (3.3). In particular, it demonstrates how Bragg peaks can become extinct when their positions happen to coincide with a node of the form factor.

The solid line shown in the diffraction pattern of Fig. 3.6 gives the best model fit to the experimental data using the above described forms of the structure and form factor and taking additionally into account instrumental influences. Figure 3.8 shows some detailed fits for dimyristoyl phosphatidylcholine (DMPC) and dipalmitoyl phosphatidylethanolamine (DPPE), both in the fluid phase. Besides the common phase, the DMPC and DPPE patterns differ markedly with respect to the number of Bragg reflections, which are four in the case of DPPE, whereas DMPC exhibits only two orders.

The global analysis model is able to fit both data sets reasonably (solid lines in Fig. 3.8) and the insets show the corresponding electron density profiles. The next step then is to derive structural parameters such as the membrane thickness d_B, bilayer separation d_W, chain length d_C, or lateral area per molecule A (Fig. 3.6) to name just the most important ones. The sectioning of d into d_B and d_W, however, is far from being trivial. The reason is the low resolution of the electron density profile due to the statistical average over all transversal lipid positions, which show significant fluctuations on the local level. It is therefore common to agree on an arbitrary definition of the extent of the lipid membrane. We usually apply [61, 113]

$$d_B = d_{HH} + 4\sigma_H \tag{3.8}$$

where $d_{HH} = 2z_H$ is the headgroup-to-headgroup distance. The hydrocarbon chain length follows from

$$d_C = z_H - d_{H1} \tag{3.9}$$

with d_{H1} being the distance between the headgroup (PO$_4$) and the polar/apolar interface of the bilayer. Usually, $d_{H1} \sim 4$ Å. The lateral area per lipid in the fluid

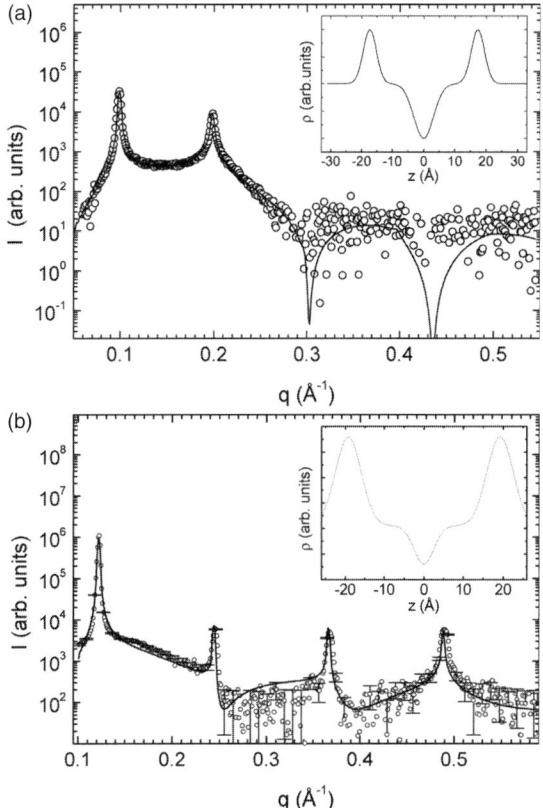

Figure 3.8 Diffraction patterns of (a) DMPC at 30 °C (from Ref. 61) and (b) DPPE at 75 °C (from Ref. 112). Solid lines correspond to fits by the global analysis model. Insets show the corresponding electron density profiles.

phase can be estimated from

$$A = \frac{V_L - V_H}{d_C} \quad (3.10)$$

where V_L is the molecular lipid volume that can be obtained from dilatometric experiments [67, 114] and V_H is the headgroup volume that is determined from the hydrocarbon chain reflections and density measurements in the gel phase [115]. $V_H = 319$ Å^3 for phosphatidylcholines (PCs) [115] and 252 Å^3 for phosphatidylethanolamines (PEs) [116]. Table 3.3 summarizes the structural parameters obtained from the analysis of the diffraction patterns shown in Fig. 3.8. The data for DMPC are compared to those obtained from Nagle's group [67]. Main differences occur for d_B and d_W, which is due to the different definition of the bilayer thickness. Nagle and co-workers usually apply a headgroup

TABLE 3.3 Structural Parameters[a] for DMPC and DPPE

Parameter	DMPC (30 °C)	DPPE (75 °C) [61, 112]
d (Å)	62.4 (62.7)	51.4
d_B (Å)	46.8 (44.2)	46.2
d_W (Å)	16.6 (18.5)	5.3
d_C (Å)	13.2 (13.1)	14.4
A (Å^2)	59.4 (59.6)	61.6
η	0.08 (0.09)	0.02

[a]The values in parentheses are results reported by Nagle and co-workers [67].

thickness d_H of 9 Å reported from neutron scattering experiments [117], whereas we use the value of 10 Å that has been estimated from space-filling models [116]. Results agree nicely if one subtracts the 2-Å difference due to the different d_H values from our d_B data.

Finally, we note that lipids often do not aggregate into MLVs but form spontaneously single shelled unilamellar vesicles (ULVs), in particular when the lipids are charged leading to an electrostatic repulsion between bilayers. ULVs of different sizes can also be prepared deliberately by extrusion (large ULVs), sonication (small ULVs), or in an alternating electric field (giant ULVs) [118]. These preparations are excluded from the present considerations.

This has no effect on the determination of the bilayer structure as the same electron density model can also be used in this case [113], which is actually one of the strongholds of this technique [62]. However, because of the absence of Bragg peaks for ULV samples, the elastic parameters cannot be obtained. This problem can be solved, however, in principle by a slight decrease of the degree of hydration using osmotic stressors such that the system exhibits again positional correlations.

3.3.2 Solid-Supported Bilayers Studied by Surface X-Ray Diffraction

In the last section, we have seen how the elastic moduli can be determined from small-angle X-ray diffraction experiments on multilamellar dispersions, emphasizing also the difficulties of disentangling B and K_C. As an alternative to randomly oriented bilayers in MLV preparations, diffraction experiments can also be performed on highly aligned lipid membranes. This can be achieved by coating a solid substrate, such as glass, mica, or silicon, with a lipid film applying various techniques (for review, see Rappolt et al. [119]). The lipid film then needs to be rehydrated, which can be done either from water vapor or by immersing the system in an aqueous solution. The first method has seen some experimental difficulties in obtaining fully hydrated lipid/water systems due to nearly unavoidable temperature gradients in sample holders, but these difficulties have recently been resolved [120]. One does not face this problem when the system is simply immersed in water [75], which has the additional advantage of being able to change, for example, the ionic strength of the aqueous phase easily [121]. On the other hand, sample loss due to a peeling off of the layers from the substrate may be a problem.

Besides the mentioned experimental difficulties that can be addressed adequately, aligned lipid systems have the clear advantage of splitting the structural information, that is averaged over 3D space in the case of MLVs, into in-plane and out-of-plane components. We apply the common convention that in-plane means structural information deriving from positional correlations that occur in the plane of the bilayer and consequently out-of-plane addresses the transverse structure of the lipid membranes. The lipid molecules are randomly oriented in the plane of the bilayer. Hence it is not possible to distinguish between x and y components and the system is said to be an in-plane powder. Consequently, the in-plane scattering vector $q_r = \sqrt{q_x^2 + q_y^2}$ and the out-of-plane scattering vector is denoted as q_z. In the past, the splitting of the scattered intensity has been capitalized on, in particular for issues concerning bilayer in-plane structure, such as in the ripple or the subgel phase [120].

Figure 3.9 gives an overview of a surface scattering experiment on aligned lipid multibilayers in the fluid phase. Clearly, the obtained scattering pattern is anisotropic, showing Bragg peaks in the q_z direction and only diffuse scattering along q_r starting at $q_r = 0$. As a consequence, a detailed analysis of the diffraction patterns yields the in-plane correlation length

$$\xi = \sqrt[4]{K_C/B} \qquad (3.11)$$

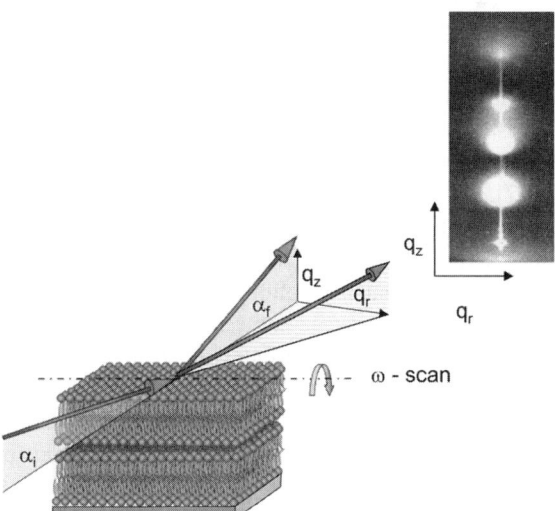

Figure 3.9 Schematic view of a surface diffraction experiment on highly aligned lipid multibilayers. The depicted diffraction pattern has been obtained by integrating the scattered intensity while rotating the substrate by the angle ω. (From Ref. 62 with permission.)

in addition to η (Eq. (3.5)). Both η and ξ are functions of the elastic parameters. Hence B and K_C can be determined directly from the analysis of diffraction patterns of solid-supported lipid films. Presently, three techniques have been developed to perform this type of analysis [63, 122–125].

Nagle's group typically uses data similar to that shown in Fig. 3.9; that is, they integrate scattered intensity by rotating the sample through the X-ray beam by some angle ω [63, 122, 126, 127]. In contrast, Salditt and co-workers separate the data into specular and off-specular contributions and analyze the diffuse scattering recorded at a fixed ω value [111, 123, 124]. The recently reported method by Li et al. [125] is in many ways similar to that of Salditt; however, it is able to analyze specular and off-specular scattering simultaneously.

The methods described so far have been restricted to X-ray studies. The reason for this is the high instrumental resolution needed to analyze the diffraction patterns, which can usually only be achieved with X-rays, simply because neutron beams are never as intense as X-ray beams and henceforth cannot be as much collimated and monochromatized. Before closing the section we want mention, however, that neutrons are superior to X-rays when it comes to measuring molecular motions by inelastic scattering. In particular, neutron spin-echo spectrometry has been applied most recently to determine the dispersion relation for membrane shape fluctuations from which naturally K_C and B follow [69].

3.3.3 Monolayer Versus Bilayer Bending

Not all lipid species form spontaneously planar bilayers, but some, like phosphatidylethanolamines (PEs), have the tendency to from nonlamellar phases such as bicontinuous cubic or inverted hexagonal mesophases [28]. Usually "nonlamellar" lipids do not form unilamellar vesicles and hence none (!) of the introduced mechanical methods of Section 3.2 are applicable. The formation of nonlamellar phases can be understood on the basis of the simple molecular shape concept of Israelachvili [27]. In short, cylindrical shaped lipid molecules such as PCs will lead to planar membranes, cone shaped lipids (e.g., lysolipids [128]) will induce convex interfaces, and wedge shaped molecules such as PEs form concave interfaces. Please note that the definition of curvature is pure convention. The sign of curvature is usually set negative for surfaces of inverse mesophases (concave interfaces) and positive for surfaces of normal oil in water aggregates (convex interfaces). Furthermore, membrane curvatures are described applying the geometrical concepts of principal curvatures c_1 and c_2 or the radii of curvature $R_1 = 1/c_1$ and $R_2 = 1/c_2$. Commonly two combinations of the principal curvatures are used: the mean curvature $H = (c_1 + c_2)/2$ and Gaussian curvature $G = c_1 \cdot c_2$. The topologically most simple phase possessing interfacial curvature is the inverted hexagonal phase H_{II}, which consists of a close packing of lipid monolayer tubes (see Fig. 3.10a). There is no Gaussian curvature since the principal interface curvature determined along the tubular axis is zero, and therefore the free energy can simply be expressed as [129, 130]

$$g_C = \frac{1}{2} K_{CP} \cdot A_P \left(\frac{1}{R_P} - \frac{1}{R_{OP}} \right)^2 \quad (3.12)$$

where K_{CP} is the bending modulus for the monolayer, A_P is the molecular area at the pivotal plane, R_P is the radius of the pivotal plane, and R_{OP} is the radius of spontaneous curvature under atmospheric conditions (cf. Fig. 3.10a). The pivotal plane is defined as a surface inside the lipid phase such that both A_P and V_P are constant when the distance between rod axes varies. Its locus may be approximately identified with the polar/apolar interface. Combining osmotic stress experiments and SAXD allows one to determine the monolayer bending modulus in a straightforward way [129, 130]. Specifically by comparing the elastic free energy of the H_{II} phase (Eq. (3.12)) under different conditions of osmotic stress Π. The plot of the osmotic work (ΠR_P^2) versus the pivotal curvature ($1/R_P$) gives concurrently the values for K_{CP} (= slope) and the spontaneous curvature ($1/R_{OP}$) (= intercept). In Fig. 3.10b such plots are shown for the osmotic work required to dehydrate the H_{II} phase of dioleoyl phosphatidylethanolamine

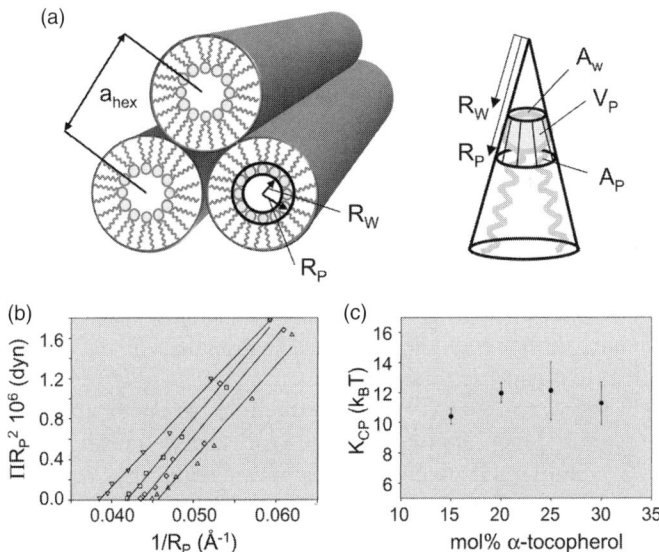

Figure 3.10 Determination of the monolayer bending modulus K_{CP} by combined SAXD and osmotic pressure experiments. (a) Scheme of the inverted hexagonal phase H_{II} and the molecular shape of the PE lipid. The following parameters are defined: a_{hex} = lattice parameter, R_P = pivotal radius, R_W = water radius, A_W = lateral area/lipid at R_W, A_P = lateral area/lipid at R_P, and V_P = pivotal volume. (b) Plot relating the osmotic work required to dehydrate the H_{II} phase of DOPE/vitamin E (α-tocopherol) mixtures as a function of the monolayer curvature $1/R_P$. (c) Deduced K_{CP} values for increasing amounts of vitamin E. (Panels (b) and (c) from Ref. 131 with permission.)

(DOPE)/vitamin E mixtures [131]. The concentration dependence of the bending modulus is shown in Fig. 3.10c. The value of the bending modulus for pure DOPE is 4.5×10^{-20} J [132]. Please note that the energy required to bend two monolayers should be twice that needed to bend only one. Hence if a bilayer consists of uncoupled monolayer sheets, then this implies that the bilayer bending modulus K_C value should have simply twice that of the monolayer modulus K_{CP}.

3.4 TEMPERATURE AND COMPOSITION BEHAVIOR OF FLUID BILAYERS

3.4.1 Temperature Dependence of Bending Rigidity

All lipids display a characteristic melting temperature where the hydrocarbon chains transform from all-*trans* into *trans–gauche* conformation [5–7]. Below the chain melting temperature, T_m, the system is in a lamellar gel-like phase denoted as L_β, or $L_{\beta'}$ if the hydrocarbon chains are tilted with respect to the lipid bilayer. Some lipids such as PCs display in a certain range of chain length and hydration an additional ripple phase $P_{\beta'}$ and the transition occurring at the temperature T_p is called the pretransition. Figure 3.4b gives an overview of the behavior of the bending rigidity across these phases. K_C displays its highest value in the gel phase, then drops across the ripple phase, and finally exhibits a value in the fluid phase that is almost one order of magnitude lower compared to the $L_{\beta'}$ phase.

Figure 3.4b displays another interesting feature. K_C goes through a pronounced minimum at the T_m, which can be related to increased density fluctuations. Although an increase of bilayer elasticity had been observed by VFA [94, 133] and ultrasound velocimetry [39, 41] and estimated from heat capacity curves [60], there was some doubt in this respect among the X-ray scattering community. In particular, it was not clear if the structural observations made in the L_α phase upon approaching the T_m could be explained by this behavior of K_C or might be due to an effect that had been overlooked.

Figure 3.11 shows the evolution of the structural parameters of DMPC in this temperature regime. Upon approaching the main phase transition, the lamellar repeat distance exhibits a distinct nonlinear increase that has been observed by several groups (see Pabst et al. [61] and references therein). This pretransitional behavior became known as "anomalous swelling" and has been a source of controversy for several years due to the difficulties of X-ray analysis techniques that rely on Bragg peaks only, upon which application it was not possible to decide whether the increase of d is due to an increase of d_B, d_W, d_C, or d_H (see Fig. 3.6 for definition of structural parameters). The resolution of this issue was of high importance to researchers in field in order to define the temperature behavior of K_C, because only if the bilayer separation increases would a drop of K_C be realistic. The coupling between bending rigidity and bilayer separation has been noticed by Helfrich [134], which in simple terms is given by the inverse

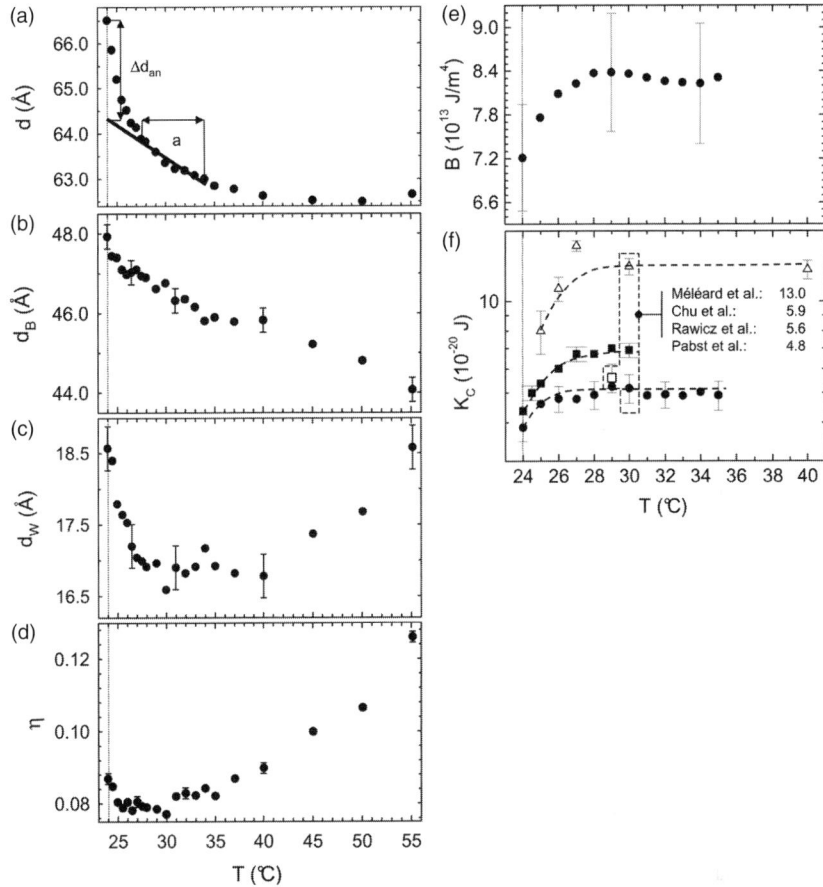

Figure 3.11 Temperature dependence of (a) the lamellar repeat distance, (b) the membrane thickness, (c) the bilayer separation, (d) the bending fluctuations, (e) the bulk modulus of interaction, and (f) the bending rigidity in the fluid phase of DMPC upon approaching the main phase transition ($T_m = 24\,°C$). Data taken from Ref. 61 (●). Panel (f) also gives the K_C values from Ref. 94 (△), Ref. 135 (■), and Ref. 78 (□) for comparison.

dependence of bending fluctuations on the bending rigidity that leads to a steric repulsion between adjacent bilayers.

The issue of anomalous swelling, therefore, called for the application of a global analysis technique in order to obtain the structural parameters in fully hydrated systems (see Section 3.3.1). Indeed, we were able to show that the pretransitional swelling is dominated by the increase of the water layer (Fig. 3.11c), whereas the bilayer thickness does increase quasilinearly over the whole temperature range studied. Figure 3.11d also shows the evolution of the fluctuation parameter, clearly demonstrating that the increase of d_W is coupled to an increase of bending fluctuations. Recall, however, that η is a function of the bending rigidity *and* the bulk modulus of compression (Eq. (3.5)). Hence the increase of η might also be due to a decrease of B only, for example, due a possible reduction of van der Waals attraction between bilayers.

Therefore it was important to explicitly show that the behavior of η is due to a decrease of K_C. As discussed in Section 3.3.1, this can be done through osmotic pressure experiments, from which one gets estimates for B that allow one to calculate K_C through Eq. (3.5). The results are shown Fig. 3.11e,f, where we clearly observe a drop of the bending rigidity and the bulk modulus of interaction. We remind the reader that the application of this technique is strictly valid only in the absence of fluctuations. Nevertheless, the overall tendency agrees surprisingly well with that observed from VFA [94] and was later confirmed using surface diffraction on highly aligned DMPC multibilayers [135] (Fig. 3.11f). Moreover, the found K_C values are actually very close to that reported by Rawicz et al. [78] and Chu et al. [135] (Fig. 3.2f and Table 3.4).

The data presented in Fig. 3.11f suggests that K_C remains constant in the L_α phase above the regime of anomalous swelling. This appears to be true for quite an extended range of temperatures, although experimental reports are scarce in this respect. Presently, there is only a single study that reports K_C for DMPC to be constant up to about 80 °C [123]. On the other hand, the d-spacing was reported to increase in a nonlinear fashion for POPC upon rising temperature starting about 50 °C above its T_m [136] (see also Fig. 3.14). This nonlinear dependence on temperature clearly indicates a drop of the bending rigidity, which was estimated to be on the order of 30% [136, 137]. If the same applies also for DMPC, we would expect to see a decrease of K_C just above 80 °C. In summary, the few experimental data available so far indicate that the bending rigidity of fluid bilayers remains constant within a temperature regime of about $T_m + 50$ °C (excluding the immediate vicinity of the T_m discussed above) and then reduces toward higher temperatures.

3.4.2 Membrane Fluctuation and Chain Length Dependence

In a recent study we paid particular attention to the chain length dependence of the fluid bilayer structure and bending rigidity of PCs [138]. For this purpose, SAXS measurements were carried out in a temperature range from 0 to 15 °C above the chain melting temperature T_m on DMPC (14:0), DPPC (16:0), distearoyl phosphatidylcholine (DSPC, 18:0), dinonadecanoyl phosphatidylcholine (DNPC, 19:0), and diarachidoyl phosphatidylcholine (DAPC, 20:0). As already mentioned, the bending modulus is expected to be proportional to the compressible membrane thickness, that is, $K_C \propto 2d_C^2$ [80] (cf. Eqs. (3.1) and (3.9)), and we would like to recall that Evans and co-workers gave experimental evidence that especially the bending moduli of saturated and monounsaturated PCs, respectively, follow this prediction very well [78] (Fig. 3.2c). Since the area compressibility moduli K_A of PCs do not alter much [67, 78], we used the common value of 0.240 N/m for all PCs studied [138, 139]. In Table 3.4 we have

TABLE 3.4 Literature Data Concerning the Bending Modulus K_C of PC/Water Systems

Lipid	K_C (10^{-20} J)	T (°C)	Method	References
DLPC	3.4	25	Entropic tension, E-field deformation	81
	5.5	30	Surface SAXS	126
DMPC	2.9	30	Surface SAXS	123, 124
	11.5	29	Vesicle fluctuation analysis	93
	4.8	30	SAXS and osmotic pressure (cf. Fig. 3.11e,f)	61
	8.0	30	SAXS and osmotic pressure	66
	5.6	29	AMP and SAXS (cf. Fig. 3.11e,f)	78, 79
	6.9	30	Surface SAXS (cf. Fig. 3.11e,f)	126, 127, 135
	7.1	29	SAXS and brush model[a]	61, 138
	9.9	40	Surface SAXS	111
	13.0	30	Vesicle fluctuation analysis (cf. Fig. 3.11e,f)	94
	13.3	30	Entropic tension, DCM	57
	5.9	30	Neutron spin echo	69
	5.8	30	NMR	72
DPPC	7.1	50	NMR	72
	13.7	50.2	Entropic tension, DCM (cf. Fig. 3.4)	57
	15.0	49.4	Vesicle fluctuation analysis (cf. Fig. 3.4)	133
	11.4	51.5	Surface SAXS on floating bilayer	147
	5.0	50	SAXS and osmotic pressure	66
	8.5	46	SAXS and brush model[a]	138
DSPC	9.5	60	SAXS and brush model[a]	138
DNPC	10.5	65	SAXS and brush model[a]	138
DAPC	10.9	73	SAXS and brush model[a]	138
POPC	8.5	30	SAXS and brush model[a]	138
	2.5	25	Entropic tension, E-field deformation	81
	3.9	24	Vesicle fluctuation analysis	56
	5.8	24	Entropic tension, E-field deformation	56
	8.5	30	Surface SAXS	127
	15.4	25	Vesicle fluctuation analysis (cf. Fig. 3.3)	90
DOPC	2.4	23	Vesicle fluctuation analysis	56
	6.1	23	Entropic tension, E-field deformation	56
	6.3	30	Surface SAXS	125
	7.3	30	Surface SAXS	63
	8.0	30	Surface SAXS	122
	8.5	18	AMP and SAXS	78
SOPC	9.2	18	AMP and SAXS	78
	3.4	21	Vesicle fluctuation analysis	56
	3.2	21	Entropic tension, E-field deformation	56

[a] The hydrocarbon core thicknesses $2d_C$ used for DMPC, DPPC, DSPC, DNPC, DAPC, and POPC were 26.7, 29.1, 30.9, 32.4, 33.0, and 29.1 Å, respectively.

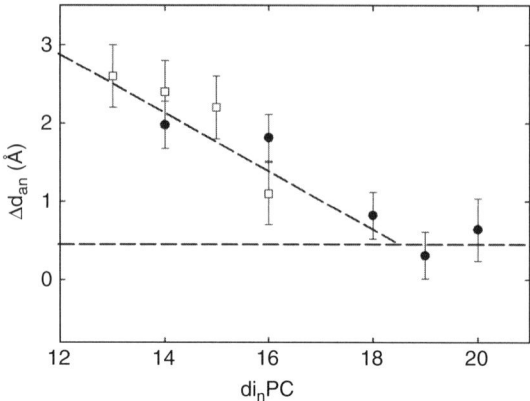

Figure 3.12 Chain length dependence of anomalous swelling component Δd_{an} (for a definition refer to Fig. 3.11a). The open squares correspond to our data [138], the solid circles to data from Korreman and Posselt [140]. (From Ref. 138 with permission.)

summarized our results and it is surprising how well these estimates compare, for instance, to the values determined by surface X-ray scattering experiments.

Note that the assumption of a constant K_A breaks down in the vicinity of the main phase transition, because it experiences in analogy to K_C a drop in this regime [60] (cf. Section 3.4.1). Hence the "brush" model is only applicable outside the anomalous swelling regime. The chain length dependence of the anomalous swelling is summarized in Fig. 3.12 [138]. In agreement with Korreman and Posselt [140] the amplitude of the swelling decreases with increasing length of the hydrocarbon chains. However, it does not vanish completely, but some residual increase of d on the order of 0.5 Å was found to remain for PCs with more than 18 hydrocarbons per chain, which can be attributed to some subtle membrane thickening that occurres in all PCs as a universal feature. Anomalous swelling defined to be due to the increase of the bilayer separation occurs only for lipids up to 18 hydrocarbons per chain (Fig. 3.13). In agreement, there is no increase of the fluctuation parameter close to T_m for chain lengths greater than 18.

3.4.3 Role of Lipid Species for Bilayer Rigidity and Shape

In the last two sections the properties of fluid PC bilayers have been discussed in minute detail, and we have demonstrated that especially their mechanical behavior in the low temperature regime is unique. Just $1-3\,°C$ above the melting point the onset of density fluctuations leads to a drop in the bilayer rigidity and causes enhanced membrane undulations, that is, to an increased interbilayer repulsion (see Section 3.4.1). These softening effects are most pronounced in short chain PCs (see Section 3.4.2). In contrast, other phospholipids such as PEs exhibit a "normal" temperature behavior: (1) the membrane thickness displays an almost

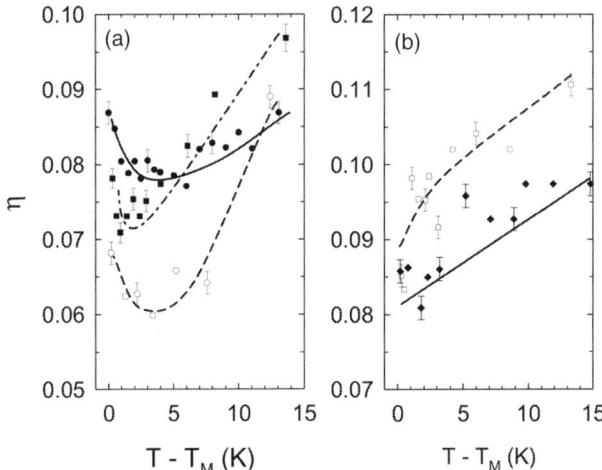

Figure 3.13 Temperature dependence of the fluctuation parameter η for (a) DMPC (●), (b) DPPC (○), (c) DSPC (■), and (d) DNPC (□, dashed line) and DAPC (♦, solid line). Lines are drawn to guide the eye. All data presented in panel (a) show an increase of fluctuations in the vicinity of T_m, which is not observed for the samples in (b). (Adapted from Ref. 138 with permission.)

linear dependence over the entire temperature regime of the L_α phase; (2) the bilayer separation is relatively small and does not increase near the melting point; and (3) the bilayer fluctuations are much smaller as compared to PC bilayer (cf. Table 3.3) and hardly change with temperature [101]. However, the PE and PC membranes differ not only in the low temperature regime, but also in their high temperature behavior.

In Fig. 3.14 the structural properties of POPC from 2 to 70 °C are compared with those of POPE from 30 to 75 °C [101]. The d-spacing of POPC decreases up to 30 °C monotonically, but then starts to increase with temperature. Remarkably, the d-value at 70 °C is almost as large as at 2 °C. A detailed examination shows that this increase is due to a strong uptake of water, which is accompanied by increased membrane fluctuations (Fig. 3.14, bottom). In principle, this is well understood, since undulation repulsion is expected to increase proportional to T^2 [141], and experimental studies showing that K_C reduces with temperature confirm this view [65, 136]. Theoretically, the undulation forces should even overwhelm the van der Waals attraction at a certain temperature and cause an unbinding of the membranes [142]. In this scenario, the bilayers are able to fluctuate freely and are no longer positionally correlated to each other. However, thermal unbinding transitions have so far only been reported for partially charged bilayer systems [143–145].

The bilayer thickness, d_B, of POPC decreases with increasing temperature. This is caused by the continuous formation of *trans–gauche* rotamers within the acyl chains. Simultaneously, the enhanced chain pressure induces the increase

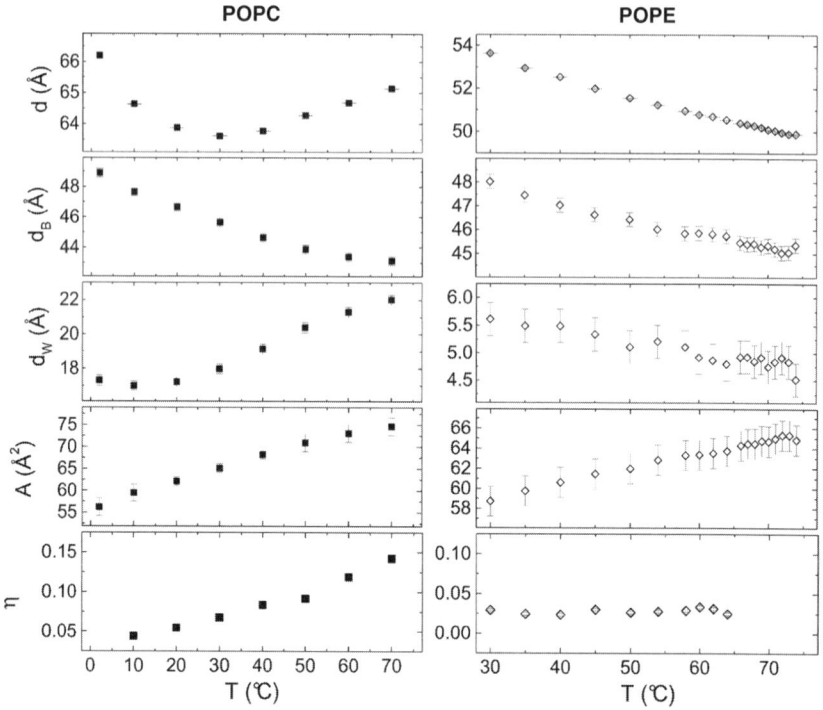

Figure 3.14 Temperature dependence of d, d_B, d_W, A, and of the Caillé parameter η of fully hydrated POPC (■) and POPE (◊) bilayers. (Panels adapted from Ref. 101 with permission.)

of the lateral area per lipid, A. In particular, we note that up to $\sim 50\,°C$ both d_B and A change almost linearly with temperature, but then display asymptotic behavior. Since only a limited number of *gauche* isomers can be generated in each hydrocarbon chain [26], the asymptotic behavior of d_B and A can be understood as a saturation effect in chain melting.

While the structural bilayer properties of POPE show a similar temperature behavior (d_B and A), the bilayer interactions are clearly different. The separation between adjacent membranes, d_W, decreases even slightly as the temperature is increased, and the bending fluctuations do not show any trend with respect to temperature (Fig. 3.14, bottom). This can be understood by the apparently strong interbilayer coupling expressed by the small bilayer separations that leads to a large B and consequently small η values (Eq. (3.5)). Nevertheless, the decrease of d_W with temperature is peculiar, because steric repulsion of bilayers usually gains influence with increasing temperature. On the other hand, it can be counteracted by the buildup of lateral membrane tension [146]. Because PEs prefer inverse curved monolayers [28] (cf. its molecular shape in Fig. 3.10a), the planar PE bilayers withstand high curvature frustration before they transform into the inverse mesophases [8, 28]; that is, the intrinsic monolayer curvature is negative in this regime and induces a high lateral tension. On this basis one may understand why undulations get suppressed in PE bilayers especially at elevated temperatures.

3.5 CONCLUSION AND OUTLOOK

In this chapter we have discussed the mechanical and structural properties of fluid bilayers and demonstrated how "macroscopic" as well as "mesoscopic" methods complement each other in the retrieval of material parameters (Sections 3.2 and 3.3). Many practical examples have been given (Figs. 3.2–3.5, 3.8, 3.10–3.14) and for sake of clarity, solely model membrane systems such as PC and PE lipid/water systems have been considered. Furthermore, we restricted ourselves to show only the influence of basic factors like temperature, chain length, and headgroup composition (Section 3.4), demonstrating that accurate characterization of fluid planar membranes is anything but trivial.

This is evident from Tables 3.2 and 3.4, which exhibit a large scatter of the reported K_C values. Moreover, there is also no correlation that allows one to discriminate data from direct (Section 3.2) and indirect (Section 3.3) methods. This is definitely unsatisfactory, but has been known since the work of Niggemann et al. [56], which demonstrated the differences in nominal equal samples applying either VFA or electric deformation. Sample preparation, sample purity, or even history all a play role; for example, Angelova et al. [85] report a measured decrease in the bending rigidity of egg-PC from 6.6 to 4.5×10^{-20} J after two weeks. Likewise, the highest value attained for egg-PC by Servuss et al. [52] might have its reasoning in an altered lipid distribution due to the extreme tubular vesicle shape [82]. On the other hand, some theoretical treatments may need to be refined. Thus experimental obscurities and/or model assumptions all together make it difficult to determine K_C on an absolute scale. Nevertheless, there is large agreement on relative changes of the bending rigidity as a function of temperature and composition of the hydrocarbon chains as shown in Section 3.4. Additionally, the rigidification of membranes by cholesterol is well accepted within the scientific community (cf. Section 3.2.2). This is encouraging, especially because membranes are systems that do no exist at constant thermodynamic conditions, but face continuous changes in temperature, pressure, pH, ionic strength, and also lipid composition. Hence we can understand a lot of the physics involved by varying these parameters, which is certainly better than being petrified by the scatter in K_C.

Apart from minor homemade problems, like lacking a convention for naming the bending modulus—at the moment you have a choice between $K_C, k_c, \kappa, \kappa_b$, or B to mention the most frequent assignments—or apart from the widespread resistance to using SI units, we have seen over the past years some progress in investigating membrane systems, especially when physical models from the liquid crystal research got integrated [24, 73]. Now we have global fitting routines at hand, allowing us to determine all important structural elements of the bilayer [64, 68, 112], and the door is wide open to tackle even elastic bilayer properties by various scattering methods [63, 69, 147]. Hand in hand with the "macroscopic" methods nearly any membrane interface from erythrocytes to biomimetic model systems can be characterized.

This opens promising possibilities for current efforts to elucidate the coupling of microscopic membrane properties to its mesoscopic features (e.g., see Refs. 15 and 62), addressing issues like how a local perturbation caused by the inclusion of a membrane protein translates into bilayer elasticity. Given the recent advances in experimental and theoretical techniques, we think that the prospects are good to find answers to this and related questions in the near future.

APPENDIX: LIST OF METHODS AND MATERIAL PARAMETERS

MPA	micropipette aspiration (K_A, K_C)
	tether formation studies (η, K_C)
VFA	vesicle fluctuation analysis (K_C)
DCM	differential confocal microscopy (K_C)
RICM	reflection interference contrast microscopy (K_A, K_C)
AFM	atomic force microscopy (E, K_C)
SAXS	small-angle X-ray scattering (K_C, B)
SANS	small-angle neutron scattering (K_C, B)
NSE	neutron spin-echo spectrometry (K_C, B)
ESR	electron spin resonance (K_C)
DLS	dynamic light scattering (η, K_A)
NMR	nuclear magnetic resonance (K_C)
DSC	differential scanning microscopy (K_C)
FPR	fluorescence photobleaching recovery (η)
	optical dynamometry studies (η, K_C)
	optical tweezers studies (η, K_A)
	electrocompression measurements (E)
v_L	lipid-specific volume (mL/g)
M_L	molecular mass of lipid (g/mol)
V_L	lipid molecular volume, $V_L = v_L M_L / 0.6022$ (Å^3)
K_B	isothermal bulk modulus (1 Pa = 1 N/m^2 = 10 dyn/cm^2)
K_A	area compressibility modulus (1 N/m (J/m^2) = 1000 dyn/cm (1000 erg/cm^2))
E	elastic modulus (Young's modulus) (1 Pa = 1 N/m^2 = 10 dyn/cm^2)
μ	surface shear modulus (1 N/m (J/m^2) = 1000 dyn/cm (1000 erg/cm^2))
K_C	bending modulus (1 J = 10^7 erg = 2.5 × 10^{20} $k_B T$)
B	bulk compression modulus of interaction (1 J/m^4 = 10^{-1} erg/cm^4)
η	Caillé parameter (a measure for membrane fluctuations)

ACKNOWLEDGMENTS

We would like to thank Peter Laggner for careful reading of the manuscript and many helpful comments. The permission to reproduce some of their experimental data by Evan Evans, John Hjort Ipsen, Chau-Hwang Lee, K.Y. Simon Ng, and Peter Rand and their co-workers is also gratefully acknowledged. Some of the research presented has been supported by the Austrian Science Funds FWF (Grants No. J2004-GEN & P17112-B10) and the European Union Grant No. INTAS-01-0105.

REFERENCES

1. E. Sackmann (1990). Molecular and global structure of membranes and lipid bilayers. *Can. J. Phys.* **68**: 999–1012.
2. E. Sackmann (1995). Biological membranes architecture and function, in *Structure and Dynamics of Membranes* (R. Lipowsky and E. Sackmann, Eds.). Elsevier, Amsterdam, pp. 1–63.
3. S. J. Singer and G. L. Nicholson (1972). The fluid mosaic model of the structure of cell membranes. *Science* **175**: 720–731.
4. L. Stryer (1988). *Biochemistry*. W. H. Freeman, San Francisco.
5. G. Cevc (Ed.) (1993). *Phospholipid Handbook*. Marcel Dekker, New York.
6. R. Lipowsky and E. Sackmann (Eds.) (1995). *Structure and Dynamics of Membranes. From Cells to Vesicles*. Elsevier, Amsterdam.
7. J. Katsaras and T. Gutberlet (Eds.) (2000). *Lipid Bilayers. Structure and Interactions*. Springer, Berlin.
8. J. M. Seddon and R. H. Templer (1995). Polymorphism of lipid water systems, in *Structure and Dynamics of Membranes* (R. Lipowsky and E. Sackmann, Eds.). North-Holland, Amsterdam, pp. 97–160.
9. T. Landh (1995). From entangled membranes to eclectic morphologies: cubic membranes as subcellular space organizers. *FEBS Lett.* **369**: 13–17.
10. M. A. Yorek (1993). Biological distribution, in *Phospholipid Handbook* (G. Cevc, Ed.). Marcel Dekker, New York
11. M. Edidin (2003). The state of lipid rafts: from model membranes to cells. *Annu. Rev. Biophys. Biomol. Struct.* **32**: 257–283.
12. K. Simons and W. L. C. Vaz (2004). Model systems, lipid rafts, and cell membranes. *Annu. Rev. Biophys. Biomol. Struct.* **32**: 263–295.
13. S. M. Gruner (1985). Intrinsic curvature hypothesis for biomembrane lipid composition: a role for nonbilayer lipids. *Proc. Natl. Acad. Sci. USA* **82**: 3665–3669.
14. J. Zimmerberg and M. M. Kozlov (2006). How proteins produce cellular membrane curvature. *Nature Rev.* **7**: 9–19.
15. P. Laggner and K. Lohner (2000). Liposome phase systems as membrane activity sensors for peptides, in *Lipid Bilayers. Structure and Interactions* (J. Katsaras and T. Gutberlet, Eds.). Springer, Berlin, pp. 233–264.

16. K. Lohner (2001). The role of membrane lipid composition in cell targeting of antimicrobial peptides, in *Development of Novel Antimicrobial Agents: Emerging Strategies* (K. Lohner, Ed.). Horizon Scientific Press, Norfolk, pp. 149–165.
17. M. Tanaka and E. Sackmann (2005). Polymer-supported membranes as models of the cell surface. *Nature* **437**: 656–663.
18. M. M. Ngundi and C. R. Taitt (2006). An array biosensor for detection of bacterial and toxic contaminants of foods. *Methods Mol. Biol.* **345**: 53–68.
19. Z. Wang, H. Shang, and G. U. Lee (2006). Nanoliter-scale reactor arrays for biochemical sensing. *Langmuir* **22**: 6723–6726.
20. A. Yaghmur, L. de Campo, S. Salentinig, L. Sagalowcz, M. E. Leser, and O. Glatter (2006). Oil-loaded monolinolein-based particles with confined inverse discontinuous cubic structure (Fd3m). *Langmuir* **22**: 517–521.
21. K. Larsson (1989). Cubic lipid–water phases: structures and biomembrane aspects. *J. Phys. Chem.* **93**: 7304–7314.
22. C. Tanford (1978). The hydrophobic effect and the organization of living matter. *Science* **200**: 1012–1018.
23. V. Luzzati (1968). X-ray diffraction studies of lipid–water systems, in *Biological Membranes* (D. Chapman, Ed.), Academic Press, New York, pp. 71–123.
24. P. G. De Gennes and J. Prost (1993). *The Physics of Liquid Crystals*. Oxford University Press, Oxford.
25. W. H. de Jeu, B. I. Ostrovskii, and A. N. Shalaginov (2003). Structure and fluctuations of smectic membranes. *Rev. Mod. Phys.* **75**: 181–235.
26. A. Seelig and J. Seelig (1974). The dynamic structure of fatty acyl chains in a phospholipid bilayer measured by deuterium magnetic resonance. *Biochemistry* **13**: 4839–4845.
27. J. Israelachvili (1991). *Intermolecular and Surface Forces*. Academic Press, London.
28. G. C. Shearman, O. Ces, R. H. Templer, and J. M. Seddon (2006). Inverse lyotropic phases of lipids and membrane curvature. *J. Phys. Condens. Matter* **18**: S1105–S1124.
29. J. F. Nagle (1980). Theory of the main lipid bilayer phase transition. *Annu. Rev. Phys. Chem.* **31**: 157–195.
30. R. Lipowsky (1995). The morphology of lipid membranes. *Curr. Opin. Struct. Biol.* **5**: 531–540.
31. T. Hianik and V. I. Passechnik (1995). *Bilayer Lipid Membranes. Structure and Mechanical Properties*. Kluwer Academic Publishers, Dordrecht.
32. S. Leibler (1989). Equilibrium statistical mechanics of fluctuating films and surfaces, in *Statistical Mechanics of Membranes and Surfaces* (D. R. Nelson, T. Piran, and S. Weinberg, Eds.), World Scientific, Singapore, pp. 1–60.
33. W. W. Webb, L. S. Barak, D. W. Tank, and E. S. Wu (1981). Molecular mobility on the cell surface. *Biochem. Soc. Symp.* **46**: 191–205.
34. R. E. Waugh (1982). Surface viscosity measurements from large bilayer vesicle tether formation. I. Analysis. *Biophys. J.* **38**: 19–27.
35. O. Alvarez and R. Latorre (1978). Voltage-dependent capacitance in lipid bilayers made from monolayers. *Biophys. J.* **21**: 1–17.

36. R. Dimova, B. Pouligny, and C. Dietrich (2000). Pretransitional effects in dimyristoylphosphatidylcholine vesicle membranes: optical dynamometry study. *Biophys. J.* **79**: 340–356.
37. T. M. Fischer (1992). Bending stiffness of lipid bilayers. 1. Bilayer couple or single-layer bending? *Biophys. J.* **63**: 1328–1335.
38. T. Hianik, M. Haburcák, K. Lohner, E. Prenner, F. Paltauf, and A. Hermetter (1998). Compressibility and density of lipid bilayers composed of polyunsaturated phospholipids and cholesterol. *Colloids Surfaces A* **139**: 189–197.
39. S. Mitaku and T. Date (1982). Anomalies of nanosecond ultrasonic relaxation in the lipid bilayer transition. *Biochim. Biophys. Acta* **14**: 211–221.
40. S. Mitaku, T. Jippo, and R. Kataoka (1983). Thermodynamic properties of the lipid bilayer transition. Pseudocritical phenomena. *Biophys. J.* **42**: 137–144.
41. S. Halstenberg, T. Heimburg, T. Hianik, U. Kaatze, and R. Krivanek (1998). Cholesterol-induced variations in the volume and enthalpy fluctuations of lipid bilayers. *Biophys. J.* **75**: 264–271.
42. L. W. Kwok and E. A. Evans (1981). Thermoelasticity of large lecithin bilayer vesicles. *Biophys. J.* **35**: 637–652.
43. E. Evans and D. Needham (1987). Physical properties of surfactant bilayer membranes: thermal transitions, elasticity, rigidity, cohesion, and colloidal interactions. *J. Phys. Chem.* **91**: 4219–4228.
44. J. O. Rädler and E. Sackmann (1993). Imaging optical thickness and separation distances of phospholipid vesicles at solid surfaces. *J. Phys. II France* **3**: 727–748.
45. R. Simson, E. Wallraff, J. Faix, J. Niewöhner, G. Gerisch, and E. Sackmann (1998). Membrane bending modulus and adhesion energy of wild-type and mutant cells of *dictyostelium* lacking talin or cortexillins. *Biophys. J.* **74**: 514–522.
46. M. F. Hildenbrand and T. M. Bayerl (2005). Differences in the modulation of collective membrane motions by ergosterol, lanosterol, and cholesterol: a dynamic light scattering study. *Biophys. J.* **88**: 3360–3367.
47. F. Nallet, D. Roux, and J. Prost (1989). Dynamic light scattering study of dilute lamellar phases. *Phys. Rev. Lett.* **62**: 276–279.
48. S. Hénon, G. Lenormand, A. Richert, and F. Gallet (1999). A new determination of the shear modulus of the human erythrocyte membrane using optical tweezers. *Biophys. J.* **76**: 1145–1151.
49. V. A. Parsegian and R. P. Rand (1995). Interaction in membrane assemblies, in *Structure and Dynamics of Membranes* (R. Lipowsky and E. Sackmann, Eds.), North-Holland, Amsterdam, pp. 643–690.
50. T. J. McIntosh, S. Advani, R. E. Burton, D. V. Zhelev, D. Needham, and S. A. Simon (1995). Experimental tests for protrusion and undulation pressures in phospholipid bilayers. *Biochemistry* **34**: 8520–8532.
51. L. Bo and R. E. Waugh (1989). Determination of bilayer membrane stiffness by tether formation from giant, thin-walled vesicles. *Biophys. J.* **55**: 509–517.
52. R. M. Servuss, W. Harbich, and W. Helfrich (1976). Measurement of the curvature-elastic modulus of egg lecithin bilayers. *Biochim. Biophys. Acta* **436**: 900–903.
53. M. B. Schneider, J. T. Jenkins, and W. W. Webb (1984). Thermal fluctuation of large cylindrical phospholipid vesicles. *Biophys. J.* **45**: 891–899.

54. M. B. Schneider, J. T. Jenkins, and W. W. Webb (1984). Thermal fluctuations of large quasi-spherical bimolecular phospholipid vesicles. *J. Phys. France* **45**: 1457–1472.
55. J. F. Faucon, M. D. Mitov, P. Méléard, I. Bivas, and P. Bothorel (1989). Bending elasticity and thermal fluctuations of lipid membranes. Theoretical and experimental requirements. *J. Phys. France* **50**: 2389–2414.
56. G. Niggemann, M. Kummrow, and W. Helfrich (1995). The bending rigidity of phosphatidylcholine bilayers: dependence on experimental method, sample cell sealing and temperature. *J. Phys. II France* **5**: 413–425.
57. C. H. Lee, W. C. Lin, and J. Wang (2001). All-optical measurements of the bending rigidity of lipid-vesicle membranes across structural phase transitions. *Phys. Rev. E* **64**: 020901.
58. Z. Shao and Q. Yang (1995). Progress in high resolution atomic force microscopy in biology. *Q. Rev. Biophys.* **28**: 195–251.
59. G. Mao, X. Liang and K. Y. S. Ng (2004). Direct force measurements of liposomes by atmomic force microscopy, in *Dekker Encyclopedia of Nanoscience and Nanotechnology*, Marcel Dekker, New York, pp. 933–942.
60. T. Heimburg (1998). Mechanical aspects of membrane thermodynamics. Estimation of the mechanical properties of lipid membranes close to the chain melting transition from calorimetry. *Biochim. Biophys. Acta* **1415**: 147–162.
61. G. Pabst, J. Katsaras, V. A. Raghunathan, and M. Rappolt (2003). Structure and interactions in the anomalous swelling regime of phospholipid bilayers. *Langmuir* **19**: 1716–1722.
62. G. Pabst (2006). Global properties of biomimetic membranes: perspectives on molecular features. *Biophys. Rev. Lett.* **1**: 57–84.
63. Y. Lyatskaya, Y. Liu, S. Tristram-Nagle, J. Katsaras, and J. F. Nagle (2001). Method for obtaining structure and interactions from oriented lipid bilayers. *Phys. Rev. E. Stat. Nonlin. Soft Matter Phys.* **63**: 011907.
64. J. Lemmich, K. Mortensen, J. H. Ipsen, T. Hønger, R. Bauer, and O. G. Mouritsen (1996). Small-angle neutron scattering from multilamellar lipid bilayers: theory, model, and experiment. *Phys. Rev. E* **53**: 5169–5180.
65. H. I. Petrache, S. Tristram-Nagle, and J. F. Nagle (1998). Fluid phase structure of EPC and DMPC bilayers. *Chem. Phys. Lipids* **95**: 83–94.
66. H. I. Petrache, N. Gouliaev, S. Tristram-Nagle, R. T. Zhang, R. M. Suter, and J. F. Nagle (1998). Interbilayer interactions from high-resolution X-ray scattering. *Phys. Rev. E* **57**: 7014–7024.
67. J. F. Nagle and S. Tristram-Nagle (2000). Structure of lipid bilayers. *Biochim. Biophys. Acta* **1469**: 159–195.
68. F. Nallet, R. Laversanne, and D. Roux (1993). Modelling X-ray or neutron scattering spectra of lyotropic lamellar phases: interplay between form and structure factors. *J. Phys. II France* **3**: 487–502.
69. M. C. Rheinstädter, W. Häußler, and T. Salditt (2006). Dispersion relation of lipid membrane fluctuations by neutron spin-echo spectrometry. *Phys. Rev. Lett.* **97**: 048103-1–048103-4.
70. J.-M. Di Meglio, M. Dvolaitzky, and D. Taupin (1985). Determination of the rigidity constant of amphiphilic film in "birefringent microemulsions": the role of cosurfactant. *J. Phys. Chem.* **89**: 871–874.

71. F. Auguste, P. Barois, L. Fredon, B. Clin, E. J. Dufourc, and A. M. Bellocq (1994). Flexibility of molecular films as determined by deuterium solid state NMR. *J. Phys. II France* **4**: 2197–2214.
72. M. F. Brown, R. L. Thurmond, S. W. Dodd, D. Otten, and K. Beyer (2002). Elastic deformation of membrane bilayers probed by deuterium NMR relaxation. *J. Am. Chem. Soc.* **124**: 8471–8484.
73. Caillé, A. (1972). Remarques sur la diffusion des rayons X dans les smectiques A. *C. R. Acad. Sci. Paris B* **274**: 891–893.
74. R. Zhang, S. Tristram-Nagle, W. Sun, R. L. Headrick, T. C. Irving, R. M. Suter, and J. F. Nagle (1996). Small-angle X-ray scattering from lipid bilayers is well described by modified Caillé theory but not by paracrystalline theory. *Biophys. J.* **70**: 349–357.
75. T. Salditt (2005). Thermal fluctuations and stability of solid-supported lipid membranes. *J. Phys. Condens. Matter* **17**: R287–R314.
76. R. Goetz, G. Gompper, and R. Lipowsky (1999). Mobility and elasticity of self-assembled membranes. *Phys. Rev. Lett.* **82**: 221–224.
77. G. Brannigan and F. L. H. Brown (2005). Composition dependence of bilayer elasticity. *J. Chem. Phys.* **122**: 074905-1–074905-11.
78. W. Rawicz, K. C. Olbrich, T. McIntosh, D. Needham, and E. Evans (2000). Effect of chain length and unsaturation on elasticity of lipid bilayers. *Biophys. J.* **79**: 328–339.
79. E. Evans and W. Rawicz (1990). Entropy-driven tension and bending elasticity in condensed-fluid membranes. *Phys. Rev. Lett.* **64**: 2094–2097.
80. L. D. Landau and E. M. Lifschitz (1989). *Elastizitätstheorie*. Akademie-Verlag, Berlin.
81. M. Kummrow and W. Helfrich (1991). Deformation of giant lipid vesicles by electric fields. *Phys. Rev. A* **44**: 8356–8360.
82. D. V. Zhelev, D. Needham, and R. M. Hochmuth (1994). A novel micropipette method for measuring the bending modulus of vesicle membranes. *Biophys. J.* **67**: 720–727.
83. F. Brochard and J. F. Lennon (1975). Frequency spectrum of the flicker phenomenon in erythrocytes. *J. Phys. France* **36**: 1035–1047.
84. F. Brochard, P. G. De Gennes, and P. Pfeuty (1976). Surface tension and deformations of membranes structures: relation to two-dimensional phase transitions. *J. Phys. France* **37**: 1099–1104.
85. M. Angelova, S. Soleau, P. Méléard, J. F. Foucon, and P. Bothorel (1992). Preparation of giant vesicles by external a.c. electric fields. Kinetics and applications. *Prog. Colloid Polym. Sci.* **89**: 127–131.
86. J. R. Henriksen and J. H. Ipsen (2002). Thermal undulations of quasi-spherical vesicles stabilized by gravity. *Eur. Phys. J. E* **9**: 365–374.
87. H.-G. Döbereiner, E. A. Evans, M. Kraus, U. Seifert, and M. Wortis (1997). Mapping vesicle shapes into the phase diagram: a comparison of experiment and theory. *Phys. Rev. E* **55**: 4458–4474.
88. H.-G. Döbereiner, G. Gompper, C. K. Haluska, D. M. Kroll, P. G. Petrov, and K. A. Riske (2003). Advanced flicker spectroscopy of fluid membranes. *Phys. Rev. Lett.* **91**: 048301-1–048301-4.
89. T. Baumgart, S. Das, W. W. Webb, and J. T. Jenkins (2005). Membrane elasticity in giant vesicles with fluid phase coexistence. *Biophys. J.* **89**: 1067–1080.

90. J. R. Henriksen, A. C. Rowat, and J. H. Ipsen (2004). Vesicle fluctuation analysis of the effects of sterols on the membrane bending rigidity. *Eur. Biophys. J.* **33**: 732–741.

91. D. Needham, T. J. McIntosh, and E. Evans (1988). Thermomechanical and transition properties of dimyristoylphosphatidylcholine/cholesterol bilayers. *Biochemistry* **27**: 4668–4673.

92. D. Needham and R. S. Nunn (1990). Elastic deformation and failure of lipid bilayer membranes containing cholesterol. *Biophys. J.* **58**: 997–1009.

93. H. P. Duwe, J. Käs, and E. Sackmann (1990). Bending moduli of lipid bilayers: modulation by solute. *J. Phys. France* **51**: 945–962.

94. P. Méléard, C. Gerbeaud, T. Pott, L. Fernandez-Puente, I. Bivas, M. D. Mitov, J. Dufourcq, and P. Bothorel (1997). Bending elasticities of model membranes: influences of temperature and sterol content. *Biophys. J.* **72**: 2616–2629.

95. I. S. Sokolnikov (1964). *Tensor Analysis: Theory and Applications to Geometry and Mechanics of Continua*. Wiley, Hoboken, NJ.

96. L. Fernandez-Puente, I. Bivas, M. D. Mitov, and P. Méléard (1994). Temperature and chain length effects on bending elasticity of phosphatidylcholine bilayers. *Europhys. Lett.* **28**: 181–186.

97. C. F. Quate (1994). The AFM as a tool for surface imaging. *Surf. Sci.* **299/300**: 980–995.

98. X. Liang, G. Mao, and K. Y. S. Ng (2004). Mechanical properties and stability measurement of cholesterol-containing liposome on mica by atomic force microscopy. *J. Colloid Interface Sci.* **278**: 53–62.

99. D. E. Laney, R. A. Garcia, S. M. Parson, and H. G. Hansma (1997). Changes in the elastic properties of cholinergic synaptic vesicles as measured by atomic force microscopy. *Biophys. J.* **72**: 806–813.

100. E. Evans (1974). Bending resistance and chemically induced moments in membrane bilayers. *Biophys. J.* **14**: 923–931.

101. M. Rappolt, P. Laggner and G. Pabst (2004). Structure and elasticity of phospholipid bilayers in the L_α phase: a comparison of phosphatidylcholine and phosphatidylethanolamine membranes, in *Recent Research Developments in Biophysics* (S. G. Pandalai, Ed.). Transworld Research Network, Trivandrum, pp. 365–394.

102. P. Laggner and H. Mio (1992). SWAX—a dual-detector camera for simultaneous small- and wide-angle X-ray diffraction in polymer and liquid crystal research. *Nucl. Instrum. Methods Phys. Res. A* **323**: 86–90.

103. H. Amenitsch, M. Rappolt, M. Kriechbaum, H. Mio, P. Laggner, and S. Bernstorff (1998). First performance assessment of the small-angle X-ray scattering beamline at ELETTRA. *J Synchrotron Radiat* **5**: 506–508.

104. S. Bernstorff, H. Amenitsch, and P. Laggner (1998). High-throughput asymmetric double-crystal monochromator of the SAXS beamline at ELETTRA. *J. Synchrotron Radiat* **5**: 1215–1221.

105. R. Hosemann and S. N. Bagchi (1962). *Direct Analysis of Diffraction by Matter*. North-Holland, Amsterdam.

106. A. Guinier (1963). *X-Ray Diffraction*. W. H. Freeman, San Francisco.

107. B. E. Warren (1969). *X-Ray Diffraction*. Addison-Wesley, Reading, MA.

108. H. I. Petrache, T. Zemb, L. Belloni, and V. A. Parsegian (2006). Salt screening and specific ion adsorption determine neutral-lipid membrane interactions. *Proc. Natl. Acad. Sci. U S A* **23**: 7982–7987.
109. H. I. Petrache, S. Tristram-Nagle, D. Harries, N. Kucerka, J. F. Nagle, and V. A. Parsegian (2006). Swelling of phospholipids by monovalent salt. *J. Lipid Res.* **47**: 302–309.
110. R. Lipowsky (1995). Generic interactions of flexible membranes, in *Structure and Dynamics of Membranes* (R. Lipowsky and E. Sackmann, Eds.). North-Holland, Amsterdam, pp. 521–602.
111. L. Mennicke, D. Constantin, and T. Salditt (2006). Structure and interaction potentials in solid-supported lipid membranes studied by X-ray reflectivity at varied osmotic pressure. *Eur. Phys. J. E* **20**: 221–230.
112. G. Pabst, M. Rappolt, H. Amenitsch, and P. Laggner (2000). Structural information from multilamellar liposomes at full hydration: full q-range fitting with high quality X-ray data. *Phys. Rev. E* **62**: 4000–4009.
113. G. Pabst, R. Koschuch, B. Pozo-Navas, M. Rappolt, K. Lohner, and P. Laggner (2003). Structural analysis of weakly ordered membrane stacks. *J. Appl. Crystallogr.* **63**: 1378–1388.
114. P. Laggner and H. Stabinger (1976). The partial specific volume changes involved in the thermotropic phase transitions of pure and mixed lecithins, in *Colliod and Interface Science* (M. Kerker, Ed.), Academic Press, New York, p. 91.
115. W. J. Sun, R. M. Suter, M. A. Knewtson, C. R. Worthington, S. Tristram-Nagle, R. Zhang, and J. F. Nagle (1994). Order and disorder in fully hydrated unoriented bilayers of gel phase dipalmitoylphosphatidylcholine. *Phys. Rev. E* **49**: 4665–4676.
116. T. J. McIntosh and S. A. Simon (1986). Area per molecule and distribution of water in fully hydrated dilauroylphosphatidylethanolamine bilayers. *Biochemistry* **25**: 4948–4952.
117. G. Büldt and H. U. Gally, A. Seelig, J. Seelig, and G. Zaccai (1978). Neutron diffraction studies on selectively deuterated phospholipid bilayers. *Nature* **271**: 182–184.
118. D. D. Lasic (1993). *Liposomes: From Physics to Applications*. Elsevier, Amsterdam.
119. M. Rappolt, H. Amenitsch, J. Strancar, C. V. Teixeira, M. Kriechbaum, G. Pabst, M. Majerowicz, and P. Laggner (2004). Phospholipid mesophases at solid interfaces: in-situ X-ray diffraction and spin-label studies. *Adv. Colloid Interface Sci.* **111**: 63–77.
120. J. Katsaras and V. A. Raghunathan (2000). Aligned lipid–water systems, in *Lipid Bilayers. Structure and Interactions* (J. Katsaras and T. Gutberlet, Eds.), Springer, Berlin.
121. H. Amenitsch, M. Rappolt, C. V. Teixeira, M. Majerowicz, and P. Laggner (2004). In situ sensing of salinity in oriented lipid multilayers by surface X-ray scattering. *Langmuir* **20**: 4621–4628.
122. Y. Liu and J. F. Nagle (2004). Diffuse scattering provides material parameters and electron density profiles of biomembranes. *Phys. Rev. E* **69**: 040901.
123. T. Salditt, M. Vogel, and W. Fenzl (2003). Thermal fluctuations and positional correlations in oriented lipid membranes. *Phys. Rev. Lett.* **90**: 178101.
124. T. Salditt, H. J. Vogel, and W. Fenzl (2004). Erratum: Thermal fluctuations and positional correlations in oriented lipid membranes [*Phys. Rev. Lett.* **90**: 178101 (2003)]. *Phys. Rev. Lett.* **93**: 169903.

125. D.-P. Li, S.-X. Hu, and M. Li (2006). Full q-space analysis of X-ray scattering of multilamellar membranes at liquid–solid interfaces. *Phys. Rev. E* **73**: 031916-1–031916-8.
126. N. Kucerka, Y. Liu, N. Chu, H. I. Petrache, S. Tristram-Nagle, and J. F. Nagle (2005). Structure of fully hydrated fluid phase DMPC and DLPC lipid bilayers using X-ray scattering from oriented multilamellar arrays and from unilamellar vesicles. *Biophys. J.* **88**: 2626–2637.
127. N. Kucerka, S. Tristram-Nagle, and J. F. Nagle (2005). Structure of fully hydrated fluid phase lipid bilayers with monounsaturated chains. *J. Membrane Biol.* **208**: 193–202.
128. N. Fuller and R. P. Rand (2001). The influence of lysolipids on the spontaneous curvature and bending elasticity of phospholipid membranes. *Biophys. J.* **81**: 243–254.
129. S. M. Gruner, V. A. Parsegian, and R. P. Rand (1986). Directly measured energy of phospholipid HII hexagonal phases. *Faraday Discuss. Chem. Soc.* **81**: 29–37.
130. R. P. Rand, N. Fuller, S. M. Gruner, and V. A. Parsegian (1990). Membrane curvature, lipid segregation, and structural transitions for phospholipids under dual-solvent stress. *Biochemistry* **29**: 76–87.
131. A. Bradford, J. Atkinson, and R. P. Rand (2003). The effect of vitamin E on the structure of membrane lipid assemblies. *J. Lipid Res.* **44**: 1940–1944.
132. Z. Chen and R. P. Rand (1997). The influence of cholesterol on phospholipid membrane curvature and bending elasticity. *Biophys. J.* **73**: 267–276.
133. L. Fernandez-Puente, I. Bivas, M. D. Mitov, and P. Méléard (1994). Temperature and chain length effects on bending elasticity of phosphatidylcholine bilayers. *Europhys. Lett.* **28**: 181–186.
134. W. Helfrich (1978). Steric interaction of fluid membranes in multilayer systems. *Z. Naturforsch.* **33a**: 305–315.
135. N. Chu, N. Kucerka, Y. Liu, S. Tristram-Nagle, and J. F. Nagle (2005). Anomalous swelling of lipid bilayer stacks is caused by softening of the bending modulus. *Phys. Rev. E* **71**: 041904-1–041904-8.
136. G. Pabst, J. Katsaras, and V. A. Raghunathan (2002). Enhancement of steric repulsion with temperature in oriented lipid multilayers. *Phys. Rev. Lett.* **88**: 128101.
137. M. Vogel, C. Munster, W. Fenzl, and T. Salditt (2000). Thermal unbinding of highly oriented phospholipid membranes. *Phys. Rev. Lett.* **84**: 390–393.
138. G. Pabst, H. Amenitsch, D. P. Kharakoz, P. Laggner, and M. Rappolt (2004). Structure and fluctuations of phosphatidylcholines in the vicinity of the main phase transition. *Phys. Rev. E* **70**: 021908-1–021908-9.
139. G. Pabst, M. Rappolt, H. Amenitsch, S. Bernstorff, and P. Laggner (2000). X-ray kinematography of temperature-jump relaxation probes the elastic properties of fluid bilayers. *Langmuir* **16**: 8994–9001.
140. S. S. Korreman and D. Posselt (2000). Chain length dependence of anomalous swelling in multilamellar lipid vesicles. *Eur. Phys. J. E* **1**: 87–91.
141. W. Helfrich (1973). Elastic properties of lipid bilayers: theory and possible experiments. *Z. Naturforsch. C* **28**: 693–703.
142. R. Lipowsky and S. Leibler (1986). Unbinding transitions of interacting membranes. *Phys. Rev. Lett.* **56**: 2541–2544.
143. M. Mutz and W. Helfrich (1989). Unbinding transition of a biological model membrane. *Phys. Rev. Lett.* **62**: 2881–2884.

144. B. Pozo-Navas, V. A. Raghunathan, J. Katsaras, M. Rappolt, K. Lohner, and G. Pabst (2003). Discontinuous unbinding of lipid multibilayers. *Phys. Rev. Lett.* **91**: 028101.
145. B. Pozo-Navas, K. Lohner, G. Deutsch, E. Sevcsik, K. A. Riske, P. Garidel, R. Dimova, and G. Pabst (2005). Composition dependence of vesicle morphology and mixing properties in a bacterial model membrane system. *Biochim. Biophys. Acta* **1716**: 40–48.
146. G. Waton and G. Porte (1993). Transient behavior and relaxations of the L3 (sponge) phase: T-jump experiments. *J. Phys. II France* **3**: 515–530.
147. J. Daillant, E. Bellet-Amalric, T. Charitat, G. Fragneto, F. Graner, S. Mora, and F. Rieutord (2005). Structure and fluctuations of a single floating lipid bilayer. *Proc. Natl. Acad. Sci. USA* **102**: 11639–11644.

CHAPTER 4

X-Ray Diffraction Studies of Lung Surfactant Membrane Structures

MARCUS LARSSON

Childrens' Hospital, Lund University Hospital, Lund, Sweden

CONTENTS

4.1	Introduction	84
4.2	X-Ray Diffraction	84
	4.2.1 Basics on Diffraction Methods	84
	4.2.2 The Solid State of Lipids	86
	4.2.3 Liquid-Crystalline Lipid Structures and Lipid–Water Phases	86
	4.2.4 Significance of the Lamellar Liquid-Crystalline Phase to Membrane Structures	88
	4.2.5 Cubic Bilayer Structures in Cell Membranes	89
	4.2.6 Minimal Surface Structures of Lipid Bilayers Can Be Described by Periodic Nodal Surfaces	89
4.3	Characterization of Exogenous Lung Surfactants	90
	4.3.1 X-Ray Studies of the Aqueous Systems of Bovine and Porcine Lung Surfactant Extracts	90
	4.3.2 Effects of Cholesterol on the Bilayer Phase Behavior in the Lung Surfactant System	92
	4.3.3 Interaction Between Lung Surfactant and Serum Albumin	97
4.4	Endogeneous Lung Surfactant	97
	4.4.1 A Surface Phase at the Air–Water Interface	97
	4.4.2 A Minimal Surface Structure Model of the Surface Phase	99
	4.4.3 On X-Ray Diffraction Data of the TM Bilayer Organization	101
4.5	Comments on Neutron Diffraction/Scattering	102
4.6	Conclusion	103
Acknowledgments		103
References		104

Structure and Dynamics of Membranous Interfaces, edited by Kaushik Nag
Copyright © 2008 John Wiley & Sons, Inc.

4.1 INTRODUCTION

Most of our present knowledge on the molecular organization of cell membranes is based on X-ray diffraction of model systems consisting of lipids, proteins, and water. During recent years, cryo-transmission electron microscopy (cryo-TEM) has been introduced, which has provided important additional information. This type of electron microscopy method allows vitrified water to be present in the samples, which is most important in studies of biomolecular organization *in vivo* [1]. Cryo-TEM has been used in examination of dispersed particles—from cell membrane assemblies to virus structures. Conventional electron microscopy (EM) is another method that has provided general insights into the ultrastructure of biomembrane organization in tissues, for example, of lung surfactant at the alveolar surface [2].

Lipids in an aqueous environment form well-defined lipid–water phases with periodic structures, which depending on the temperature are crystalline or liquid-crystalline. Lung surfactant is the lipid–protein membrane system that lines the alveolar spaces toward air. It is a beautiful example of a coherent membrane structure providing fundamental physiological functions. Lung surfactant can therefore be used as a model system for general biomembrane studies, such as lipid–protein interaction, interfacial energy, and rheological properties. The lung surface is quite difficult to study directly. Therefore a combination of different experimental methods is needed.

The lung surfactant system has evolved with terrestrial animals in order to facilitate oxygen and carbon dioxide transport between air and an aqueous environment by reducing collapsing and destabilizing forces in the terminal air spaces [3]. Lung surfactant in mammals is secreted from the alveolar type II cells. This complex membrane system consists of lipid/protein bilayers that self-assemble into two types of arrangements: lamellar bodies and tubular myelin [4]. The main lipid component is dipalmitoyl phosphatidylcholine (DPPC) [5] and about 13 mol% cholesterol is also present [6]. There are four proteins specific for lung surfactant. Two are cationic and hydrophobic: SP-B and SP-C. These are solubilized in the bilayer in the lamellar bodies (LBs) [7]. The other two, SP-A and SP-D, are anionic and hydrophilic and both possess immunological functions in the innate defense [8]. Furthermore, SP-A interacting with membrane-bound SP-B and calcium ions forms a scaffold that spans the lung surfactant bilayers into a more hydrated organization—tubular myelin (TM) [9]. The functional organization of lung surfactant has been the subject of intensive research during the latest decades and it can serve as an illustrative example of both possibilities and limitations of X-ray diffraction methods.

4.2 X-RAY DIFFRACTION

4.2.1 Basics on Diffraction Methods

The *International Tables for X-ray Crystallography* [10] is the most important source of information covering this field. Alexander's textbook on X-ray diffraction is useful also for lipid applications, even if it is mainly focused on polymers [11].

The characteristic feature of a crystal is periodicity and the existence of a smallest structure unit, the *unit cell*, which is defined by its three axes: a, b, and c. The atomic positions within the unit cell are defined by the x, y and z coordinates along the a, b, and c axes, respectively. The angles between these axes are termed alpha (between b and c), beta (between a and c), and gamma (between a and b). The unit cell in lipids usually contains two or four molecules and reflects the fact that, also in the crystalline state, the molecules form bimolecular unit layers as in cell membranes.

The whole crystal is obtained by repetition of the unit cell in three dimensions, and the resulting corners of the unit cells in space are called *lattice points*. Planes through these points—*lattice planes*—are identified by their indices (*hkl*), which is equal to the number of intersections along the a, b, and c axes, respectively. If, for example, a set of all parallel lattice planes cuts the a axis four times, the h index of these planes is 4.

When X-ray light interacts with periodic lattices of atoms, diffraction will take place when the Bragg equation is fulfilled:

$$n\lambda = 2d \sin \Theta$$

where d is the distance between the lattice planes that we consider, n is an integer defining any multiple of the wavelength of the X-ray radiation λ, and the angle between the X-ray beam and this particular set of lattice planes is Θ. There is a simple reflection analogy that is useful in order to understand Bragg's law. Thus diffraction takes place when the difference in length that the X-rays have to travel at "reflection" from adjacent planes is equal to one wavelength, or a multiple of the wavelength. All lattice planes through a crystal can therefore give diffraction at the directions in relation to the incident beam with $n = 1, 2, 3\ldots$. By measuring the position of the reflections, the geometry and symmetry of the crystal lattice can be determined, and therefore also the dimensions of the unit cell.

It is more complicated to determine the complete crystal structure—the positions of all the atoms in the unit cell. This requires recording of all individual X-ray reflections from a single crystal. The procedure will briefly be summarized here. Every atom in the unit cell will contribute to the reflection intensities of a particular lattice plane, which is identified by its indices (*hkl*). The sum of all atomic contributions gives the value of the so-called structure factor $F(hkl)$. If we know the structure factor, the electron density within the unit cell can be calculated by a Fourier summation. Only the absolute value of the structure factor can be directly determined experimentally, not the phase angle of this complex number. The determination of the phase angles is thus a critical step in X-ray crystallography. The X-ray data discussed in this chapter, however, is focused on liquid-crystalline phases, where single-crystal structural determinations are impossible. Rather, we are concerned with periodicity and structure unit geometry. In order to follow phase transitions induced by temperature, a very powerful method is the recording of the X-ray diffraction pattern versus temperature. A useful feature in such recordings is the possibility to differentiate between the order/disorder on atomic distances from the long-range periodicity. This will be illustrated in Sections 4.3 and 4.4.

The *experimental setup* is simply an optical bench along a collimated X-ray beam, usually the radiation used is copper K-α. In the studies described later the beam is linear after collimation and penetrates a sample with a thickness of about 1 mm (usually a glass capillary). Also, point focus geometry is used. The primary beam is collected in a beam stopper and the scattered or diffracted radiation is recorded versus the diffraction angle (using photographic film or electronic position-sensitive detectors). The experimental limitation is to record the diffraction at very low diffraction angles, which requires high precision of the collimation geometry, related to the strength of the primary beam.

4.2.2 The Solid State of Lipids

Knowledge of the crystal structure of lipids has been fundamental in the understanding of lipid molecular arrangement in all states of order and also in the analysis of disordered structures like cell membranes. In the solid state, lipid structures usually consist of infinite bilayers with the hydrocarbon chains fully extended and close packed, with the polar end groups forming surface planes outwards. The hydrocarbon chain close-packing arrangement can be identified from the X-ray diffraction, and there are two alternatives that dominate in membrane lipid crystals. One is a hexagonal chain packing, which is characterized by one dominant short-spacing diffraction line at about 4.1 Å. The other chain packing arrangement is also termed according to its symmetry—the orthorhombic packing. It gives two strong diffraction lines at about 3.8 and 4.2 Å [12, 13].

Similar bilayer units with crystalline chains can also be formed by cell membranes. In early studies of lipid–water phases, such as the lamellar liquid-crystalline phase of dipalmitoyl phosphatidylcholine, crystallization of the bilayers was achieved by cooling. This phase transition was termed the "main transition" and the structure after crystallization was described as a "gel phase." We will come back to this behavior in discussion of lung surfactant.

4.2.3 Liquid-Crystalline Lipid Structures and Lipid–Water Phases

Luzzati and co-workers reported in 1960 the structures of the liquid-crystalline phases in soap–water systems, and their pioneering work is the basis for X-ray studies of lipids and membranes even today [14]. A crystal of a lipid can be described as a stack of crystalline lipid bilayers. Such a crystal can melt directly when it is heated, or it can sometimes pass a state where the hydrocarbon chains of the bilayers become disordered as in liquid paraffin although the gross structure remains unchanged.

When the X-ray diffraction pattern during such a transition is followed, the lattice dimension corresponding to the bilayer thickness is observed to shift a bit, which corresponds to the high molecular disorder in the liquid-crystalline phase. The short lattice spacings ("short-spacings") reflect the close packing of hydrocarbon chains—the spacings around 4 Å that we just discussed. They are well defined in the crystalline phase but disappear at the transition into a liquid-crystalline phase, due to lack of order at such short distances. It was in fact the experimentally proven liquid-like nature of the hydrocarbon chains, in spite of crystallographic order over long distances, that was major evidence for the structure models presented by Luzzati and co-workers.

The long-range order is analyzed in the small-angle region (by SAXS—small-angle X-ray scattering). A limited region in the wide-angle region defines order or disorder of the hydrocarbon chains (WAXS—wide-angle X-ray scattering). Typical lipid–water phases are identified from their X-ray diffraction pattern in the wide-angle region by a broad and diffuse X-ray scattering band around 4.5 Å, corresponding to the average distance between the carbon atoms in the chains [14].

The Lamellar Liquid-Crystalline Phase This phase—also called L_α phase—consists of lipid bilayers alternating with water layers [14]. Within the bilayer, the molecules possess translational freedom and are therefore highly disordered. Thus there is only periodicity in one dimension of this phase, and the dimension of the structure unit is the water layer thickness + the lipid bilayer thickness. The X-ray spacings therefore exhibit the following ratios:

$1/1 : 1/2 : 1/3 : 1/4 : 1/5 : 1/6 \ldots$ (corresponding to the first, second, etc. reflection order)

The intensity of the X-ray reflections is reduced by an exponential function of the temperature factor and the diffraction angle. The temperature factor is high in the lamellar phases. Therefore, due to high thermal disorder in the L_α phase, only two to four spacings are usually observed. By plotting the diffraction spacing versus composition, however, it is still possible to identify this phase. The thickness of the lipid bilayer can then be obtained as well as the water layer variations. Examples on this will be demonstrated later.

The Hexagonal Phases There are two types of such lipid–water phases. In order to illustrate the structure, let us consider a cylindrical micelle of soap molecules in water. When the concentration is increased, the micelles will successively adopt a parallel arrangement and ultimately pack in the same way, as a bundle of any set of cylindrical rods will close-pack, with every cylindrical rod surrounded by six others [14]. This is the structure of the ordinary hexagonal phase, termed H_I.

The other hexagonal phase is just the inversion of H_I with regard to water and lipid localization. This phase is termed H_{II}. There is thus a hexagonal arrangement of water cylinders with lipid bilayers in between.

In both cases the X-ray diffraction pattern corresponds to two-dimensional periodicity with the characteristic ratios between the three first observed reflection spacings:

$$1/1 : 1/\sqrt{3} : 1/\sqrt{4} : 1/\sqrt{7}$$

Usually only the first three reflections are observed. From the spacings the radius of the cylinders and the distances between cylinders can be calculated (from volume ratios of lipid:water). According to the phase diagram, it is often obvious if the hexagonal phase is H_{II} or H_I, as an H_{II} phase can coexist with excess of water, whereas an H_I phase by dilution with water forms a micellar solution (sometimes via a cubic phase) [13].

Cubic Lipid–Water Phases with the Lipid Bilayer Forming a Minimal Surface   There is finally a third general type of lipid–water phase, which consists of infinite bilayers that are curved in space into so-called minimal surfaces. The mathematical condition is that the bilayer is as convex as it is concave in all positions. The bilayer is free from self-intersections, and it separates two congruent water channel systems. These cubic phases are therefore both lipid and water continuous [15].

The X-ray reflections that characterize such a cubic phase show the following possible spacings:

$$1/1 : 1/\sqrt{2} : 1/\sqrt{3} : 1/\sqrt{4} : 1/\sqrt{5} : 1/\sqrt{6} : 1/\sqrt{8} : 1/\sqrt{9} : 1/\sqrt{10} : 1/\sqrt{11}, \ldots$$

Usually many reflections are observed and this is probably related to the three-dimensional periodicity, making such phases quite stiff.

There are three types of such minimal surface-based cubic phases, which have been seen in lipid–water systems, and all three occur in cell membranes. The actual type can be identified by certain absences among the observed X-ray reflections. It should be mentioned that cubic phases also exist as a cubic arrangement of micellar aggregates. These micellar aggregates can be either lipids in a water environment, or water aggregates in a lipid matrix [37].

4.2.4 Significance of the Lamellar Liquid-Crystalline Phase to Membrane Structures

Unilamellar and multilamellar vesicles (also termed liposomes) represent colloidal dispersions of the lamellar liquid-crystalline phase in excess of water (above the limit of swelling of the lamellar phase). This lipid bilayer conformation dominates cell membranes *in vivo*. Cholesterol is a common component in mammalian cell membranes. A structural change of the bilayer structure takes place at a certain concentration of cholesterol in relation to the bilayer phospholipids. Above this concentration, the membrane is transformed into a more ordered form; this phase segregation within bilayers/membranes induced by cholesterol is discussed in Section 4.3.

Furthermore, the lipid molecules may crystallize in a lamellar liquid-crystalline lipid–water phase, although the lamellar arrangement of water layers and bilayers persists. Such phase transitions in saturated phosphatidylcholines, such as dipalmitoyl phosphatidylcholine (DPPC) that dominates in lung surfactant, involve another crystallized bilayer phase with a periodicity also along the bilayers (known as ripple bilayers). This complex behavior of DPPC in water was first fully explained by Janiak et al. [16].

Lung surfactant (the lipids, SP-B, and SP-C) is secreted from the alveolar type II cells as a spherically concentric arrangement of lipid bilayers (SP-A is secreted in a separate pathway by the type II cells). These colloidal particles termed lamellar bodies (LBs) have the same bilayer conformation at physiological temperature as in lamellar liquid-crystalline phases.

4.2.5 Cubic Bilayer Structures in Cell Membranes

In connection with the early studies of the cubic structure in well-defined lipid–water phases, it was demonstrated that the membrane system in the so-called prolamellar body of plants was fully consistent with the cubic minimal surface structure of lipid–water phases [17]. Chloroplasts, the photosynthetic machinery in plants, are transformed in darkness into etioplasts, where the stacks of thylakoid membranes have been reorganized into such prolamellar bodies with cubic membranes. Many hundreds of examples of similar cubic bilayer conformations in cell membrane assemblies have been observed [15].

4.2.6 Minimal Surface Structures of Lipid Bilayers Can Be Described by Periodic Nodal Surfaces

Periodic minimal surfaces are very complicated to describe analytically, but von Schnering and Nesper [18] discovered simple nodal surface equations that approximate the minimal surface structure within a few percent. An example is the cubic bilayer structures observed in lipid–water systems corresponding to the gyroid minimal surface (one of the cubic phases mentioned earlier). It can be obtained by the following nodal surface equation [20]:

$$\cos \pi x \sin \pi y + \sin \pi x \cos \pi z + \cos \pi y \sin \pi z = 0$$

Another example is the tetragonal minimal surface CLP ("crossing layers of parallels").

The conformation of such a lipid bilayer can be obtained by the equation

$$\cos \frac{\pi}{4}(x - y)e^{(1/40)\cos \pi z} - \cos \frac{\pi}{4}(x + y) = 0$$

The structure is shown in Fig. 4.1 and was generated by the computer program Mathematica 2.2 for Macintosh (Wolfram Research Inc., USA).

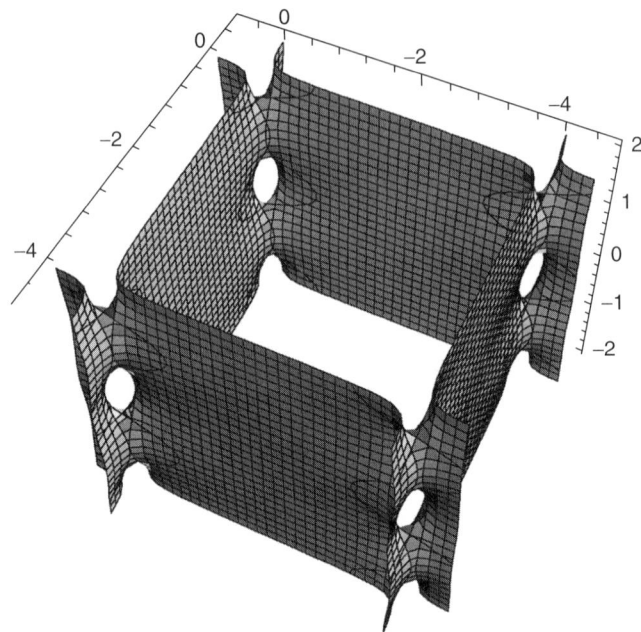

Figure 4.1 The nodal surface representation of the CLP minimal surface bilayer structure. The structure was calculated on a Macintosh computer using the program Mathematica 2.2.

This surface corresponds to the proposed structure of the lipid bilayer in the alveolar surface phase [19], an important functional organization covered in this Chapter (see Section 4.4). Andersson et al. [20] have introduced a method for description of periodic structures with finite periodicity for the description of biological structures. Using this method, a nodal surface description of a thin film of the CLP phase, which can model the alveolar lining, can be calculated.

4.3 CHARACTERIZATION OF EXOGENOUS LUNG SURFACTANTS

It will now be demonstrated how X-ray diffraction can be applied in the analysis of structure and aqueous swelling behavior of lung surfactant membrane phases; see Refs. 21–23.

4.3.1 X-Ray Studies of the Aqueous Systems of Bovine and Porcine Lung Surfactant Extracts

The X-ray diffraction curves versus temperature are used in order to characterize the structures and to find out if there are any phase transitions.

A gel phase with a crystalline bilayer shows ideal swelling that is a linear swelling against the volume fraction of water. The partial specific volume of the phospholipids is rather close to that of water. Therefore the spacings that we observe follows the equation

$$d = d_l[1 + c/(100 - c)]$$

where d_l is the thickness of the bilayer and c is the weight concentration of water in percent. From a series of measurements versus concentration, the lipid bilayer thickness can be obtained by extrapolation to zero concentration. The concentration showing the limit of swelling is obtained by the intercept between the curve with increasing d values versus c and the line with constant d versus c (when the swelling limit has been reached).

In the lamellar liquid-crystalline phase there is usually some deviation from a linear swelling behavior, which may reflect increasing disorder within the lipid bilayer with increasing water content.

A Bovine Lung Surfactant Extract The diffraction curves at a series of temperatures of the bovine surfactant preparation BLES [21] formed by equilibration with 50% aqueous phase consisting of physiological saline solution (0.15 M) are shown in Fig. 4.2. As is evident from the diffraction in the wide-angle region, there is a transition at 20–25 °C from a gel phase with crystalline chains into an L_α phase with disordered chains (a diffraction peak at 4.2 Å in the wide-angle region disappears in this temperature range).

If we select the curve at 5 °C, the thickness of the lipid bilayer can be calculated by the formula above. It is 44.3 Å assuming that the partial specific volumes of

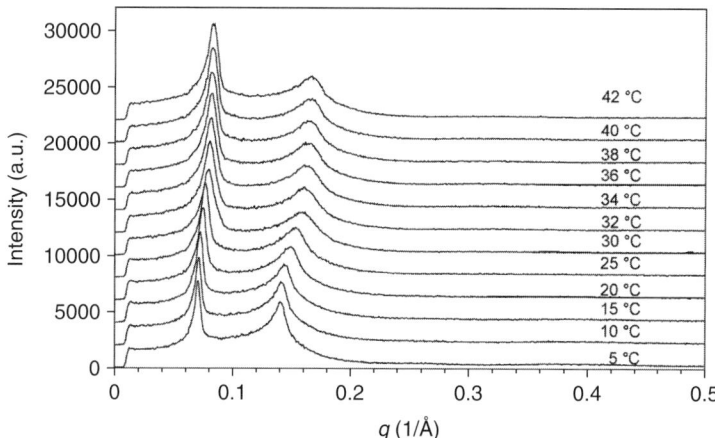

Figure 4.2 X-ray diffraction versus temperature in the small-angle region of a sample of 50 wt% bovine lung surfactant extract equilibrated with 50 wt% of saline solution, reproduced with permission from [21].

the lipid bilayer and water are the same. A sample equilibrated in the presence of excess of the same aqueous phase at the same temperature shows a spacing of 98.6 Å. We can again apply the formula and then calculate the limit of swelling c:

$$98.6 = 44.3[1 + c/(100 - c)]$$

which gives a swelling limit of 55% in the gel phase.

If the series of diffraction curves in Fig. 4.2 are compared, small and regular changes with temperature can be seen within each phase, whereas in the range 20–25 °C, involving the transition from the gel phase into the lamellar liquid-crystalline phase, the changes are more pronounced. Similar calculations can be performed at other temperatures. In the L_α phase, however, it is more complicated to obtain exact bilayer dimensions, as many swelling curves at different concentrations are needed in order to determine the deviation from linear swelling.

A Porcine Lung Surfactant Extract By mixing a series of aqueous samples of the pharmacy grade porcine surfactant extract HL-10 [22], with increasing proportions of a water phase, examination of the samples in the polarizing microscope after equilibration was used to determine the limit of swelling. With water and with physiological electrolyte solutions, the swelling limit was in the range 55–58 wt% (above this limit a free-water phase was observed). The small-angle diffraction curves at maximum swelling in pure water and in physiological saline solution at 42 °C are shown in Fig. 4.3. The wide-angle diffraction curves above and below the bilayer melting transition are also shown. The crystallization of the bilayer takes place at about 34–36 °C on cooling, as shown by the occurrence of a peak at 4.2 Å, corresponding to the lateral packing of the hydrocarbon chains.

These diffraction curves reflect two types of lamellar liquid-crystalline phases, with different bilayer organizations. In saline solution, the structure unit is identical to that of the usual L_α phase, with one lipid bilayer + one water layer. In distilled water, however, the structure unit consists of two bilayers. The observed structure is proposed to be due to colocalization of the surfactant proteins SP-B and SP-C in one gap between the bilayers, whereas the other gap is assumed to be free from proteins. This structure is shown in Fig. 4.4. Beside X-ray data, cryo-TEM was used in deriving this structure [22].

4.3.2 Effects of Cholesterol on the Bilayer Phase Behavior in the Lung Surfactant System

Cholesterol is known to constitute an important component of mammalian membranes. A breakthrough in the understanding of its physical functions came via monolayer studies by McConnell and co-workers [24], and by X-ray diffraction studies of aqueous phosphatidylcholine phases [25]. Thus cholesterol was observed to induce phase separation of a cholesterol-rich phase and a cholesterol-poor phase with an immiscibility gap between 8 to 20 mol%

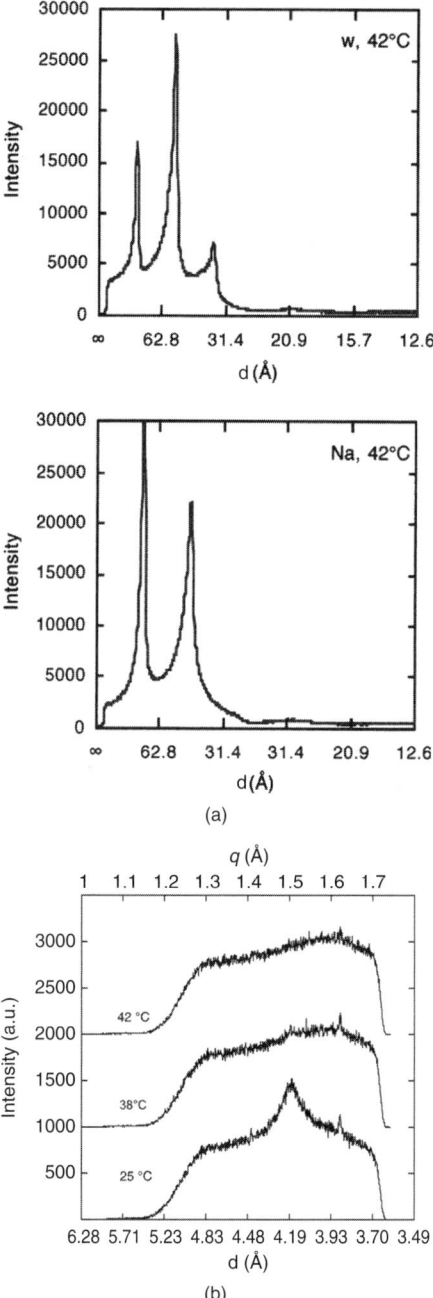

Figure 4.3 (a) X-ray diffraction curves in the small-angle region at 42 °C of porcine lung surfactant extract HL-10 at maximal swelling in water (top) and in physiological saline solution (bottom). (b) The wide-angle region of the same sample as in (a) showing the crystallization transition at cooling. (From Ref. 22 with permission.)

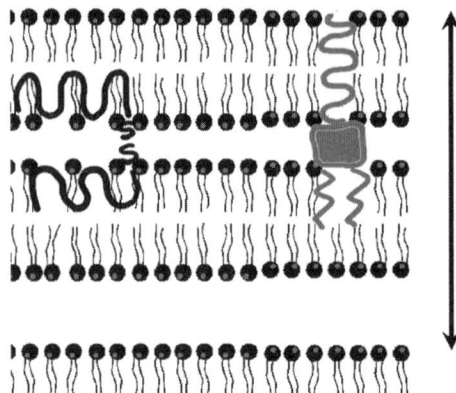

Figure 4.4 Proposed structure model of the double-bilayer lamellar liquid-crystalline structure of a porcine lung surfactant extract, HL-10, in distilled water. The tentative location of the cationic surfactant proteins, SP-B (to the left) and SP-C (to the right), attaching the apposing bilayers, are indicated.

cholesterol. The latter authors also showed that the structure of the segregated bilayers is well organized, with stripes of each phase alternating laterally in the bilayer [25]. In biomembranes, these two phases both correspond to lamellar liquid-crystalline phases, but the high cholesterol containing phase is more ordered (the acyl chains have less freedom to move chaotically), and it has therefore been termed $L_\alpha(o)$. This mechanism of phase segregation within the lipid bilayer in simple systems has a corresponding behavior in biomembranes. The $L_\alpha(o)$ bilayer domains are termed lipid rafts or caveolae when they are curved [26]. The coexisting bilayer conformations will most likely have important physiological functions; membrane-associated proteins like SP-B and SP-C may be preferably located in the $L_\alpha(o)$ domains.

Lung surfactant bilayers *in vivo* consist of about 13.3 mol% cholesterol [6], whereas the extracts used clinically are almost free from cholesterol as a consequence of the standard extraction procedures used. The effect of addition of 13 mol% cholesterol to such cholesterol-free surfactant preparations was therefore examined by X-ray diffraction [23]. The effect on the bilayer structure, as demonstrated in Fig. 4.5, is obvious when we compare the samples without cholesterol (Fig. 4.5a) and with cholesterol added (Fig. 4.5b). A remarkable effect of the cholesterol-induced phase segregation into stripes of an L_α phase structure alternating with stripes of $L_\alpha(o)$ is the loss of a diffraction peak at 4.2 Å below the temperature at which the L_α phase must be expected to be crystalline. Curves recorded every 5 °C from 5 to 42 °C indicate that 20–25 °C is the melting transition interval, and scattering curves below and at the end of this transition range are shown in Fig. 4.5b. The total lack of any pronounced diffraction at 4.2 Å in the segregated bilayer structure indicates that the L_α stripes are so small that line broadening effects reduce their diffraction intensity.

Furthermore, the addition of cholesterol results in a reduction of the bilayer melting transition temperature from about 34–36 °C to about 20–25 °C. It might be assumed that the cholesterol-poor phase is the continuous phase, which is mainly responsible for the melting transition. A preferred association of disaturated PC species to the cholesterol-poor phase would then be expected to show such reduction of the bilayer melting transition.

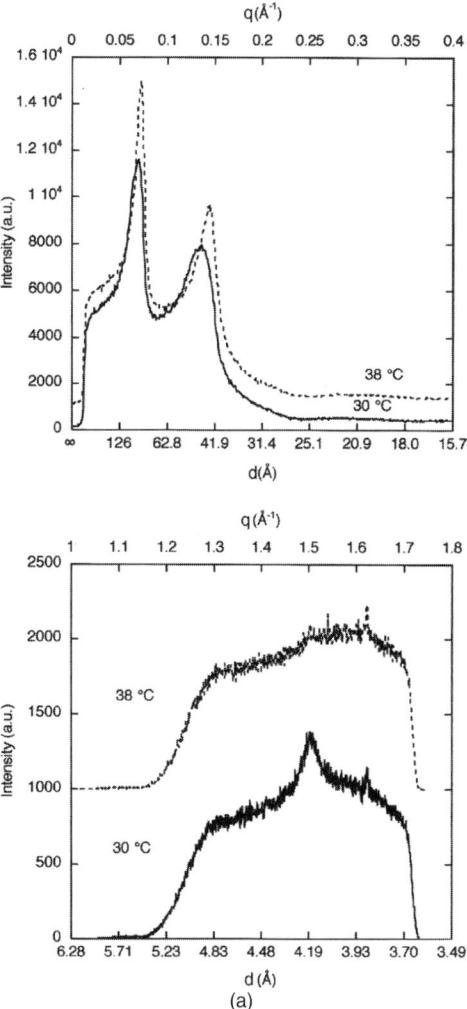

Figure 4.5 (a) X-ray diffraction curves of a porcine lung surfactant extract without cholesterol, swollen in Ringer solution 40:60 (w/w). The small-angle region is shown above and the wide-angle region below. (b) X-ray diffraction curves of a porcine lung surfactant extract (the same as in (a)) with the addition of 13 mol% cholesterol, swollen in Ringer solution 40:60 (w/w). The small-angle region is shown above and the wide angle region below. (From Ref. 23 with permission.)

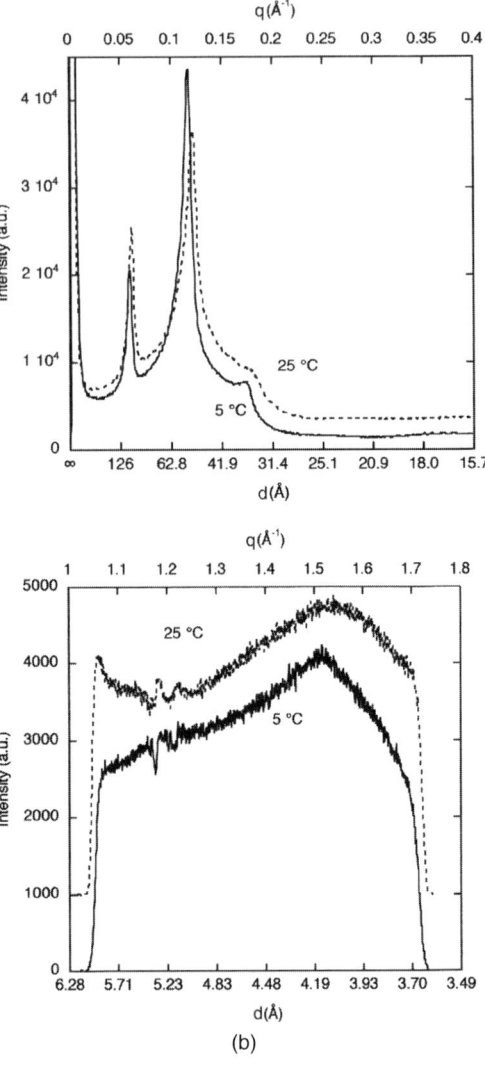

Figure 4.5 (*Continued*)

Cholesterol phase segregation has been proposed to have an important role in lung surfactant bilayers, as the X-ray diffraction patterns of native lung surfactant lavage indicate a similar bilayer segregation as discussed earlier [27]. It should be expected that the domains of cholesterol-rich bilayers, where SP-B and SP-C may be preferentially localized, contribute to the unfolding of LBs and the further development into the TM structure. In this process, the association of the hydrophilic and anionic surfactant protein SP-A to the cationic and hydrophobic proteins SP-B and SP-C drives the transformation.

4.3.3 Interaction Between Lung Surfactant and Serum Albumin

A study of the swelling properties of the bovine surfactant preparation BLES discussed previously has also been performed in the presence of serum albumin in the aqueous medium during swelling of the bilayer [21]. Plasma leakage into the alveolar spaces is a well-known inactivator of surfactant function and an important pathogenetic process in lung disease. Albumin is the most abundant protein in plasma and has been implicated in the inactivation of surfactant [21]. The swelling behavior (with albumin present) determined as described earlier was analyzed in two alternative structure models. In the first alternative, the albumin molecules are assumed to be closely associated to the bilayer. In the second alternative, these molecules are regarded as "free" from bilayer association and solved in the aqueous layer.

The results support a model where the albumin molecules are integrated into the bilayer surface zone. In both alternatives there is an increased hydrocarbon chain disorder, which is evident as the bilayer is thinner than in the albumin-free system. This important effect is likely due to electrostatic interaction between cationic SP-B and SP-C and albumin, which is anionic at physiological pH.

4.4 ENDOGENEOUS LUNG SURFACTANT

Finally, we will consider the structure of the air–water interface lining the lung alveoli, with particular attention to recent X-ray diffraction data. The fluid layer covering the alveoli has for a long time been studied extensively by EM, identifying the two types of bilayer structures that we now are familiar with: LBs and TM. The LBs represent a well-known dispersed state, which at present is described as multilamellar vesicles or liposomes. The TM bilayer organization, on the other hand, is a three-dimensional organization, which according to electron micrographs has been regarded as a square arrangement of intersecting bilayers [28]. The EM textures of TM bilayers in lung washings from rabbit illustrate this characteristic texture, shown in Fig. 4.6, where also the scaffolding protein SP-A is clearly observed. From the Fourier transform of a uniform region, the perfect symmetry is illustrated [27].

4.4.1 A Surface Phase at the Air–Water Interface

About half a century ago, Clements introduced the monolayer model of the alveolar surface, which still seems to be generally accepted [29]. A continuous lung surfactant monomolecular film was proposed to form the lung surface toward air, explaining the low interfacial tension. During the last decade, this concept has been criticized, for example, by Dorrington and Young [30], with arguments derived from key phenomena in lung physiology and surface chemistry.

The introduction of cryo-transmission electron microscopy (cryo-TEM) in studies of the alveolar surface showed the existence of a uniform organized

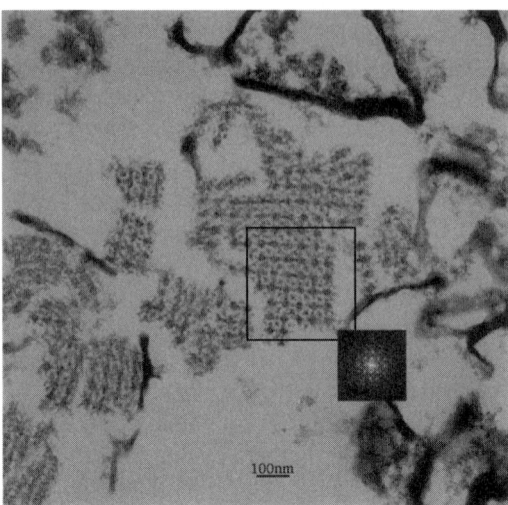

Figure 4.6 Electron micrograph of a TM sample (from the large aggregate fraction of rabbit lung lavage). The Fourier transform from the boxed region is also shown. (From Ref. 27 with permission.)

bilayer phase. This coherent *surface phase* was identified in directly deposited surface samples from freshly opened lungs. The cryo-TEM textures showed a bilayer repetition of about 500 Å, like that of TM. This indicated that the alveolar surface film is uniform and coherent with all the characteristics of a true phase. The same samples were analyzed in the polarizing microscope, confirming the existence and liquid-crystalline nature of this surface phase [31]. Considering these observations of direct deposits from the lung surface, it was proposed that the observed coherent and uniform phase, not a monolayer, constitutes the alveolar lining [19]. Acceptance has been slow, however, as the generally established monolayer model was challenged.

In fact, there is also on the macroscopic level striking evidence for the existence of two separate phases lining the alveoli: a lipid bilayer surface phase and a subphase of water. The turbid phase is birefringent and homogeneous as seen in the polarizing microscope [31, 32]. A glass capillary brought into contact with a freshly opened rabbit lung surface shows the presence of two phases: a turbid phase and a smaller amount of a water phase, located beneath the surfactant phase (unpublished observation). Similar phase separation is seen with reconstituted TM and also in the large aggregate fraction of rabbit lung lavage [32]. The actual amount of free water at the alveolar surface is hard to assess by the above experiment as some tissue damage occurs during the experimental procedure. However, it seems likely the LBs will require some amount of free water for the transition into the TM-type structure. That water will thus be rapidly incorporated into the developing TM surface phase.

As far as I know, no *equilibrium* monolayer of lipids and proteins has been described that exhibits a low surface tension similar to what is physiologically required at the alveolar surface. A phase forming an interface, however, is a different situation. A theoretical analysis of the monolayer model versus the surface phase model has been reported by Kashchiev and Exerowa [33]. Their analysis clearly favors the surface phase model.

Schurch et al. [34] have reported EM observations of regions of a multilayered alveolar surface structure. Such a localized bilayer structure, however, is quite different from the continuous *surface phase* discussed here. It may be mentioned that during recent years many reports on the alveolar surface structure describe a monolayer with a *"monolayer-associated reservoir."* Whatever that might be, it will have the surface tension of the monolayer and will in physical aspects be *different from a uniform bilayer phase* forming the surface toward air.

4.4.2 A Minimal Surface Structure Model of the Surface Phase

The observed properties of the surface phase indicated a continuous bilayer structure, and it was therefore natural to consider the possibility of a minimal surface structure consistent with the ultrastructural evidence [19]. It was found that there is a minimal surface structure termed CLP that is fully compatible with the experimental results. A strict mathematical analysis of this minimal surface was first reported by Lidin and Hyde [35], and the nodal surface description that was introduced as an approximation is shown in Fig. 4.1. The structure of the center of the bilayer following the CLP minimal surface is shown in Fig. 4.7 with the principal arrangement of the scaffolding protein SP-A indicated. Here, the topology is of central interest, and the reader should note that the CLP structure forces SP-A to arrange in a pairwise fashion with perpendicular pairs in adjacent layers along the c axis.

The general assumption in the literature seems to be that SP-A, in its role to scaffold TM, interacts four units at a time, forming a cross-like structure. However, SP-A has a strong tendency toward linear association into fibrils [36]. The cross-like structure formed by four SP-A's has, to my knowledge, only been observed in the context of TM (i.e., lipids, SP-B, and calcium present). Thus limitations in the resolution along the c axis (i.e., sample thickness) may induce projection artifacts with two perpendicular SP-A pairs appearing like a cross. The formation of SP-A dimers as the functional unit, however, seems more likely, with binding taking place at the CRD domain of each SP-A.

The structure on the molecular and lipid bilayer level involving the surfactant proteins is shown schematically in two adjacent cross sections along the c axis of the tetragonal structure in Figs. 4.8a and 4.8b, respectively. The scaffolding SP-A molecules, which are electrostatically associated to SP-B, and the lipids at the maximally curved bilayer regions (the catenoids) are drawn in the figure as molecular pairs within the tubular diagonal. These SP-A pairs associate to the bilayer-embedded SP-B and have a crucial role in formation of the TM structure by spanning the bilayer into the CLP type of structure. Of course, projection

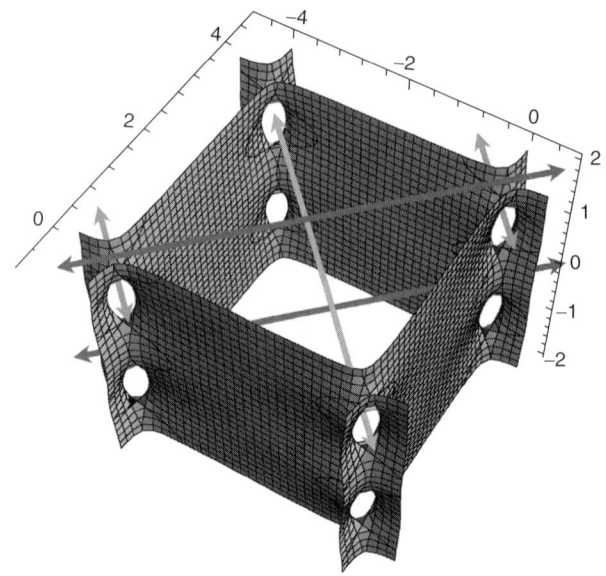

Figure 4.7 The CLP minimal surface model of the bilayer structure in the alveolar surface phase. The lines through the catenoid connections between adjacent tubuli indicate the SP-A network (cf. Fig. 4.8).

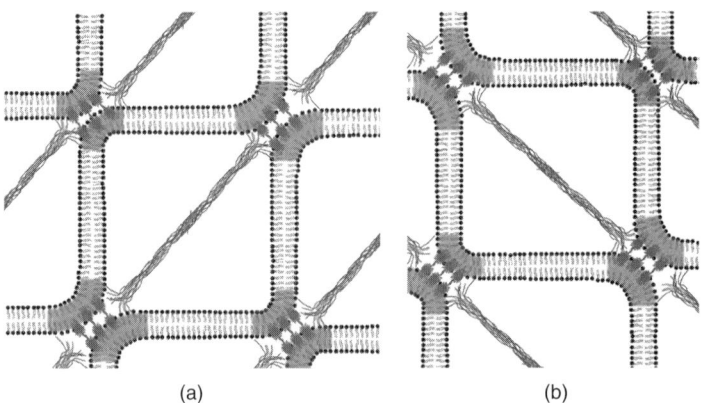

Figure 4.8 (a) A cross section through the xy plane at $z = 0$ or 1, showing catenoid openings between tubuli in the CLP bilayer model of the TM structure. SP-B proposed to associate apposing bilayers in the corners is indicated, as well as the associated SP-A pair along the diagonal of the tubular space. The shaded part of the bilayer indicates that it is a cholesterol-rich region with $L_\alpha(o)$ conformation. (b) The same structure as in (a), but here the cross section at $z = \frac{1}{2}$ is shown with the SP-A molecules oriented perpendicular to those shown in (a). Thus these SP-A molecules form parallel rod systems in each plane $z = 0.5, 1.5, \ldots$, and the direction of the linear SP-A nets alternate between perpendicular orientations from one z-plane of proteins to the next.

of Figs. 4.8a and 4.8b on top of each other will produce the familiar cross-like shape usually seen in EM images of SP-A. The shaded areas in the curved regions indicate the hypothetical location of $L_\alpha(o)$ regions (which will harbor SP-B and SP-C).

The periodicity of the CLP structure is defined by the tetragonal axis of about 500 Å, and the proposed arrangement of the SP-A protein along the diagonals (attached to SP-B and SP-C in the catenoid openings) is also shown. Thus the repetition period exhibited by the SP-A protein network according to the CLP model will be about 350 Å ($500/\sqrt{2}$ Å), as shown in Fig. 4.9. Although the SP-A network will appear as an X-shape when viewed along the c axis (in typical EM sections), each SP-A pair (see Fig. 4.6) is located in separate planes.

Such a protein network has a high electron density. Therefore the X-ray diffraction can be assumed to be dominated by the contribution from this protein network.

4.4.3 On X-Ray Diffraction Data of the TM Bilayer Organization

X-ray diffraction data of TM (from the large aggregate fraction of lung lavage from rat) was recently reported by Quinn and co-workers [37].

A well-defined diffraction pattern showing a one-dimensional periodicity with a unit length of 335 Å was observed. Seven orders of reflections were observed, which is remarkable in a liquid-crystalline bilayer phase. Only the minimal

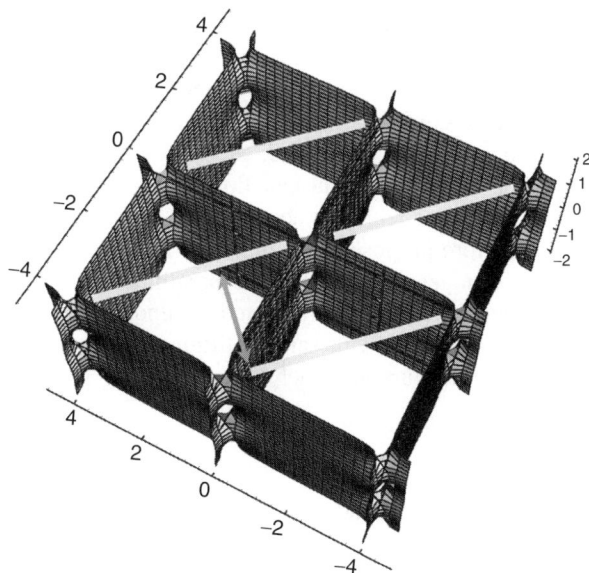

Figure 4.9 The CLP model of the surface phase with the SP-A rod system in one z-plane indicated. The distance between these rods in this plane is indicated by d. This distance should be expected to dominate the diffraction pattern.

surface type of cubic phases reported earlier have exhibited a similar large number of diffraction spacings. Quinn also shows a micrograph of this TM by conventional EM. The texture is quite irregular and a bilayer period of 600 Å is given. Most observations, however, report a value of 500 Å. It seems reasonable to consider these new diffraction data in relation to the structure model of TM presented here.

As pointed out in the CLP structure description just given, the scaffolding protein network is expected to contribute strongly to the X-ray diffraction from this organization. The arrangement of the protein network (in one plane along the c axis) is illustrated in Fig. 4.9. It is more rigid than lipid bilayer structures, and the proteins have higher electron density. The distance between the rods in each cross-section plane where they are located, as seen in Fig. 4.9, is about 350 Å (the diagonal of the 500-Å tetragonal cell divided by 2). From this we can assume that the X-ray data by Quinn is consistent with the model of TM presented here, even if it is no proof.

The traditional TM structure model consists of a square pattern of intersecting bilayers, with tetragonal two-dimensional periodicity, and a liquid-like disorder in the third dimension—the c direction. According to EM ultrastructural evidence, a complex of four SP-A units is assumed to be located in the same a, b plane. Such a structure will show dominating diffraction spacings corresponding to the a- and b-axes interplanar distance of about 500 Å, as well as spacings corresponding to the (110) interplanar distance of about 350 Å. The lack of diffraction spacings related to the tetragonal axis (500 Å) is therefore not consistent with the traditional TM model.

Finally, it is interesting to consider the expected mechanical properties of the CLP phase model of the TM structure. As the entire surface is one bilayer continuum without intersections, successive changes by mechanical deformation from the minimal surface conformation of the bilayer result in a concertina-like movement [19]. As the structure is ordered, such deformations will, to a certain extent, be elastic. The bilayer material can flow freely through this membrane continuum to adopt the shape and area according to changes in geometry. Such a mechanism was therefore proposed to explain the low energy requirements for alveolar surface area changes involved in tidal breathing.

In conclusion, X-ray diffraction has furthered the understanding of the internal organization of TM. Cryo-TEM with sample tilting and computer-aided reconstruction may be a fruitful future approach to analyze lung surfactant membrane organization.

4.5 COMMENTS ON NEUTRON DIFFRACTION/SCATTERING

Neutron diffraction can be applied to the study of lipid–water systems in a similar way as X-ray scattering and diffraction. The strong scattering difference between deuterium and hydrogen makes it possible to go further in the structure analysis using neutron diffraction as compared to X-ray studies. This technique can reveal structural details in a lipid–protein bilayer of a liquid-crystalline phase, such as localization and orientation of proteins in the bilayer, by substituting heavy water for water and perdeuterated lipids for lipids.

Neutron reflectometry is a related method particularly suited for studying surface films in the thickness range from a few nanometers to a few hundred nanometers. This technique has recently been applied to membranes of pharmacy grade lung surfactant extracts, conclusively confirming that lung surfactant forms a coherent multilayered surface phase at the air–water interface [38]. Furthermore, this technique allowed determination of the protein concentration in the interfacial layer, and it was found that there is a strong enrichment of surfactant proteins SP-B and SP-C in the first two layers toward air. This observation may be taken as an indication of the mechanism initiating the surface phase formation. At exposure of this surfactant dispersion to air, the driving force to reduce the surface tension is likely to initially result in formation of a monolayer of SP-B and SP-C, as these proteins have an equilibrium distribution between location within lipid bilayers and solution in the water phase. The slower process of unfolding of the vesicular bilayers at the air–water interface will then start a domino-like buildup of multilayers into a surface phase.

4.6 CONCLUSION

As has been shown in this chapter, certain features and/or clinical problems related to lung surfactant can be studied using X-ray diffraction techniques. This implicitly requires the bulk equilibrium structures of lung surfactant to be of liquid-crystalline order. Cryo-TEM of directly deposited alveolar surface films exhibit liquid-crystalline order on the scale of thousands of nanometers. Furthermore, polarizing microscopy with its possibility to identify birefringent structures on a larger scale has revealed birefringence in directly deposited alveolar surfactant films, as well as striking kinetic phenomena in the surface phase of exogenous surfactants [22]. Neutron reflectometry has recently revealed the structure of the air–water surface phase, with enrichment of proteins in the first two layers. Taken together, this indicates that lung surfactant should be regarded as a continuous liquid-crystalline system at the alveolar interface, a *surface phase*. What to me seems crucial is to acknowledge that all surfactant material in the human lung, be it the healthy surface film or a life-saving exogenous surfactant, must be formed spontaneously at the alveolar surface, and thus must be regarded as an equilibrium structure, where breathing of course regularly deforms the organization.

ACKNOWLEDGMENTS

I am grateful to Kåre Larsson for introducing me to the use of X-ray diffraction for studies of the structure of lipid–water phases. The author is supported by an ALF-Grant, Lund University Hospital, Sweden.

REFERENCES

1. J. Dubochet, M. Adrian, J. J. Chang, J. Lepault, and A.W McDowall (1987). Cryoelectron microscopy of vitrified specimen, in *Cryotechniques in Biological Electron Microscopy* (R.A. Steinbrecht and K. Zierold, Eds.). Springer-Verlag, Berlin, pp. 114–131.
2. J. Gil (1971). Ultrastructure of lung fixed under physiologically defined conditions. *Arch. Intern. Med.* **127**:896–902.
3. J. H. Power, I. R. Doyle, K. Davidson, and T. E. Nicholas (1999). Ultrastructural and protein analysis of surfactant in the Australian lungfish *Neoceratodus forsteri*: evidence for conservation of composition for 300 million years. *J. Exp. Biol.* **20218**:2543–2550.
4. R. J. Sanderson and A. E. Vatter (1977). A mode of formation of tubular myelin from lamellar bodies in the lung. *J. Cell. Biol.* **74**(3):1027–1031.
5. M. C. Kahn, G. J. Anderson, W. R. Anyan, and S. B. Hall (1995). Phosphatidylcholine molecular species of calf lung surfactant. *Am. J. Physiol.* **269**(5 Pt 1):567–573.
6. R. Veldhuizen, K. Nag, S. Orgeig, and F. Possmayer (1998). The role of lipids in pulmonary surfactant. *Biochim. Biophys. Acta* **1408**(2-3):90–108.
7. J. Johansson and T. Curstedt (2005). New synthetic surfactants—basic science. *Biol. Neonate* **87**(4):332–337.
8. F. van Iwaarden and L. M. G. van Golde (1995). In *Surfactant Therapy for Lung Disease* (B. Robertson and H.W. Taeusch, Eds.). Marcel Dekker New York, pp. 75–88.
9. Y. Suzuki, Y. Fujita, and K. Kogishi (1989). Reconstitution of tubular myelin from synthetic lipids and proteins associated with pig pulmonary surfactant. *Am. Rev. Respir. Dis.* **140**(1):75–81.
10. T. Hahn (Ed.) (2002). *International Tables for Crystallography*, Vol. A, 5th ed. Kluwer Academic Publishers, Dordrecht.
11. L. E. Alexander (1969). *X-Ray Diffraction Methods in Polymer Science*. Wiley-Interscience, London.
12. D. Small (1986). *Handbook of Lipid Research, Volume 4. The Physical Chemistry of Lipids*. Plenum Press, New York.
13. K. Larsson (1994). *Lipid Handbook* (F. G. Gunstone, J. L. Harwood, and F. B. Padley, Eds.). Chapman and Hall, London.
14. V. Luzzati, T. Gulik-Krzywicki, and A. Tardieu (1968). Polymorphism of lecithins. *Nature* **218**(5146):1031–1034.
15. S. T. Hyde, S. Andersson, K. Larsson, Z. Blum, T. Landh, S. Lidin, and B. Ninham (1997). *Language of Shape*. Elsevier, Amsterdam.
16. M. J. Janiak, D. M. Small, and G. G. Shipley (1976). Nature of the thermal pretransition of synthetic phospholipids: dimyristolyl- and dipalmitoyllecithin. *Biochemistry* **15**(21):4575–4580.
17. I. Lindstedt and C. Liljenberg (1990). On the periodic minimal surface structure of the plant prolamellar body. *Physiol. Plant* **80**:1–4.

18. H. G. von Schnering and R. Nesper (1991). Nodal surfaces of Fourier series: fundamental invariants of structured matter. *Z. Phys. B Condensed Matter* **83**:407–412.

19. M. Larsson (2002). *A Surface Phase Model of the Alveolar Surface Lining: Ultrastructural Analysis and In Vivo Applications*. Thesis, Lund University, Lund, Sweden. ISBN 91-628-5448-8.

20. S. Andersson, K. Larsson, M. Larsson, and M. Jacob (1999). *Biomathematics: Mathematics of Biostructures and Biodynamics*. Elsevier, Amsterdam.

21. M. Larsson, T. Nylander, K. M. W. Keough, and K. Nag (2006). An X-ray diffraction study of alterations in bovine lung surfactant bilayer structures induced by albumin. *Chem. Phys. Lipids* **144**(2):137–145.

22. M. Larsson, J. J. Haitsma, B. Lachmann, K. Larsson, T. Nylander, and P. Wollmer (2002). Enhanced efficacy of porcine lung surfactant extract by utilization of its aqueous swelling dynamics. *Clin. Physiol. Funct. Imaging* **22**:39–48.

23. M. Larsson, K. Larsson, T. Nylander, and P. Wollmer (2003). The bilayer melting transition in lung surfactant bilayers: the role of cholesterol. *Eur. Biophys. J.* **31**(8):633–636.

24. D. J. Recktenwald and H. M. McConnell (1981). Phase equilibria in binary mixtures of phosphatidylcholine and cholesterol. *Biochemistry* **20**(15):4505–4510.

25. W. Knoll, G. Schmidt, K. Ibel, and E. Sackmann (1985). Small-angle neutron scattering study of lateral phase separation in dimyristoylphosphatidylcholine–cholesterol mixed membranes. *Biochemistry* **24**:5240–5246.

26. K. Simons and E. Ikonen (2000). How cells handle cholesterol. *Science* **290**(5497):1721–1725.

27. M. Larsson, O. Terasaki, and K. Larsson (2004). A solid state transition in the tetragonal lipid bilayer structure at the lung alveolar surface. *Solid State Sci.* **5**:109–114.

28. M. C. Williams (1977). Conversion of lamellar body membranes into tubular myelin in alveoli of fetal rat lungs. *J. Cell Biol.* **72**:260–277.

29. J. A. Clements (1957). Surface tension of lung extracts. *Proc. Soc. Exp. Biol. Med.* **95**:170–172.

30. K. L. Dorrington and J. D. Young (2001). Development of the concept of a liquid pulmonary alveolar lining layer. *Br. J. Anaesth.* **86**(5):614–617.

31. M. Larsson, K. Larsson, and P. Wollmer (2002). The alveolar surface is lined by a coherent liquid-crystalline phase. *Prog. Colloid Polym. Sci.* **120**:28–34.

32. M. Larsson, J. F. van Iwaarden, J. J. Haitsma, B. Lachmann, and P. Wollmer (2003). Human SP-A and a pharmacy-grade porcine lung surfactant extract can be reconstituted into tubular myelin—a comparative structural study of alveolar surfactants using cryo-transmission electron microscopy. *Clin. Physiol. Funct. Imaging* **23**(4):199–203.

33. D. Kashchiev and D. Exerowa (2001). Structure and surface energy of the surfactant layer on the alveolar surface. *Eur. Biophys. J.* **30**:34–41.

34. S. Schurch, F. H. Y. Green, and H. Bachhofen (1998). Formation and structure of surface films: captive bubble surfactometry. *Biochim. Biophys. Acta* **1408**:180–202.

35. S. Lidin and S. T. Hyde (1987). A construction algorithm for minimal surfaces. *J. Physique* **48**:1585–1590.

36. N. Palanyiar et al. (1999). Filaments of surfactant protein A specifically interact with corrugated surfaces of phospholipid membranes. *Am. J. Physiol.* **276**(4 Pt 1): L631–L641.
37. K. Larsson, P. Quinn, K. Sato, and F. Tiberg (2006). *Lipids: Structure, Physical Properties and Functionality*, Vol. 19. The Oily Press Lipid Library, Bridgwater, UK.
38. D. Follows, R. K. Thomas, F. Tiberg, and M. Larsson (2007). Multilayers at the surface of solutions of exogenous lung surfactant: direct observation by neutron reflection. *Biochim. Biophys. Acta* **1768**(2):228–235.

CHAPTER 5

Neutron and X-Ray Scattering from Isotropic and Aligned Membranes

J. KATSARAS

Canadian Neutron Beam Centre, National Research Council, Chalk River Laboratories, Chalk River, Ontario, Canada; Biophysics Interdepartmental Group, Guelph-Waterloo Physics Institute, University of Guelph, Guelph, Ontario, Canada; and Department of Physics, Brock University, St. Catherines, Ontario, Canada

J. PENCER

Canadian Neutron Beam Centre, National Research Council, Chalk River Laboratories, Chalk River, Ontario, Canada; and Department of Physics, St. Francis Xavier University, Antigonish, Nova Scotia, Canada

M.-P. NIEH, T. ABRAHAM, and N. KUČERKA

Canadian Neutron Beam Centre, National Research Council, Chalk River Laboratories, Chalk River, Ontario, Canada

THAD A. HARROUN

Department of Physics, Brock University, St. Catherines, Ontario, Canada

CONTENTS

5.1	Introduction	108
5.2	Instrumentation	109
	5.2.1 The Neutron	109
	5.2.2 Production of Neutrons	110
	5.2.3 X-Rays	112
	5.2.4 Production of X-Rays	113
5.3	Types of Scattering	115
5.4	Model Membranes	117
	5.4.1 Isotropic Membranes	118
	5.4.2 Aligned Membranes	121
References		130

Structure and Dynamics of Membranous Interfaces, edited by Kaushik Nag
Copyright © 2008 John Wiley & Sons, Inc.

5.1 INTRODUCTION

Neutron and X-ray scattering are two of the most powerful structural determination techniques presently available. Although many are familiar with the exploits of atomic resolution X-ray crystallography—in particular, protein crystallography—these techniques also have applications in polymer science, colloid chemistry, and materials science. With regard to structural biology, the various neutron and X-ray scattering techniques complement crystallographic studies that require hard-to-obtain high quality crystals of macromolecules. Moreover, small-angle neutron (SANS) and X-ray scattering (SAXS) are two structural techniques capable of studying biomolecules and their assemblies in solution, including molecules whose crystallization conditions have yet to be determined.

Neutron and X-ray scattering are similar in that both techniques are capable of providing dynamical and structural information. However, whereas X-rays are scattered primarily by electrons, neutrons are fundamental particles scattered primarily by their interaction with atomic nuclei. Although the amplitude of X-ray scattering increases in a simple way with atomic mass, neutron scattering amplitudes depend in a complex manner on the nucleus' mass, spin, and energy levels. Since neutrons interact differently with the different nuclei, including with the various isotopes of elements, this allows for the powerful and commonly used method of contrast variation. In structural biology the classic example is the substitution of hydrogen (^{1}H) for its heavier isotope deuterium (^{2}H). It should be pointed out that contrast variation is often more easily exploited with neutron rather than X-ray scattering.

In biological systems, the ubiquitous cell membrane surrounds cells and separates their contents from the external environment. It is accepted that the lipid bilayer is the underlying structure of most, if not all, biomembranes. The self-assembly of lipids in solution into model membranes for laboratory study is based on three pivotal experimental findings. (1) Using an apparatus developed by Agnes Pockels [1], Langmuir proposed that fatty acid molecules form a monolayer by orienting themselves in such a fashion that their carboxyl groups are in contact with water, while their hydrocarbon chains stick out in the air away from the water [2]. (2) Using lipids extracted from plasma membranes and a Langmuir trough, Gorter and Grendel [3] reported that lipids can form bilayers, since the surface area of the monolayer formed from the lipids extracted from a cell's plasma membrane was approximately equal to twice the surface area of the cell itself. (3) Mueller et al. [4] reconstituted a lipid bilayer from the lipids extracted from a cow's brain.

Over the decades that followed, the validation of the lipid bilayer and its association with the life sciences [5] and biotechnologys [6, 7] is of central interest to a wide variety of investigators including biochemists, biologists, biophysicists, bioengineers and technologists, electrochemists, physiologists, pharmacologists, surface and colloid scientists, and others working on ultrathin films and membrane phenomena.

5.2 INSTRUMENTATION

5.2.1 The Neutron

The neutron is an electrically neutral, subatomic, elementary particle, whose existence had been postulated by Ernest Rutherford [8] and was subsequently discovered in 1932 by the English physicist James Chadwick [9]. With the exception of hydrogen (^{1}H), the neutron is found in all atomic nuclei, has a mass slightly larger than a proton—but approximately 1840 times greater than that of an electron—a nuclear spin of $\frac{1}{2}$, and a magnetic moment [10] of $-1.91 eh/(2m_p)$. Neutrons are only stable when bound by an atomic nucleus, while free neutrons are unstable with a mean lifetime of approximately 900 s, and decay into a proton, an electron, and an antineutrino [11].

Although the neutron's interaction with atomic nuclei is weak, the scattering "power" (cross section) of an atom is not related to its atomic number. As such, neighboring elements in the periodic table can have substantially different scattering cross sections. Moreover, the interaction of a neutron with the nucleus of an atom allows for an element's isotopes to be differentiated. The classic example of this is the isotopic substitution of ^{1}H for ^{2}H (deuterium), commonly used in the study of polymeric and biologically relevant materials. As a result of their intrinsic properties neutrons are used as follows:

1. The weak interaction of neutrons with atomic nuclei makes them a highly penetrating probe. This feature allows the construction of enclosed, windowless, and complex sample environments, and enables the measurement of bulk processes under realistic conditions without significant attenuation of the neutron beam by the sample container.
2. Since the scattering ability of an atom is not strongly related to its atomic number, neutrons are used extensively to locate "light," low atomic number atoms among "heavier" atoms. In the case of polymeric materials, neutrons are used to precisely locate hydrogen atoms, otherwise invisible to X-rays.
3. ^{1}H and ^{2}H atoms scatter neutrons differently. ^{1}H has a negative coherent scattering length (e.g., $b_{coh} = -3.74 \times 10^{-15}$ m) lending it "contrast" when surrounded by other, positive scattering length atoms. For biological samples, intrinsically rich in hydrogen, judicious substitution of ^{2}H ($b_{coh} = +6.67 \times 10^{-15}$ m) for ^{1}H provides a powerful method for selectively tuning the "contrast" of a given macromolecule. One can therefore accentuate, or nullify, the scattering from particular parts of a macromolecular complex by selective deuteration. This powerful technique is commonly referred to as "contrast variation" [12].
4. Neutron energies are similar to the energies of atomic and electronic processes in the millielectron volt (meV) to electron volt (eV) range. This allows for the study of various dynamic properties (i.e., translations, rotations, vibrations, and lattice modes) exhibited by molecules, and eV transitions within the electronic structure of materials.

5. Neutrons possess a magnetic moment (spin-half particles), and spin-polarized neutron beams are ideally suited to the study of magnetic structures (short- and long-range) and short-wavelength magnetic fluctuations [10]. The cross sections for magnetic scattering are of the same magnitude as those for nuclear scattering.

5.2.2 Production of Neutrons

Neutron beams with intensities suitable for scattering experiments are presently being produced either by nuclear reactors, where the fission of uranium nuclei (^{235}U) results in neutrons of energies between 0.5 and 3 MeV, or by spallation sources, where accelerated charged subatomic particles strike a heavy metal target, expelling neutrons from the target nuclei (Fig. 5.1) [13, 14].

In reactors, fission takes place when a fissile nucleus of similar mass captures a neutron. Upon capture, the nucleus splits into two nearly equal-mass fragments producing, on average, 2.5 high energy neutrons per fission. Since the probability of a fast (high energy) fission neutron interacting with a fissile nucleus is small, to sustain the chain reaction, neutrons must be slowed down or "thermalized" by passing through a moderator. In practice, moderators such as H_2O, D_2O, graphite, or beryllium are used, and the resulting "thermal" neutrons exhibit a peak energy flux centered at ~1.2 Å. Thermal neutrons can further be slowed down by interaction with another, colder moderator, for example liquid hydrogen (~20 K) [15, 16]. These "cold" neutrons, with their Maxwellian distribution now shifted towards lower energies ($\lambda \sim 5$–10 Å), can then be transported over many meters to various experimental stations by total reflection from ^{58}Ni-coated optically flat glass guides [17]. Recently, supermirrors made of, for example, Ni/Ti multilayers, can increase the effective total reflection critical angle θ_c by up to a factor of 3, compared to pure Ni [18, 19]. They do so by utilizing not only the total external reflection component, but also the constructive interference (Bragg reflection) from the successive layers of Ni, effectively extending the incident angle of total external reflection.

The desired energy neutrons are selected through Bragg scattering from a single crystal or pass through a velocity selector whose rotating vanes, coated with a neutron-absorbing material, allow only selected energy neutrons to pass. These neutrons then impinge on the sample, which are then scattered and are usually detected by a ^{3}He-filled two-dimensional (2D) detector. Cold neutrons are ideally suited for the study of systems that commonly self-assemble into larger structural units, such as polymeric and biologically relevant materials.

Another technology for producing neutrons is accelerator-based, pulsed neutron sources (Fig. 5.1B), whereby an ion source produces negative hydrogen (H$^-$) ions, which are collected and accelerated to form a pulsed beam with energies on the order of kiloelectron volts (keV) to megaelectron volts (MeV) [20]. This beam is then delivered to a linear accelerator, which accelerates the H$^-$ beam to energies approaching 1 gigaelectron volts (GeV), such as in the case of the Spallation Neutron Source (Oak Ridge, TN). The H$^-$ ions are stripped of their electrons, producing protons (H$^+$) that are bunched and further intensified using an accumulator ring structure, such as a synchrotron. Once bunched, the protons are extracted from the ring in short (10^{-6} s), intense, proton pulses at a rate of 50–60 times per second. The individual pulses are delivered to a metal target (e.g., Pb, Ta, Hg, U), releasing neutrons from the target material's nuclei. The characteristic neutron spectra emanating from reactor- and accelerator-based

(a)

Figure 5.1 (A) The National Research Universal (NRU) reactor, a steady-state neutron source, is located at Chalk River Laboratories. Pictured are staff standing on top of the reactor core cover plate, in front of the fuel rod extraction crane. It is a heavy water (D$_2$O) moderated and cooled reactor and is the major world source for medical isotopes (e.g., ^{99}Mo, ^{125}I, ^{131}I, and ^{192}Ir). Presently, the reactor operates at 125 MW and uses 20% enriched ^{235}U fuel. NRU was Canada's third nuclear reactor and went critical in 1957, and the first ever reactor with online refueling capability. (B) Schematic of the Spallation Neutron Source (SNS), an accelerator-based source located at Oak Ridge, TN and completed at a cost of $\sim$\$1.4 billion. Accelerated H$^-$ ions produced by an ion source (a) are stripped of their electrons and delivered to the Linac (Linear accelerator) (b), producing a 1-GeV H$^+$ ion beam. The H$^+$ ions are bunched and intensified by the accumulator ring (c), which are then delivered to the liquid mercury target (d) housed in the target station (e), which houses the various neutron scattering instruments (f). The SNS chose mercury as the target for the proton pulses for the following reasons: (1) unlike solid materials, liquid mercury does not experience radiation damage; (2) mercury is a high atomic number material resulting in many spallation neutrons ($\sim$20–30 neutrons/mercury atom); and (3) compared to a solid target, a liquid target at room temperature is better at dissipating heat and withstanding shock effects.

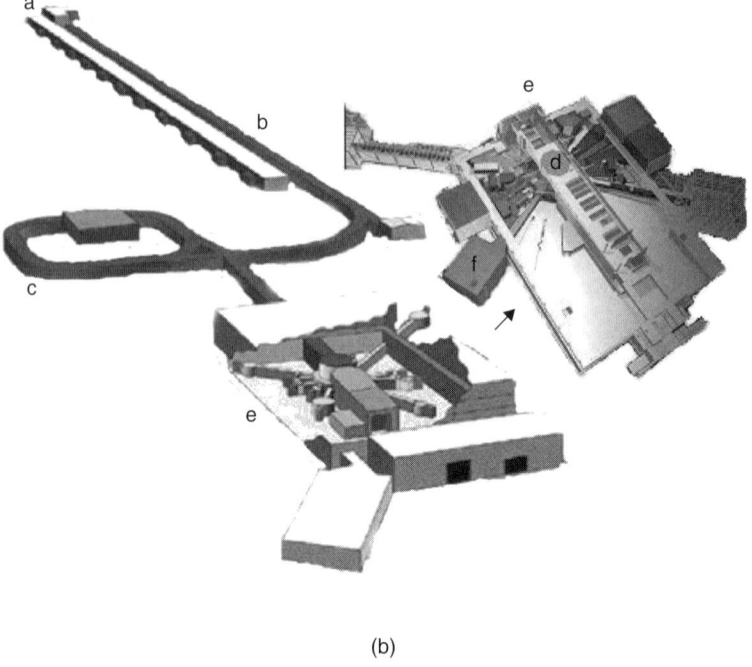

(b)

Figure 5.1 (*Continued*)

neutron sources differ considerably. Spallation neutron spectra possess both a high energy "slowing" component and a thermalized component. Spallation neutrons are thermalized in a fashion similar to those produced by reactor-based neutron sources, by passing through a vessel of hydrogen-rich material at a fixed temperature.

Compared to reactor sources, the biggest advantage of spallation sources is that the production of neutrons results in much less heat, translating into increased neutron fluxes. Despite this, since neutrons at spallation sources are produced in pulses, the time-averaged flux of even the most powerful pulsed sources is less than that of high flux reactor sources. However, judicious use of large detector arrays and time-of-flight techniques, which can utilize the wide range of neutron wavelengths present in each pulse, can exploit the high brightness and can, for certain experiments, more than compensate for the time-averaged flux disadvantage.

5.2.3 X-Rays

X-rays are electromagnetic waves whose wavelengths range from about 10^{-8} to 10^{-12} m and were discovered in 1895 by Wilhelm Conrad Roentgen by discharging electrical current in an evacuated glass tube commonly known as a "Crookes tube." Unlike neutrons, which scatter from atomic nuclei, X-rays interact vigorously with the atom's electron cloud. The atomic form factor is a measure of the scattering strength of an element, and for very small scattering angles, scales with Z, the number of electrons in the element, which is also the atomic number. Since the electron cloud makes up most of the atom's volume and shape, the X-ray form factor strongly depends on the angle of photon scatter, unlike neutrons, whose scatter from the nucleus can be treated as a point source in the Born approximation [21].

5.2.4 Production of X-Rays

Over the sixty years since their discovery, the most common means of producing X-rays were by the impact of high energy electrons, accelerated over several thousand volts in an evacuated glass tube, onto a metal anode. Rapidly decelerating electrons lose their kinetic energy, which is converted into radiation and heat. This type of X-ray radiation, with a range of energies from a few keV to a maximum energy corresponding to that of the accelerated electron beam, is referred to as bremsstrahlung radiation [21].

X-rays are also produced when the high energy incoming electron interacts with the target's K-shell electrons, imparting enough energy to knock the electron out of its K-shell. When L- and M-shell electrons cascade down to fill the K-shell vacancies, the energy released results in the intense characteristic K_α and K_β X-ray photons. Each target element has its own characteristic K_α and K_β X-rays. However, this process of producing X-rays is very inefficient as only $\sim 0.1\%$ of electrons produce K-shell vacancies.

A significant advance in X-ray production came with the development of the rotating anode. In this case, the metal anode is fashioned as a rotating wheel that shows its circumference to the electron beam, thereby increasing the effective area bombarded by the electrons by factor of $2\pi r$, where r is the radius of the rotating wheel. The wheel is typically water-cooled internally, and since the heat is better dissipated, more electrons can impinge on the anode with a resultant tenfold increase in X-ray flux over a conventional sealed tube with a fixed anode. A further improvement in X-ray flux came as a result of synchrotrons.

Synchrotron radiation is produced when a charged particle traveling at high speed undergoes acceleration, and was first observed (1947) at General Electric's laboratories by Elder and co-workers [22, 23]. At first, this radiation was considered a nuisance and wasted energy in particle physics research. However, in the 1960s it was recognized that this intense light could be used to carry out a variety of experiments. Today, these machines are commonly referred to as first-generation synchrotrons and are operated in a "parasitic" mode whereby condensed-matter physicists exploited the synchrotron light, while high-energy physicists used the electron beam.

In the 1970s and 1980s synchrotrons were constructed solely for their ability to generate synchrotron radiation. These so-called second-generation synchrotrons were large circular rings where charged particles, such as electrons or positrons were guided around the ring through a series of bending magnets at nearly the speed of light. As the magnets alter the electrons' path, the electrons are accelerated toward the center of the ring, emitting synchrotron radiation. Examples of second-generation synchrotron sources are Brookhaven's National Synchrotron Light Source (NSLS, Upton, NY), commissioned in 1984, and The Photon Factory (Tsukuba, Japan).

The latest, third-generation synchrotrons, such as the Advanced Photon Source (APS, Chicago, IL), Spring 8 (Hyogo, Japan), and the European Synchrotron Radiation Facility (ESRF, Grenoble, France) are $\sim 10^{10}$ times brighter (photons/s/mm^2/mrad^2) than first-generation sources and $\sim 10^{15}$ times brighter than conventional sealed tubes and rotating anodes (Fig. 5.2). This increased brightness has been achieved through the use of insertion devices known as

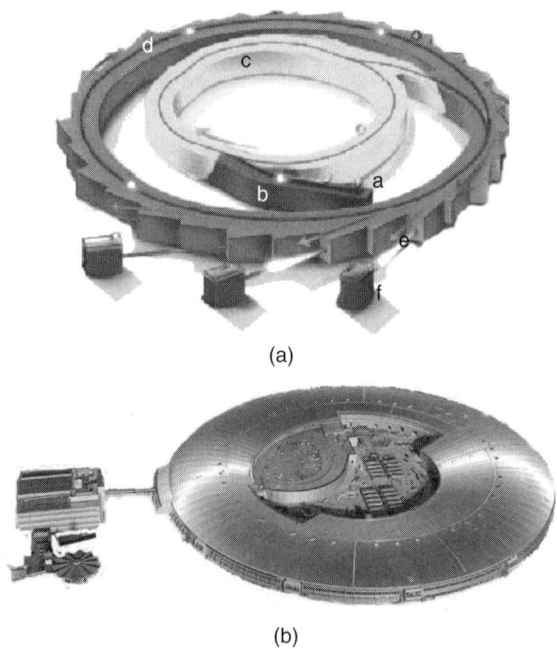

Figure 5.2 (A) Schematic of a third-generation synchrotron source. (a) An electron gun produces electrons that are accelerated by a (b) Linac to nearly the speed of light ($\sim$300,000 km/s). (c) The electrons from the Linac are then further accelerated to 99.999994% of the speed of light, after which they are transported to the (d) storage ring, where they circulate for hours in a high vacuum environment. Bending magnets direct the electrons around the ring—X-rays are emitted when electrons moving near the speed of light are forced to change direction—and (e) insertion devices (i.e., wigglers and undulators) are placed in the straight sections of the storage ring, producing the brightest synchrotron light. (f) The experimental station is where X-rays interact with the sample and are detected. (B) The Diamond synchrotron located at the Harwell Chilton science campus.

wigglers and undulators. Wigglers consist of a series of high field magnets placed above and below the electron beam along a straight section of the storage ring. The magnets cause the electrons to undulate with a tighter radius, increasing the light emitted by the electrons at shorter wavelengths. Undulators are similar to multipole wigglers, but with a larger number of poles. The effect is that strong interference occurs between the radiation from consecutive magnets, which results in a spectral profile with a peak at a specific wavelength. The energy of the emitted X-rays is tuned by changing the gap between the magnet poles.

5.3 TYPES OF SCATTERING

Coherent scattering is the result of interference between the wavefunctions of scattered particles arising from correlations among the atoms within the sample. It can be either elastic or inelastic [10]. Elastic scattering occurs when the neutron or X-ray scatters without any change in kinetic energy, while inelastic scattering involves an exchange in kinetic energy with the sample. In addition, neutrons exhibit isotropic incoherent scattering, the result of correlations between the same nuclei at time zero and at a later time t. Neutron incoherent scattering generally contributes to the general background intensity, but can be used in the determination of dynamical information. In the case of X-rays, Compton scattering is sometimes referred to as incoherent scattering, but it is not analogous to neutron incoherent scattering. For the purposes of this chapter we limit ourselves to coherent, elastic scattering.

The wavevector $\mathbf{k}$, points in the direction of the neutron or X-ray beam and has magnitude $|\mathbf{k}| = 2\pi/\lambda$, where λ is the wavelength. The scattering vector $\mathbf{Q}$, is the difference between the incident and final wavevectors, $\mathbf{k}_i - \mathbf{k}_f$, and is schematically represented for a crystalline material in Fig. 5.3. From the figure, which depicts equally spaced planes of atoms separated by a distance d, Bragg's law ($\lambda = 2d \sin \theta$) can easily be derived. To observe a diffraction peak, the Laue condition $\mathbf{Q} = \mathbf{G}$, where $\mathbf{G}$ is a reciprocal-lattice vector, must be met.

Analytically, one can derive Bragg's law from the magnitude of the scattering vector $|\mathbf{Q}|$. From Fig. 5.3 we observe that $|\mathbf{Q}| = 2k \sin \theta$, where 2θ is the angle between the incident beam and diffracted beam. Since $k = 2\pi/\lambda$ and the magnitude of the reciprocal-lattice vector $|\mathbf{G}| = 2\pi/d$, carrying out the appropriate substitutions results in $\lambda = 2d \sin \theta$. Simply, this is the condition for constructive interference of waves with incident angle θ on a set of equidistant planes separated by a distance d. From this condition, it is possible to derive the concept of the structure factor for a crystal. The structure factor represents the cumulative scattering of the electrons or nuclei in a real sample and the effects of their arrangement within the sample.

The structure factor $\mathbf{F}$ is the Fourier transform of the arrangement of the atoms in the sample, $\rho(x,y,z)$. The function $\mathbf{F}$ is in general, a complex function with both real and imaginary parts. The imaginary part is related to the "phase" of $\mathbf{F}$.

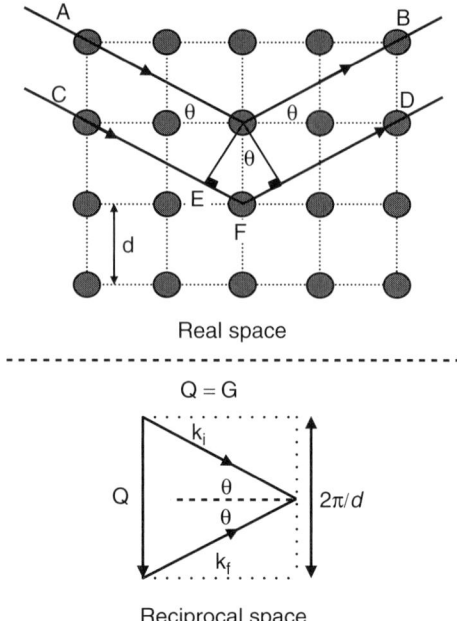

Figure 5.3 The equivalence of Bragg's law (real space) and the Laue condition (reciprocal space) for a two-dimensional square lattice. In the Bragg condition, neutrons or X-rays are specularly reflected from planes of atoms separated by a distance d. The requirement for constructive interference between two waves resulting from the different planes is that the pathlength difference is an integer number of the wavelength (λ), leading to the well-known expression $\lambda = 2d \sin \theta$, where 2θ is the angle between the incident beam and diffracted beam. The Laue condition requires that $\mathbf{Q} = \mathbf{G}$ (constructive interference). Since $|\mathbf{G}| = 2\pi/d$, $|\mathbf{Q}| = 2k \sin \theta$ and $k = 2\pi/\lambda$, it then follows that $\lambda = 2d \sin \theta$, commonly known as Bragg's law.

However, the intensity of the detected scattered radiation, I, is proportional to the magnitude of $\mathbf{F}$, $I = |\mathbf{F}|^2$, meaning the phase of $\mathbf{F}$ is not detected. In crystallography this is the widely known "phase problem." Through various techniques, one can determine the phases of $\mathbf{F}$, and inverse Fourier transform to reconstruct the arrangement of atoms in the sample.

Small-angle neutron scattering (SANS) probes structures on the order of 10^{-9} to $\sim 10^{-6}$ m and is a technique uniquely suitable for the study of isotropic materials in solution [24]. As such, SANS is sensitive not only to local length scales, but also length scales describing the gross morphology of the system. Deuterated materials, including deuterated solvents, are used to enhance neutron contrast [14]. The neutron wavelength, λ, and the scattering angle, θ, determine the length scale, d, being probed ($d \sim \lambda/\theta$).

5.4 MODEL MEMBRANES

In biological systems, plasma membranes surround cells and function as an interface between the cell's interior and exterior environments. Phospholipids, cholesterol, and a variety of integral and peripheral proteins are the chief components of these membranes.

Lipids constitute as much as 50% of the mass of eukaryotic cellular membranes, while the remaining mass is accounted for by proteins [25]. The most common lipids have a hydrophilic polar headgroup and two hydrophobic fatty acid chains that can differ in length and degree of unsaturation (presence or absence of inter carbon double bonds). Normally, the shorter of the two tails is saturated (no double bonds) while the longer hydrocarbon chain contains one or more double bonds. Difference in the length, saturation of the hydrocarbon chains, and chemical makeup of the headgroups all conspire to influence the packing of the lipid molecules, resulting in a variety of structural phases when pure lipids are placed in solution (Fig. 5.4) [25, 26].

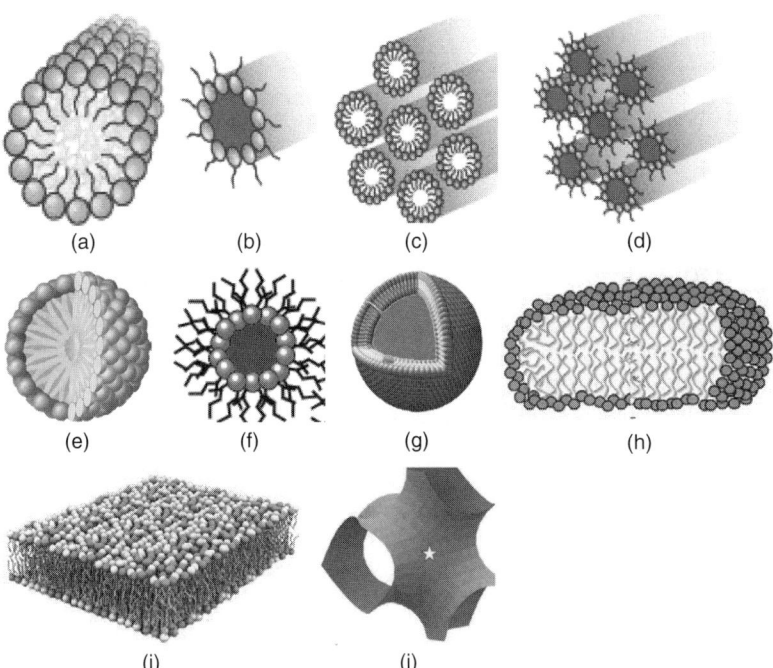

Figure 5.4 Example morphologies adopted by lipids: (a) prolate micelle, (b) inverse prolate micelle, (c) hexagonal, (d) inverted hexagonal, (e) micelle, (f) inverted micelle, (g) unilamellar vesicle, (h) bilayered micelle, (e) bilayer, and (j) cubic.

5.4.1 Isotropic Membranes

Spontaneously Forming Unilamellar Vesicles Phospholipids share many of the characteristics exhibited by surfactants. However, since phospholipids are biological in origin, they are generally considered to be a promising group of materials suitable for the engineering of biocompatible systems for medical purposes. Unilamellar vesicles (ULVs) made of lipids and other biocompatible molecules are used to enhance the efficacy of various pharmaceuticals by protecting encapsulated material and by acting as targeted carriers [27]. These ULVs are usually produced through sonication of multilamellar vesicles (MLVs), resulting in nonuniform size or polydisperse ULVs, or through multistage extrusion of MLVs. However, these mechanically produced ULVs are generally unstable and revert over a period of time (e.g., hours to days) to their native MLV conformation [28]. In addition, the efficacy of polydisperse ULVs in treating disease is compromised as both polydispersity and instability lead to highly variable circulation half-lives in the patient. Generally, ULVs produced by mechanical means are costly, have a limited "shelf life," and are not amenable to industrial scale production.

Over the last decade or so, spontaneously forming ULVs have been observed in cationic–anionic surfactant mixtures [29–33] and cationic surfactant systems [34]. Although some of these are believed to be thermodynamically stable, with low size distributions, issues concerning biocompatibility and biodegradability must be considered for them to be relevant to medical applications. Spontaneously formed ULVs have previously been found in phospholipid mixtures composed of long- and short-chain lipids [35, 36]. However, their stability as a function of total lipid concentration (c_{lp}) and polydispersity was seldom studied.

There has also been a great deal of scientific activity in lipid systems forming bilayered micelles, commonly referred to as "bicelles" [37, 38]. Although bicelles are often formed using a variety of surfactants [39–41], a number of groups have demonstrated biomimetic bicelles composed of long- (e.g., dimyristoyl phosphatidylcholine (DMPC)) and short-chain (e.g., dihexanoyl phosphatidylcholine (DHPC)) zwitterionic phospholipids doped with either paramagnetic ions, charged lipids/surfactants, or both [42–44].

Nieh and co-workers [45] have brought together these three topics of inquiry, by discovering the presence of spontaneously formed, biocompatible, reasonably monodisperse and stable ULVs in bicelle mixtures consisting of the lipids DMPC and DHPC, and the anionic lipid dimyristoyl phosphatidylglycerol (DMPG) [45]. The system's phase diagram is dominated by three liquid-crystalline morphologies. Below a critical temperature T_c, comparable to the chain melting transition temperature ($T_M = 23\,°C$) of DMPC, the mixture exhibits an isotropic dispersion of disk-like micelles over a wide range of total lipid concentration (c_{lp}). At temperatures above T_c, a lamellar phase is found for $c_{lp} \geq 2$ wt%, whereas at lower concentrations, monodisperse ULVs are obtained. These results raised several questions regarding the thermodynamics of the system. Do higher concentration lamellae undergo a complete unbinding transition to form ULVs at lower concentrations? Are the ULVs formed thermodynamically stable, and what dictates their size? To answer these questions, a series of SANS and polarized optical microscopy (POM) experiments were devised, whereby the formation of the various morphologies from different dilution and temperature pathways was explored [46].

Figure 5.5 shows a 25 wt% sample, at 45 °C, being diluted to a final concentration of 0.1 wt%. Samples with $c_{lp} \geq 2.5$ wt% exhibited Bragg maxima, typical of a lamellar phase with a unit cell size of d. The unit cell changed with decreasing lipid concentration, from 104 to 1348 Å, as a function of c_{lp}^{-1}. In contrast, SANS patterns for $c_{lp} \leq 1.25$ wt% did not exhibit Bragg maxima, while the scattering intensity decayed as q^{-2} over an extended q-range. This is consistent with the form factor of locally flat, sheet-like objects such as bilayers making up a ULV. The fact that these scattering curves were absent of any oscillations indicates that the ULVs were highly polydisperse. Fitting with a model of noninteracting spherical ULVs resulted in an average radius $<R>$ larger than 1200 Å with an absolute minimum polydispersity (standard deviation of radius/mean radius) $p \sim 0.25$. From these observations it was clear that at c_{lp} below 2.5 wt% the lamellae completely unbind, meaning the lamellar phase may not be thermodynamically stable.

No sharp peaks were observed after these samples (1.25 and 2.5 wt%) were cooled down to 10 °C. The SANS data were best described by the bilayered micelle morphology and the best fit was obtained using a combination of the core-shell-discoidal (CSD) model and the Hayter–Penfold structure factor, which

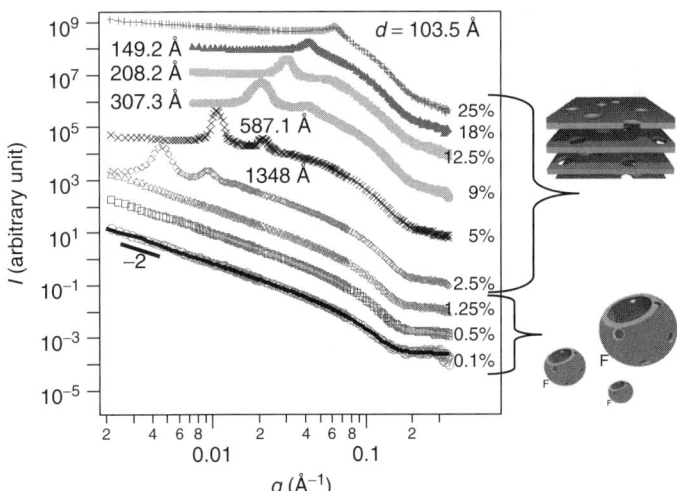

Figure 5.5 SANS profiles of DMPC/DHPC ULVs doped with DMPG diluted at 45 °C. The profiles for $c_{lp} \geq 2.5$ wt% are characteristic of multibilayers. The scattered intensity for $c_{lp} \leq 1.25$ wt% follows a q^{-2} dependence, indicative of isolated bilayers. The solid line is a fit to the data from the 0.1-wt% sample using a model of noninteracting polydisperse ULVs.

resulted in a disk core radius, R, of 590 and 220 Å for the 1.25- and 2.5 wt% samples, respectively. Moreover the bilayered micelles for both lipid concentrations had the same bilayer thickness (42 Å).

On reheating to 45 °C, the lamellar morphology was recovered for all of the samples with $c_{lp} \geq 2.5$ wt%. However, reheating had a very different effect on the 1.25-wt% sample. The scattering pattern showed an oscillatory behavior as a function of q, instead of the monotonic decay initially seen at 45 °C. This was a clear indication that the initial polydisperse ULVs, when reformed from the bilayered micelle morphology, transformed into monodisperse ULVs. However, it is interesting to note that recooling a 0.1-wt% sample containing monodisperse ULVs to 10 °C did not result in bilayered micelles, but rather the data was best fit using a model of monodisperse oblate ellipsoids.

The major conclusions of the study by Nieh et al. [45] (schematically shown in Fig. 5.6) can be summarized as follows:

1. Diluting the lamellar morphology at 45 °C eventually leads to the lamellae completely unbinding. The individual unbound lamellae form highly polydisperse ULVs.
2. For a 1.25-wt% sample when cooled from 45 to 10 °C, polydisperse ULVs adopt a uniform size bicelle morphology, which when reheated reform into monodisperse ULVs.

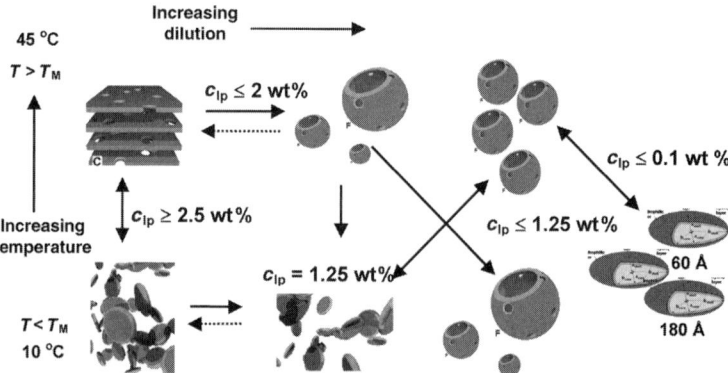

Figure 5.6 Schematic of morphological transformations exhibited by DMPC/DHPC/DMPG lipid mixtures. Diluting below a critical lipid concentration, at a temperature $T > T_c$, results in bilayers unbinding and forming polydisperse ULVs. On cooling below T_c, 1.25-wt% polydisperse ULVs transform into an isotropic bicellar solution, which on reheating to $T > T_c$ gives rise to monodisperse ULVs. For $c_{lp} \leq 0.5$ wt%, polydisperse ULVs are trapped and cannot transform into bicelles at $T = 10$ °C. Monodisperse ULVs can also be obtained by diluting the bicellar phase to a c_{lp} of ~ 1.25 wt% and heating above T_c. However, in the case of very dilute mixtures, (i.e., $c_{lp} \leq 0.1$ wt% and $T < T_c$), bicelles are not recoverable. Instead, oblate ellipsoids are formed. The dashed lines indicate plausible transformations, which were not proved experimentally. For details the reader is referred to Ref. 46.

3. When 0.1-wt% monodisperse ULVs are cooled down to 10 °C, they do not reform bicelles, but instead transform into oblate ellipsoids.
4. For $c_{lp} \leq 0.5$ wt%, the ULV morphology at 45 °C remains unchanged even when the sample is cooled down to 10 °C.

It was concluded that the various results indicated that ULVs are most likely the thermodynamically stable morphology for temperatures greater than those corresponding to the T_M of DMPC and c_{lp} less than the unbinding c_{lp}. Finally, the ULV size distribution depended on the equilibrium precursor morphology. Recently, the ULVs were formulated to carry bi-modal (optical and MRI) contrast payloads in sufficient concentrations to allow for signal detection in vivo.

5.4.2 Aligned Membranes

Being amphiphilic molecules, lipids self-assemble into a variety of morphologies when in contact with water. Dipalmitoyl phosphatidylcholine (DPPC) is one of the most extensively studied and best understood lipids, which when completely hydrated, exists in four distinct lamellar morphologies. On decreasing temperature they occur in the following sequence: L_α, $P_{\beta'}$, L'_β, and $L_{c'}$. The L_α phase contains no long-range order within the plane of the bilayer, and the hydrocarbon chains are in a disordered "melted" state. On the other hand, both the periodically modulated lamellar $P_{\beta'}$ and gel $L_{\beta'}$ phases are characterized by a 2D ordering of their hydrocarbon chains in the bilayers, with no interbilayer correlations. It has also been shown that the $L_{\beta'}$ phase in DMPC, in fact, consists of three phases with differing directions of the molecular tilt with respect to the 2D bond direction [47, 48]. The lowest temperature $L_{c'}$ phase was first observed calorimetrically by Chen et al. [49] in fully hydrated DPPC bilayers after a few days of incubation at ∼0 °C.

When compared to powder or liposomal preparations, aligned membranes have allowed a variety of techniques of extracting unambiguous structural information. With regard to scattering, because the signal is not isotropic, as is the case for liposomal preparations, aligned samples permit clear differentiation between in-plane and out-of-plane structures. Moreover, data collection is accelerated and the sample quantity required is generally a fraction of that needed for nonaligned samples. Although the use of aligned samples, for all of the reasons stated, is highly desirable, their use was previously limited as they could not be fully hydrated. However, that problem was overcome, opening the doorway for their use in biologically relevant studies [50, 51].

X-Ray Diffraction from Aligned $L_{c'}$ DPPC Multibilayers Until 1980, DPPC MLV suspensions were known to exhibit only two thermotropic phase transitions: the sharp main $L_{\beta'}$–L_α transition centered at ∼41 °C and the broad $P_{\beta'}$–$L_{\beta'}$ pretransition centered at ∼35 °C. However, a new transition was observed by Chen et al. [49] using differential scanning calorimetry (DSC), after storing DPPC at ∼0 °C for several days. Since then, numerous investigations were conducted in order to determine the structural features characteristic of this morphology.

The structural changes accompanying the $L_{\beta'}$–$L_{c'}$ transition, commonly referred to as the subtransition, had been well documented using powder X-ray diffraction [52, 53]. Compared to $L_{\beta'}$ bilayers, it was generally accepted that $L_{c'}$ bilayers exhibited a decreased lamellar periodicity and the diffraction patterns contained a number of "additional" reflections, supposedly the result of a more ordered system [52]. It was not until the production of highly aligned samples that the structural details of $L_{c'}$ DPPC bilayers were elucidated [54, 55].

Figure 5.7 shows the diffraction from aligned $L_{c'}$ DPPC bilayers. From the visual examination of the diffraction pattern it is reasonably simple to identify the reflections corresponding to structure parallel and perpendicular to the plane of the bilayer. This would not be possible from a powder diffraction pattern resulting from a dispersion of MLVs. Of importance is that the 4.2- and 4.5-Å reflections, which lie in a straight line with the 3.9-Å reflection, and also parallel with the c^* axis, are the result of the form factor of finite length hydrocarbon chains. Consequently, these reflections do not arise from lattice spacings and, as such, could not be surmised from powder patterns. Moreover, because the hydrocarbon chains in DPPC are tilted, the 3.9- and 3.8-Å repeat spacings do not directly correspond to the spacing of lattice planes; only their projections (e.g., 4.65 and 4.4 Å) onto the axis perpendicular to c^* do. Again, from powder patterns all of this information regarding chain tilt becomes obscured.

From analysis of the diffraction pattern presented in Fig. 5.7, it was concluded that the $L_{c'}$ phase of DPPC is described by a 2D molecular lattice containing two lipid molecules. Moreover, the molecular lattices are positionally correlated across a single bilayer, but this positional correlation does not extend to adjacent bilayers (i.e., there are no out-of-plane correlations).

X-Ray Diffraction from Aligned $P_{\beta'}$ DPPC Multibilayers

Hydrated disaturated phosphatidylcholines often exhibit an intriguing thermodynamic phase known as $P_{\beta'}$, or ripple morphology, in which the time-averaged bilayers are locally flat, and which can be formed by cooling from the fluid L_α phase.

Evidence that the $P_{\beta'}$ phase is indeed rippled and not flat has accumulated over 25 years from freeze-fracture electron microscopy (FFEM) and X-ray diffraction. [56–60] The most precise structure of the $P_{\beta'}$ phase was determined for DMPC multibilayers using a synchrotron data set taken at $T = 18\,°C$ and 25-wt% water, corresponding to a relative humidity RH ∼98% [61]. This DMPC ripple profile has an asymmetric, "sawtooth" shape, rather than a smooth symmetric or simple sinusoidal shape (Fig. 5.8).

Although there was a general agreement that the asymmetric ripple pattern exists, it also became clear that there was an additional ripple morphology, as imaged by FFEM [56–58, 60] and deduced from powder X-ray diffraction. [62, 63]. After considerable uncertainty lasting over a decade regarding the conditions for seeing different ripple patterns, it was noted that the $P_{\beta'}$ phase can differ depending on whether it is formed on cooling from the L_α phase or on

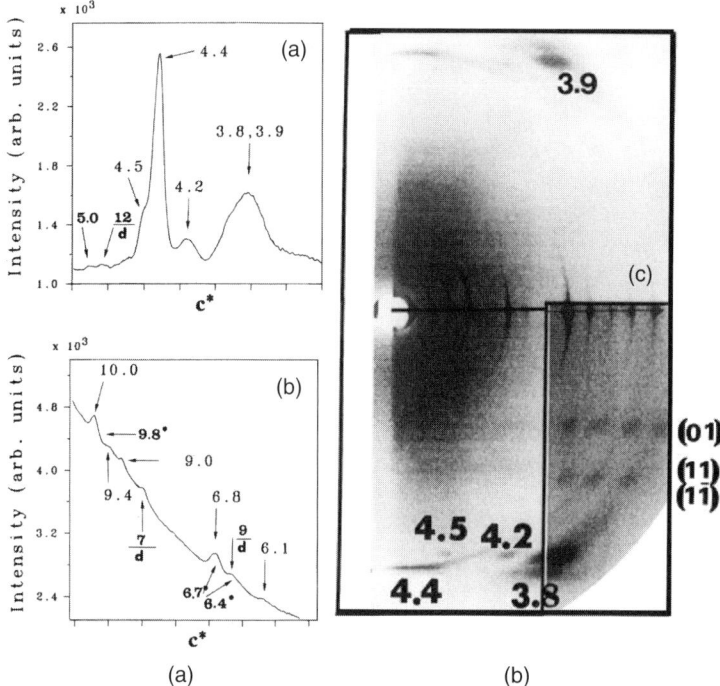

Figure 5.7 (A) X-ray powder diffraction patterns of the (a) wide-angle region showing reflections arising from the hydrocarbon chains and their corresponding repeat spacings, and (b) the 10.0- and 6.8-Å regions. The broad reflection in (a) contains both the 3.8- and 3.9-Å reflections seen in the aligned pattern (B). From this diffraction pattern the positional relationship between the various Bragg reflections (i.e., those arising from in-plane vs. out-of-plane structure) are clearly seen. The 10.0- and 6.8-Å lattice line reflections, arising from the ordering of DPPC molecules in two dimensions, are labeled (01) and (1$\bar{1}$), (11), respectively. Data were collected with an 18-kW Rigaku Rotaflex RU300 rotating anode generator and a 2D MarResearch imaging plate detector having a plate diameter of 180 mm and pixel size of 150 μm × 150 μm. For complete details see Refs. 54 and [55].

heating from the $L_{\beta'}$ phase. The consensus, at least in the case of DPPC bilayers, was that upon heating from the $L_{\beta'}$ phase the $P_{\beta'}$ phase formed consisted of short wavelength ($\lambda \sim 145$ Å) asymmetric ripples, while the ripple phase formed upon cooling from the L_α phase was a mixture consisting of long wavelength ($\lambda \sim 260$ Å) ripples coexisting with the usual short wavelength ripples [58, 64]. However, coexistence over an extended range of temperature generally violates the Gibbs phase rule. Coexistence was disputed when previously supporting X-ray data were reanalyzed, and along with powder neutron diffraction data concluded that the analysis was consistent with pure long wavelength rippled bilayers [65].

The solution of the vapor pressure paradox [51] allowed for the construction of a sample environment capable of fully hydrating aligned lipid multibilayers

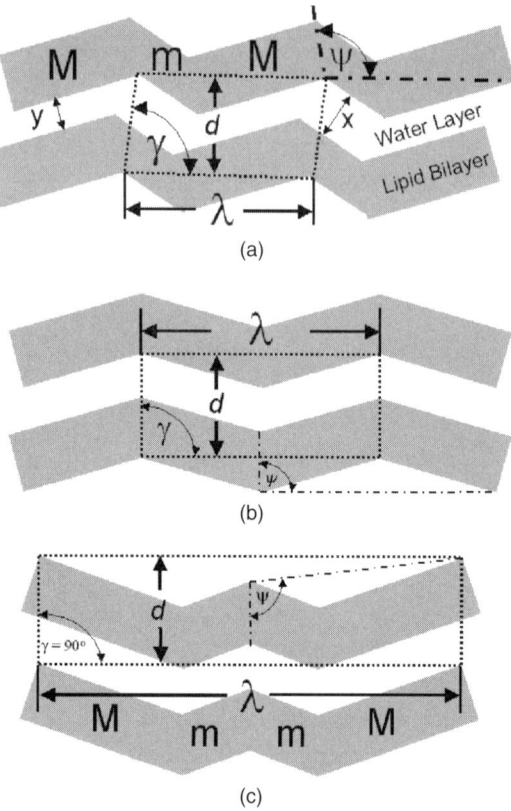

Figure 5.8 Schematic drawing showing (A) asymmetric (MmMm) rippled bilayers separated by a layer of water. The unit cell, shown by the dotted parallelogram, has dimensions d, λ, and a monoclinic angle γ. ψ is a structural angle made between the two adjacent "kinks" in the bilayer and the horizontal. (B) Symmetric ripple whereby M and m are indistinguishable. For this type of bilayer, a γ of 90° is expected, and $\gamma = \psi$. (C) Schematic drawing of the symmetric, W-shaped ripple having a repeat of MmmM or mMMm. As was the case of the symmetric ripple (B) a W-shaped ripple has a γ of 90°, however, unlike the symmetric ripple, $\gamma \neq \psi$. For further details the reader is referred to Ref. 66.

[66] (Fig. 5.9). One outstanding issue that could be resolved using fully hydrated aligned samples so that their bilayer repeat spacings were indistinguishable from those bilayers immersed in water [50] was whether or not DPPC bilayers formed, upon cooling, a mixture of short and long wavelength rippled bilayers or simply a single population of long wavelength rippled bilayers [62, 63, 65]. The experiments were carried out at the Cornell High Energy Synchrotron Source (CHESS, Itahca, NY) using monochromatic X-rays ($\lambda = 1.38$ Å) and a 2D charge coupled device (CCD) detector. The flux at the sample was determined to be $\sim 10^{12}$ photons mm^{-2} s^{-1} [67].

From the 2D diffraction patterns (Fig. 5.10), it is clear that upon heating both chiral and racemic DPPC bilayers form one population of rippled bilayers, as all reflections can be indexed to one unique unit cell. However, upon cooling, the situation is very different. The data in Fig. 5.10 supports the commonly accepted notion by Yao et al. [62] and Matuoka et al. [63]. that there are two distinct, coexisting populations of ripples differing in d, ripple wavelength, and ripple symmetry. Moreover, direct evidence for the existence of the symmetric ripple was provided, for the first time, using synchrotron radiation.

X-Ray Scattering from Lα Lipid Bilayers It is difficult to obtain the structure of a fully hydrated L_α phase bilayer as bilayer fluctuations degrade the Bragg signal, which in the traditional crystallographic approach is used to determine the membrane's structure. Employing stacks of aligned multibilayers, a new method of analysis was proposed whereby the focus is on the diffuse scattering (in the plane of the bilayer, Q_r direction) taking place across the entire Q_z range (Fig. 5.11A), instead of on the discrete Bragg reflections [68, 69]. Compared to traditional Bragg scattering, this method has access to data over a greater range in Q_z, thereby providing better real-space resolution. However, in contrast to the

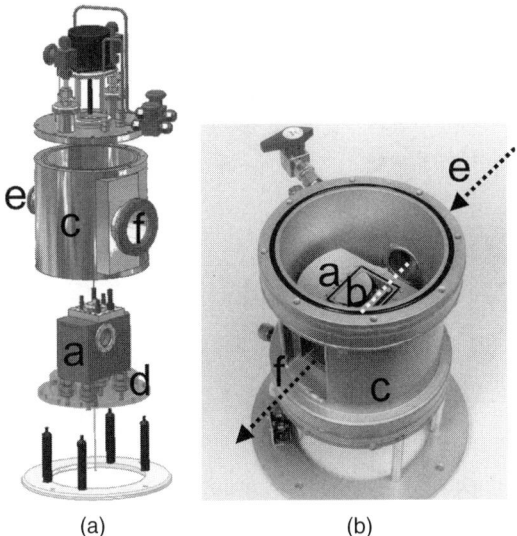

(a)   (b)

Figure 5.9 Schematic of X-ray sample environment capable of attaining 100 % RH. (A) Exploded view and (B) perspective view of the assembled sample environment. The temperature controlled sample cell (a) is a sealed unit (shown with sealing panel removed) and contains the water-saturated evaporative sponge used in hydrating the sample and the cylindrical sample support (b) enabling the collection of multiple Bragg reflections. The sample chamber is thermally isolated from the independently temperature controlled outer "jacket" (c) via the use of an acrylic base (d). Entrance and exit windows are made of aluminized mylar (e) or kapton (f), respectively. For complete details the reader is referred to Ref. 67.

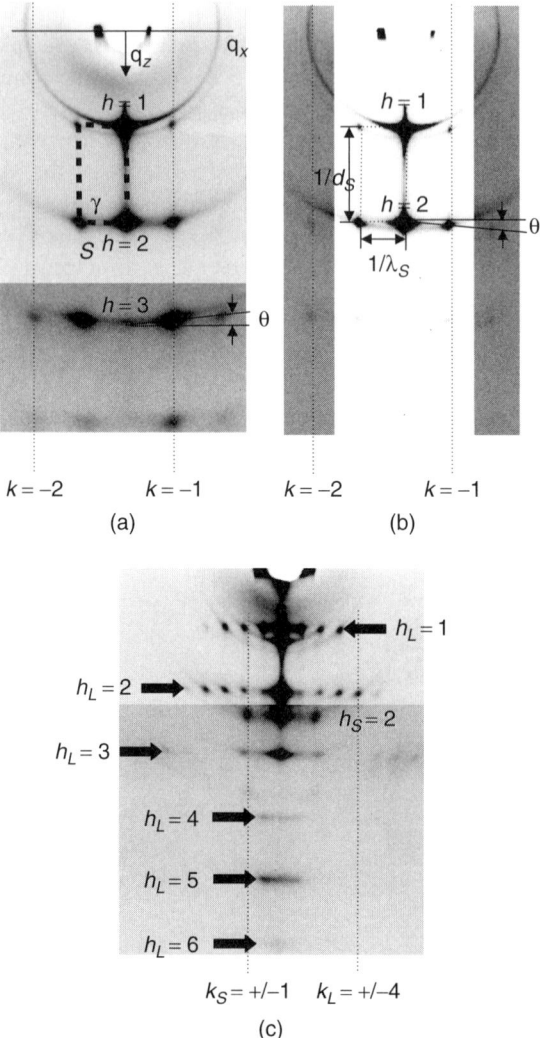

Figure 5.10 Two-dimensional diffraction patterns of oriented (A) l-DPPC and (B) dl-DPPC ripple phases formed on heating from the $L_{\beta'}$ phase. (C) Two-dimensional diffraction pattern of $P_{\beta'}$ dl-DPPC rippled bilayer phases formed on cooling from the disordered L_α phase. The coexistence of two rippled bilayer populations is depicted by the indices (h_S, k_S) for the short ripple phase and the indices r for the long ripple phase. The bold arrows identify the family of peaks due to the long ripple. From the diffraction patterns the various lattice parameters are directly obtainable. Cooling L_α phase l-DPPC bilayers also results in two rippled bilayer populations (data not shown). For complete description of experimental data the reader is referred to Ref. 67 from which this figure was adapted.

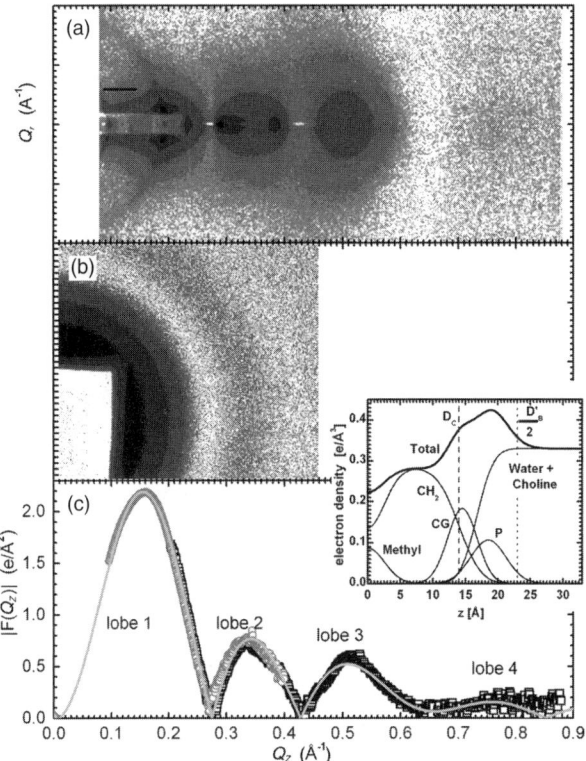

Figure 5.11 Gray-scale X-ray diffraction plot of the log of background-subtracted intensity from aligned multibilayers (A) and isotropic ULVs (B) of fully hydrated DPPC in the fluid phase. (C) Analysis of the diffuse scattering data from aligned multibilayers resulted in the high Q_z data (squares), while ULVs provided the low Q_z data (circles). The solid curve is the fit to the data using the hybrid model of electron density distribution across the bilayer. Inset to the figure shows such a model of one-dimensional electron density and its decomposition into various sub molecular components (shown for only one-half of the bilayer). The vertical broken lines show the hydrocarbon thickness (D_C) and half of the estimated bilayer thickness ($D'_B/2$).

usual crystallographic approach of Bragg scattering, this method of analysis does not extend to data in the low Q_z, as in this region of Q_z the diffuse scattering is weak relative to the very strong scattering exhibited by the first and second Bragg reflections.

An alternative approach is to utilize isotropic ULVs (Fig. 5.11B). Scattering data from ULVs directly provide the continuous form factor, which is the Fourier transform of the scattering density distribution across the bilayer. However, a dilute system of unoriented bilayers results in a rapidly decaying scattered intensity, which, although it contains good counting statistics at low Q_z, exhibits rather poor counting statistics in the high Q_z region where the scattering from the bilayer, on average, is near the level of the background. To summarize, aligned multibilayers provide excellent high Q_z data (i.e., short to medium length scale features), while scattering data from ULVs are better at resolving medium to long length scale structural features (i.e., low Q_z).

Kučerka et al. [70] have recently combined the scattering data from fully hydrated ULVs and aligned multibilayers into a single scattering curve covering an extended Q_z range(Fig. 5.11C). Among other lipids, the structure of a fully hydrated L_α phase DPPC bilayer was presented in terms of a hybrid model (inset to Fig. 5.11) [71] based on a functional form obtained from molecular dynamics simulations. Using a model approach to analyze the data allows not only for the determination of the bilayer's total one-dimensional electron density profile, but also for the location of its individual submolecular components. Thus high quality experimental data covering an extended range in reciprocal space is better able to resolve structural features, including biomolecules added to the membrane.

Neutron Diffraction from Aligned Lα Multibilayers Aligned lipid bilayers have proved useful for elucidating their liquid-crystalline structures at high and low temperatures, but have revealed important structural information under physiologically relevant conditions as well. Numerous experiments have used neutron diffraction to locate individual membrane components along the normal of the bilayer plane. The technique relies on the specific deuterium labeling of molecular groups and the Fourier transform reconstruction of the bilayer profile from the diffraction data. The unique ability of neutrons to distinguish between hydrogen and deuterium atoms provides the signal from the deuterium label, which remains once all of the unlabeled sample data has been subtracted away.

A first application of the technique was provided by Zaccaï et al. [72], who identified the various molecular components of a DPPC bilayer [73, 74]. The molecular components of bilayers formed, for example, from dioleoyl phosphatidylcholine (DOPC) [75, 76] and phosphatidylinositol [77] lipids have also been measured. The location within the bilayer matrix of several membrane-bound molecules has also been determined. They include anesthetics [78], squalene [79], selectively labeled protein residues [80, 81], and cholesterol in disaturated and mixed saturated–monounsaturated PC bilayers [82, 83].

Cholesterol has greatly reduced solubility within polyunsaturated fatty acid PC lipids, leading to questions of cholesterol's orientation and location within PUFA lipids. Harroun et al. [84] employed neutron diffraction with specifically deuterium-labeled cholesterol incorporated into PC bilayers with increasing degrees of unsaturation. The deuterium atom replacements were located on the first steroid ring, near the hydroxyl oxygen atom. This end of the molecule usually is directed outward to the aqueous environment, and the long axis of the molecule stands upright in the acyl chain matrix.

After subtracting the neutron scattering length density profile of the unlabeled cholesterol experiments, the remaining signal shown in Fig. 5.12 is the time- and

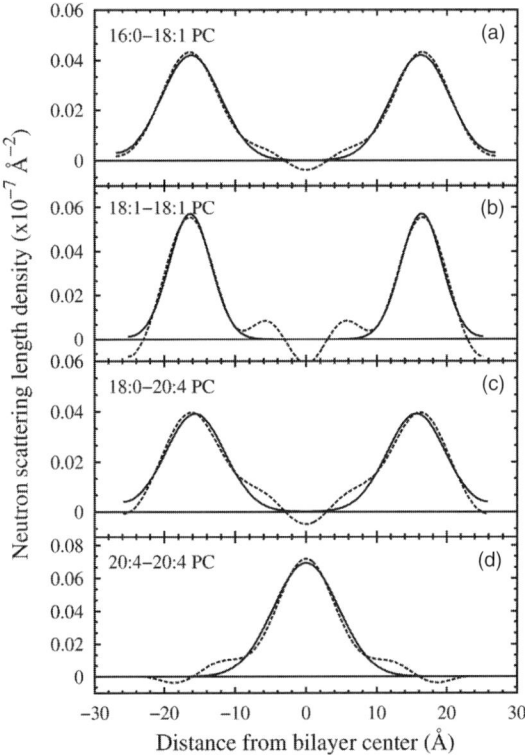

Figure 5.12 The time- and sample-averaged center of mass of the hydroxyl group of cholesterol located within bilayers of increasing degrees of acyl-chain unsaturation. The origin of the abscissa is the center of the bilayer. The calculated SLD profiles have a spatial resolution of 10 Å. The dashed line is the measured data, and the solid line is a fit with a single Gaussian function. Fitting is performed in reciprocal space by taking the difference in the measured form factors.

sample-averaged center of mass distribution of the deuterium atoms attached to the steroid ring of the cholesterol, relative to the center of a PC bilayer. For lipids with at least one acyl chain of no or single double bonds, the hydroxyl end of cholesterol is located near the lipid–aqueous interface, approximately 1.6 nm from the center of the bilayer. This was expected, since cholesterol has a relatively high solubility in these lipids. However, in the di- arachidonyl (20:4–20:4 PC) bilayer, the center of mass shifts to the center of the bilayer. This surprising result must be interpreted as the reorientation of the molecule such that the hydroxyl group end now is located at the bilayer center. This is the consequence of either the cholesterol laying flat within the bilayer, or the complete inversion of the molecule. The inability of the PUFA acyl chains to accommodate cholesterol in its normal upright position must have physiological implications, which have not yet been explored.

REFERENCES

1. A. Pockels (1891). Surface tension. *Nature* **43**: 437.
2. I. Langmuir (1917). The constitution and fundamental properties of solids and liquids. II. Liquids. *J. Am. Chem. Soc.* **39**: 1848.
3. E. Gorter and F. Grendel (1925). On bimolecular layers of lipids on the chromocytes of the blood. *J. Exp. Med.* **41**: 439.
4. P. Mueller, D. O. Rudin, H. T. Tien, and W. C. Wescott (1962). Reconstitution of cell membrane structure *in vitro* and its transformation into an excitable system. *Nature* **194**: 979.
5. S. J. Singer and G. L. Nicolson (1972). The fluid mosaic model of the structure of cell membranes. *Science* **175**: 720.
6. T. Allen, C. Hansen, F. Martin, C. Redemann, and A. Yau-Young (1991). Liposomes containing synthetic lipid derivatives of poly(ethylene glycol) show prolonged circulation half-lives *in vivo*. *Biochim. Biophys. Acta* **1066**: 29.
7. M. Woodle and D. Lasic (1992). Sterically stabilized liposomes. *Biochim. Biophys. Acta* **1113**: 171.
8. E. Rutherford (1920). Nuclear constitution of atoms. *Proc. R. Soc. London A* **97**: 374.
9. J. Chadwick (1932). The existence of a neutron. *Proc. R. Soc. London A* **136**: 692.
10. H. Dachs (1978). Principles of neutron diffraction, in *Topics in Current Physics: Neutron Diffraction* (H. Dach, Ed.), Springer-Verlag, Berlin.
11. W. Mampe, P. Ageron, C. Bates, J. M. Pendlebury, and A. Steyerl (1989). Neutron lifetime measured with stored ultracold neutrons. *Phys. Rev. Lett.* **63**: 593.
12. B. Jacrot (1976). The study of biological structures by neutron scattering from solution. *Rep. Prog. Phys.* **39**: 911.
13. J. Katsaras, M.-P. Nieh, T. A. Harroun, M. Chakrapani, and M. J. Watson (2004). Neutron scattering from biologically relevant materials. *Phys. Canada* **60**: 93.
14. T. A. Harroun, G. D. Wignall, and J. Katsaras (2006). Neutron scattering for biology, in *Neutron Scattering in Biology: Techniques and Applications* J. Fitter, T. Gutberlet, and J. Katsaras, Eds.). Springer-Verlag, Berlin.
15. R. E. Williams, J. M. Rowe, and P. Kopetka (1979). The liquid hydrogen moderator at the NIST reactor, in *Proceedings of the International Workshop on Cold Moderators for Pulsed Neutron Sources* (J. M. Carpenter and E. B. Iverson, Eds.). Argonne National Laboratory, p. 79.
16. R. E. Williams and J. M. Rowe (2002). Developments in neutron beam devices and an advanced cold source for the NIST research reactor. *Physica B* **311**: 117.
17. H. Aschauer, A. Fleischmann, C. Schanzer, and E. Steichele (2000). Neutron guides at the FRM-II. *Physica B* **283**: 323.
18. Ch. Rehm and M. Agamalian (2002). Flux gain for a next-generation neutron reflectometer resulting from improved supermirror performance. *Appl. Phys. A* **77** (Suppl.): S1483.
19. M. Senthil Kumar, P. Böni, and D. Clemens (1998). Mechanical and structural properties of Ni/Ti multilayers and films: an application to neutron supermirrors. *J. Appl. Phys.* **84**: 6940.
20. N. Watanabe (2003). Neutronics of pulsed spallation neutron sources. *Rep. Prog. Phys.* **66**: 339.

21. J. Als-Nielsen and D. McMorrow (2001). *Elements of Modern X-Ray Physics*. Wiley, Hoboken, NJ.
22. F. R. Elder, A. M. Gurewitsch, R. V. Langmuir, and H. C. Pollock (1947). Radiation from electrons in a synchrotron. *Phys. Rev.* **71**: 829.
23. F. R. Elder, A. M. Gurewitsch, R. V. Langmuir, and H. C. Pollock (1947). A 70-MeV synchrotron. *J. Appl. Phys.* **18**: 810.
24. J. K. Krueger and G. D. Wignall (2006). Small-angle neutron scattering from biological molecules, in *Neutron Scattering in Biology: Techniques and Applications* (J. Fitter, T. Gutberlet, and J. Katsaras, Eds.). Springer-Verlag, Berlin, p. 127.
25. R. B. Gennis (1989). *Biomembranes: Molecular Structure and Function*. Springer-Verlag, New York.
26. P. Yeagle (1992). *The Structure of Biological Membranes*. CRC Press, Boca Raton, FL.
27. G. Storm and D. J. A. Crommelin (1998). Liposomes: quo vadis? *Pharm. Sci. Technol. Today*, **1**: 19.
28. I. Stanish and A. Singh (2001). Highly stable vesicles composed of a new chain-terminus acetylenic photopolymeric phospholipid. *Chem. Phys. Lipids.* **112**: 99.
29. E. W. Kaler, A. K. Murthy, B. E. Rodriguez, and J. A. N. Zasadzinski (1989). Spontaneous vesicle formation in aqueous mixtures of single-tailed surfactant. *Science* **245**: 1371.
30. E. W. Kaler, K. L. Herrington, A. K. Murthy, and J. A. N. Zasadzinski (1992). Phase behavior and structures of mixtures of anionic and cationic surfactants. *J. Phys. Chem.* **96**: 6698.
31. M. Villeneuve, S. Kaneshina, T. Imae, and M. Aratono (1999). Vesicle–micelle equilibrium of anionic and cationic surfactant mixture studied by surface tension. *Langmuir* **15**: 2029.
32. M. Bergstrom, J. S. Pedersen, P. Schurtenberger, and S. U. Egelhaaf (1999). Small-angle neutron scattering (SANS) study of vesicles and lamellar sheets formed from mixtures of an anionic and a cationic surfactant. *J. Phys. Chem. B* **103**: 9888.
33. M. Bergstrom and J. S. Pedersen (2000). A small-angle neutron scattering study of surfactant aggregates formed in aqueous mixtures of sodium dodecyl sulfate and didodecyldimethylammonium bromide. *J. Phys. Chem. B* **104**: 4155.
34. M. I. Viseu, K. Edwards, C. S. Campos, and S. M. B. Costa (2000). Spontaneous vesicle formed in aqueous mixtures of two cationic amphiphiles. *Langmuir* **16**: 2105.
35. N. E. Gabriel and M. F. Roberts (1984). Spontaneous formation of stable unilamellar vesicles. *Biochemistry* **23**: 4011.
36. M. Ollivon, S. Lesieur, C. Grabrielle-Madelmont, and M. Paternotre (2000). Vesicle reconstitution from lipid-detergent mixed micelles. *Biochim. Biophys. Acta.* **1508**: 34.
37. C. R. Sanders 2nd and G. C. Landis (1995). Reconstitution of membrane proteins into lipid-rich bilayered mixed micelles for NMR-studies. *Biochemistry* **34**: 4030.
38. J. Katsaras, R. L. Donaberger, I. P. Swainson, D. C. Tennant, Z. Tun, R. R. Vold, and R. S. Prosser (1997). Rarely observed phase transitions in a novel lyotropic liquid crystal system. *Phys. Rev. Lett.* **78**: 899.
39. C. R. Sanders 2nd and J. H. Prestegard (1990). Magnetically orientable phospholipid bilayers containing small amounts of bile salt analogue, CHAPSO. *Biophys. J.* **58**: 447.

40. J. Chung and J. H. Prestegard (1993). Characterization of field-oriented aqueous liquid crystals by NMR diffusion measurements. *J. Phys. Chem.* **97**: 9837.
41. C. R. Sanders 2nd and G. C. Landis (1995). Reconstitution of membrane proteins into lipid-rich bilayered mixed micelles for NMR-studies. *Biochemistry* **34**: 4030.
42. R. S. Prosser, S. A. Hunt, J. A. DiNatale, and R. R. Vold (1996). Magnetically aligned membrane model systems with positive order parameter: switching the sign of Szz with paramagnetic ions. *J. Am. Chem. Soc.* **118**: 269.
43. R. S. Prosser, J. S. Hwang, and R. R. Vold (1998). Magnetically aligned phospholipid bilayers with positive ordering: a new model membrane system. *Biophys. J.* **74**: 2405.
44. J. A. Losonczi and J. H. Prestegard (1998). Improved dilute bicelle solutions for high resolution NMR of biological macromolecules. *J. Biomol. NMR* **12**: 447.
45. M.-P. Nieh, T. A. Harroun, V. A. Raghunathan, C. J. Glinka, and J. Katsaras (2004). Spontaneously formed monodisperse biomimetic unilamellar vesicles: the effect of charge, dilution, and time. *Biophys. J.* **86**: 2615.
46. M.-P. Nieh, V. A. Raghunathan, S. R. Kline, T. A. Harroun, C.-H. Huang, J. Pencer, and J. Katsaras (2005). Spontaneously formed unilamellar vesicles with path-dependent size distribution. *Langmuir* **21**: 6656.
47. E. B. Sirota, G. S. Smith, C. R. Safinya, R. J. Plano, and N. A. Clark (1988). X-ray scattering studies of aligned, stacked surfactant membranes. *Science* **242**: 1406.
48. J. Katsaras, D. S.-C. Yang, and R. M. Epand (1992). Fatty acid chain tilt angles and directions in dipalmitoyl phosphatidylcholine bilayers. *Biophys. J.* **63**: 1170.
49. S. C. Chen, J. M. Sturtevant, and B. J. Gaffney (1980). Scanning calorimetric evidence for a third phase transition in phosphatidylcholine bilayers. *Proc. Natl. Acad. Sci. USA* **77**: 5060.
50. J. Katsaras (1997). Highy aligned lipid membrane systems in the physiologically relevant "excess water" condition. *Biophys. J.* **73**: 2924.
51. J. Katsaras (1998). Adsorbed to a rigid substrate, DMPC multibilayers attain full hydration in all mesophases. *Biophys. J.* **75**: 2157.
52. M. J. Ruocco and G. G. Shipley (1982). Characterization of the sub-transition of hydrated dipalmitoylphosphatidylcholine bilayers. X-ray diffraction study. *Biophys. Biochim. Acta* **684**: 59.
53. J. Stümpel, H. Eibl, and A. Nicksch (1983). X-ray analysis and calorimetry on phosphatidylcholine model membranes. The influence of length and position of acyl chains upon structure and phase behaviour. *Biochim. Biophys. Acta* **727**: 246.
54. V. A. Raghunathan and J. Katsaras (1995). Structure of the $L_{c'}$ phase in a hydrated lipid multilamellar system. *Phys. Rev. Lett.* **74**: 4456.
55. J. Katsaras and V. A. Raghunathan (1995). Evidence for a two-dimensional lattice in subgel phase DPPC bilayers. *Biochemistry* **34**: 4684.
56. B. R. Copeland and H. M. McConnel (1980). The rippled structure in bilayer membranes of phosphatidylcholine and binary mixtures of phosphatidylcholine and cholesterol. *Biochim. Biophys. Acta* **599**: 95.
57. A. Hicks, M. Dinda, and M. A. Singer (1987). The ripple phase of phosphatidylcholines: effect of chain-length and cholesterol. *Biochim. Biophys. Acta* **903**: 177.
58. J. A. N. Zasadzinski (1988). Effect of stereo configuration on ripple phases ($P_{\beta'}$) of dipalmitoylphosphatidylcholine. *Biochim. Biophys. Acta* **946**: 235.

59. H. W. Meyer, B. Dobner, and K. Semmler (1996). Pretransition-ripples in bilayers of dipalmitoylphosphatidylcholine: undulation or periodic segments? A freeze-fracture study. *Chem. Phys. Lipids* **82**: 179.
60. J. T. Woodward and J. A. N. Zasadzinski (1997). High-resolution scanning tunneling microscopy of fully hydrated ripple-phase bilayers. *Biophys. J.* **72**: 964.
61. D. C. Wack and W. W. Webb (1989). Synchrotron X-ray study of the modulated lamellar phase $P_{\beta'}$ in the lecithin–water system. *Phys. Rev. A* **40**: 2712.
62. H. Yao, S. Matuoka, B. Tenchov, and I. Hatta (1991). Metastable ripple phase of fully hydrated dipalmitoylphosphatidylcholine as studied by small angle X-ray scattering. *Biophys. J.* **59**: 252.
63. S. Matuoka, S., H. Yao, S. Kato, and I. Hatta (1993). Condition for the appearance of the metastable $P_{\beta'}$ phase in fully hydrated phosphatidylcholines as studied by small-angle X-ray diffraction. *Biophys. J.* **64**: 1456.
64. B. G. Tenchov, H. Yao, and I. Hatta (1989). Time-resolved X-ray diffraction and calorimetric studies at low scan rates. I. Fully hydrated dipalmitoylphosphatidylcholine (DPPC) and DPPC/water/ethanol phases. *Biophys. J.* **56**: 757.
65. P. C. Mason, B. D. Gaulin, R. M. Epand, G. D. Wignall, and J. S. Lin (1999). Small angle neutron scattering and calorimetric studies of large unilamellar vesicles of the phospholipid dipalmitoylphosphatidylcholine. *Phys. Rev. E* **59**: 921.
66. J. Katsaras and M. J. Watson (2000). Sample cell capable of 100% relative humidity suitable for X-ray diffraction of aligned lipid multibilayers. *Rev. Sci. Instrum.* **71**: 1737.
67. J. Katsaras, S. Tristram-Nagle, Y. Liu, R. L. Headrick, E. Fontes, P. C. Mason, and J. F. Nagle (2000). Clarification of the ripple phase of lecithin bilayers using fully hydrated, aligned samples. *Phys. Rev. E.* **61**: 5668.
68. Y. Lyatskaya, Y. Liu, S. Tristram-Nagle, J. Katsaras, and J. F. Nagle (2001). Method for obtaining structure and interactions from oriented lipid bilayers. *Phys. Rev. E* **63**: 011907.
69. Y. Liu and J. F. Nagle (2004). Diffuse scattering provides material parameters and electron density profiles of biomembranes. *Phys. Rev. E.* **69**: 040901.
70. N. Kučerka, Y. Liu, N. Chu, H. I. Petrache, S. Tristram-Nagle, and J. F. Nagle (2005). Structure of fully hydrated fluid phase DMPC and DLPC lipid bilayers using X-ray scattering from oriented multilamellar arrays and from unilamellar vesicles. *Biophys. J.* **88**: 2626.
71. N. Kučerka, S. Tristram-Nagle, and J. F. Nagle (2006). Closer look at structure of fully hydrated fluid phase DPPC bilayers. *Biophys. J. Lett.* **90**: L83.
72. G. Zaccaï, J. K. Blasie, and B. P. Schoenborn (1975). Neutron diffraction studies on the location of water in lecithin bilayer model membranes. *Proc. Natl. Acad. Sci. USA* **72**: 376.
73. G. Büldt, G., H. U. Gally, J. Seelig, and G. Zaccai (1979). Neutron diffraction studies on phosphatidylcholine model membranes. *J. Mol. Biol.* **134**: 673.
74. G. Zaccai, G. Büldt, A. Seelig, and J. Seelig (1979). Neutron diffraction studies on phosphatidylcholine model membranes. *J. Mol. Biol.* **134**: 693.
75. M. C. Wiener and S. H. White (1991). Fluid bilayer structure determination by the combined use of X-ray and neutron diffraction. 1. Fluid bilayer models and the limits of resolution. *Biophys. J.* **59**: 162.

76. M. C. Wiener and S. H. White (1991). Fluid bilayer structure determination by the combined use of X-ray and neutron diffraction. 2. "Composition-space" refinement method. *Biophys. J.* **59**: 174.
77. J. P. Bradshaw, J. P., R. J. Bushby, C. C. D. Giles, and M. R. Saunders (1999). Orientation of the headgroup of phosphatidylinositol in a model biomembrane as determined by neutron diffraction. *Biochemistry* **38**: 8393.
78. S. R. Wassall, M. R. Brzustowicz, S. R. Shaikh, V. Cherezov, M. Caffery, and W. Stillwell (2004). Order from disorder, corralling cholesterol with chaotic lipids. The role of polyunsaturated lipids in membrane raft formation. *Chem. Phys. Lipids* **132**: 79.
79. T. Hauss, S. Dante, N. A. Dencher, and T. H. Haines (2002). Squalane is in the midplane of the lipid bilayer: implications for its function as a proton permeability barrier. *Biochim. Biophys. Acta* **1556**: 149.
80. J. P. Bradshaw, M. J. M. Darkes, T. A. Harroun, J. Katsaras, and R. M. Epand (2000). Oblique membrane insertion of viral fusion peptide probed by neutron diffraction. *Biochemistry* **39**: 6581.
81. S. Dante, T. Hauss, and N. A. Dencher (2002). β-Amyloid 25 to 35 is intercalated in anionic and zwitterionic lipid membranes to different extents. *Biophys. J.* **83**: 2610.
82. A. Léonard, A., C. Escriv, M. Laguerre, E. Pebay-Peyroula, W. Néri, T. Pott, J. Katsaras, and E. J. Dufourc (2001). Location of cholesterol in DMPC membranes. A comparative study by neutron diffraction and molecular mechanics simulation. *Langmuir* **17**: 2019.
83. D. L. Worcester, D. L. and N. P. Franks (1976). Structural analysis of hydrated egg lecithin and cholesterol bilayers. *J. Mol. Biol.* **100**: 359.
84. T. A. Harroun, J. Katsaras, and S. R. Wassall (2006). Cholesterol hydroxyl group is found to reside in the center of a polyunsaturated lipid membrane. *Biochemistry* **45**: 1227.

PART II
DYNAMICS AND MOLECULAR EVENTS AT MEMBRANE INTERFACES

CHAPTER 6

Interaction of Plasma Proteins with Phospholipids at Interfaces

CHIA-LIN YIN, DORCAS, ANNA DUDEK, and CHIEN-HSIANG CHANG

Department of Chemical Engineering, National Cheng Kung University, Tainan, Taiwan

CONTENTS

6.1	Introduction	137
6.2	Dynamic Tension Behavior Analysis	139
6.3	Dynamic Monolayer Behavior Analysis	140
6.4	Brewster Angle Microscopy	140
6.5	Infrared Reflection–Absorption Spectroscopy	141
6.6	Mixed Monolayers Containing Albumin	142
6.7	Mixed Monolayers Containing Fibrinogen	151
6.8	Mixed Monolayers Containing γ-Globulins	157
6.9	Conclusion	159
Acknowledgments		160
References		160

6.1 INTRODUCTION

Dynamic surface properties of mixed phospholipid/protein monolayers at air–liquid interfaces have attracted attention from various disciplines because many biological membrane characteristics can be derived from monolayer studies. For example, peripheral membrane proteins play an important role in cell functions, and phospholipid monolayers at air–liquid interfaces have been used extensively as model membrane systems to provide information on the interactions between peripheral membrane proteins and membranes. One of the advantages to use monolayers as model membrane systems is that the monolayer parameters, such as molecular density, at air–liquid interfaces can be controlled precisely.

Structure and Dynamics of Membranous Interfaces, edited by Kaushik Nag
Copyright © 2008 John Wiley & Sons, Inc.

Moreover, several techniques can be applied at interfaces to investigate the mixed monolayer behavior of phospholipids with proteins. The model membrane systems of phospholipids and proteins are thus useful to obtain information on various membranous processes [1].

The importance of investigating mixed phospholipid/protein monolayer behavior at air–liquid interfaces is also obvious when one considers the mixture is relevant to proper interfacial functions of lung surfactant [2]. The specific interfacial properties of lung surfactant at the interface of air–alveolar lining layer include the abilities to reduce dynamic surface tension effectively, to respread readily at the expanded interface following dynamic compression, and to adsorb fast from the subphase to the interface [3, 4]. It is well accepted that dipalmitoyl phosphatidylcholine (DPPC) is the major phospholipid ingredient of lung surfactant, being responsible for the surface tension lowering activity or the dynamic interfacial properties of lung surfactant.

The specific proteins existing in lung surfactant, such as SP-B and SP-C, show excellent ability to improve the adsorption and respreading of phospholipid components [5]. However, plasma proteins, which can be introduced into alveolar space as a result of lung injury, possess an inhibitory effect on the surface activity of lung surfactant and contribute to the causes of adult or acute respiratory distress syndrome [6, 7]. This difference is apparently related to their surface properties at air–liquid interfaces. Therefore it is not surprising that the dynamic surface properties of mixed phospholipid/protein monolayers relevant to lung surfactant systems at air–liquid interfaces have been studied extensively.

Mixed spread monolayers of DPPC with bovine albumin or fibrinogen have been investigated in terms of surface pressure–area characteristics, and it has been shown that the plasma proteins did not seem to affect the maximum surface tension lowering ability of DPPC [8]. However, although serum albumin in the subphase stabilized the surface activity of spread DPPC monolayers, the adsorbed monolayers formed by mixtures of DPPC and albumin exhibited marked inhibition of DPPC surface activity [9].

For mixed spread monolayers of DPPC with albumin, a considerable change in the monolayer composition occurred due to the protein-residue rejection upon interface compression. When the mixed monolayers were compressed to reach very high surface pressures, DPPC collapsed and left the interface irreversibly [10]. In addition, the hysteresis isotherms of mixed spread plasma protein/pig lung surfactant monolayers indicated that the protein molecules might leave the interface and carry along surfactant lipids with them upon interface compression [11].

It has been reported that plasma proteins seemed to inhibit lung surfactant functions by preventing the phospholipids from adsorbing onto the interface, possibly by competition for space at the interface rather than by direct molecular interactions between proteins and phospholipids [11, 12]. Moreover, fibrinogen seemed to be the most deleterious to surface refining of pulmonary surfactant [11, 13] and a more potent inhibitor of surfactant TA than albumin [14].

There is no doubt that general principles involved in the inhibitory effect of plasma proteins on the dynamic surface activity of DPPC or lung surfactant systems can be derived from past reports. Nevertheless, recent mixed monolayer studies especially with sophisticated experimental designs and/or in situ microscopic/spectroscopic techniques do provide more detailed information about the interactions of plasma proteins with phospholipids at interfaces. Thus this chapter focuses on the recent progress in understanding the dynamic interfacial properties of mixed monolayers of phospholipids, in particular, DPPC, with plasma proteins at air–liquid interfaces.

6.2 DYNAMIC TENSION BEHAVIOR ANALYSIS

Pulsating bubble surfactometry is generally applied to investigate the dynamic adsorption of lung surfactant or mixed lipid/protein systems at air–liquid interfaces. A commercial pulsating bubble surfactometer (PBS) (General Transco, Inc., USA) is available to be used for measuring dynamic surface tensions of samples under pulsating-area conditions [15, 16]. Measurements can be controlled at a specific temperature by the internal heater of the PBS. To perform an experiment, a bubble with a radius of 0.40 mm is formed first in an aqueous sample contained in a special-designed sample chamber, and the surface-active molecules start adsorbing onto the bubble surface to reduce the surface tension at the air–liquid interface. The pressure difference across the air–liquid interface can be monitored with a pressure transducer, and the surface tension is calculated from the pressure difference and bubble radius by using the Young–Laplace equation. After a steady tension value is reached, the volume displacement of the sample can be controlled with a pulsating rod, and the bubble radius then oscillates between 0.40 and 0.55 mm. A pulsation rate up to 100 cycles/min can be applied in the experiments. During the pulsation, the pressure differences across the air–liquid interface are measured and the corresponding dynamic surface tensions are calculated. The steady tension value can be confirmed by following the surface tension variation with time after the pulsation is finished, and then a repeated experiment can be performed with the same sample.

If a surface pulsation experiment is started after the equilibrium surface tension of the sample is reached, the dynamic surface tension increases first as the bubble area is expanded due to the decreased surface concentration of surface-active molecules at the interface. Thus, during the surface expansion cycle, the surface-active molecules in the bulk phase adsorb onto and replenish the interface. When the bubble area is compressed back to its original size, the surface concentration is increased and the surface tension is decreased. Hence the excess molecules move from the interface to the bulk phase through desorption during the surface compression cycle. If the desorption process is not fast enough, the molecules accumulate at the interface and the surface concentration becomes higher than the equilibrium surface concentration, resulting in dynamic surface tensions lower than the equilibrium surface tension.

Thus it is sometimes observed that the minimum surface tension obtained in a PBS measurement is lower than the equilibrium surface tension, which is a result of molecule adsorption hysteresis between the expansion and compression stages of an air–liquid interface [17].

6.3 DYNAMIC MONOLAYER BEHAVIOR ANALYSIS

To study mixed phospholipid/protein monolayer behavior at air–liquid interfaces, a widely applied method is to use the Langmuir film balance to record changes of surface pressure as a function of the molecular area at constant temperature or the surface pressure–area (π–A) isotherm. The surface pressure (π) is defined as the surface tension difference of pure subphase (γ_0) with the subphase containing a monolayer (γ) and can be calculated from the relationship $\pi = \gamma_0 - \gamma$. In a Langmuir film balance, an aqueous subphase is contained in a so-called Langmuir trough with one or two movable barriers. An organic solvent sample containing monolayer-forming materials can be spread at the air–liquid interface usually by a microsyringe. A period of 10–20 min is allowed for solvent evaporation, and the spread monolayer is then formed at the interface. Compression or expansion of the monolayer at the interface can be carried out with the computer-controlled movable barriers, and the corresponding surface pressure variations are usually measured with the Wilhelmy plate method. A surface pressure–area (π–A) isotherm can thus be constructed.

An analysis of the π-A isotherm can provide information about the molecular packing at the interface and the phase states of the two-dimensional monolayer. When an interface compression–expansion process is applied, the π-A hysteresis isotherm can be recorded. This allows one to study the mechanisms of monolayer respreading and/or dynamic adsorption behavior of molecules at the interface during cyclic area changes [18, 19]. Moreover, a Langmuir film balance is generally incorporated with in situ microscopic or spectroscopic techniques to investigate the dynamic monolayer behavior at air–liquid interfaces.

6.4 BREWSTER ANGLE MICROSCOPY

Brewster angle microscopy (BAM) is a microscopic technique based on the characteristics of reflectivity of light at an interface under the Brewster angle condition. A commercial instrument is available (NFT, Germany) to be used for studying the monolayer morphology at air–liquid interfaces in situ. BAM has generally been applied to investigate lipid–protein interactions in lung surfactant and cell membrane systems. When a beam of p-polarized light is incident on the pure water surface with the Brewster angle, no reflection occurs. For the air–water interface, the Brewster angle is approximately $53°$. However, the presence of a monolayer at the interface will disturb the Brewster angle condition, causing some reflection. The reflected intensity depends strongly on the monolayer properties, such as surface concentration, thickness, and optical parameters. The reflected light can be detected with a CCD camera. The optical anisotropy resulting from the differences in molecular orientation in the monolayer can be analyzed with the incorporation of an analyzer in the path of the reflected beam. A detailed discussion about the principle and application of BAM has been given by Lheveder et al. [20].

BAM has been used to provide direct visualization of the monolayer phase transitions at air–liquid interfaces [21–23]. It has also been applied to compare the interfacial properties of Gibbs (adsorbed) monolayers with those of spread (Langmuir) monolayers [24, 25]. In addition, the information of gray level or relative reflectivity obtained in a BAM image can be used to determine the thickness of monolayer regions, even when the refractive index of the monolayer is unknown [26, 27].

6.5 INFRARED REFLECTION–ABSORPTION SPECTROSCOPY

Infrared reflection–absorption spectroscopy (IRRAS) is a widely applied in situ technique in biological interface studies. The theory and applications of IRRAS have been discussed by Dluhy [28] and by Tolstoy et al. [29]. The reflectance–absorbance (RA) intensity is equal to $-\log_{10}(R/R_o)$, where R and R_o are the reflectivities of the monolayer-covered and pure subphase surfaces, respectively. The intensities of vibrational bands for a monolayer on the water surface are negative when showed as the absorbance–wavelength relationship, and the basis for the negative absorbance in the monolayer spectra has been explored [30, 31].

An IRRAS analysis of monolayer behavior at air–liquid interfaces is conducted using an infrared (IR) spectrometer usually with a liquid-nitrogen-cooled mercury cadmium telluride (MCT) detector. A commercial monolayer/grazing angle accessory with a removable Teflon Langmuir trough is available to be used with an IR spectrometer to obtain the IR spectrum of a monolayer at the air–liquid interface. The IR spectrum can be taken using polarized or unpolarized light with a controlled incidence angle.

IRRAS has been applied successfully to study the mixed monolayer and competitive adsorption behavior of lipids and proteins at air–liquid interfaces in situ, especially for lung surfactant systems. In the cases of phospholipid monolayers at the air–water interface, the spectrum analysis is mainly based on the peak positions and intensities of CH_2 symmetric (v_s-CH_2 at $\sim$2850 cm^{-1}) and antisymmetric (v_a-CH_2 at $\sim$2920 cm^{-1}) stretching vibrational bands as well as carbonyl group (C=O) stretching of polar region vibrations. It has been reported that the wavenumber corresponding to the maximum in the v_a-CH_2 band is strongly related to the packing and conformation of the molecular hydrocarbon chains in the monolayer, and a lower wavenumber is a characteristic of highly ordered conformation of the acyl chains [32, 33]. The changes in peak intensity are connected with the variations of surface concentration [34]. Another parameter that can be evaluated from the spectrum is the full width at half-maximum

(FWHM) of the peak. FWHM is proportional to degree of rotational mobility and flexibility within the layer and is increased with the decrease in the hydrocarbon chain order [29]. There are many reports demonstrating that the secondary structures of proteins at interfaces can also be detected by IRRAS [35–38].

Major absorption bands coming from amide groups of proteins are amide A (N—H symmetric stretching), amide I (C=O stretching), and amide II (N—H in plane scissoring and C—N symmetric stretching) heavily mixed with amide III (N—H in plane scissoring). For the analysis of proteins at air–liquid interfaces, amide I and amide II with frequencies in the ranges of 1600–1700 and 1510–1580 cm^{-1}, respectively, are usually used [29].

6.6 MIXED MONOLAYERS CONTAINING ALBUMIN

Dynamic surface tension responses for mixed dipalmitoyl phosphatidylcholine (DPPC)/albumin dispersions under both constant-area and pulsating-area conditions have been reported to demonstrate the inhibitory effect of albumin on the dynamic surface activity of DPPC [39, 40]. With the presence of albumin in a DPPC dispersion, the adsorption of DPPC molecules or the formation of a compact DPPC monolayer at the air–liquid interface was inhibited with the extent depending on the albumin bulk concentration. When the albumin concentration was 10 or 100 ppm, the equilibrium surface tension and minimum surface tension achieved during interface pulsation were significantly higher than that obtained for a pure 1000-ppm DPPC dispersion, showing the strong influence of albumin on the surface tension lowering properties of DPPC. Furthermore, the dynamic tension responses were dominated by albumin and the ability of DPPC molecules to adsorb at the interface was disrupted. Apparently, an albumin concentration of 10 or 100 ppm was enough to inhibit the adsorption of DPPC with a concentration of 1000 ppm.

The dynamic tension data obtained under a constant-area condition also showed that the adsorption of albumin at 100 or 1000 ppm was much faster than that of a sonicated 1000-ppm DPPC dispersion, suggesting that the formation of an adsorbed DPPC monolayer could be strongly inhibited by albumin. The dominant adsorption of albumin at the interface for a 1000-ppm/1000-ppm DPPC/albumin dispersion has been confirmed by an infrared reflection–absorption spectroscopy (IRRAS) analysis [40]. By measuring the time-dependent reflectance–absorbance (RA) intensities of amide I band from albumin and v_a-CH_2 band from DPPC at the interface, it was shown that the adsorption of albumin molecules was much faster than that of DPPC aggregates. Furthermore, the adsorption of DPPC did not cause significant repulsion of preadsorbed albumin molecules from the interface. Albumin apparently had an advantage over DPPC when competing for space at interfaces by adsorbing rapidly and possibly irreversibly, thus interfering with the DPPC transport to the interface and inhibiting the formation of a compact DPPC monolayer at the interface.

It is well known that the dynamic surface tension behavior of a DPPC dispersion is strongly affected by the preparation protocol [41–44]. With certain preparation protocols, a 1000-ppm DPPC dispersion, probably containing mostly vesicles, could show faster and more extensive adsorption, resulting in much lower dynamic surface tensions, as low as 1 mN/m, under a pulsating-area condition [45]. However, when the 1000-ppm DPPC dispersion with improved dynamic adsorption characteristics was added with 1000-ppm albumin, albumin still controlled the tension behavior at both constant-area and pulsating-area conditions and thus inhibited the dynamic surface activity of DPPC. The intensity of amide I band from albumin in an IRRAS analysis also provided direct evidence on the dominant adsorption of albumin at interfaces.

When DPPC molecules were spread onto the interface with a preadsorbed albumin layer, the surface tension was reduced and was different from that of a DPPC monolayer alone, implying that DPPC molecules were able to enter the protein layer by the spreading approach [40]. With an IRRAS analysis, higher absolute RA value and lower frequency response of DPPC v_a-CH$_2$ band, corresponding to highly ordered hydrocarbon chains, were detected after the DPPC spreading. This result indicated that DPPC could enter the protein layer with the spreading mechanism and remain at the interface. Moreover, the absolute RA value of albumin amide I band was found to decrease with increasing spread DPPC surface concentration, suggesting the induced desorption of albumin molecules by the formation of a closely packed DPPC monolayer at the interface.

The interface compression–expansion hysteresis curves of a mixed spread DPPC/adsorbed albumin layer on a 1000-ppm albumin subphase are demonstrated in Fig. 6.1. During the first compression stage, a prominent plateau at the surface pressure ~58 mN/m was detected. It has been proposed that, at lower surface pressures, where the pure lipid was expanded, homogeneous lipid/protein monolayers might be obtained. However, the monolayers could then demix at higher surface pressures, where the pure lipid was condensed [46, 47]. Therefore the observed surface pressure transition in the compression curve of a mixed DPPC/albumin monolayer could be explained by the extensive exclusion of albumin molecules from the mixed monolayer due to the close packing of insoluble DPPC molecules. A similar plateau has been observed in the compression curve of a mixed DPPC/surfactant protein-C (SP-C) layer at the air–liquid interface, and it was suggested that material was removed upon compression but remained to associate with the monolayer [48].

Upon further compression, a characteristic collapse plateau of a DPPC monolayer was observed in the compression isotherm (Fig. 6.1). Thus most of the albumin molecules seemed to be preferentially expelled from the monolayer, resulting in a condensed phase mainly consisting of DPPC with a collapse plateau similar to that of a pure DPPC monolayer. Similar behavior has been found for mixed DPPC/β-lactoglobulin monolayers by synchrotron X-ray experiments at grazing incidence [49].

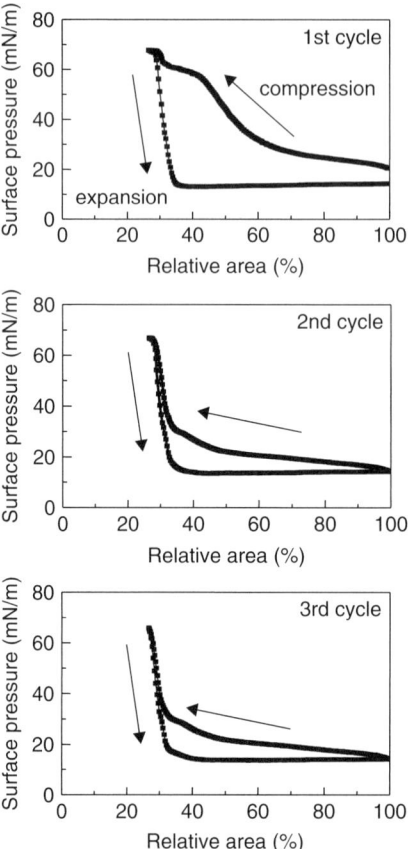

Figure 6.1 Consecutive surface pressure–relative area hysteresis curves of a mixed DPPC/albumin layer on a 1000-ppm albumin subphase at 25 °C with an initial area per DPPC molecule of 1.0 nm^2. Data were obtained at an area compression–expansion rate of 18 cm^2/min with the working area of the Langmuir trough being 70 × 7 cm^2. The relative area represents the ratio of the actual interfacial area during the compression–expansion cycle to the initial interfacial area.

During the interface expansion stage, the surface pressure of the mixed monolayer decreased sharply and then increased gradually until the end of the expansion stage, which suggested the reentry of surface-active materials (mostly albumin) at the interface, especially at larger interface areas (Fig. 6.1). Keough et al. [11] have also found that proteins might compete with surfactants for surface space under conditions when the layers were not highly compressed. The formation of a protein-dominated layer at the interface may prevent the phospholipid adsorption [12]. In addition, the increasingly insignificant collapse characteristic detected during the following cycles indicated that fewer free DPPC molecules were available at the interface.

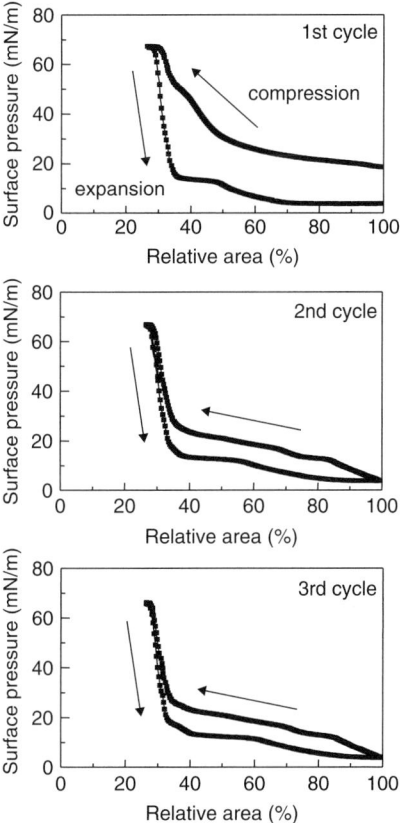

Figure 6.2 Consecutive surface pressure–relative area hysteresis curves of a mixed DPPC/albumin layer on a 10-ppm albumin subphase.

With a lower albumin subphase concentration of 10 ppm, the readsorption of albumin molecules at the expanded interface became less significant, as judged from the lower surface pressures obtained at the end of each cycle (Fig. 6.2). Nevertheless, the albumin readsorption could still be detected by Brewster angle microscopy (BAM), especially at lower surface pressures (Fig. 6.3). Moreover, the characteristic kidney shape of DPPC domains observed by BAM during the following interface compression stage was somewhat distorted, indicating that some protein molecules might still be available at the interface and thus a partly disturbed ordered DPPC phase might result due to the interactions between the two compounds (Fig. 6.4) [50, 51]. The loss of free DPPC molecules at interfaces during consecutive compression–expansion cycles could also be confirmed by the BAM observations (Fig. 6.5).

For DPPC molecules spread at the interface to form a mixed monolayer with preadsorbed albumin molecules on a 1000-ppm albumin subphase, an IRRAS

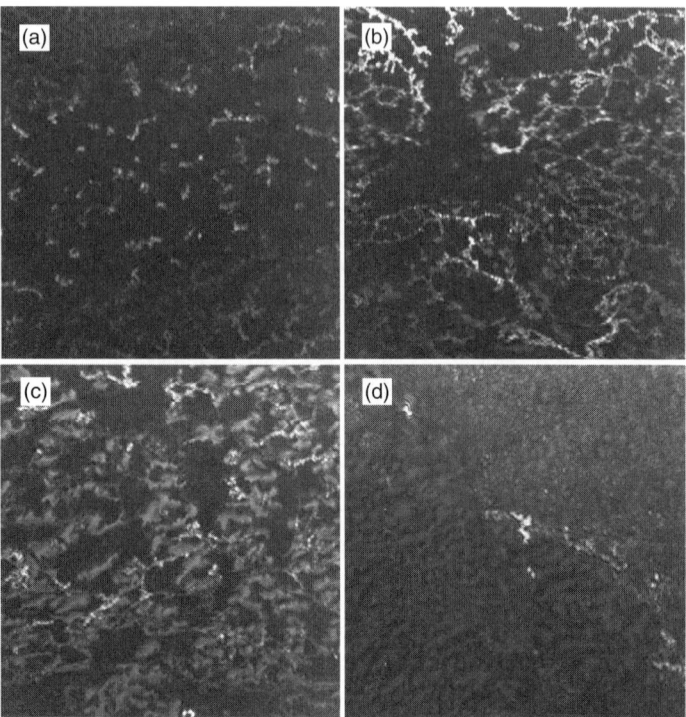

Figure 6.3 A mixed DPPC/albumin layer on a 10-ppm albumin subphase was compressed to pass the DPPC collapse point and then was expanded to a relative area of 50%. The albumin readsorption could be demonstrated by the morphology changes with time observed in a BAM analysis. The corresponding elapsed times for the BAM images are (a) 0.5 min, (b) 17 min, (c) 25 min, and (d) 97 min. The images show a section of the air–liquid interface with a size of 430 μm × 430 μm.

analysis showed that a higher absolute RA value (0.0034) of the v_a-CH$_2$ band in comparison with that (0.0028) of a pure DPPC monolayer was detected, evidently attributed to the closely packed DPPC molecules due to the presence of preadsorbed albumin molecules at the interface (Fig. 6.6a). It has been reported that a lower wavenumber is a characteristic of highly ordered conformation of the acyl chains [32, 33, 52]. Thus the close packing of spread DPPC molecules in the mixed layer at the interface was also suggested by the lower v_a-CH$_2$ wavenumber of ∼2921 cm^{-1} (Fig. 6.6b), as compared with 2923 cm^{-1} for a pure DPPC monolayer.

At the end of the first interface compression stage, the absolute v_a-CH$_2$ RA value (0.0045) of the mixed DPPC/albumin layer was lower than that (0.0049) for a pure DPPC monolayer, implying that extra free DPPC molecules were removed from the interface upon interface compression as a result of the presence of albumin (Fig. 6.6a). In addition, the absolute amide I RA value for the

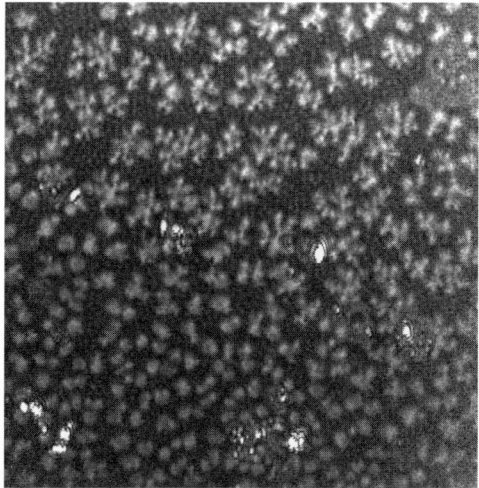

Figure 6.4 The BAM image of a mixed DPPC/albumin layer on a 10 ppm albumin subphase during the second interface compression stage at a surface pressure of ~14.5 mN/m.

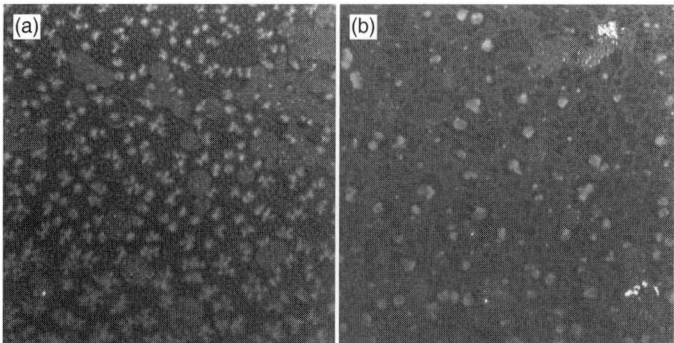

Figure 6.5 The loss of free DPPC molecules in a mixed DPPC/albumin layer on a 10-ppm albumin subphase during consecutive compression–expansion cycles could be confirmed by the decreasing numbers of DPPC domains observed by BAM in the (a) second and (b) third interface compression stages at the same relative area.

albumin molecules at the interface always decreased upon interface compression, suggesting significant expulsion of albumin molecules from the interface (Fig. 6.7). The albumin molecules were expelled apparently due to the presence of insoluble DPPC molecules at the interface.

During the following interface expansion stage, the albumin adsorption is clearly demonstrated by the increased absolute amide I RA value upon interface expansion (Fig. 6.7). Moreover, the wavenumber corresponding to the

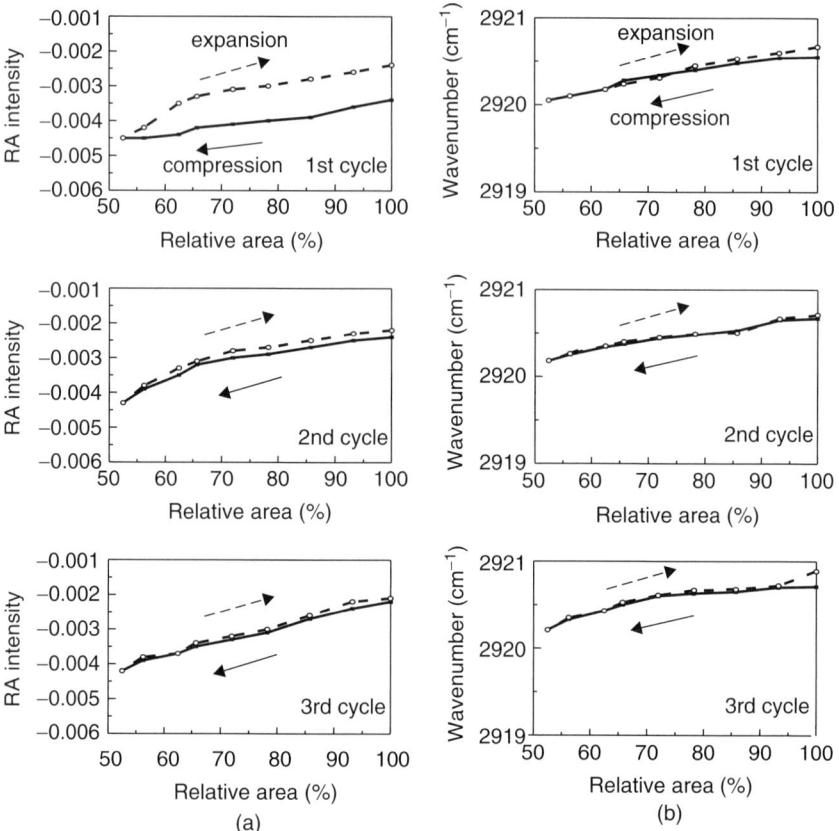

Figure 6.6 Consecutive (a) ν_a-CH$_2$ RA intensity–relative area and (b) ν_a-CH$_2$ wavenumber–relative area hysteresis curves of a mixed DPPC/albumin layer on a 1000-ppm albumin subphase at room temperature with an initial area per DPPC molecule of 0.704 nm^2. The mixed layer was continuously compressed and then expanded by a barrier at a rate of 0.00875 nm^2/(DPPC molecule)-min. The IR spectra were obtained using unpolarized light with an angle of incidence of 40°, as measured perpendicular to the liquid surface.

maximum in the ν_a-CH$_2$ band for DPPC remained essentially unchanged at 2920–2921 cm^{-1} during the interface expansion stage, implying that the free DPPC molecules were always in a highly ordered state (Fig. 6.6b). Because it is well known that DPPC has poor respreading ability, the close packing of free DPPC molecules was apparently related to the significant adsorption of albumin molecules to the interface upon interface expansion. Under the circumstances, more albumin molecules were available at the interface and occupied part of the space at the interface, enhancing the conformational order of the hydrocarbon chains of the remaining free DPPC molecules.

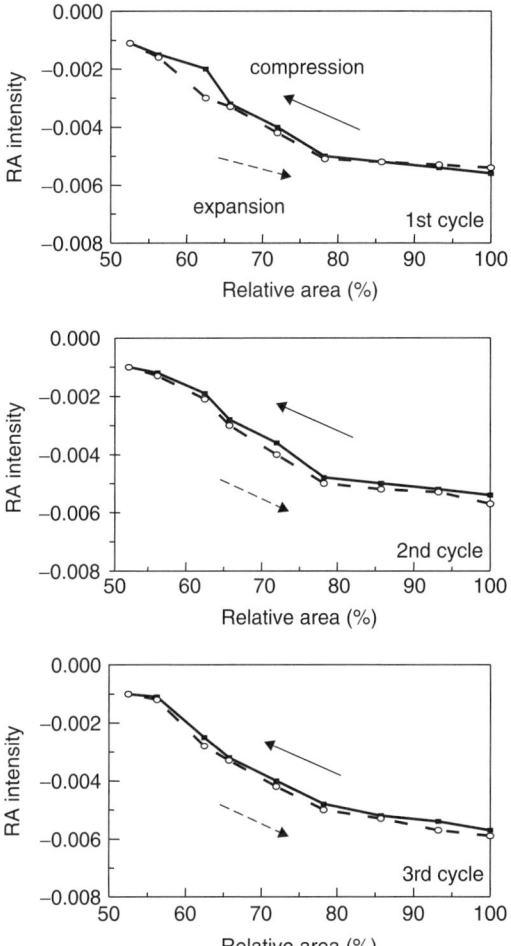

Figure 6.7 Consecutive amide I RA intensity–relative area hysteresis curves of a mixed DPPC/albumin layer on a 1000-ppm albumin subphase.

During the second and third interface compression–expansion cycles, the decreased absolute v_a-CH$_2$ RA values detected at the ends of the cycles suggested that some of the remaining free DPPC molecules after the first cycle were further removed with albumin during the cycles (Fig. 6.6a). The intensity changes might seem insignificant, but the variations in the surface concentration of free DPPC molecules would be pronounced, if one considered the albumin molecules adsorbed at the interface and occupied certain interface space upon interface expansion. The significant albumin adsorption during the interface expansion stage can be realized from the absolute amide I RA values detected at the end of each cycle (Fig. 6.7).

Albumin molecules have been introduced into the subphase with a prespread DPPC monolayer at the interface, and the albumin adsorption on the monolayer or monolayer penetration by albumin has been suggested by the surface tension changes [45]. However, the resulting tension variations also implied that the monolayer was still mainly composed of DPPC and the substantial adsorption of albumin was somewhat prevented. The inhibited adsorption of albumin was confirmed from the essentially unchanged IRRAS spectrum as before the albumin introduction. The albumin molecules might approach the interfacial region but did not displace the DPPC molecules from the interface. When a DPPC dispersion was introduced dropwise or with the Trurnit method onto the surface of a DPPC/albumin mixture, a significant drop in the surface tension suggested that once such a DPPC monolayer was formed at the interface, albumin could no longer adsorb with a significant surface concentration.

For a mixed dilauroyl phosphatidylcholine (DLPC)/albumin system, the dynamic surface tension and adsorption behavior at air–liquid interfaces have been investigated by Phang and Franses [53]. At short adsorption times, the surface layer of a 1000-ppm/1000-ppm DLPC/albumin mixture mainly consisted of fast-adsorbed albumin molecules as suggested by the surface tension responses. At longer adsorption times, the tension data approached those of a pure DLPC dispersion, implying DLPC molecules had adsorbed and probably replaced most or all of the albumin molecules at the interface. These results were also inferred from the IRRAS spectrum being about the same as that of a pure DLPC dispersion. The expulsion process of albumin would take longer if the DLPC concentration was decreased or would become faster if the albumin concentration was decreased. It is interesting to note that the 1000-ppm albumin could not inhibit the dynamic surface activity of DLPC at constant-area conditions, but it did slightly inhibit DLPC from achieving the lowest surface tension under a pulsating-area condition due to its fast adsorption characteristic and comparatively slow expulsion by DLPC. With a lower albumin concentration, the inhibition of DLPC dynamic surface activity became insignificant, because fewer albumin molecules were available in the subphase and then only minor albumin adsorption occurred during the interface expansion stage.

When a DLPC dispersion was injected into the subphase with a preadsorbed albumin layer at the interface to yield final concentrations of 1000-ppm DLPC and 1000-ppm albumin in the subphase, higher protein surface concentrations at the interface were inferred from the RA intensity variations of the amide I band in an IRRAS analysis. However, after a period of adsorption time, it appeared that the protein bands were unable to be detected and albumin was no longer available in a significant amount in the monolayer. Moreover, the IRRAS spectra were similar to that of an aqueous DLPC dispersion, showing that DLPC indeed expelled, mostly or completely, the adsorbed albumin layer from the interface. A hypothesis of the expulsion mechanism has been proposed as follows. After DLPC was injected into the albumin subphase, DLPC started adsorbing and possibly bound to albumin, making it more hydrophobic. Hence more albumin molecules adsorbed to the interface. However, DLPC molecules continued adsorbing and finally displaced or expelled albumin molecules from the interface.

An egg lecithin sample has been spread on a nearly clean surface of [^{14}C]albumin subphase, and then the surface concentration of [^{14}C]albumin was detected with a radiotracer technique [54]. It was found that the surface concentration of albumin increased rapidly, suggesting that the lecithin monolayer enhanced the albumin adsorption at short times. For an egg lecithin sample spread at the interface with a preadsorbed albumin layer, the albumin adsorption was also initially enhanced. Subsequently, some albumin molecules were apparently desorbed at long times. The initially enhanced protein adsorption by a spread lecithin monolayer may be due either to hydrophobic interactions between the lecithin chains and albumin or to electrostatic interactions of albumin with the lecithin headgroups [46]. Moreover, spreading of more lecithin molecules resulted in faster and nearly complete desorption of albumin at the interface with no adsorption enhancement. For example, spreading of a closely packed lecithin monolayer caused eventually desorption of most of the previously adsorbed albumin layer, apparently because of the surface pressure and steric exclusion effects produced by the spread lecithin monolayer.

6.7 MIXED MONOLAYERS CONTAINING FIBRINOGEN

For a 1000-ppm/750-ppm DPPC/fibrinogen mixture, the dynamic surface tension behavior under a constant-area condition has been shown similar to that of a fibrinogen solution alone, but with higher tension values probably due to the formation of DPPC/fibrinogen aggregates [38]. The species present in the adsorbed layer at the interface have been directly probed by IRRAS. The lipid bands, v_a-CH$_2$, v_s-CH$_2$, and C=O, and the protein amide I and II bands were observed, suggesting that both fibrinogen and DPPC were available at the interface. Although DPPC molecules were found at the interfacial region, fibrinogen still controlled the surface tension behavior, implying that the adsorbed DPPC monolayer was incomplete with a low surface concentration.

The inhibitory effect of fibrinogen on the formation of a compact adsorbed DPPC monolayer from a DPPC dispersion at a pulsating air–liquid interface has been demonstrated by a dynamic surface tension analysis [55]. The minimum and maximum surface tensions achieved by a 1000-ppm/1000-ppm DPPC/fibrinogen mixture during the interface pulsation were much higher than that obtained for a pure DPPC dispersion, showing strong fibrinogen-induced inhibition on the dynamic adsorption ability of DPPC. Furthermore, the dynamic surface tension curve obtained for the mixture almost overlapped with that obtained for a pure fibrinogen solution, indicating that formation of the adsorbed monolayer at the interface was dominated by fibrinogen and the DPPC ability of adsorbing onto the interface was greatly depressed. A reasonable explanation for the inactivation of DPPC dynamic surface activity by fibrinogen may be similar to that proposed for the mixed DPPC/albumin system [40].

A study about the fibrinogen-dominated dynamic tension behavior under a pulsating-area condition has also been reported for a mixture of 1000-ppm/750-ppm DPPC/fibrinogen by Kim and Franses [38]. When DPPC molecules were spread onto the interface in a closely packed state with a preadsorbed fibrinogen layer on a 750-ppm fibrinogen subphase, the surface tension reduction suggested that a DPPC monolayer did form and the fibrinogen-induced inhibition on the surface tension lowering ability of DPPC was insignificant. The corresponding IRRAS analysis showed that after the DPPC monolayer was formed, the amide I and II bands of fibrinogen disappeared and the usual lipid bands, v_a-CH$_2$, v_s-CH$_2$, and C=O, appeared. This result indicated that fibrinogen molecules were expelled from the interface by the DPPC monolayer. Most of the DPPC molecules, however, remained at the interface.

A detailed analysis of the mixed DPPC/fibrinogen monolayer behavior at air–liquid interfaces under continuous compression–expansion conditions has been performed by Kuo et al. [19]. When DPPC molecules were spread in a gaseous state on a 10-ppm fibrinogen subphase, the rise in the initial surface pressure was detected and was probably due to the enhanced adsorption of protein molecules by a lecithin monolayer [54] or due to the interactions between a lecithin monolayer and adsorbed protein molecules at the interface [56, 57].

During the first interface compression stage, a surface pressure transition was observed, at which the DPPC molecules were in a state corresponding to liquid-condensed state. The observed surface pressure transition could be explained by the extensive exclusion of fibrinogen molecules from the mixed monolayer due to the close packing of insoluble DPPC molecules. Upon further compression, a collapse characteristic of DPPC was detected in the isotherm, confirming that the fibrinogen molecules were preferentially expelled from the interface and a condensed monolayer mainly consisting of DPPC resulted. During the following interface expansion stage, unlike that observed for a pure DPPC monolayer, the surface pressure never dropped to zero, implying that some fibrinogen molecules entered the interface and were involved in the rise of surface pressure at the expanded interface.

During the interface compression stage of the second cycle, the characteristic liquid-expanded/liquid-condensed phase transition of a DPPC monolayer could be detected in the isotherm, suggesting that DPPC and fibrinogen acted as separate components in the mixed monolayer, which has been proposed before [8]. For mixed monolayers of DPPC with hydrophilic proteins, it has been demonstrated that no interaction was found between DPPC and proteins, and thus the ideal mixing behavior was exhibited [58]. Moreover, the phase transition occurred at a much larger interface area in comparison with that observed for a DPPC monolayer alone, indicating that limited space was available for the remaining free DPPC molecules after the first cycle. This is very likely to be attributed to the entry of protein molecules during the prior interface expansion stage [8, 50]. During the expansion stage of the second cycle, the surface pressure–area curve was similar to that shown in the first cycle, except that higher surface pressures resulted at larger interface areas, which implied that more fibrinogen molecules entered the interface at larger interface areas upon interface expansion. A similar trend was observed for the third compression–expansion cycle. Furthermore, the increasingly insignificant collapse characteristic detected in the cycles suggested that fewer free DPPC molecules were available at the interface.

With a higher fibrinogen subphase concentration of 1000 ppm, the DPPC collapse characteristic disappeared during the first interface compression stage of the mixed DPPC/fibrinogen monolayer, implying that an appreciable number of DPPC molecules were removed with the excluded fibrinogen molecules [19]. It has also been reported that in mixed monolayers of lung surfactants with soluble plasma proteins, plasma proteins left the interface by accompaniment with surfactant lipids [11]. In addition, during the first interface expansion stage, the surface pressure was always greater than zero and increased steadily at larger interface areas. This obviously resulted from the significant penetration of fibrinogen molecules into the expanded monolayer on a 1000-ppm fibrinogen subphase.

Because a compressed fibrinogen layer alone did not reach surface pressures higher than 40 mN/m, the presence of free DPPC molecules at the interface could be inferred from the higher surface pressures detected during the second interface compression stage. However, the lower maximum surface pressure in comparison with that shown in the first cycle indicated that some of the remaining free DPPC molecules after the first cycle were further removed with fibrinogen during the second interface compression stage. Furthermore, during the second interface expansion stage, more fibrinogen molecules entered the interface and thus higher surface pressures were achieved at larger interface areas. The comparatively significant penetration behavior of fibrinogen at the interface would result in protein replenishment at the interface. For the third interface compression–expansion cycle, similar trends, such as a slightly decreased maximum surface pressure during area compression and more pronounced entry behavior of fibrinogen during area expansion, were also observed.

The relative numbers of the free DPPC molecules existing at the interface during each compression stage have been estimated from the hysteresis curves. It was found that, with a higher fibrinogen subphase concentration, fewer free DPPC molecules were available at the interface. It was also indicated that after fibrinogen molecules were expelled from the interface during the first interface compression stage, part of the originally spread DPPC molecules were also lost from the interface. The induced removal of DPPC by the exclusion of fibrinogen during area compression apparently resulted in the increasingly insignificant collapse characteristic of the mixed monolayer in the following cycles.

When DPPC molecules were spread at the interface with a preadsorbed fibrinogen layer on a 70-ppm fibrinogen subphase, a higher absolute v_a-CH_2 RA value (0.0035) in comparison with that (0.0028) of a pure DPPC monolayer was detected in an IRRAS analysis (Fig. 6.8a). This result was evidently attributed to the induced close packing of spread DPPC molecules by the presence of preadsorbed fibrinogen molecules at the interface. The close packing of DPPC molecules in the mixed monolayer at the interface was also suggested by the lower v_a-CH_2 wavenumber of $\sim$2920 cm^{-1} (Fig. 6.8b), as compared with 2923 cm^{-1}

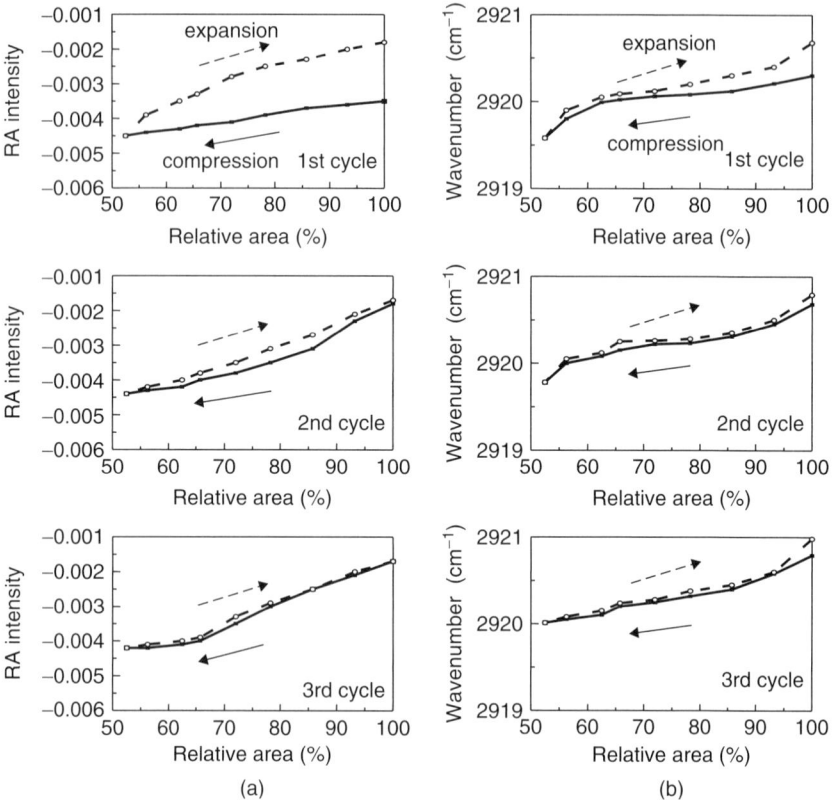

Figure 6.8 Consecutive (a) ν_a-CH_2 RA intensity–relative area and (b) ν_a-CH_2 wavenumber–relative area hysteresis curves of a mixed DPPC/fibrinogen layer on a 70-ppm fibrinogen subphase.

for a pure DPPC monolayer. A similar experiment has been performed by Kim and Franses [38] with DPPC molecules spread at a higher surface concentration, and the preadsorbed fibrinogen molecules were almost expelled from the interface.

For the mixed DPPC/fibrinogen layer, the absolute ν_a-CH_2 RA value (0.0045) detected at the end of the first interface compression stage was lower than that (0.0049) for a pure DPPC monolayer, implying that extra DPPC molecules were removed from the interface upon interface compression due to the presence of fibrinogen (Fig. 6.8a). During the first interface compression stage, the absolute amide I RA value for the fibrinogen molecules at the interface always decreased, indicating that significant expulsion of fibrinogen molecules from the interface occurred upon interface compression (Fig. 6.9). However, the absolute amide I RA value detected at the end of the interface compression stage showed that

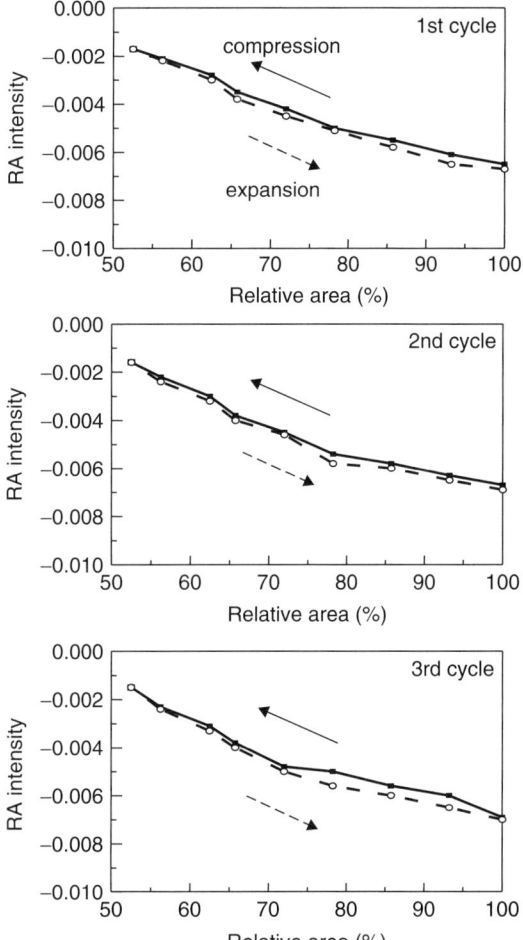

Figure 6.9 Consecutive amide I RA intensity–relative area hysteresis curves of a mixed DPPC/fibrinogen layer on a 70-ppm fibrinogen subphase.

some fibrinogen molecules still existed at the interface even though the rest of free DPPC molecules were closely packed.

During the following interface expansion stage, the absolute v_a-CH$_2$ RA value decreased gradually to a value (0.0018) close to that (0.0017) observed for a pure DPPC monolayer (Fig. 6.8a). However, a smaller number of free DPPC molecules were available if one took account of the fibrinogen adsorption. The fibrinogen adsorption was clearly demonstrated by the increased absolute amide I RA value upon interface expansion (Fig. 6.9). Moreover, the wavenumber corresponding to the maximum in the v_a-CH$_2$ band for DPPC was essentially unchanged at ~2920 cm^{-1} during the interface expansion stage, which indicated that the free

DPPC molecules were always in a highly ordered state (Fig. 6.8b). The close packing of DPPC molecules upon interface expansion apparently resulted from the significant adsorption of fibrinogen molecules occupying part of the space at the interface.

During the second and third cycles, the absolute v_a-CH$_2$ RA values detected at the ends of the cycles were slightly lower than that detected in the first cycle, suggesting that some of the remaining free DPPC molecules after the first cycle were further removed with fibrinogen during the cycles (Fig. 6.8a). Furthermore, the significant fibrinogen adsorption during the interface expansion stage could be inferred from Fig. 6.9, in which the absolute amide I RA value of fibrinogen increased upon interface expansion.

With a higher fibrinogen concentration of 700 ppm, a similar analysis has been reported and a greater loss of DPPC with more pronounced readsorption of fibrinogen during the interface compression–expansion cycles has been found [59]. In addition, detailed experiments have been designed by Kim and Franses [38] to investigate whether the fibrinogen will adsorb on the polar group of DPPC at the air–liquid interface, or whether fibrinogen will penetrate the DPPC monolayer. When a liquid-expanded DPPC monolayer was formed by spreading DPPC molecules onto the air–liquid interface and then fibrinogen was injected into the subphase to yield a bulk concentration of approximately 750 ppm, an IRRAS analysis showed that the protein amide I and II bands were observed 1 min after the fibrinogen injection and disappeared 10 min later. Apparently, protein molecules adsorbed in the first few minutes and then were expelled from the interface, which was similar to the Vroman effect [60, 61]. The fibrinogen might approach the interfacial region and penetrate into the DPPC monolayer, but it could not remain at the interface and was eventually expelled from the interface. The expulsion of fibrinogen by DPPC would occur if the lipid could produce a lower surface tension or a higher surface pressure than fibrinogen.

With a closely packed DPPC Langmuir monolayer available at the interface, the surface pressure of the monolayer showed little effect of fibrinogen injection in the subphase and the IRRAS spectrum was essentially unchanged as before the fibrinogen introduction [38]. Fibrinogen, therefore, did not adsorb significantly on the DPPC monolayer, and it could not displace the DPPC layer. That is, under the circumstances, fibrinogen did not inhibit the surface tension lowering properties of DPPC.

When a DPPC dispersion was introduced onto the surface of a DPPC/fibrinogen mixture by the Trurnit method, the surface tension variations demonstrated that the dispersion spread on the interface and formed an aqueous layer on top of the original aqueous surface. Because this layer did not contain initially any fibrinogen, it allowed the quick and uninhibited formation of a DPPC monolayer, which in turn prevented the fibrinogen adsorption. In addition, a DPPC dispersion has been introduced dropwise onto the surface of a DPPC/fibrinogen mixture, and the surface tension change was similar to that observed for the case with the Trurnit method [38].

For a mixed DLPC/fibrinogen system, if fibrinogen was allowed to adsorbed for 1 h and then a DLPC dispersion was injected into the fibrinogen subphase to yield a mixture of ~730-ppm fibrinogen and 970-ppm DLPC, the surface tension behavior approached that of a DLPC dispersion alone at longer times, suggesting that DLPC had adsorbed and probably expelled most or all of the fibrinogen layer from the interface [62]. The expulsion was apparently driven by the lowering of the system surface free energy, which was reduced more by a complete DLPC monolayer than by a fibrinogen layer. The corresponding IRRAS analysis indicated that the absolute RA value of fibrinogen amide I band increased temporarily, then decreased, and finally dropped to zero. Moreover, the absolute RA value of the DLPC lipid v_a-CH$_2$ band increased first and remained nearly steady. These results showed that DLPC could expel the adsorbed fibrinogen layer from the aqueous interface.

6.8 MIXED MONOLAYERS CONTAINING γ-GLOBULINS

Concentration-dependent effects of γ-globulins on the equilibrium and dynamic surface tension behavior of a 1000-ppm DPPC dispersion are demonstrated in Fig. 6.10. The addition of γ-globulins clearly resulted in a pronounced inhibitory effect on the ability of DPPC molecules to adsorb at the air–liquid interface. Even with the presence of 10-ppm γ-globulins in a 1000-ppm DPPC dispersion, γ-globulin molecules were found to play an important role in the dynamic adsorption at the air–liquid interface. However, DPPC molecules also took part in the dynamic surface tension behavior of the mixture.

When the γ-globulin concentration was increased to a higher concentration of 1000 ppm, the equilibrium surface tension of the mixture was significantly higher than that measured in the absence of γ-globulins (Fig. 6.10). Furthermore, the dynamic tension responses were dominated by γ-globulins and the ability of DPPC molecules to adsorb at the interface was disrupted. All these changes showed the strong influence of γ-globulins on the formation of an adsorbed DPPC monolayer at the interface and suggested that a γ-globulin concentration of 1000 ppm was enough to inhibit the dynamic adsorption of DPPC with a concentration of 1000 ppm.

The roles of γ-globulins in the dynamic monolayer behavior of the mixed DPPC/γ-globulins system have been investigated by monitoring the surface pressure variations during interface compression–expansion cycles [18]. When DPPC molecules were spread at the interface of air–aqueous phase containing 10-ppm γ-globulins, the initial surface pressure increased slightly, which suggested the formation of a mixed lipid/protein layer at the interface.

During the interface compression stage of the first cycle, the characteristic liquid-expanded/liquid-condensed phase transition of a DPPC monolayer was observed in the isotherm, and DPPC and γ-globulins possibly acted as separated components in the mixed monolayer. Similar behavior has been found for the mixed monolayers of DPPC with albumin or fibrinogen [8, 19]. When the interface area was decreased until a DPPC monolayer was in a condensed solid-like

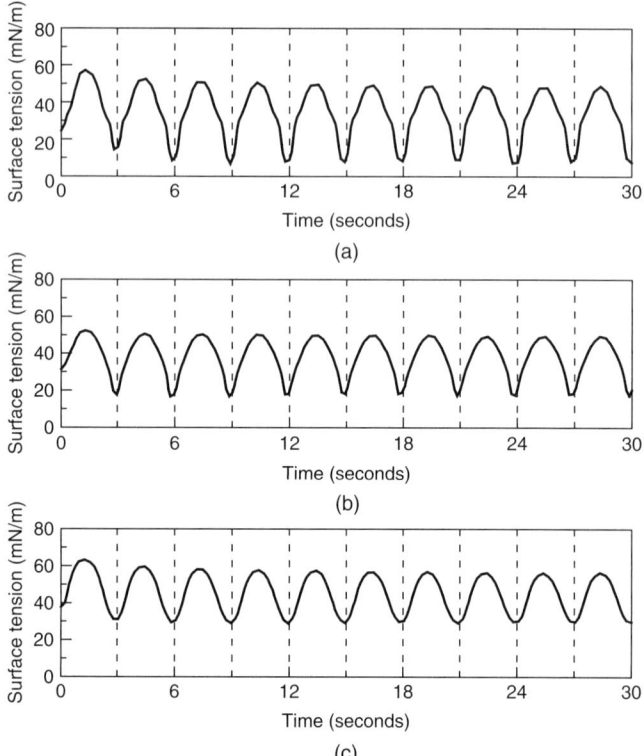

Figure 6.10 Dynamic surface tension behavior of (a) 1000-ppm DPPC, (b) 1000-ppm/10-ppm DPPC/γ-globulins, and (c) 1000-ppm/1000-ppm DPPC/γ-globulins dispersions under a pulsating-area condition of 20 cycles/min at 37 °C.

state, there was a surface pressures transition at ~60 mN/m. This transition could be explained by the forced repulsion of the protein molecules from the mixed monolayer at higher surface pressures due to the presence of the closely packed phospholipid molecules [46, 47]. When the interface area or the mixed monolayer was compressed further, a characteristic DPPC collapse state at surface pressures of ~70 mN/m was reached, which was contributed by a DPPC-rich layer. During the first interface expansion stage, the surface pressure variations suggested that the monolayer was composed mainly of DPPC with detectable entry of γ-globulins at the interface.

During the interface compression stage of the second cycle, the surface pressure was always higher than that for a pure DPPC monolayer at the same interface area, implying the reentry of protein molecules at lower surface pressures during prior interface expansion. The γ-globulin molecules were forced to expel from the overcrowded interface during the interface compression stage, and they might just accumulate immediately below the interface and closely associate with the residual monolayer at the interface. Thus it would be much easier for them to readsorb onto the interface during the subsequent interface expansion stage. A similar trend was also observed for the third interface compression–expansion cycle.

When the γ-globulin subphase concentration was increased to 100 ppm, spread DPPC molecules caused a significant rise in surface pressure, apparently due to a higher surface concentration of the preadsorbed γ-globulin layer at the interface. The transition at surface pressures of ~60 mN/m, presumably associating with the exclusion of γ-globulin molecules, occurred at a slightly larger interface area than that for the case of 10-ppm γ-globulin subphase. This was obviously due to a larger number of γ-globulin molecules adsorbing at the interface before interface compression, and thus the γ-globulin molecules were squeezed out from the monolayer at a larger interface area. The excluded γ-globulin molecules seemed to remove an appreciable number of DPPC molecules from the mixed monolayer, as suggested by the comparatively insignificant collapse plateau detected at the end of the interface compression stage. Similar behavior has also been found for the mixed monolayers of DPPC with albumin or fibrinogen [19]. The expansion curve of the mixed monolayer was similar to that for a DPPC monolayer alone, but the surface pressure did not drop to zero as observed for a pure DPPC monolayer, apparently caused by the more pronounced penetration of γ-globulins into the expanded monolayer on a 100-ppm γ-globulin subphase.

Some of the remaining DPPC molecules after the first cycle were removed with γ-globulins during the second interface compression stage, as suggested by the insignificant collapse behavior. The exclusion of γ-globulins associated by the removal of DPPC evidently resulted in the depletion of DPPC molecules at the interface. Moreover, the significant penetration behavior of γ-globulins at the interface, indicated by the surface pressure variations, represented a process for protein replenishment in the monolayer. Similar behavior was also observed for the third cycle, with a slightly decreased maximum surface pressure during the interface compression stage and more pronounced reentry behavior of γ-globulins during the interface expansion stage.

6.9 CONCLUSION

Based on the dynamic surface tension measurements, especially under pulsating-area conditions, the concentration-dependent inhibitory effects of plasma proteins on the dynamic adsorption or monolayer formation of DPPC at air–liquid interfaces have clearly been demonstrated. It appears that the adsorption time scale of DPPC is larger than that of plasma proteins. Consequently, plasma proteins have an advantage over DPPC to compete for the space at interfaces by adsorbing rapidly, thus inhibiting the formation of a compact adsorbed DPPC monolayer at interfaces. However, the dynamic interfacial properties of a phospholipid dispersion is strongly affected by the dispersion methodology or the dispersion status, and it is therefore difficult to conduct a fair comparison for the available data in the literature. A systematic study on the plasma protein ability to inhibit the monolayer formation or the dynamic adsorption of DPPC at interfaces is then necessary in order to take the dispersion characteristics into account. Moreover, recent studies of the dispersion effects on the adsorbed phospholipid monolayer formation at interfaces may suggest possible approaches to improve the formulation of artificial lung surfactants containing insoluble phospholipids such as DPPC.

Current evidence from dynamic monolayer behavior studies supports that the inhibition of DPPC dynamic surface activity at the cyclic interface by plasma proteins may be explained by the partial exclusion of plasma proteins associated with the removal of DPPC at the interface during interface compression and by the significant reentry of plasma proteins during the subsequent interface expansion. In a continuous interface compression–expansion process, the induced loss of free DPPC molecules by plasma protein expulsion from the interface and the dominant readsorption of plasma protein molecules to the interface will result in pronounced depletion and thus an incomplete monolayer of DPPC at the interface. However, the roles of plasma protein structures in the protein-induced DPPC removal still remain to be studied. The abilities of in situ IRRAS technique to probe the mixed phospholipid/plasma protein monolayer composition and possibly to perform the conformational analysis of protein secondary structures may offer a means to resolve this issue. Moreover, in situ BAM technique can provide information on the monolayer characteristics, such as orientational order of the monolayer domains and phase transitions in monolayers, at the air–liquid interfaces. Consequently, it may be expected that IRRAS and BAM will be used extensively to study the dynamic and structural characteristics of mixed phospholipid/plasma protein monolayers at interfaces.

Furthermore, it has been reported in a recent paper that serum was much more potent than albumin in reducing the surface activity of bovine lipid extract surfactant, probably because serum could perturb the structure–function properties of the lung surfactant extract [63]. This result indicates that other nonprotein or protein–lipid combined components of serum may play an important role in the lung surfactant inactivation. The possible synergistic effect of plasma proteins with other components in serum on the dynamic surface activity of phospholipids or lung surfactant systems thus seems to be an interesting subject to study.

ACKNOWLEDGMENTS

Work done in the authors' laboratory was supported by the National Science Council, Taiwan. The authors are indebted to Professors Y.-M. Yang and Y.-L. Lee for helpful discussions about this chapter.

REFERENCES

1. J. Boucher, E. Trudel, M. Méthot, P. Desmeules, and C. Salesse (2007). Organization, structure and activity of proteins in monolayers. *Colloids Surf. B: Biointerfaces* **58**: 73–90.

2. J. Goerke (1998). Pulmonary surfactant: functions and molecular composition. *Biochim. Biophys. Acta* **1408**: 79–89.
3. R. H. Notter (1989). Physical chemistry and physiological activity of pulmonary surfactants, in *Surfactant Replacement Therapy* (D. L. Shapiro and R. H. Notter, Eds.). Alan R. Liss, New York, pp. 19–70.
4. K. M. W. Keough (1992). Physical chemistry of pulmonary surfactant in the terminal spaces, in *Pulmonary Surfactant: From Molecular Biology to Clinical Practice* (R. Robertson, L. M. G. van Golde, and J. J. Batenburg, Eds.). Elsevier, Amsterdam, pp. 109–164.
5. J. A. Whitsett and J. Baatz (1992). Hydrophobic surfactant proteins SP-B and SP-C: molecular biology, structure and function, in *Pulmonary Surfactant: From Molecular Biology to Clinical Practice* (R. Robertson, L. M. G. van Golde, and J. J. Batenburg, Eds.). Elsevier, Amsterdam, pp. 55–85.
6. M. Ikegami, A. Jobe, H. Jacobs, and R. Lam (1984). A protein from airways of premature lambs that inhibits surfactant function. *J. Appl. Physiol.* **57**: 1134–1142.
7. A. Gunther and W. Seeger (1995). Resistance to surfactant inactivation, in *Surfactant Therapy for Lung Disease* (B. Robertson and H. W. Taeusch, Eds.). Marcel Dekker, New York, pp. 269–292.
8. S. A. Tabak, and R. H. Notter (1977). Effect of plasma proteins on the dynamic π–A characteristic of saturated phospholipids films. *J. Colloid Interface Sci.* **59**: 293–300.
9. G. Colacicco and M. K. Basu (1978). Effect of serum albumin on dynamic force–area curve of dipalmitoyl lecithin. *Respir. Physiol.* **32**: 265–280.
10. S. Taneva, I. Panaiotov, and L. Ter-Minassian-Saraga (1984). Effect of surface pressure on mixed dipalmitoyllecithin–serum albumin monolayer composition. *Colloids Surf.* **10**: 101–111.
11. K. M. W. Keough, C. S. Parson, P. T. Phang, and M. G. Tweeddale (1988). Interactions between plasma proteins and pulmonary surfactant: surface balance studies. *Can. J. Physiol. Pharmacol.* **66**: 1166–1173.
12. B. A. Holm, G. Enhorning, and R. H. Notter (1988). A biophysical mechanism by which plasma protein inhibits surfactant activity. *Chem. Phys. Lipids* **49**: 49–55.
13. W. Seeger, G. Stohr, H. R. D. Wolf, and H. Neujof (1985). Alternation of surfactant function due to protein leakage: special interaction with fibrin monomer. *J. Appl. Physiol.* **58**: 326–338.
14. T. Fuchimukai, T. Fujiwara, A. Takahashi, and G. Enhorning (1987). Artificial pulmonary surfactant inhibited by proteins. *J. Appl. Physiol.* **62**: 429–437.
15. G. Enhorning (1977). Pulsating bubble technique for evaluating pulmonary surfactant. *J. Appl. Physiol.* **43**: 198–203.
16. C.-H. Chang and E. I. Franses (1994). An analysis of the factors affecting dynamic tension measurements with the pulsating bubble surfactometer. *J. Colloid Interface Sci.* **164**: 107–113.
17. C.-H. Chang and E. I. Franses (1994). Dynamic tension behavior of aqueous octanol solutions under constant-area and pulsating-area conditions. *Chem. Eng. Sci.* **49**: 313–325.
18. C.-H. Chang, S.-D. Yu, T.-K. Chuang, and C.-N. Liang (2000). Roles of γ-globulin in the dynamic interfacial behavior of mixed dipalmitoyl phosphatidylcholine/γ-globulin monolayers at air/liquid interfaces. *J. Colloid Interface Sci.* **22**: 461–468.

19. R.-R. Kuo, C.-H. Chang, Y.-M. Yang, and J.-R. Maa (2003). Induced removal of dipalmitoyl phosphatidylcholine by the exclusion of fibrinogen from compressed monolayers at air/liquid interfaces. *J. Colloid Interface Sci.* **257**: 108–115.
20. C. Lheveder, J. Meunier, and S. Hénon (2000). Brewster angle microscopy, in *Physical Chemistry of Biological Interfaces* (A. Baszkin and W. Norde, Eds.). Marcel Dekker, New York, pp. 559–575.
21. G. A. Overbeck, D. Hönig, and S. Möbius (1993). Visualization of first- and second-order phase transitions in eicosanol monolayers using Brewster angle microscopy. *Langmuir* **9**: 555–560.
22. V. Melzer and D. Vollhardt (1996). Formation of condensed phase patterns in adsorption layers. *Phys. Rev. Lett.* **76**: 3770–3773.
23. M. M. Hossain, M. Yoshida, and T. Kato (2000). Higher-order structure formation in adsorbed monolayers at aqueous solution surfaces studied by Brewster angle microscopy. *Langmuir* **16**: 3345–3348.
24. V. Melzer, D. Vollhardt, G. Brezesinski, and H. Mohwald (1998). Similarities in the phase properties of Gibbs and Langmuir monolayers. *J. Phys. Chem. B* **102**: 591–597.
25. M. M. Hossain, M. Yoshida, K. Iimura, N. Suzuki, and T. Kato (2000). Phase transition in adsorbed monolayers of 2-hydroxyethyl laurate at the air–water interface. *Colloids Surf. A: Physicochem. Eng. Aspects* **171**: 105–113.
26. M. N. G. de Mul and J. A. Mann, Jr. (1998). Determination of the thickness and optical properties of a Langmuir film from the domain morphology by Brewster angle microscopy. *Langmuir* **14**: 2455–2466.
27. J. M. Rodríguez Patino, C. Carrera Sánchez, and M. R. Rodríguez Niño (1999). Is Brewster angle microscopy a useful technique to distinguish between isotropic domains in β-casein-monoolein mixed monolayers at the air–water interface? *Langmuir* **15**: 4777–4788.
28. R. A. Dluhy (2000). Infrared spectroscopy of biophysical monomolecular films at interfaces: theory and applications, in *Physical Chemistry of Biological Interfaces* (A. Baszkin and W. Norde, Eds.), Marcel Dekker, New York, pp. 711–747.
29. V. P. Tolstoy, I. V. Chernyshova, and V. A. Skryshevsky (2003). *Handbook of Infrared Spectroscopy of Ultrathin Films*. Wiley, Hoboken, NJ.
30. R. A. Dluhy and D. G. Cornell (1985). In Situ measurement of the infrared spectra of insoluble monolayers at the air–water interface. *J. Phys. Chem.* **89**: 3195–3197.
31. R. A. Dluhy (1986). Quantitative external reflection infrared spectroscopic analysis of insoluble monolayers spread at the air–water interface. *J. Phys. Chem.* **90**: 1373–1379.
32. M. L. Mitchell and R. A. Dluhy (1988). In Situ FTIR investigation of phospholipid monolayer phase transitions at the air–water interface. *J. Am. Chem. Soc.* **110**: 712–718.
33. R. Mendelsohn, J. W. Brauner, and A. Gericke (1995). External infrared reflection–absorption spectrometry monolayer films at the air–water interface. *Annu. Rev. Phys. Chem.* **46**: 305–334.
34. C. R. Flach, A. Gericke, and R. Mendelsohn (1997). Quantitative determination of molecular chain tilt angles in monolayer films at the air/water interface: infrared reflection/absorption spectroscopy of behenic acid methyl ester. *J. Phys. Chem. B* **101**: 58–65.

35. K. K. Chittur (1998). FTIR/ATR for protein adsorption to biomaterial surfaces. *Biomaterials* **19**: 357–369.
36. X. Bi, S. Taneva, K. M. W. Keough, R. Mendelsohn, and C. R. Flach (2001). Thermal stability and DPPC/Ca^{2+} interactions of pulmonary surfactant SP-A from bulk-phase and monolayer IR spectroscopy. *Biochemistry* **40**: 13659–13669.
37. M. D. Lad, F. Birembaut, R. A. Frazier, and R. J. Green (2005). Protein–lipid interactions at the air/water interface. *Chem. Phys.* **7**: 3478–3485.
38. S. H. Kim and E. I. Franses (2006). Competitive adsorption of fibrinogen and dipalmitoylphosphatidylcholine at the air/aqueous interface. *J. Colloid Interface Sci.* **295**: 84–92.
39. C.-C. Cheng and C.-H. Chang (2000). Retardation effect of tyloxapol on inactivation of dipalmitoyl phosphatidylcholine surface activity by albumin. *Langmuir* **16**: 437–441.
40. X. Wen and E. I. Franses (2001). Adsorption of bovine serum albumin at the air/water interface and its effect on the formation of DPPC surface film. *Colloids Surf. A: Physicochem. Eng. Aspects* **190**: 319–332.
41. S. Y. Park, C.-H. Chang, D. J. Ahn, and E. I. Franses (1993). Dynamic surface tension behavior of hexadecanol spread and adsorbed monolayers. *Langmuir* **9**: 3640–3648.
42. S. Y. Park and E. I. Franses (1995). Hexadecanol microstructures of crystallites in aqueous dispersions and of Langmuir–Blodgett monolayers. *Langmuir* **11**: 2187–2194.
43. C.-H. Chang, K. A. Coltharp, S. Y. Park, and E. I. Franses (1996). Surface tension measurements with the pulsating bubble method. *Colloids Surf. A: Physicochem. Eng. Aspects* **114**: 185–197.
44. S. Y. Park, S. C. Peck, C.-H. Chang, and E. I. Franses (1996). The roles of dispersed surfactant particles on the dynamic tension behavior of aqueous surfactant system, in *Dynamic Properties of Interfaces and Association Structures* (V. Pillai and D. Shah, Eds.). AOCS Press, Champaign, IL, pp. 1–22.
45. S. H. Kim and E. I. Franses (2005). New protocols for preparing dipalmitoylphosphatidylcholine dispersions and controlling surface tension and competitive adsorption with albumin at the air/aqueous interface. *Colloids Surf. B: Biointerfaces* **43**: 256–266.
46. D. G. Cornell and R. J. Carroll (1985). Miscibility in lipid–protein monolayers. *J. Colloid Interface Sci.* **108**: 226–233.
47. M. R. Rodríguez Niño, P. J. Wilde, D. C. Clark, and J. M. Rodríguez Patino (1998). Surface dilational properties of protein and lipid films at the air–water interface. *Langmuir* **14**: 2160–2166.
48. S. G. Taneva and K. M. W. Keough (1994). Dynamic surface properties of pulmonary surfactant proteins SP-B and SP-C and their mixture with dipalmitoylphosphatidylcholine. *Biochemistry* **33**: 14660–14670.
49. J. Zhao, D. Vollhardt, G. Brezesinski, S. Siegel, J. Wu, J. B. Li, and R. Miller (2000). Effect of protein penetration into phospholipid monolayers: morphology and structure. *Colloids Surf. A: Physicochem. Eng. Aspects* **171**: 175–184.
50. H. Zhang, X. Wang, G. Cui, and J. Li (2000). Stability investigation of the mixed DPPC/protein monolayer at the air–water interface. *Colloids Surf. A: Physicochem. Eng. Aspects* **175**: 77–82.

51. X. Wang, H. Zhang, G. Cui, and J. Li (2001). Structure characterization and stability of mixed lipid/protein monolayer at the air/water interface. *J. Mol. Liquids* **90**: 149–156.

52. A. Gericke and H. Hühnerfuss (1993). In Situ investigation of saturated long-chain fatty acids at the air/water interface by external infrared reflection–absorption spectrometry. *J. Phys. Chem.* **97**: 12899–12908.

53. T. L. Phang and E. I. Franses (2004). Expulsion of bovine serum albumin from the air/water interface by a sparingly soluble lecithin lipid. *J Colloid Interface Sci.* **275**: 477–487.

54. D. Cho, G. Narsimhan, and E. I. Franses (1997). Interactions of spread lecithin monolayers with bovine serum albumin in aqueous solution. *Langmuir* **13**: 4710–4715.

55. Y.-L. Liu and C.-H. Chang (2002). Inhibitory effects of fibrinogen on the dynamic tension-lowering activity of dipalmitoyl phosphatidylcholine dispersions in the presence of tyloxapol. *Colloid Polym. Sci.* **280**: 683–687.

56. M. Ivanova, I. Panaiotov, T. Trifonova, M. Echkenazi, G. Konstantinov, and R. Ivanova (1984). Interaction of antilipid A immunoglobulin G and normal immunoglobulin G, incorporated in a lipid A monolayer. *Colloids Surf.* **10**: 269–282.

57. J. M. Rodríguez Patino and M. R. Rodríguez Niño (1995). Protein adsorption and protein–lipid interactions at the air–aqueous solution interface. *Colloids Surf. A: Physicochem. Eng. Aspects* **103**: 91–103.

58. T. Mita (1989). Lipid–protein interaction in mixed monolayers from phospholipids and proteins. *Bull. Chem. Soc. Jpn.* **62**: 3114–3121.

59. C.-L. Yin and C.-H. Chang (2006). Infrared spectroscopy analysis of mixed DPPC/fibrinogen layer behavior at the air/liquid interface under a continuous compression–expansion condition. *Langmuir* **22**: 6629–6634.

60. L. Vroman and A. L. Adams (1986). Adsorption of proteins out of plasma and solutions in narrow spaces. *J. Colloid Interface Sci.* **111**: 391–402.

61. E. F. Leonard and L. Vroman (1991). Is the Vroman effect of importance in the interaction of blood with artificial materials? *J. Biomater. Sci. Polym. E* **3**: 95–107.

62. T. L. Phang, S. J. McClellan, and E. I. Franses (2005). Displacement of fibrinogen from the air/aqueous interface by dilauroylphosphatidylcholine lipid. *Langmuir* **21**: 10140–10147.

63. K. Nag, A. Hillier, K. Parsons, and M. F. Garcia (2007). Interactions of serum with lung surfactant extract in the bronchiolar and alveolar airway models. *Respir. Physiol. Neurobiol.* **157**: 411–424.

CHAPTER 7

Monitoring of Membrane-Associated Protein Binding and of Enzyme Activity in Monolayers at the Air-Water Interface by Infrared Spectroscopy

SYLVAIN BUSSIÈRES, JULIE BOUCHER, PHILIPPE DESMEULES, and MICHEL GRANDBOIS

Unité de Recherche en Ophtalmologie, Centre de Recherche du CHUQ, Département d'Ophtalmologie, Faculté de médecine, Université Laval, Québec, Québec, Canada

BERNARD DESBAT

CBMN, UMR-CNRS, Université Bordeaux I, Pessac, France

CHRISTIAN SALESSE

Unité de Recherche en Ophtalmologie, Centre de Recherche du CHUQ, Département d'Ophtalmologie, Faculté de Médecine, Université Laval, Québec, Québec, Canada

CONTENTS

7.1	Introduction	166
7.2	Monitoring of Phospholipase A_2 Monolayer Binding and Hydrolysis by Infrared Spectroscopy	168
7.3	Monitoring of Lecithin Retinol Acyltransferase Monolayer Binding and Hydrolysis by Infrared Spectroscopy	174
7.4	Monitoring of Binding of an Acylated Protein onto Phospholipid Monolayers by Surface Pressure Measurements	177
7.5	Monitoring of Binding of an Acylated Protein onto Phospholipid Monolayers by Infrared Spectroscopy	181
7.6	Conclusion	183
	Acknowledgments	184
	References	184

Structure and Dynamics of Membranous Interfaces, edited by Kaushik Nag
Copyright © 2008 John Wiley & Sons, Inc.

7.1 INTRODUCTION

A large number of processes involving different types of proteins occur at the membrane interface. For example, G-protein coupled receptors are membrane proteins that activate membrane-associated G-proteins as the first step of a transduction cascade that involves several additional membrane-associated proteins (for a review, see Ref. 1). Visual phototransduction, where most reactions take place at the membrane interface, is a good example of such a transduction cascade (for reviews, see Refs. 2 and 3). Indeed, after the activation of transducin, a membrane-associated G-protein, by the membrane protein rhodopsin, transducin alpha subunit activates an additional membrane-associated protein called phosphodiesterase. The calcium-dependent inhibition of rhodopsin is also regulated by two other membrane-associated proteins. These proteins represent only a few examples of the processes regulated at the membrane interface, which thus illustrates the complexity and the number of reactions occurring at this interface. Therefore it is important to improve our understanding of the interactions taking place between membrane-associated proteins and membranes.

Several enzymatic processes also take place at the membrane interface. Hydrolysis of membrane components by phospholipases, for example, involves first the binding of these enzymes followed by membrane hydrolysis [4]. The enzymatic transformation of hydrophobic compounds is also occurring at the membrane interface. For example, the enzymatic reduction, oxidation, and esterification of vitamin A during the regeneration of the chromophore of rhodopsin all take place at the membrane interface by the concerted action of retinol dehydrogenase and lecithin retinol acyltransferase, which are membrane-associated proteins (for reviews, see Refs. 5 and 6).

Different strategies are used by proteins to drive their attachment to membranes such as the presence of one α-helix at both the N and C terminals (Fig. 7.1a), one α-helix at either the N or C terminal (Fig. 7.1b), a fatty acylation (Fig. 7.1c), or a small hydrophobic stretch located at the protein surface (Fig. 7.1d). It has been known for many years that proteins can be posttranslationally modified by

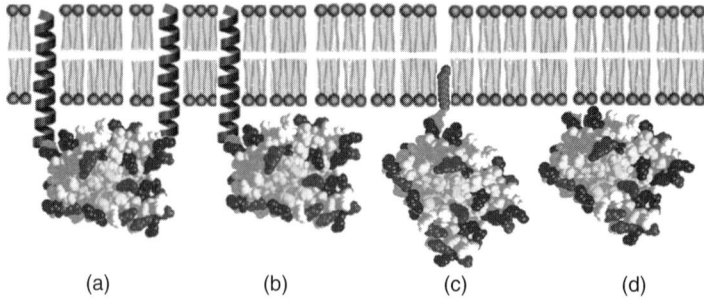

Figure 7.1 Schematic diagram of the different strategies used by membrane-associated proteins to bind membranes. Protein binding to membranes via (a) N- and C-terminal α-helices, (b) one terminal α-helix, (c) covalent acylation, and (d) a small protein stretch.

the covalent attachment of a fatty acyl chain such as myristoylation, prenylation, or palmitoylation (for a review, see Ref. 7). However, the widespread occurrence and the significance of these modifications have only recently been appreciated. Indeed, it is now widely demonstrated that, for example, many viral and cellular proteins are N-terminally acylated by myristic acid (14:0) and other fatty acids (i.e., C12:0, C14:1, C14:2, C16:0) (for reviews, see Refs. 8–10). Such a hydrophobic modification has been shown to play a key role in protein targeting to membranes [7]. However, it has been argued that the presence of one acylation is barely enough to allow membrane binding [11–13]. Hydrophobic residues could also play an important role in this binding [14]. Moreover, several lines of evidence suggest that other types of proteins bind membranes by virtue of their N- and/or C-terminal α-helices [15]. Finally, other enzymes whose activity is taking place at the membrane interface, such as venom phospholipases, are not acylated and do not bear N- or C-terminal attachments [4]. It has thus been postulated that small hydrophobic stretches are involved in their membrane binding although charge–charge interactions can also be involved. The intricate interaction between these proteins and membranes is still not well understood and minor contributions to membrane binding arising from amino acids have not yet been clarified.

Model membranes are useful to improve our understanding of protein binding to membranes. Proteins injected into the buffer subphase underneath lipid monolayers represent good model membrane systems (for reviews, see Refs. 16–19). It consists in spreading lipids from an organic solvent on a buffer subphase. The surface density of lipids and hence the lateral pressure can be controlled by use of compression barriers (Fig. 7.2). A single type of lipid or mixtures of lipids that more appropriately mimic the natural membranes can be used to study the influence of the polar headgroup and the fatty acyl chains on protein binding. After the injection of proteins into the subphase (Fig. 7.3a), one can work at a fixed surface pressure or molecular area and respectively monitor molecular area or surface pressure increase as a function of time to study kinetics of protein binding. The organization, structure, and dynamics of these films can be studied by a large number of modern methods of investigation including, for example, fluorescence microscopy, Brewster angle microscopy, X-ray and neutron reflectivity, ellipsometry, and rheology as well as infrared, absorption, and fluorescence spectroscopy (for a review, see Ref. 16). Films can also be transferred onto solid substrates and studied by atomic force microscopy to get higher resolution information.

In this chapter, three different types of proteins have been used to demonstrate the usefulness of the monolayer model membrane system and to improve our understanding of the binding of membrane-associated proteins and their action onto phospholipid monolayers: (1) phospholipase A_2 that binds the membrane through a small hydrophobic stretch (see Fig. 7.1d) and then hydrolyzes the *sn*-2 fatty acyl chain of membrane phospholipids; (2) lecithin retinol acyltransferase (LRAT) that binds membranes and then hydrolyzes the *sn*-1 fatty acyl chain of membrane phospholipids; the N- and C-terminal α-helices of

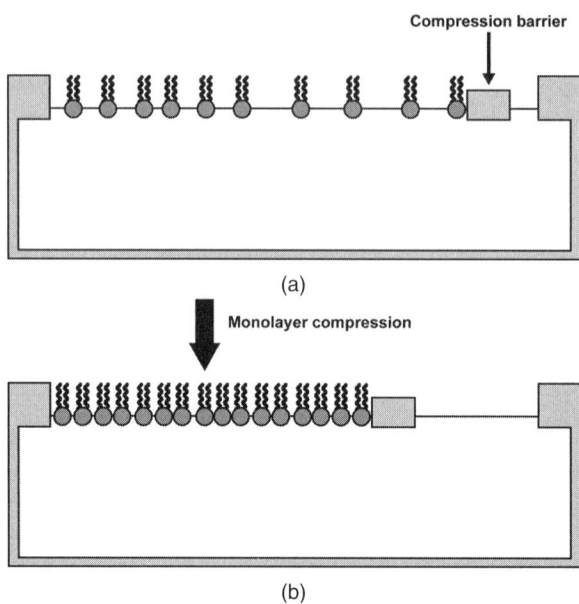

Figure 7.2 Schematic diagram of the monolayer method. (a) Phospholipids are spread from an organic solvent and then (b) compressed using the compression barriers.

LRAT (Fig. 7.1a) have been removed to find out if it could nevertheless bind membranes; and (3) acylated recoverin (Fig. 7.1c) that changes conformation in a calcium-dependent manner, which results in the extrusion or occlusion of its myristoyl moiety from/in its hydrophobic pocket in the presence or absence of calcium, respectively.

7.2 MONITORING OF PHOSPHOLIPASE A_2 MONOLAYER BINDING AND HYDROLYSIS BY INFRARED SPECTROSCOPY

To obtain molecular information on the dynamics, orientation, organization, and secondary structure of proteins during their adsorption onto phospholipid monolayers, we have used polarization modulation infrared reflection absorption spectroscopy (PM-IRRAS), which allows measurement of infrared spectra in monolayers at the lipid–buffer interface (Fig. 7.3b). This method overcomes the water absorption limitation using rapid polarization modulation of the incident beam, which allows the observation of infrared absorption bands of phospholipids and proteins at the monomolecular level [20, 21]. Furthermore, it is possible to extract information from the infrared spectra on conformation and orientation of proteins and phospholipids in monolayers [22–30] as well as the kinetics of hydrolysis of monolayers by enzymes [31–34].

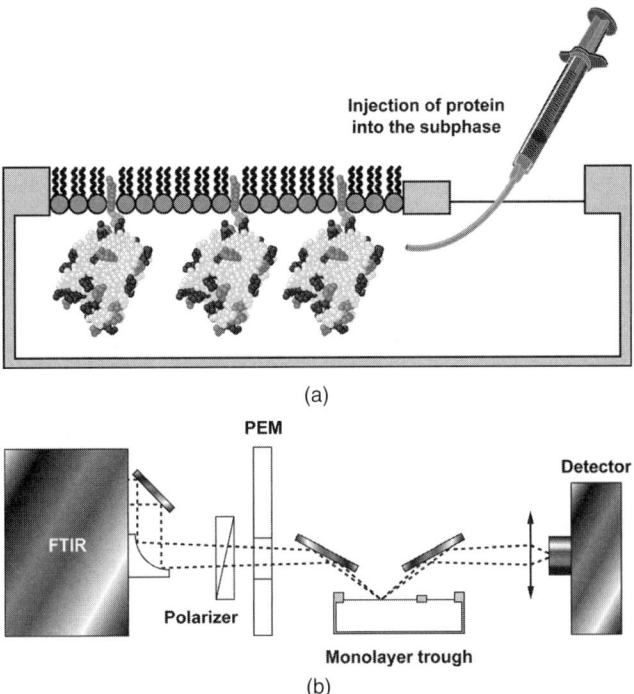

Figure 7.3 (a) Schematic diagram of protein injection into the monolayer subphase and binding of an acylated protein onto this monolayer. The acylated protein shown here is recoverin. The proposed orientation has been deduced from the simulations of the PM-IRRAS spectra shown in the inset of Fig. 7.10. (b) Schematic diagram of the PM-IRRAS setup used in the present studies. The infrared light from the Fourier-transformed infrared spectrometer (FTIR) is s- and p-polarized using the photoelastic modulator (PEM), reflected by the monolayer surface and then focused on the detector. The optical setup used for the PM-IRRAS measurements has been described previously [78].

Phospholipase A_2 (PLA$_2$) is an interfacially activated enzyme that catalyzes regio- and stereospecific hydrolysis of the *sn*-2 acyl ester linkage of phospholipids [4]. PLA$_2$ is indeed inactive when its substrate is not organized in membranes or model membranes. During the course of dipalmitoyl phosphatidylcholine (DPPC; 16:0/16:0-PC) monolayer hydrolysis by PLA$_2$, two insoluble hydrolysis products are formed: palmitic acid (16:0) and lysopalmitoyl phosphatidylcholine (16:0-PC). When PLA$_2$ is injected into the subphase of the 16:0/16:0-PC monolayer, no change in surface pressure can be observed, which suggests that it barely protudes into the monolayer, which corresponds to the situation depicted in Fig. 7.1d. However, infrared spectroscopy allowed researchers to follow the kinetics of monolayer hydrolysis by PLA$_2$. PM-IRRAS spectra of a 16:0/16:0-PC monolayer at the air–water interface in the 1000–1800-cm^{-1} region measured at different extents of monolayer hydrolysis by PLA$_2$ are shown in Fig. 7.4 [33]. The three most intense bands in the spectrum of pure 16:0/16:0-PC before PLA$_2$ injection into the subphase have been attributed to the νC=O ester stretching band at 1733 cm^{-1}, the δ-CH$_2$ bending mode at 1468 cm^{-1}, and the ν_a-P=O at 1228 cm^{-1} (spectrum 1, Fig. 7.4). The assignment of these bands is in agreement with previous data [31]. The presence of the broad negative infrared band in the 1700–1640-cm^{-1} region has previously been documented [20, 35, 36]. In brief, the water absorption band is centered at approximately 1640 cm^{-1}, which corresponds to the position where the PM-IRRAS spectrum changes abruptly, showing a positive band followed immediately by a negative band. This behavior is due to an abrupt variation of the refractive index of the aqueous subphase in this range of frequency, which has been confirmed by other reports [31, 37, 38].

Upon monolayer hydrolysis by PLA$_2$, a decrease in the intensity as well as a broadening of the C=O ester band at 1733 cm^{-1} is observed (spectra 2–6, Fig. 7.4). The decrease of this band, which was also observed by Gericke and Hühnerfuss [31], is due to the hydrolysis of the *sn*-2 acyl ester linkage of 16:0/16:0-PC by PLA$_2$. The residual ν-C=O ester band at the end of the hydrolysis (spectrum 6, Fig. 7.4) is due to the *sn*-1 acyl ester linkage of 16:0-PC hydrolysis product that remains at the surface. The broadening of this band can be interpreted by the formation of an increasingly disordered monolayer during hydrolysis [25].

An additional important change in spectra 2–6 of Fig. 7.4 is observed upon monolayer hydrolysis by PLA$_2$. Indeed, the appearance of two new bands at 1537 and 1575 cm^{-1} can be seen in this figure. The use of the asymmetric phospholipid 6:0/16:0-PC (1-caproyl-2-palmitoyl-*sn*-3-phosphatidylcholine) was very useful for the assignment of these bands because its hydrolysis by PLA$_2$ produces an insoluble fatty acid (palmitic acid, 16:0) and a soluble 6:0-PC (lysocaproyl phosphatidylcholine) [33, 39]. The attribution of these infrared bands can thus be directly confirmed in the presence of PLA$_2$. Figure 7.5 shows the PM-IRRAS spectra before (spectrum 1) and after (spectrum 2) the injection of PLA$_2$ into the subphase monolayer. Before PLA$_2$ injection, a ν-C=O ester stretching band at 1731 cm^{-1} can be seen in spectrum 1 (Fig. 7.5) as observed in the case of the 16:0/16:0-PC monolayer before hydrolysis (spectrum 1, Fig. 7.4). At the end of hydrolysis (spectrum 2, Fig. 7.5), this band disappears almost completely due to the solubilization into the subphase of the 6:0-PC species, which contains the ν-C=O ester groups. In addition, two new bands can be seen at 1575 and 1537 cm^{-1}. The position of these two bands corresponds very well to that of the doublet observed after 16:0/16:0-PC monolayer hydrolysis (spectrum 6, Fig. 7.4) [33].

The precise attribution of these infrared bands has been achieved by measuring PM-IRRAS spectra of palmitic acid (16:0) hydrolysis product. At alkaline pH, the carboxylic acid of fatty acids is known to be deprotonated and palmitate salts are readily formed. The results from spectrum 6 in Fig. 7.4 and spectrum 2 in Fig. 7.5 suggest that the doublet observed at 1537 and 1575 cm^{-1} after a

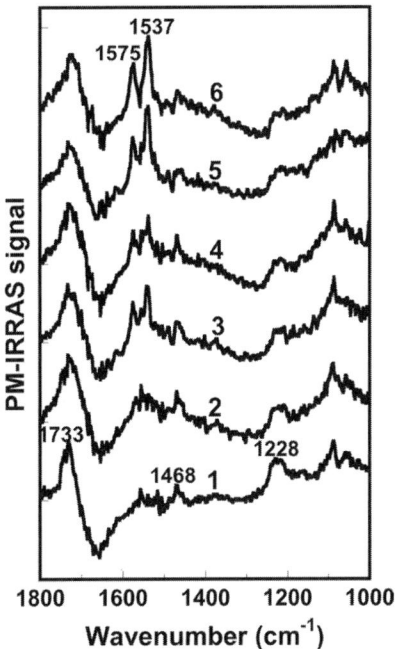

Figure 7.4 PM-IRRAS spectra of 16:0/16:0-PC monolayer hydrolysis by PLA$_2$ (0.4 units/mL) at 8 mN/m. Spectra 1–6: (1) pure 16:0/16:0-PC monolayer before and (2) 10 min, (3) 25 min, (4) 50 min, (5) 70 min, and (6) 90 min after PLA$_2$ injection into the subphase. Subphase contains 150-mM NaCl, 5-mM CaCl$_2$, 10-mM Tris, pH 8.9, $T = 25$ °C. *Naja naja* venom PLA$_2$ and 16:0/16:0-PC were purchased from Sigma and used without further purification. For all monolayer measurements, the water used was filtrated on Millipore Nanopure System. Its resistivity and surface tension were higher than 18 MΩ·cm and 71 mN/m, respectively. Surface pressure was measured by the Wilhelmy method using a filter paper. Phospholipid monolayers were spread by using hexane:ethanol (9:1).

large extent of 16:0/16:0-PC hydrolysis by PLA$_2$ can be assigned to the formation of an ionic salt of palmitic acid. However, in principle, these vibrations could be assigned to either a sodium- or a calcium-palmitate complex or both. Indeed, the subphase used contains 5-mM CaCl$_2$, 100-mM NaCl, and 10-mM Tris at pH 8.9. In order to make a clear assignment of these bands observed after monolayer hydrolysis by PLA$_2$, spectra of pure palmitic acid were measured on different subphases including each individual component of the buffer used for monolayer hydrolysis by PLA$_2$. The spectrum of palmitic acid on CaCl$_2$ (spectrum 4, Fig. 7.5) or on the same buffer as that used in Fig. 7.4 (spectrum not shown) shows two well-defined bands at 1538 and 1573 cm^{-1}. There is a very good correlation between the position of these bands at 1538 and 1573 cm^{-1} (spectrum 4, Fig. 7.5) and those observed at 1537 and 1575 cm^{-1} after a large

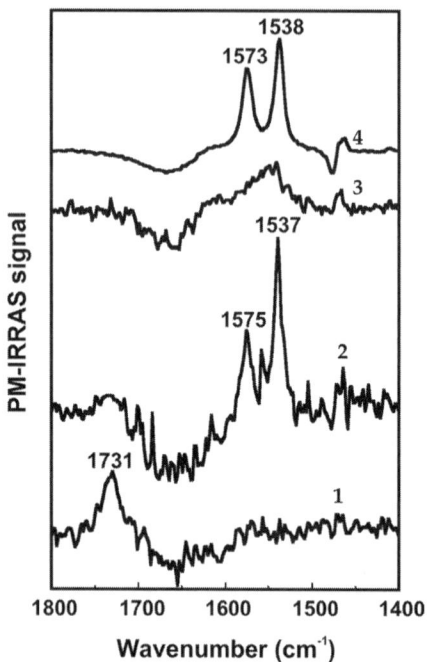

Figure 7.5 PM-IRRAS spectra of 6:0/16:0-PC monolayer at 8 mN/m. The asymmetric phospholipid 1-caproyl-2-palmitoyl-*sn*-3-phosphatidylcholine(6:0/16:0-PC) was kindly provided as described [33]. Spectra 1 and 2: (1) pure 6:0/16:0-PC monolayer before and (2) 1 h after PLA_2 injection (0.4 units/mL) into the same subphase as in Fig. 7.4. PM-IRRAS spectra of a monolayer of pure palmitic acid (16:0) spread on different subphases. Spectra 3 and 4: (3) 10-mM Tris, pH 8.9 at a surface pressure of 8 mN/m and (4) 5-mM $CaCl_2$ at a surface pressure of 25 mN/m.

extent of hydrolysis of 16:0/16:0-PC (spectrum 6, Fig. 7.4) and 6:0/16:0-PC (spectrum 2, Fig. 7.5) by PLA_2. The assignment of these bands to the formation of a calcium–palmitate complex is also supported by Ahn and Franses [40] and Marshbanks et al. [41], who have measured the infrared spectrum of LB films of stearic acid spread on a subphase containing 0.1-mM calcium (without sodium) at pH 8 by ATR spectroscopy. Ahn and Franses [40] have observed a doublet in the 1540–1573-cm^{-1} region, whereas Marshbanks et al. [41] have shown a spectrum with well-defined bands at 1538 and 1574 cm^{-1}. These bands can only be attributed to the formation of a calcium–palmitate complex. This assignment is further supported by the spectrum of solid calcium–palmitate reported by Clarke [42], which also shows the presence of a doublet at 1540 and 1580 cm^{-1}. The two infrared bands at 1537 and 1575 cm^{-1} observed in Fig. 7.4 have thus been assigned to the v_a-COO^- vibration doublet of calcium–palmitate. These two well-defined bands at 1538 and 1573 cm^{-1} are absent in the spectra of palmitic acid on a subphase containing either NaCl (spectrum not shown) or pure Tris buffer (spectrum 3, Fig. 7.5). Instead, a broad band located at 1562 cm^{-1} is observed for the sodium–palmitate complex or in the presence of the pure Tris buffer. This broadening suggests a disordering of this monolayer in the presence of sodium or Tris. In addition, this assignment is supported by the observation by Clarke [42] of a single band at 1560 cm^{-1} in the infrared spectrum of solid sodium–palmitate.

The amide I and II bands of PLA$_2$ cannot be observed directly during monolayer hydrolysis, which further supports its proposed organization depicted in Fig. 7.1d, where only a small hydrophobic stretch of the protein penetrates the monolayer to perform hydrolysis of the *sn*-2 phospholipid acyl chain. However, the presence of PLA$_2$ at the monolayer level can be demonstrated when the spectrum of the phospholipid monolayer is subtracted from the same monolayer in the presence of PLA$_2$. Indeed, an amide I band at 1660 cm^{-1} and an amide II band at 1542 cm^{-1} can be seen in Fig. 7.6. The amide I band is widely used to determine the secondary structure of proteins [43, 44]. For example, the position of α-helices (1648–1657 cm^{-1}), β-sheets (1623–1641 and 1674–1695 cm^{-1}), turns (1662–1686 cm^{-1}), and disordered structures (1642–1657 cm^{-1}) have previously been attributed [43, 44]. The position of the amide I band suggests that this PLA$_2$ contains a large number of α-helices.

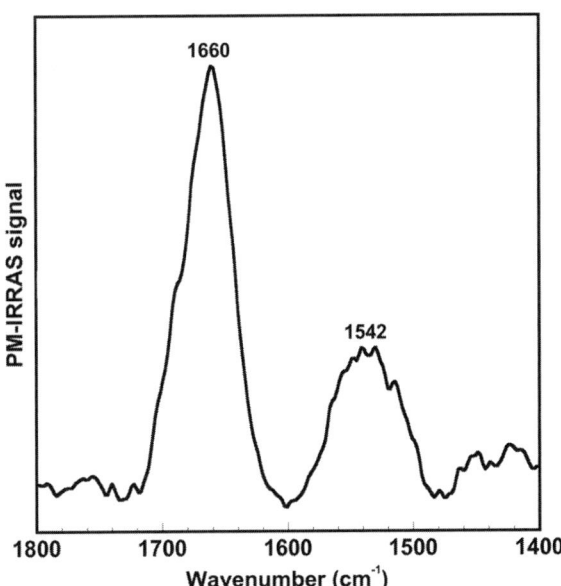

Figure 7.6 Difference PM-IRRAS spectrum: spectrum of PLA$_2$ adsorbed to a mixed monolayer of 6:0/16:0-PC, 16:0-PC, and 16:0 (20:40:40)—spectrum of the same mixed monolayer in absence of PLA$_2$. Subphase concentration of PLA$_2$ is 1 unit/mL. Subphase is the same as in Fig. 7.4.

7.3 MONITORING OF LECITHIN RETINOL ACYLTRANSFERASE MONOLAYER BINDING AND HYDROLYSIS BY INFRARED SPECTROSCOPY

Lecithin retinol acyltransferase (LRAT) is a 25.3-kDa protein [45] that has been shown to be expressed in several tissues including testis, liver, intestine, and the retinal pigment epithelium (RPE) [46]. It is likely involved in the mobilization and storage of vitamin A [47]. It catalyzes hydrolysis of the sn-1 acyl chain of phospholipids and transfers this acyl group to all-trans retinol [48–51] to generate all-trans retinyl esters. The analysis of the amino acid sequence of LRAT suggests two possible transmembrane α-helices, in the N- and C-terminal regions (Fig. 7.1a) at positions 9–31 and 195–222 [45]. However, recent data suggest that only the C-terminal α-helix is necessary for membrane targeting [52], which rather suggests that the configuration depicted in Fig. 7.1b is valid for LRAT. Until now, the full sequence of LRAT with its N- and C-terminal α-helices could not yet be successfully solubilized after its overexpression [53] presumably because of the hydrophobicity of its N- and C-terminal segments. A recombinant truncated form of LRAT (tLRAT) (amino acids 31–196) has thus been produced, where these transmembrane regions have been removed [53]. The active site of LRAT has three essential residues (C161, Y154, and H60), all of which are conserved in the LRAT family of enzymes [54]. This active site is located within the sequence of tLRAT, which was thus shown to be fully active [53]. In order to hydrolyze membrane phospholipids, this active site of tLRAT must interact with membranes. However, information on the membrane binding of tLRAT in the absence of its hydrophobic segments is lacking. Phospholipid monolayers were thus used to determine the extent of binding of this recombinant tLRAT as well as its hydrolytic activity in monolayers.

Binding of tLRAT onto phospholipid monolayers can be observed directly by the measurement of the increase in surface pressure after the injection of the protein into the monolayer subphase as shown in Fig. 7.7a. The adsorption of tLRAT onto monolayers of 1,2-dioleoyl-sn-glycero-3-phosphatidylethanolamine (DOPE, 18:1/18:1-PE), 1,2-dioleoyl-sn-glycero-3-phosphatidylserine (DOPS, 18:1/18:1-PS), and 1,2-dioleoyl-sn-glycero-3-phosphatidylcholine (DOPC, 18:1/18:1-PC) at an initial surface pressure of 15 mN/m is shown in Fig. 7.7a. These phospholipids were selected because they show only one physical state at room temperature and the oleoyl fatty acyl chains are one of the three main fatty acyl chains transferred to all-trans retinol by LRAT [51]. Moreover, the use of the same fatty acyl chains for these three phospholipids allowed evaluation of the preference of tLRAT for a phospholipid headgroup. It can be seen that the kinetics and extent of adsorption of tLRAT is much larger in the presence of 18:1/18:1-PS than in the presence of 18:1/18:1-PC or 18:1/18:1-PE (Fig. 7.7a), which suggests that charge–charge interactions are involved in the membrane binding of tLRAT. Moreover, the adsorption maximum is reached within less than 7 min. The extent of adsorption of tLRAT is very similar in the presence of 18:1/18:1-PE and 18:1/18:1-PC. Monolayer hydrolysis cannot

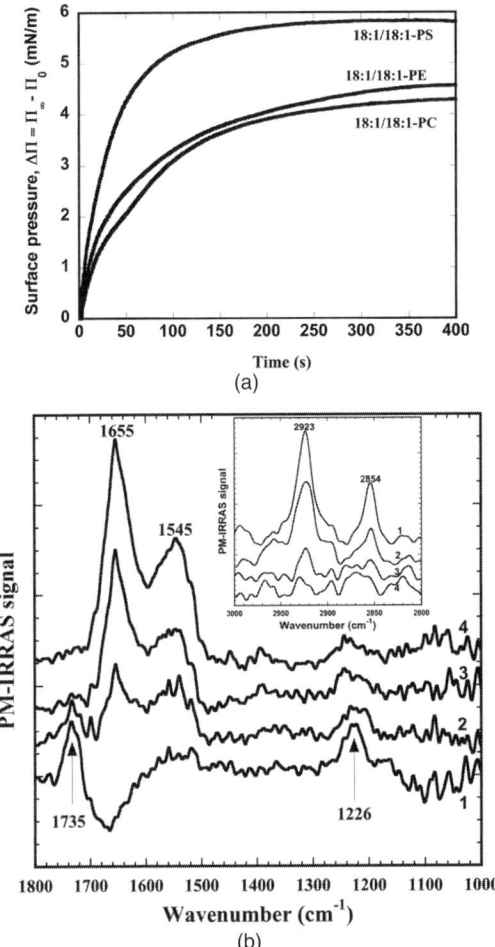

Figure 7.7 The truncated form of the human LRAT gene has been cloned in the vector pET11a as described previously [53]. Overexpression was achieved using *Escherichia coli* Bl21 (DE3) pLysS. Purification was performed with a His-Trap column. The SDS-PAGE electrophoresis and Western blot analyses of purified tLRAT allowed researchers to estimate the purity of tLRAT to more than 95%. (a) Effect of different phospholipid headgroups (18:1/18:1-PC, 18:1/18:1-PE, and 18:1/18:1-PS; Avanti Polar Lipids) on the adsorption of tLRAT at an initial surface pressure of 15 mN/m. A sample of 0.5 μg of tLRAT was injected into the subphase containing 500 μL of 50-mM phosphate buffer at pH 7. (b) PM-IRRAS spectra in the 1000–1800-cm^{-1} region of a 14:0/14:0-PC monolayer at an initial surface pressure of 15 mN/m before (spectrum 1) and 2–11 min (spectrum 2), 21–30 min (spectrum 3), and 36–45 min (spectrum 4) after tLRAT injection into the subphase. A sample of 50 μg of tLRAT was injected into the subphase containing 25 mL of 50-mM carbonate buffer at pH 9.2. (Inset) PM-IRRAS spectra in the 2800–3000-cm^{-1} region of a 14:0/14:0-PC monolayer at an initial surface pressure of 15 mN/m before (spectrum 1) and 2–11 min (spectrum 2), 21–30 min (spectrum 3), and 36–45 min (spectrum 4) after tLRAT injection into the same subphase as described above.

readily be observed using these phospholipids because the hydrolyzed oleic acid is insoluble in the buffer subphase. Therefore infrared spectroscopy has been used to follow hydrolysis of phospholipid monolayers by tLRAT in a manner similar to PLA_2.

Figure 7.7b shows the PM-IRRAS spectra in the 1000–1800-cm^{-1} region of a pure dimyristoyl phosphatidylcholine (DMPC; 14:0/14:0-PC) monolayer at 15 mN/m before (spectrum 1) and after (spectra 2–4) the injection of tLRAT into the monolayer subphase [55]. Surface pressure increased from 15 to a stable value of 20 mN/m within the 9-min measuring period of the first spectrum after tLRAT injection into the subphase (spectrum 2, Fig. 7.7b). The 14:0/14:0-PC has been chosen to allow observation of monolayer hydrolysis by tLRAT. Indeed, it has previously been shown that both hydrolysis products resulting from PLA_2 activity on a 14:0/14:0-PC monolayer are readily soluble into the monolayer subphase (myristic acid (14:0) and lysomyristoyl phosphatidylcholine (14:0-PC)) [56]. Given that the same hydrolysis products are obtained from the tLRAT enzymatic activity, one could expect to observe a decrease in the intensity of the typical infrared bands of 14:0/14:0-PC upon monolayer hydrolysis. As can be seen in Fig. 7.7b (spectrum 1), the v-C=O ester stretching band and the v_a-P=O band of pure 14:0/14:0-PC are located at 1735 and 1226 cm^{-1}, respectively, which is consistent with previous reports [33] as well as with the bands observed for 16:0/16:0-PC (spectrum 1, Fig. 7.4). The intensity of these two bands decreases after the injection of tLRAT into the subphase. The first spectrum measured soon after tLRAT injection into the subphase (spectrum 2, Fig. 7.7b) shows a significant decrease of the intensity of the C=O ester and P=O bands. Longer adsorption times lead to a much larger extent of monolayer hydrolysis as demonstrated by the further decrease in the intensity of these 14:0/14:0-PC infrared bands (spectra 3 and 4, Fig. 7.7b). The hydrolysis of the 14:0/14:0-PC monolayer is further confirmed by the observation of a similar decrease in the intensity of the antisymmetric and symmetric CH_2 stretching bands of the 14:0/14:0-PC monolayer (inset of Fig. 7.7b). The position of these bands, which are located, respectively, at 2923 and 2854 cm^{-1}, is consistent with previous observations [57]. It is noteworthy to mention that the calcium–palmitate bands at 1575 and 1537 cm^{-1} observed during PLA_2 monolayer hydrolysis (Figs. 7.4 and 7.5) can not be seen in Fig. 7.7b . This can be explained by the fact that no calcium was present in the subphase in our experiments with tLRAT. Moreover, LRAT is known to form a covalent bond with the hydrolyzed fatty acyl chain until it is transferred to retinol [54], which would thus prevent formation of the calcium–palmitate complex. In parallel to monolayer hydrolysis, the adsorption of tLRAT onto the 14:0/14:0-PC monolayer can be observed directly by the appearance of the amide I and II bands of tLRAT at 1655 and 1545 cm^{-1}, respectively (spectra 2–4, Fig. 7.7b). The position of the amide I band of tLRAT suggests that it contains a significant amount of α-helices [58, 59]. The direct observation of these typical protein bands contrasts with the fact that subtraction of the phospholipid monolayer was necessary to observe PLA_2 amide bands (Fig. 7.6). These data thus further demonstrate that tLRAT significantly penetrates the phospholipid monolayer even though its

N- and C-terminal α-helices have been removed. In addition, it can be seen that the more tLRAT is adsorbed, the larger is the intensity of the amide I and II bands, and the more hydrolyzed is the monolayer. In future experiments, we will attempt to overexpress a tLRAT where only the N-terminal α-helix will be removed given that only the C-terminal α-helix is necessary for membrane targeting [52]. It can be postulated that very different kinetics of adsorption should be obtained when this C-terminal α-helix is present.

7.4 MONITORING OF BINDING OF AN ACYLATED PROTEIN ONTO PHOSPHOLIPID MONOLAYERS BY SURFACE PRESSURE MEASUREMENTS

Recoverin is a 23-kDa calcium-binding protein from retinal rod cells of vertebrates [60] that is involved in the visual phototransduction cascade. It is a member of the EF-hand superfamily of proteins. Only two of its four EF-hand motifs, a helix–loop–helix of 29 residues, bind Ca^{2+} [61–63]. This structural feature was called EF-hand because this motif was originally described for the helices five (E) and six (F) of the carp parvalbumin [61]. Recoverin is N-acylated predominantly by an amino-terminal myristoyl group [60, 64–66]. Structural analyses of myristoylated recoverin gave rise to the concept of the Ca^{2+}–myristoyl switch [62, 67, 68], where the occlusion and extrusion of the myristoyl group of recoverin is induced by Ca^{2+} ions. Indeed, binding of two Ca^{2+} ions by recoverin induces the extrusion of its myristoyl group [60, 67, 69] whereas the dissociation of Ca^{2+} results in its sequestration in a hydrophobic cleft [63, 68, 70]. Myristoylated and nonmyristoylated recoverin have been overexpressed and purified. Their kinetics of binding onto phospholipid monolayers have then been assayed by performing measurements of surface pressure as a function of time. Kinetics is started as soon as myristoylated or nonmyristoylated recoverin is injected into the subphase in the presence or the absence of calcium (Fig. 7.8) [71]. It can be seen that the slowest kinetics of adsorption is observed with myristoylated recoverin in the absence of calcium (in the presence of EGTA). In this case, the adsorption of myristoylated recoverin onto a 14:0/14:0-PC monolayer leads to a surface pressure increase from 5 to 12 mN/m within 6500 seconds. However, the adsorption of myristoylated recoverin onto the 14:0/14:0-PC monolayer is drastically accelerated by the presence of calcium ions, which leads to an increase of surface pressure from 6 to 16.5 mN/m within only 200 seconds (see inset of Fig. 7.8). Interestingly, the $\Pi-t$ adsorption isotherm of the nonmyristoylated recoverin onto a 14:0/14:0-PC monolayer is also accelerated by the presence of calcium. However, this effect is smaller than in the case of myristoylated recoverin in the presence of calcium (see inset of Fig. 7.8). Indeed, an increase from 6 to 14.5 mN/m within 1000 seconds is observed for recoverin in the absence of its myristoyl moiety. This result suggests that nonmyristoylated recoverin underwent a conformational change following calcium binding, which exposed amino acids that increased its affinity for the phospholipid monolayer. The trend of these curves has been fitted successfully with a

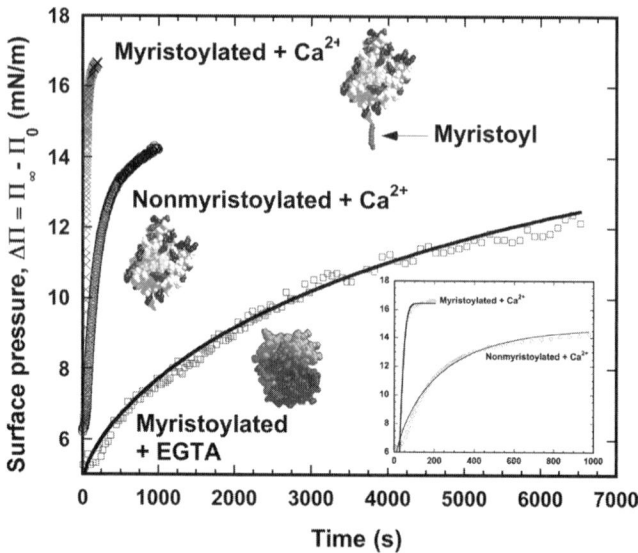

Figure 7.8 Bovine recoverin has been cloned from bovine retina into the pET11a plasmid (Novagen). Nonmyristoylated and myristoylated recoverin were expressed and purified essentially as reported by Ray et al. [79] and modified by Desmeules et al. [80] using the *E. coli* strain BL21 (DE3) pLysS and a single-step procedure using the calcium-dependent binding of recoverin to low-substituted phenyl sepharose. The purity of recoverin was at least 99% as judged by gel electrophoresis and Coomassie blue staining. Myristoylation of recoverin has been quantified by reverse phase high pressure liquid chromatography using a hydrophobic column Jupiter 5μ C4 300 Å (Phenomenex). The trough used in these experiments was made of glass because no adsorption was observed at the air–water interface when using Teflon® covered troughs due to the irreversible binding of recoverin to this hydrophobic material. $\Pi-t$ adsorption isotherms are shown for myristoylated and nonmyristoylated recoverin onto a 14:0/14:0-PC monolayer ($\Pi_0 = 5$ mN/m) in the presence (myristoylated + Ca^{2+} and nonmyristoylated + Ca^{2+}) and absence of calcium (myristoylated + EGTA) into the subphase. The time zero corresponds to the injection of recoverin into the subphase. In all cases, the final concentration of recoverin was 50 nM. The subphase was 1-mM Hepes, pH 7.5, 100-mM NaCl, and 1-mM $CaCl_2$ (or 1-mM EGTA). (Inset) To better appreciate the kinetics of adsorption of myristoylated and nonmyristoylated recoverin in the presence of calcium, the x axis has been extended to 1000 s. The stretched exponential used to fit the experimental data is $\Pi_t = \Pi_\infty - \Pi_0 e^{-(kt)\beta}$, where Π_t is the surface pressure of the monolayer at time t, Π_∞ is the surface pressure of the monolayer at equilibrium, Π_0 is the initial surface pressure of the monolayer, k is the rate coefficient, and β is the exponential scaling factor.

stretched exponential (Fig. 7.8) that allowed researchers to quantify and compare the rate of adsorption of recoverin under these different conditions. The rate coefficients have been calculated for myristoylated recoverin in the presence (0.028 s^{-1}) or in the absence (0.00017 s^{-1}) of calcium as well as for nonmyristoylated recoverin in the presence of calcium (0.0048 s^{-1}). These rate constant ratios indicate that the adsorption of myristoylated recoverin in the presence of calcium is 165 times faster than in the absence of calcium (0.028 s^{-1}/0.00017 s^{-1}), whereas the adsorption of nonmyristoylated recoverin in the presence of calcium is 30 times faster than myristoylated recoverin in the absence of calcium (0.0048 s^{-1}/0.00017 s^{-1}). It is important to stress that the observed large and quick increase in surface pressure following recoverin injection into the subphase does not result from protein denaturation upon adsorption at the lipid–water interface. Indeed, as shown in Section 7.5, the position and shape of the amide I and II bands of recoverin adsorbed onto phospholipid monolayers are very similar to those of recoverin in solution (compare spectrum 6 with spectra 1–5, Fig. 7.10), which clearly demonstrates that it is not denatured in our experiments.

These results can be interpreted on the basis of the NMR and X-ray diffraction structural data obtained for myristoylated and nonmyristoylated recoverin in the presence and the absence of calcium [63, 67, 68]. In the absence of calcium, the NMR study in solution has shown that the myristoyl group of calcium-free myristoylated recoverin is buried inside the protein in contact with a deep hydrophobic cleft formed by residues F23, W36, Y53, F56, F57, and Y86 [63, 67, 68]. The presence of this hydrophobic cleft has also been demonstrated for the calcium-free nonmyristoylated recoverin by X-ray crystallography [63]. Consequently, recoverin in its calcium-free state has less peripheral hydrophobic residues in contact with the solvent than the calcium-bound state. Indeed, it has been shown that Ca^{2+} binding to recoverin leads to the unclamping and extrusion of the myristoyl group but also to a 45° rotation of the N-terminal domain relative to the carboxy-terminal domain, thus exposing hydrophobic residues F23, W36, Y53, F56, F57, and Y86 to the solvent [67, 69, 72]. Furthermore, NMR and circular dichroism studies have shown that nonmyristoylated recoverin undergoes a calcium-modulated structural change very similar to that of myristoylated recoverin upon calcium binding [67, 69, 73]. The results presented in Fig. 7.8 thus suggest that the higher rate (30-fold) of adsorption of nonmyristoylated recoverin onto the 14:0/14:0-PC monolayer in the presence of calcium compared to the calcium-free myristoylated recoverin can be explained by the interaction of the exposed hydrophobic region of recoverin with the phospholipid monolayer in the presence of calcium. These data thus strongly suggest that the increased rate of adsorption of myristoylated recoverin compared to nonmyristoylated recoverin in the presence of calcium (0.028 s^{-1}/0.0048 s^{-1} ≈ 5.8-fold) arises solely from the myristoyl moiety of recoverin. In others words, our results indicate that the increase in surface hydrophobicity of recoverin reduces the kinetic barrier of its adsorption to phospholipid monolayers.

The phospholipid headgroup and acyl chain preference of myristoylated recoverin has also been measured by the increase in surface pressure as a function of time. The extent of monolayer binding by myristoylated recoverin can be determined by plotting $\prod_\infty - \prod_0$ (surface pressure at equilibrium − initial surface pressure) as a function of the initial surface pressure ($\prod_0$) as shown in the inset

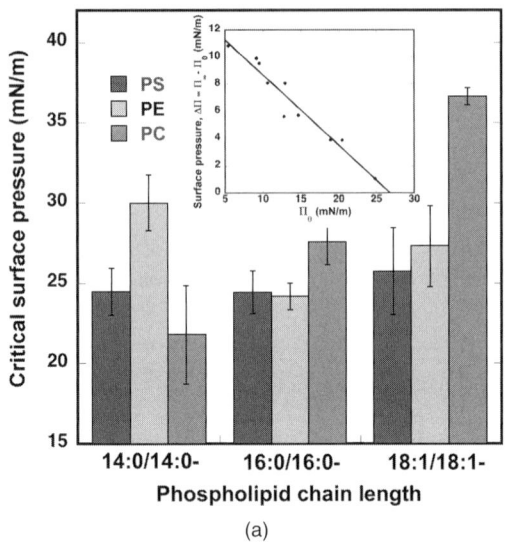

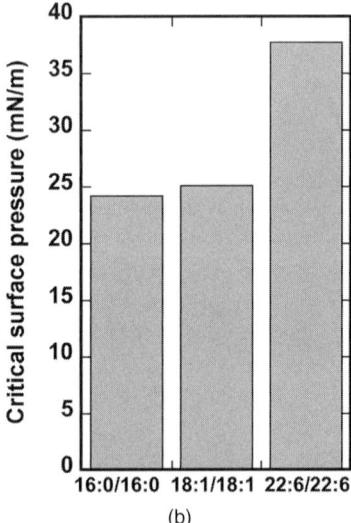

Figure 7.9 Histograms of the critical surface pressure of binding of myristoylated recoverin onto different phospholipid monolayers. (a) The phospholipids used are 14:0/14:0-PC, 16:0/16:0-PC, and 18:1/18:1-PC, -PE, and -PS (PC, phosphatidylcholine; PE, phosphatidylethanolamine; PS, phosphatidylserine). (Inset) Critical surface pressure has been determined by extrapolating the plot of $\prod_\infty - \prod_0$ as a function of $\prod_0$, where the curve reaches the x axis. The subphase was 1-mM Hepes, pH 7.5, 100-mM NaCl, 1-mM $CaCl_2$, and 5-mM β-mercaptoethanol. The final concentration of recoverin was 52 nM. All measurements were performed using the 500-µL multiwell trough from Kibron. Phospholipids were spread using a chloroform solution for PC and PE, whereas PS was solubilized in hexane: ethanol (9:1). (b) The phospholipids used are 16:0/16:0-PE, 18:1/18:1-PE, and 22:6ω3/22:6ω3−PE. Other conditions are the same as in (a).

of Fig. 7.9a. The critical surface pressure is the intercept with the x axis, which corresponds to the maximum surface pressure of protein binding [74]. Figure 7.9a shows a histogram of the critical surface pressure of binding of myristoylated recoverin onto different phospholipids including phosphatidylcholine (PC), phosphatidylserine (PS), and phosphatidylethanolamine (PE) with different fatty acyl chains (14:0/14:0-PC, -PS, -PE; 16:0/16:0-PC, -PS, -PE; 18:1/18:1-PC, -PS, -PE). It can be seen that for the 14:0/14:0 species, myristoylated recoverin shows a larger extent of binding in the presence of phosphatidylethanolamine whereas phosphatidylcholine and phosphatidylserine show similar extents of binding. In contrast, for the 16:0/16:0 and 18:1/18:1 species, myristoylated recoverin shows a larger extent of binding in the presence of PC. In fact, the extent of binding of myristoylated recoverin increases with PC chain length, which is not true for PE and PS. This can be explained by the fact that PE and PS form hydrogen bonds between their individual polar headgroups that create a stronger barrier for the myristoyl group of recoverin to penetrate the membrane [75]. In addition, a stronger effect of chain length can be observed for PE with a phospholipid bearing very long fatty acyl chains. Indeed, as can be seen in Fig. 7.9b, a much larger monolayer binding is observed for myristoylated recoverin in the presence of didocosahexaenoyl PE ($22:6\omega3/22:6\omega3$-PE). In fact, the critical surface pressure obtained with this phospholipid (37.8 mN/m) is in the range presumed for membrane lateral pressure (30–35 mN/m) [76]. Docosahexaenoyl ($22:6\omega3$) comprises up to 50% rod photoreceptor fatty acids [77]. The observation of a larger extent of recoverin binding in the presence of this polyunsaturated phospholipid suggests that photoreceptor membrane fluidity favors recoverin binding.

7.5 MONITORING OF BINDING OF AN ACYLATED PROTEIN ONTO PHOSPHOLIPID MONOLAYERS BY INFRARED SPECTROSCOPY

Typical spectra recorded during myristoylated recoverin adsorption onto a 14:0/14:0-PC monolayer are presented in Fig. 7.10 [71]. Two broad bands stand out in each spectrum corresponding, respectively, to the amide I (centered at 1655 cm^{-1}) and amide II (at approximately 1550 cm^{-1}) bands of recoverin. It can be seen that the intensity of the amide I band increases with surface pressure (Fig. 7.10). The position and shape of the amide I band (1655 cm^{-1}) of myristoylated recoverin in the presence of calcium reveal a major proportion of α-helices [24, 38, 44], which is in agreement with its known NMR structure in solution [68, 70]. Moreover, circular dichroism has shown that myristoylated as well as nonmyristoylated recoverin contains 65% α-helices [68].

The infrared spectrum of myristoylated recoverin in aqueous solution has been measured using the stage of the Golden Gate by attenuated total reflection (ATR) (spectrum 6, Fig. 7.10). The Golden Gate is a versatile infrared sampling system that is made of type IIa diamond, which has been selected for its high sensitivity as a single reflection ATR. The position of the amide I band (1646 cm^{-1})

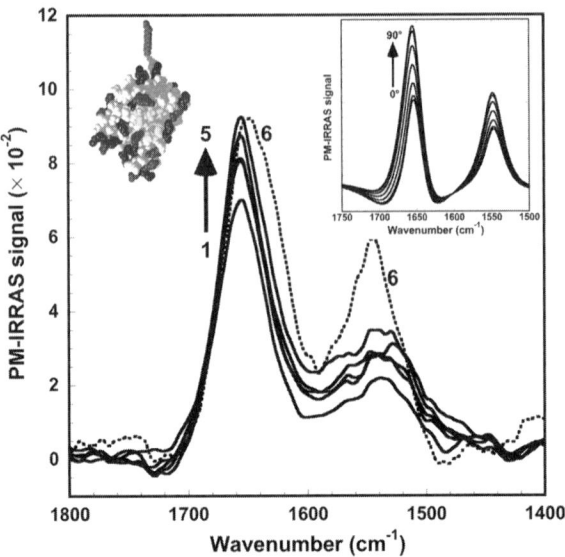

Figure 7.10 PM-IRRAS spectra of myristoylated recoverin during its adsorption onto a 14:0/14:0-PC monolayer in the presence of calcium ($\Pi_0 = 5$ mN/m). Spectra 1 (5–11.2 mN/m), 2 (11.5–12.7 mN/m), 3 (12.7–12.8 mN/m), 4 (12.8–12.9 mN/m), and 5 (12.9–13 mN/m). Spectrum 6 has been measured using the stage of the Golden Gate with a drop (5 μL) of a solution of recoverin at a concentration of 2.5 μg/mL. Each PM-IRRAS spectrum was obtained after 9 min of acquisition. The subphase was 1-mM Hepes, pH 7.5, 100-mM NaCl, and 1-mM CaCl$_2$. The final concentration of recoverin was 50 nM. These measurements were performed with a polarizer positioned in front of the detector as described earlier [81]. (Inset) Simulated spectra of myristoylated recoverin at different orientations (from theta = $0°$ to $90°$ with respect to the z axis) on the basis of its NMR structure [70]. The z axis has been taken as the normal to α-helix 2 of recoverin (as well as several other α-helices). The orientation of recoverin deduced from the simulation of the spectra is also shown in the figure. The plane of the monolayer is perpendicular to the main axis of recoverin.

together with the observation of an increased bandwidth on the low frequency part indicates a slightly larger disordering of recoverin secondary structure in solution compared to when it is bound to lipid monolayers. Moreover, it has been previously shown that information can be obtained on the orientation of an α-helical peptide from the measurement of the amide I/amide II ratio [25]. The discrepancy between the ratio of the amide I/amide II bands observed in the ATR spectrum (1.6; spectrum 6, Fig. 7.10) and that in the PM-IRRAS spectra (2.7; spectra 1–5, Fig. 7.10) suggests that recoverin has an anisotropic organization at the phospholipid–buffer interface in contrast to being randomly oriented in solution. Simulations of spectra 1–5 of recoverin have been performed (inset of Fig. 7.10) in order to estimate its orientation by comparing the experimental amide I/amide II ratio with those of the simulated spectra. It can be seen that the spectrum with an amide I/amide II ratio that most closely corresponds to the experimental value of 2.7 (spectra 1–5, Fig. 7.10) is that where recoverin has an orientation of 0° (Fig. 7.10)—that shown in Fig. 7.3a when recoverin is adsorbed onto the phospholipid monolayer. In addition, such a high amide I/amide II ratio suggests that most of the α-helices of recoverin are preferentially oriented parallel to the water surface, which fits well with the orientation of recoverin shown in Fig. 7.3a as well as with the model proposed by Valentine et al. [72], who have estimated the orientation of myristoylated recoverin bound to bicelles.

7.6 CONCLUSION

Membrane-associated proteins play essential functions in different cellular processes. It is thus very important to improve our understanding of the mechanism of their binding to membranes. These proteins can bind membranes by means of a small hydrophobic stretch as in the case of venom PLA_2, by N- and/or C-terminal α-helices as for LRAT, or by a covalently attached fatty acylation like in the case of recoverin. PLA_2 binding has no effect on phospholipid monolayer surface pressure, which supports its postulated binding mechanism. Nevertheless, it efficiently hydrolyzes phospholipid monolayers to produce calcium–palmitate and lysophospholipid hydrolysis products as demonstrated by the measurement of infrared spectra during PLA_2 monolayer hydrolysis. Injection of a truncated form of LRAT (tLRAT) into the phospholipid monolayer subphase induces a significant increase in surface pressure even though its N- and C-terminal α-helices have been removed. Moreover, a stronger binding can be observed in the presence of phosphatidylserine, which suggests that charge–charge interactions are involved because this phospholipid bears a negative charge at neutral pH. Its binding and hydrolytic action on phospholipid monolayers have also been monitored by infrared spectroscopy. Indeed, the presence of the tLRAT amide I and II bands demonstrated that it protudes into the monolayer. Moreover, the decrease in the intensity of the bands corresponding to the soluble hydrolysis products of 16:0/16:0-PC allowed researchers to confirm the hydrolytic activity of tLRAT. Finally, the kinetics of monolayer binding by myristoylated recoverin in the presence of calcium was much faster than nonmyristoylated recoverin in the presence of calcium as well as myristoylated recoverin in the absence of calcium. These measurements allowed researchers to draw conclusions on the importance of hydrophobic residues and the myristoyl moiety on recoverin monolayer binding. Moreover, the preferential binding of phosphatidylcholine and the long polyunsaturated 22:6ω3 fatty acids from critical surface pressure measurements suggests that the presence of these phospholipids in photoreceptor membranes should favor recoverin binding. The extensive binding of myristoylated recoverin was further confirmed by infrared spectroscopy and simulations of these spectra allowed estimation of its orientation relative to the monolayer surface.

ACKNOWLEDGMENTS

The authors are indebted to the Natural Sciences and Engineering Research Council of Canada (NSERC) for financial support. CS is a Chercheur boursier national of the Fonds de la Recherche en Santé du Québec (FRSQ). PD holds a joint scholarship from the CIHR and the Gimbel Eye Foundation. CS also acknowledges support from the Centre National de la Recherche Scientifique during his sabbatical leave at the Université de Bordeaux (France).

REFERENCES

1. P. L. Yeagle and A. D. Albert (2006). G-protein coupled receptor structure. *Biochim. Biophys. Acta* **1768**: 808–824.
2. I. M. Pepe (2001). Recent advances in our understanding of rhodopsin and phototransduction. *Prog. Retin. Eye Res.* **20**: 733–759.
3. Y. Shichida and H. Imai (1998). Visual pigment: G-protein-coupled receptor for light signals. *Cell. Mol. Life Sci.* **54**: 1299–1315.
4. H. Van den Bosch (1980). Intracellular phospholipases A. *Biochim. Biophys. Acta* **604**: 191–246.
5. T. D. Lamb and E. N. Pugh, Jr. (2004). Dark adaptation and the retinoid cycle of vision. *Prog. Retin. Eye Res.* **23**: 307–380.
6. J. K. McBee, K. Palczewski, W. Baehr, and D. R. Pepperberg (2001). Confronting complexity: the interlink of phototransduction and retinoid metabolism in the vertebrate retina. *Prog. Retin. Eye Res*. **20**: 469–529.
7. T. Magee and M. C. Seabra (2005). Fatty acylation and prenylation of proteins: what's hot in fat. *Curr. Opin. Cell. Biol.* **17**: 190–196.
8. J. T. Dunphy and M. E. Linder (1998). Signalling functions of protein palmitoylation. *Biochim. Biophys. Acta* **1436**: 245–261.
9. M. D. Resh (1999). Fatty acylation of proteins: new insights into membrane targeting of myristoylated and palmitoylated proteins. *Biochim. Biophys. Acta* **1451**: 1–16.
10. M. D. Resh (2004). Membrane targeting of lipid modified signal transduction proteins. *Subcell. Biochem.* **37**: 217–232.
11. A. V. Finkelstein and J. Janin (1989). The price of lost freedom: entropy of bimolecular complex formation. *Protein Eng.* **3**: 1–3.
12. D. Murray, N. Ben-Tal, B. Honig, and S. McLaughlin (1997). Electrostatic interaction of myristoylated proteins with membranes: simple physics, complicated biology. *Structure* **5**: 985–989.
13. J. R. Silvius and M. J. Zuckermann (1993). Interbilayer transfer of phospholipid-anchored macromolecules via monomer diffusion. *Biochemistry* **32**: 3153–3161.
14. S. Grenier, P. Desmeules, A. K. Dutta, A. Yamazaki, and C. Salesse (1998). Determination of the depth of penetration of the alpha subunit of retinal G protein in membranes: a spectroscopic study. *Biochim. Biophys. Acta* **1370**: 199–206.
15. M. Liden and U. Eriksson (2006). Understanding retinol metabolism: structure and function of retinol dehydrogenases. *J. Biol. Chem.* **281**: 13001–13004.

16. P. Dynarowicz-Latka, A. Dhanabalan, and O. N. Oliveira, Jr. (2001). Modern physicochemical research on Langmuir monolayers. *Adv. Colloid Interface Sci.* **91**: 221–293.
17. V. M. Kaganer, H. Mohwald, and P. Dutta (1999). Structure and phase transitions in Langmuir monolayers. *Rev. Modern Phys.* **71**: 779–820.
18. R. Maget-Dana (1999). The monolayer technique: a potent tool for studying the interfacial properties of antimicrobial and membrane-lytic peptides and their interactions with lipid membranes. *Biochim. Biophys. Acta* **1462**: 109–140.
19. J. Boucher, E. Trudel, M. Methot, P. Desmeules, and C. Salesse (2007). Organization, structure and activity of proteins in monolayers. *Colloids Surf. B Biointerfaces* **58**: 73–90.
20. D. Blaudez, T. Buffeteau, J. C. Cornut, B. Desbat, N. Escafre, M. Pezolet, and J. M. Turlet (1994). Polarization modulation FTIR spectroscopy at the air–water interface. *Thin Solid Films* **242**: 146–150.
21. T. Buffeteau, B. Desbat, and J. M. Turlet (1991). Polarization modulation FT-IR spectroscopy of surfaces and ultra-thin films: experimental procedure and quantitative analysis. *Appl. Spectrosc.* **45**: 380–389.
22. E. Bellet-Amalric, D. Blaudez, B. Desbat, F. Graner, F. Gauthier, and A. Renault (2000). Interaction of the third helix of *Antennapedia* homeodomain and a phospholipid monolayer, studied by ellipsometry and PM-IRRAS at the air–water interface. *Biochim. Biophys. Acta* **1467**: 131–143.
23. S. Castano, B. Desbat, and J. Dufourcq (2000). Ideally amphipathic beta-sheeted peptides at interfaces: structure, orientation, affinities for lipids and hemolytic activity of (KL)(m)K peptides. *Biochim. Biophys. Acta* **1463**: 65–80.
24. S. Castano, B. Desbat, M. Laguerre, and J. Dufourcq (1999). Structure, orientation and affinity for interfaces and lipids of ideally amphipathic lytic $L_i K_j (i = 2j)$ peptides. *Biochim. Biophys. Acta* **1416**: 176–194.
25. I. Cornut, B. Desbat, J. M. Turlet, and J. Dufourcq (1996). In situ study by polarization modulated Fourier transform infrared spectroscopy of the structure and orientation of lipids and amphipathic peptides at the air–water interface. *Biophys. J.* **70**: 305–312.
26. C. R. Flach, F. G. Prendergast, and R. Mendelsohn (1996). Infrared reflection–absorption of melittin interaction with phospholipid monolayers at the air/water interface. *Biophys. J.* **70**: 539–546.
27. J. Gallant, B. Desbat, D. Vaknin, and C. Salesse (1998). Polarization-modulated infrared spectroscopy and X-ray reflectivity of photosystem II core complex at the gas–water interface. *Biophys. J.* **75**: 2888–2899.
28. H. Lavoie, B. Desbat, D. Vaknin, and C. Salesse (2002). Structure of rhodopsin in monolayers at the air–water interface: a PM-IRRAS and X-ray reflectivity study. *Biochemistry* **41**: 13424–13434.
29. A. Meister, C. Nicolini, H. Waldmann, J. Kuhlmann, A. Kerth, R. Winter, and A. Blume (2006). Insertion of lipidated Ras proteins into lipid monolayers studied by infrared reflection absorption spectroscopy (IRRAS). *Biophys. J.* **91**: 1388–1401.
30. F. Wu, A. Gericke, C. R. Flach, T. R. Mealy, B. A. Seaton, and R. Mendelsohn (1998). Domain structure and molecular conformation in annexin V/1,2-dimyristoyl-*sn*-glycero-3-phosphate/Ca^{2+} aqueous monolayers: a Brewster angle microscopy/infrared reflection–absorption spectroscopy study. *Biophys. J.* **74**: 3273–3281.

31. A. Gericke and H. Hühnerfuss (1994). IR reflection absorption spectroscopy: a versatile tool for studying interfacial enzymatic processes. *Chem. Phys. Lipids* **74**: 205–210.

32. M. Grandbois, B. Desbat, D. Blaudez, and C. Salesse (1999). Polarization-modulated infrared absorption spectroscopy measurement of phospholipid monolayer hydrolysis by phospholipase C. *Langmuir* **15**: 6594–6597.

33. M. Grandbois, B. Desbat, and C. Salesse (2000). Monitoring of phospholipid monolayer hydrolysis by phospholipase A_2 by use of polarization-modulated Fourier transform infrared spectroscopy. *Biophys. Chem.* **88**: 127–135.

34. X. Wang, S. Zheng, Q. He, G. Brezesinski, H. Mohwald, and J. Li (2005). Hydrolysis reaction analysis of L-alpha-distearoylphosphatidylcholine monolayer catalyzed by phospholipase A_2 with polarization-modulated infrared reflection absorption spectroscopy. *Langmuir* **21**: 1051–1054.

35. D. Blaudez, T. Buffeteau, J. C. Cornut, B. Desbat, N. Escafre, M. Pézelet, and J. M. Turlet (1993). Polarization-modulated FT-IR spectroscopy of a spread monolayer at the air–water interface. *Appl. Spectrosc.* **47**: 869–874.

36. D. Blaudez, J. M. Turlet, J. Dufourcq, D. Bard, T. Buffeteau, and B. Desbat (1996). Investigations at the air/water interface using polarization modulation IR spectroscopy. *J. Chem. Soc. Faraday. Trans.* **92**: 525–530.

37. R. A. Dluhy, N. A., Wright, and P. R. Griffiths (1988). In situ measurement of the FT-IR spectra of phospholipid monolayers at the air/water interface. *Appl. Spectrosc.* **42**: 138–141.

38. C. R. Flach, J. W. Brauner, J. W. Taylor, R. C. Baldwin, and R. Mendelsohn (1994). External reflection FTIR of peptide monolayer films *in situ* at the air/water interface: experimental design, spectra-structure correlations, and effects of hydrogen–deuterium exchange. *Biophys. J.* **67**: 402–410.

39. K. M. Maloney, M. Grandbois, D. W. Grainger, C. Salesse, K. A. Lewis, and M. F. Roberts (1995). Phospholipase A_2 domain formation in hydrolyzed asymmetric phospholipid monolayers at the air/water interface. *Biochim. Biophys. Acta* **1235**: 395–405.

40. D. J. Ahn and E. I. Franses (1992). Orientations of chain axes and transition moments in Langmuir–Blodgett monolayers determined by polarized FTIR-ATR spectroscopy. *J. Phys. Chem.* **96**: 9952–9959.

41. T. L. Marshbanks, J. A. Ahn, and E. I. Franses (1994). Transport and ion exchange in Langmuir–Blodgett films: water transport and film microstructure by attenuated total reflectance Fourier transform infrared spectroscopy. *Langmuir* **10**: 276–285.

42. P. E. Clarke (1994). Infrared spectroscopy. Introduction to surfactant analysis. *Blackie Academic & Professional* 266–276.

43. A. Barth and C. Zscherp (2002). What vibrations tell us about proteins. *Q. Rev. Biophys.* **35**: 369–430.

44. E. Goormaghtigh, V. Cabiaux, and J. M. Ruysschaert (1994). Determination of soluble and membrane protein structure by Fourier transform infrared spectroscopy. III. Secondary structures. *Subcell. Biochem.* **23**: 405–450.

45. A. Ruiz, A. Winston, Y.-H. Lim, B. A. Gilbert, R. R. Rando, and D. Bok (1999). Molecular and biochemical characterization of lecithin retinol acyltransferase. *J. Biol. Chem.* **274**: 3834–3841.

46. A. Ruiz and D. Bok (2000). Molecular characterization of lecithin-retinol acyltransferase. *Methods Enzymol*. **316**: 400–413.
47. A. Ruiz, M. H. Kuehn, J. L. Andorf, E. Stone, G. S. Hageman, and D. Bok (2001). Genomic organization and mutation analysis of the gene encoding lecithin retinol acyltransferase in human retinal pigment epithelium. *Invest. Ophthalmol. Vis. Sci.* **42**: 31–37.
48. R. J. Barry, F. J. Canada, and R. R. Rando (1989). Solubilization and partial purification of retinyl ester synthetase and retinoid isomerase from bovine ocular pigment epithelium. *J. Biol. Chem.* **264**: 9231–9238.
49. P. N. MacDonald and D. E. Ong (1988). Evidence for a lecithin-retinol acyltransferase activity in the rat small intestine. *J. Biol. Chem.* **263**: 12478–12482.
50. R. R. Rando (2001). The biochemistry of the visual cycle. *Chem. Rev.* **101**: 1881–1896.
51. J. C. Saari and D. L. Bredberg (1989). Lecithin:retinol acyltransferase in retinal pigment epithelial microsomes. *J. Biol. Chem.* **264**: 8636–8640.
52. A. R. Moise, M. Golczak, Y. Imanishi, and K. Palczewski (2007). Topology and membrane association of lecithin: retinol acyltransferase (LRAT). *J. Biol. Chem.* **282**: 2081–2090.
53. D. Bok, A. Ruiz, O. Yaron, W. J. Jahng, A. Ray, L. Xue, and R. R. Rando (2003). Purification and characterization of a transmembrane domain-deleted form of lecithin retinol acyltransferase. *Biochemistry* **42**: 6090–6098.
54. L. Xue and R. R. Rando (2004). Roles of cysteine 161 and tyrosine 154 in the lecithin-retinol acyltransferase mechanism. *Biochemistry* **43**: 6120–6126.
55. S. Bussières, T. Buffeteau, B. Desbat, R. Breton, and C. Salesse (2006). Secondary structure of a truncated form of lecithin-retinol acyltransferase in solution and evidence for its binding and hydrolytic action in monolayers. *Biochim. Biophys. Acta—Biomembranes* **1778**: 1324–1334.
56. D. W. Grainger, A. Reichert, H. Ringsdorf, C. Salesse, D. E. Davies, and J. B. Lloyd (1990). Mixed monolayers of natural and polymeric phospholipids: structural characterization by physical and enzymatic methods. *Biochim. Biophys. Acta* **1022**: 146–154.
57. R. A. Dluhy, K. E. Reilly, R. D. Hunt, M. L. Mitchell, A. J. Mautone, and R. Mendelsohn (1989). Infrared spectroscopic investigations of pulmonary surfactant. Surface film transitions at the air-water interface and bulk phase thermotropism. *Biophys J.* **56**: 1173–1181.
58. H. Lavoie, B. Desbat, D. Vaknin, and C. Salesse (2002). Structure of rhodopsin in monolayers at the air–water interface: a PM-IRRAS and X-ray reflectivity study. *Biochemistry* **41**: 13424–13434.
59. F. Wu, C. Flach, B. Seaton, T. Mealy, and R. Mendelsohn (1999). Stability of annexin V in ternary complexes with Ca^{2+} and anionic phospholipids: IR studies of monolayer and bulk phases. *Biochemistry* **38**: 792–799.
60. J. B. Ames, T. Tanaka, M. Ikura, and L. Stryer (1995). Nuclear magnetic resonance evidence for Ca^{2+}-induced extrusion of the myristoyl group of recoverin. *J. Biol. Chem.* **270**: 30909–30913.
61. R. D. Burgoyne (2004). The neuronal calcium-sensor proteins. *Biochim. Biophys. Acta* **1742**: 59–68.

62. A. M. Dizhoor, S. Ray, S. Kumar, G. Niemi, M. Spencer, D. Brolley, K. A. Walsh, P. P. Philipov, J. B. Hurley, and L. Stryer (1991). Recoverin: a calcium sensitive activator of retinal rod guanylate cyclase. *Science* **251**: 915–918.

63. K. M. Flaherty, S. Zozulya, L. Stryer, and D. B. McKay (1993). Three-dimensional structure of recoverin, a calcium sensor in vision. *Cell* **75**: 709–716.

64. A. M. Dizhoor, C. K. Chen, E. Olshevskaya, V. V. Sinelnikova, P. Phillipov, and J. B. Hurley (1993). Role of the acylated amino terminus of recoverin in Ca^{2+}-dependent membrane interaction. *Science* **259**: 829–832.

65. A. M. Dizhoor, L. H. Ericsson, R. S. Johnson, S. Kumar, E. Olshevskaya, S. Zozulya, T. A. Neubert, L. Stryer, J. B. Hurley, and K. A. Walsh (1992). The NH_2 terminus of retinal recoverin is acylated by a small family of fatty acids. *J. Biol. Chem.* **267**: 16033–16036.

66. S. Zozulya and L. Stryer (1992). Calcium–myristoyl protein switch. *Proc. Natl. Acad. Sci. USA* **89**: 11569–11573.

67. J. B. Ames, R. Ishima, T. Tanaka, J. I. Gordon, L. Stryer, and M. Ikura (1997). Molecular mechanics of calcium–myristoyl switches. *Nature* **389**: 198–202.

68. T. Tanaka, J. B. Ames, T. S. Harvey, L. Stryer, and M. Ikura (1995). Sequestration of the membrane-targeting myristoyl group of recoverin in the calcium-free state. *Nature* **376**: 444–447.

69. R. E. Hughes, P. S. Brzovic, R. E. Klevit, and J. B. Hurley (1995). Calcium-dependent solvation of the myristoyl group of recoverin. *Biochemistry* **34**: 11410–11416.

70. J. B. Ames, T. Tanaka, L. Stryer, and M. Ikura (1994). Secondary structure of myristoylated recoverin determined by three-dimensional heteronuclear NMR: implications for the calcium–myristoyl switch. *Biochemistry* **33**: 10743–10753.

71. P. Desmeules, S. E. Penney, B. Desbat, and C. Salesse (2006). Determination of the contribution of the myristoyl group and hydrophobic amino acids of recoverin on its dynamics of binding to lipid monolayers. *Biophys. J.* **93**: 2069–2082.

72. K. G. Valentine, M. F. Mesleh, S. J. Opella, M. Ikura, and J. B. Ames (2003). Structure, topology, and dynamics of myristoylated recoverin bound to phospholipid bilayers. *Biochemistry* **42**: 6333–6340.

73. M. Kataoka, K. Mihara, and F. Tokunaga (1993). Recoverin alters its surface properties depending on both calcium-binding and N-terminal myristoylation. *J. Biochem. (Tokyo)* **114**: 535–540.

74. M. T. Kennedy, H. Brockman, and F. Rusnak (1997). Determinants of calcineurin binding to model membranes. *Biochemistry* **36**: 13579–13585.

75. H. Hauser, I. Pascher, R. H. Pearson, and S. Sundell (1981). Preferred conformation and molecular packing of phosphatidylethanolamine and phosphatidylcholine. *Biochim. Biophys. Acta* **650**: 21–51.

76. D. Marsh (1996). Lateral pressure in membranes. *Biochim. Biophys. Acta* **1286**: 183–223.

77. C. Salesse, F. Boucher, and R. M. Leblanc (1984). An evaluation of purity criteria for bovine rod outer segment membranes. *Anal. Biochem.* **142**: 258–266.

78. D. Blaudez, T. Buffeteau, J. C. Cornut, B. Desbat, N. Escafre, M. Pezolet, and J. M. Turlet (1993). Polarization modulation FT-IR spectroscopy of surfaces and ultra-thin films: experimental procedure and quantitative analysis. *Appl. Spectrosc.* **45**: 380–389.

79. S. Ray, S. Zozulya, G. A. Niemi, K. M. Flaherty, D. Brolley, A. M. Dizhoor, D. B. McKay, J. Hurley, and L. Stryer (1992). Cloning, expression, and crystallization of recoverin, a calcium sensor in vision. *Proc. Natl. Acad. Sci. USA* **89**: 5705–5709.
80. P. Desmeules, S. E. Penney, and C. Salesse (2006). Single-step purification of myristoylated and nonmyristoylated recoverin and substrate dependence of myristoylation level. *Anal. Biochem.* **349**: 25–32.
81. J. Saccani, T. Buffeteau, B. Desbat, and D. Blaudez (2003). Increasing detectivity of polarization modulation infrared reflection–absorption spectroscopy for the study of ultrathin films deposited on various substrates. *Appl. Spectrosc.* **57**: 1260–1265.

CHAPTER 8

Chirality and Dipolar Interactions of Membrane Mimetic Amphiphilic Molecules

NILASHIS NANDI[*,1] and K. THIRUMOORTHY[1]

Chemistry Department, Birla Institute of Technology and Science, Pilani, Rajasthan, India

DIETER VOLLHARDT

Max Planck Institute of Colloids and Interfaces, Potsdam, Germany

CONTENTS

8.1	Introduction	191
8.2	Chirality in Membrane Mimetic Systems: Experimental Observation	192
8.3	Chirality in Membrane Mimetic Systems: Theoretical Studies	204
8.4	Dipolar Interaction in Membrane Mimetic Systems: Experimental Observation	209
8.5	Dipolar Interaction in Membrane Mimetic Systems: Theoretical Studies	213
8.6	Conclusion	220
Acknowledgments		220
References		220

8.1 INTRODUCTION

Membranes have been the subject of intensive studies for a long time due to their important biological role as cell constituents that are absolutely essential

*Corresponding author
[1]Present address: Chemistry Department, Kalyani University, Kalyani, 741235, West Bengal, India; E-mail: nilashisnandi@yahoo.com.

Structure and Dynamics of Membranous Interfaces, edited by Kaushik Nag
Copyright © 2008 John Wiley & Sons, Inc.

for all living beings. Membranes have an amphiphilic nature and are generally composed of lipid molecules with proteins and carbohydrates as the other principal components. Various classes of lipids such as fatty acids, esters of fatty acids (acylglycerolic and phospholipid amphiphiles), isoprenoids (sterol, sterol esters), and glycolipids are present in mammalian, plant, and bacterial cell membranes. The central part of a membrane is a bilayer composed of amphiphiles, which is the critical component of all biological membranes. Due to the complexity of the membrane structure, often bilayers and monolayers are used as simpler model systems to understand the physicochemical forces and factors controlling membrane dynamics and function. These systems are self-aggregated systems with the principal driving factor being a balance between hydrophobic effect and headgroup repulsion. However, recent experimental and theortical studies indicate that several subtle molecular factors play a major role in determining the morphology of the self-aggregating systems composed of amphiphiles. For example, both chirality and polarity of a given amphiphile molecule influence the aggregate structure of monolayers and bilayers. Since almost all biological lipids are chiral as well as polar in nature, their role in membrane morphology and function is worth exploring.

It is well known that the structure–function relationship plays an important role in biological systems. Active and functional biological molecules have unique structure and a change of structure results in variation of the functionality of the molecule or may even lead to malfunction. Despite the fact that the structure–function relationship is an accepted principle, the molecular understanding of this correlation is far from achieved for many biological molecules. Why biological lipids are overwhelmingly chiral is an unanswered question. Similarly, the role of polarity of lipids in bipolar membranes is yet to be explored in detail. It might have an important role in determining the relative position of transmembrane proteins (which generally have large dipole moment) and controlling the functionality of the same as an efficient receptor or part of the membrane channel. Understanding of chirality and polarity effects in bilayer and monolayer models may lead to better understanding of membrane–ligand interactions in various chemoreception processes.

Recent molecular studies have focused on the properties of self-aggregating systems such as monolayers and bilayers, which are important and well-known membrane models. This chapter focuses on the effect of molecular chirality and polarity of lipids and glycerolic amphiphiles, which are useful to study as membrane mimetic molecules. In the following, we review recent experimental observations and molecular understanding of the effect of chirality and polarity on the aggregate morphology of amphiphiles like lipids and glycerolic headgroup amphiphiles.

8.2 CHIRALITY IN MEMBRANE MIMETIC SYSTEMS: EXPERIMENTAL OBSERVATION

Langmuir monolayers and bilayers are successful membrane model systems to correlate the molecular chirality with mesoscopic structure. The experimental results on the morphology of the condensed phase domains and the two-dimensional lattice structure obtained from surface pressure–area per molecule ($\pi - A$) isotherms, Brewster angle microscopy (BAM), and grazing incidence X-ray diffraction (GIXD) provide an excellent basis for molecular interpretations. When the monolayer of a chiral amphiphile is composed of a pure enantiomer, the molecules within the corresponding domains develop mutual intermolecular azimuthal orientation. Whereas the molecular tilt from the normal remains unchanged, the continuous or sudden variation of the azimuthal orientation is a feature of enantiomerically pure domains. The intermolecular orientation varies along the length and width of the domain. This mutual orientation is propagated over the mesoscopic length scale and the domains or fragments of domains become curved. In general, heterochiral racemic domains do not show anisotropic shape development. This indicates that, in the case of heterochiral racemic systems, mutual intermolecular azimuthal orientation is either absent or canceled over the mesoscopic length scale. However, in cases where the racemic domains develop curved arms with opposite handedness, chiral separation occurs. This happens when the interaction between the same enantiomers is favored over the two different enantiomers, designated as homochirality; whereas the opposite case, where the interaction of the two different enantiomers is more favored compared to the interaction between the same type of enantiomers, is denoted as heterochirality. Chiral discrimination is observed in 1,2-dipalmitoyl-*sn*-3-glycero-phosphatidylcholine (DPPC) and other lipid systems. We discuss chirality-dependent features observed in lipid and glycerolic amphiphiles from $\pi - A$ isotherms, optical (BAM), and X-ray diffraction (GIXD) experiments. Information on different length scales (macroscopic to microscopic) are available from these studies and provide a detailed understanding of the aggregates.

The effect of molecular chirality is manifested in monolayers, when the molecular separation is relatively small. It is well known that stereochemical effects on the monolayer phase behavior are difficult to detect when the intermolecular separation is large [1]. Consequently, chiral effects cannot be expected in the fluid (gaseous or liquid expanded) phase. With increasing temperature or decreasing surface pressure, the rotational disorder of the molecules increases rapidly. They rotate more freely about their long axis. As a result, the intermolecular interaction becomes gradually insensitive to any asymmetry present about the short axis of the respective molecule and the molecule becomes effectively symmetric to the neighboring molecules. In such a case, chiral interactions are relatively unimportant. It is expected that the chiral discrimination will gradually disappear as the system approaches the free rotating limit. Thus chiral interaction is more dominant at lower temperature or higher surface pressure where spatial or orientational correlations between the molecules are present, as recently pointed out [2].

General conclusions on the phase behavior of amphiphilic monolayers can be drawn considering the $\pi - A$ isotherms at different temperatures. Usually $\pi - A$ isotherms are recorded using a computer-interfaced film balance with a Wilhelmy-type pressure measuring system. Neglecting details, three general single-phase states can be distinguished, namely, the *fluid* (gaseous, liquid-expanded (LE)), *condensed* (liquid-condensed (LC)), and *solid* (S) phase. Three characteristic shapes of a generalized $\pi-A$ isotherm representative of respective temperatures or alkyl chain length can be measured (Fig. 8.1). The most interesting type of $\pi-A$ isotherm shows a nonhorizontal plateau region after a kink point in the isotherm characteristic for a first-order phase transition from the fluid to the condensed phase. The plateau represents a two-phase coexistence between the fluid and the condensed phase. The lower the temperature, the lower the surface pressure of the plateau, but the more extended the area per molecule covered by the plateau. The transition to the solid monolayer phase is usually indicated by a kink in the steep part of the isotherm. At low enough temperatures the two-phase coexistence region already exists at a surface pressure $\pi \sim 0$; that is, this is the second general type of isotherm. In this case, the $\pi-A$ isotherm does not allow the exact determination of the first-order phase transition point. Oppositely, at high enough temperatures, a continuous increase of the surface pressure with decreasing area per molecule suggests that over the whole area range no phase transition to the condensed or solid phases occurs. This third type of isotherm is not of interest for studies of the monolayer characteristics.

Phospholipid amphiphiles constitute an important class of chiral compounds due to their presence in biological membranes. As a result, monolayers of homologous dialkanoyl phosphatidylcholine (DMPC, DPPC), dialkanoyl phosphatidic acid (DMPA, DPPA), dialkanoyl phosphatidylethanolamine (DMPE), and so on have been investigated in detail by several techniques. Usually phospholipids have two alkyl chains and the chiral carbon atom is in close proximity to the headgroup. The temperature at which the characteristic features of the phase diagram of phospholipids appear is dependent on the alkyl chain length and the headgroup [3]. This suggests that the characteristic features of the phase diagram

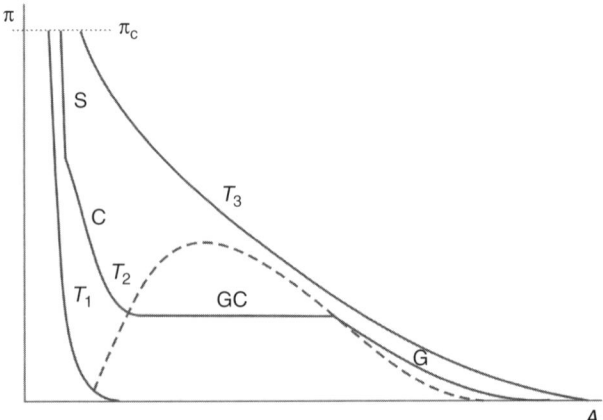

Figure 8.1 Generalized surface pressure–area ($\pi-A$) isotherm at various temperatures. The gaseous (G), condensed (C), and solid (S) phases and coexistence regions are shown.

are dependent on the chirality of the phospholipid molecules, since the variation in the alkyl chain length and nature of the headgroup change the stereochemical arrangement of the groups about the chiral center. However, detailed studies are not yet available to quantify this correlation. Monolayers of phospholipids with more than two chains have also been studied [4–8]. The isotherms of a triple-chain phosphatidylcholine have been investigated in order to address the important issue of estimation of the relative importance of the interactions arising from the head and those arising from the alkyl chain regions of the amphiphile [4]. It is observed that the transition pressure π_c corresponding to the transition from the fluid (LE) phase to the condensed (LC) phase of the enantiomeric and racemic 3-O-hexadecyl-2-(2′-hexadecylstearoyl)glycerol-sn-1-phosphocholine and racemic 1-O-hexadecyl-2-(2′-hexadecylstearoyl)glycerol-3-phosphocholine, is independent of chirality. However, a very small change in area and a pronounced change in the slope at pressures above π_c were observed for the enantiomers but not for the racemates. It was suggested that this could be an indication of a transition between two condensed states in the enantiomeric monolayer.

Two optical methods most extensively used in Langmuir monolayer studies are Brewster angle microscopy (BAM) and fluorescence microscopy. Reflection spectroscopy is a useful technique for studying organized molecular assemblies at interfaces [9–12]. The BAM technique gathers detailed information about the molecular orientation by direct visualization of the monolayer. When p-polarized light is used and incidence takes place at the Brewster angle, no light is reflected from the pure air–water interface [13–16]. However, changes in the molecular density and/or refractive index by the condensed phase of a monolayer on the aqueous subphase leads to a measurable change in the reflectivity and allows one to visualize and record the monolayer morphology. In a typical experimental setup, a Brewster angle microscope (Fig. 8.2) is mounted to a computer-interfaced film balance [17]. The light source of the Brewster angle microscope is a He–Ne laser. The spatial resolution of the microscope is in the range of a few micrometers. BAM images are taken with a CCD camera and images are formed in a video unit and can be stored there. The distortion caused due to incidence at the Brewster angle can be corrected by digital image processing software. If the

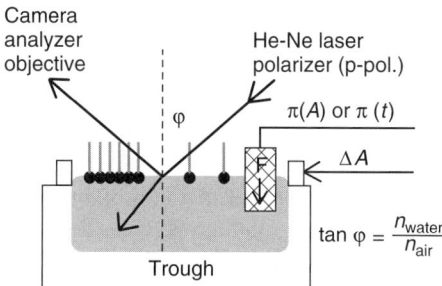

Figure 8.2 Principle of Brewster angle microscopy (BAM). Positions of polarizer, analyzer, and the Langmuir monolayer are shown.

BAM image were recorded as a mirror image, then the sense of the curvature of the corresponding domain in the enantiomeric monolayers would be opposite to that of the BAM image. The correct sense of the domain can be obtained by using corresponding image processing software.

BAM studies of monolayers revealed an enormous variety of domains in the condensed phase. The structures of the corresponding molecules have a profound influence in driving these domain shapes as well as the molecular orientation within a particular domain. Many of the molecules are chiral and in many cases the chirality has a major influence on the morphology, which is clearly detected by BAM. As pointed out earlier, the effect of chiral discrimination is unobserved in $\pi - A$ isotherm measurements of enantiomeric and racemic amphiphiles, but clear discrimination is revealed from the direct visualization of the domain shape and inner texture by BAM in several chiral monolayers. The texture gives a direct indication of the average molecular orientation in the monolayer. However, a characterization of the orientational order in a condensed phase requires that the observed textures occur under well-defined conditions. BAM can also distinguish between the jumps of orientations and a continuous change in orientation. Careful observation must be made of whether the lines that separate the dark and bright areas remain at a fixed position or are shifted, respectively, when the analyzer is rotated. Observing a sequence of images using a video rather than observing single images can achieve this. It may be noted that an external probe is not used in the BAM technique. Thus careful investigation using BAM can reveal many microscopic details of the monolayer without any external perturbation. BAM is advantageous compared to the fluorescence technique, which requires dye molecules as a probe to investigate the monolayer morphology.

Fluorescence microscopy is another extensively used method to study monolayers [3, 18, 19]. In this method a fluorescent probe (a dye molecule) is added to the monolayer, which is excited using a high pressure mercury lamp. Then the monolayer state is observed with a microscope mounted with a camera. A variety of probes have been used. Due to the presence of dye molecules, the effect of trace components on the monolayer properties cannot be ignored in fluorescence studies.

A number of studies of the effect of chirality on the morphology of biomimetic monolayers were focused on the model phospholipids DPPC and DMPE. Before discussing in detail some results of phospholipid monolayers, we summarize how chirality is apparent in the domain morphology of a phospholipid (Fig. 8.3). In equilibrium, phospholipid monolayers form compact domains of different shape and inner texture. As the relaxation time for reaching equilibrium is rather high, nonequilibrium structures in the form of dendritic or fractal-like structures are observed under the corresponding experimental conditions. The chirality of the enantiomeric forms of compact domains can be apparent in the form of clockwise or counterclockwise curvatures of defect lines, domain arms (e.g., triskelions), or regions of the same azimuthal orientation. The curved defect lines can meet within the domain or at their edge. In all cases, the two enantiomers show curvatures of opposite sense. The corresponding nonequilibrium structures with dendritic

Figure 8.3 (a) Representative equilibrium domains of phospholipids: (left) L-DPPC monolayers with curved domain arms; (middle) L-DPP(Me)$_2$E monolayers with curved defect lines; and (right) L-DPPE monolayers with curvature in the brightness change within the domains. (b) Representative nonequilibrium domains of phospholipids: (left) L-DPPC monolayer far from equilibrium with fractal-like (disordered) growth without chiral influence; (middle, right) L-DPPE and DL-DPPE monolayers not so far from equilibrium with dendritic (ordered) growth with chiral influence; curved axes and arms for enantiomeric monolayers (L-DPPE), straight axes and arms for racemic monolayers (DL-DPPE).

growth develop curvatures of the same sense. In the fractal-like structures the effect of chirality cannot be observed, because they are formed at the beginning of rapid growth where the patterns are obviously nonordered. In the corresponding racematic mixtures of different phospholipids, curvatures are not formed but rather straight dendrite arms or noncurved compact domains are formed. If the formation of the condensed phase domain occurs far from equilibrium, the growth kinetics can be so rapid that the stable phase does not have time to reach its lowest energy state on the microscopic level. A metastable microstructure results and the growth patterns formed under these nonequilibrium conditions are mainly affected by the complicated interplay between the microscopic interfacial dynamics, such as surface kinetics, surface tension, and crystalline anisotropy, and external driving forces, such as supersaturation and undercooling [17]. In highly supersaturated monolayers, the macroscopic dynamics is determined by the diffusion field, which tends to drive the formation of fractal-like objects. Consequently, the shapes of growing condensed phase domains can depend substantially on the compression rate of the monolayer. For example, the shapes of the DMPE domains are fractal-like at high compression rates as expected for diffusion-limited aggregation, while only at very low compression rates are compact domains evolved [17, 20]. It would be very interesting to study the effect of the molecular chirality on the growth kinetics or the relaxation of the condensed phase structure.

The comparison of the BAM images of the enantiomeric phospholipids L-DPPE and L-DPPC monolayers reveals very different domain textures (Fig. 8.4). In equilibrium, the L-DPPE monolayers form compact domains with a different number of segments (five to eight) separated by chirally curved lines where the molecular orientation jumps. The lines join together in a center located inside the circular domain or at its edge when the shape of the domain is bean shaped. Under nonequilibrium conditions at high compression rates, the molecule orientation in the branched domains is similar to that in the compact domains. The texture of the condensed phase domains of DPPC monolayers is strongly affected by the compression rates. At very low compression rates, equilibrium shapes evolve in the form of triskelions curved oppositely for the two enantiomeric forms and noncurved arms for the racemic mixture. The lines running through the arms of the enantiomeric domains mark a strong continuous change in the orientation. In the enantiomeric DPPC domains, the orientation changes gradually within each arm of the triskelions [20–22]. This difference in the outer shape of the domain and the inner structure is striking. Differences in the molecular structure and corresponding differences in the chiral features of the molecules are responsible for the observed shapes. Epifluorescence microscopic studies of DPPC have established that the handedness of the highly ordered domains is directly related to the absolute configuration of the enantiomer [23]. It was observed that the condensed phase domains of the monolayers composed of 75% R-DPPC and 25% S-DPPC are triskelion shaped and the arms of the domains are oriented in an anticlockwise (left-handed) fashion. Domains composed of 75% S-DPPC and 25% R-DPPC are also triskelion shaped and the arms are oriented in a clockwise (right-handed) fashion. The domains of racemic

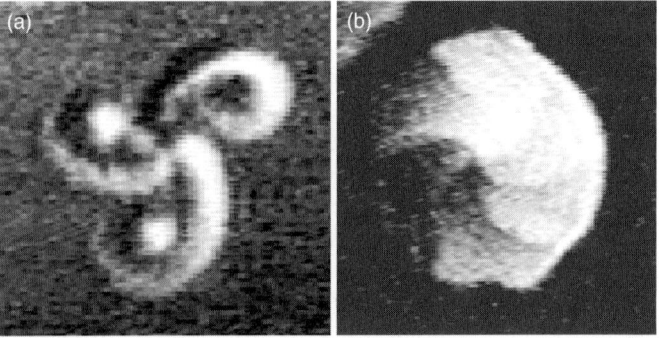

Figure 8.4 BAM images of the enantiomeric (a) DPPC and (b) DPPE condensed phase domains. Image size 200×200 μm.

DPPC (equimolar mixture of *R*-DPPC and *S*-DPPC) show development of arms without curvature. The arms of the domains are oriented in specific directions, that is, a chiral object in mesoscopic length scale. Clearly, this mesoscopic chirality is triggered by microscopic chirality present at the molecular level. Theoretical studies on the effect of molecular chirality and the curvature of the enantiomeric domains of DPPC are discussed later.

An interesting comparative study of the morphological features of phospholipid monolayers with closely similar chemical structure was carried out using BAM [22]. Four phospholipids of the same chain length are only different with respect to the number of methyl groups at the nitrogen of the headgroup. The morphological features of dipalmitoyl phosphatidyl-*N*-monomethylethanolamine (DPP(Me)E), dipalmitoyl phosphatidyl-*N*-dimethylethanolamine (DPP(Me)$_2$E) are studied and they are compared with those of monolayers of dipalmitoyl phosphatidylethanolamine (DPPE), dimyristoyl phosphatidylethanolamine (DMPE), and dipalmitoyl phosphatidylcholine (DPPC) (Fig. 8.5). The properties of the condensed phase domains of DPP(Me$_2$)E and DPP(Me)E are between those of DPPC and DPPE or DMPE. In nonequilibrium, fractal-like or dendritic domains

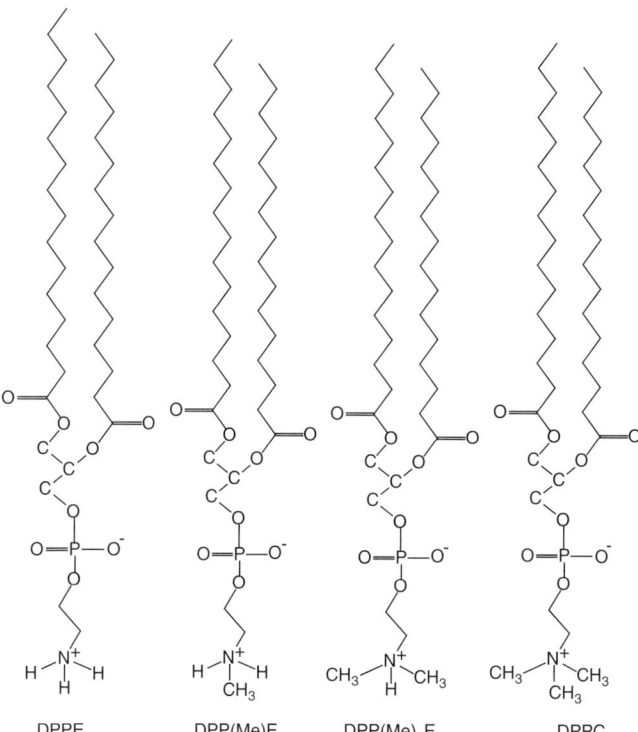

Figure 8.5 Effect of the headgroup variation of chiral phospholipids on the texture of the equilibrium domains. Four lipids, DPPE, DPP(Me)E, DPP(Me)$_2$E, and DPPC, are shown.

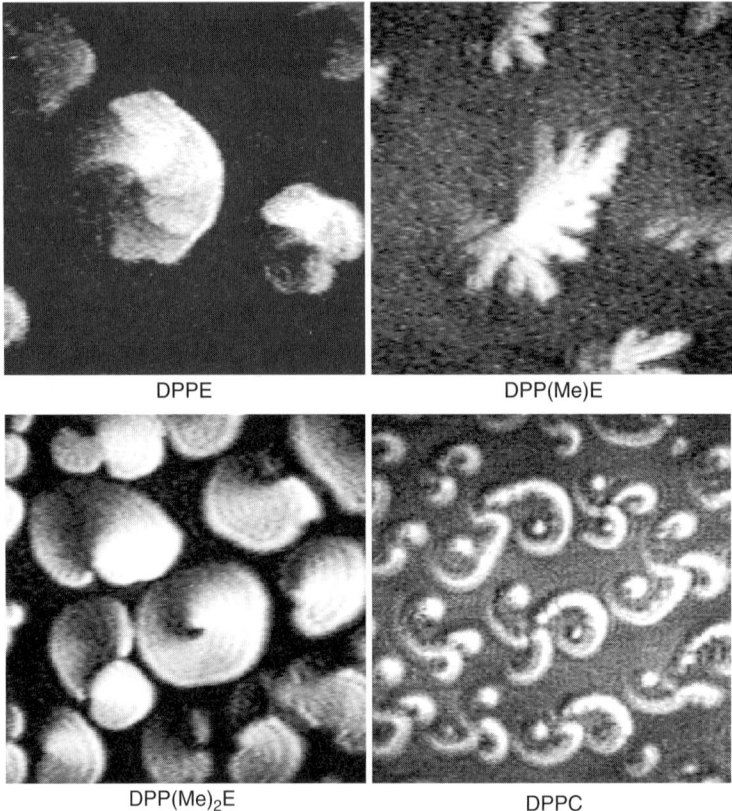

Figure 8.5 (*Continued*)

are formed depending on the compression rate. The fractal-like domains of DPP(Me)$_2$E or DPPC do not reflect chirality as they are grown obviously too far from equilibrium, whereas the dendritic domains of DPP(Me)E and DPPE are specifically curved, indicating the influence of chirality. However, in the racemic mixtures of phospholipids (e.g., in DL-DPPE monolayers), straight dendritic arms are observed. The absence of a curvature in the dendritic arms of racemates points out that the curvature is an effect of the chirality and not a characteristic of the observed texture.

Compact phospholipid domains can be obtained at very low compression rates or as a result of a long-time shape relaxation after stopping the monolayer compression. Under these conditions close to equilibrium, chirality affects the domain texture in different ways, but curvatures corresponding to the enantiomeric forms are always observed in the domain texture [23]. Enantiomeric DPPC domains form triskelions with curved arms and a remarkable continuous change in the orientation within the arms. It is interesting to note that domain shapes with relatively straight arms (without curvature) are formed in monolayers of the racemic DL-DPPC mixture [24]. Compact domains are subdivided into segments by curved lines at which the orientation jumps are observed in enantiomeric L-DMPE monolayers. In the compact L-DPP(Me$_2$)E domains the orientation changes by 360° around a point inside or at the edge of the domain and no jump in the orientation is observed.

Growth of chiral domains in dimyristoyl phosphatidic acid (DMPA) monolayers containing 1 mol% cholesterol has been studied [25]. The shape is spiral. The domain shape is dependent on pH. Ionic strength and pH are controlled by addition of NaCl and NaOH, respectively. The lipid spirals are observed in a narrow range of pH~11. At lower pH (~10.5) the spirals convert into coffee bean shapes. These effects may be related to an increase or decrease in electrostatic forces, which leads to a change in the boundary/area ratio. It is argued that the lipids have an excess charge or a dipole moment normal to the surface and the surface electrostatic repulsion is driving the thin, elongated shape. The line tension competes to reduce the boundary line and the effect of cholesterol is to decrease the line tension and thereby favor the elongated shape. Clearly, the formation of spiral shapes is additionally driven by the chirality of the molecules.

The role of the headgroup size and the concomitant symmetry in crystalline domains responsible for a given domain shape in L-α-dimyristoyl phosphatidic acid (DMPA) monolayers containing 1 mol% cholesterol have also been studied [26]. DMPA at high pH has a double charged headgroup, which is expected to occupy a large area. In condensed monolayer phases, the alkyl chains are tilted. In this respect, the arrangement of the molecules is similar to that of DPPC monolayers. It is suggested that the symmetry is reduced from hexagonal symmetry, as observed in lipids with small headgroups, and the domain shape becomes elongated rather than symmetric [27]. Such elongation is not to be expected when the chain tilt is small or zero. This is the case for the DMPA monolayers with cholesterol at pH 8 and DMPE monolayers with cholesterol. Based on this proposition, the formation of the elongated domain shape is essentially driven by the chain tilt, which in turn can be controlled by the presence of ions in the solution. Also, the effect of chirality is to bend the elongated domains. The possibilities of chiral diffusion and specific molecular interactions like those between CO and phosphate groups are suggested to be the reasons for such bending. The permanent dipole moment in the headgroup region has an in-plane dipole moment, which slightly deviates from the preferred growth direction of the crystal (presumably the x direction) due to the asymmetric arrangement of the headgroups. The asymmetric arrangement of the headgroups should cause the molecular arrangement to change from rectangular to trapezoidal shapes. As a result, the neighboring molecular dipoles would be nonparallel and would have a nonzero angle between them with respect to the preferred growth direction.

In many experimental studies of the morphology of lipid monolayers, added substances are present that could have substantial influence on domain morphology. Cholesterol is an important constituent of the biological membrane and studies of the lipid monolayer structure and phases are often made including cholesterol as one of the components. One important reason to include cholesterol in such studies is to understand the lipid distribution and interaction in biological membranes. Due to the chirality of the cholesterol molecule, the results of the structural or morphological studies, in such two-component systems containing an amphiphilic component and cholesterol, are difficult to interpret unambiguously. Explicitly, it is difficult to conclude whether the observed morphological feature is arising from the amphiphile or the added substance. The effect of cholesterol on the monolayer packing properties of dipalmitoyl phosphatidylcholine (DPPC) monolayers and 1-palmitoyl-2-oleoyl phosphatidylcholine (POPC) was studied [27]. The introduction of cholesterol to these monolayers decreased the average molecular area occupied by the phosphatidylcholine molecules by inducing order into the acyl chain. The effect is more prominent in DPPC than POPC monolayers. The closer packing of the alkyl chains in the presence of cholesterol could drive stronger chiral interaction compared to the monolayer where cholesterol is not added. It is unclear, however, whether or not this condensing effect triggers the stronger chiral shapes in the presence of cholesterol.

Optical studies of the monolayers draw our attention to the fact that in many cases the macroscopic methods, like the $\pi - A$ isotherm measurements, fail to detect the chirality effects present at mesoscopic or microscopic levels. A bewildering variety of inner structures can be observed in domains composed of different classes of molecular structure also clearly revealed by BAM. The results of BAM evidently point out the decisive role of molecular chirality in determining the molecular orientation and the resultant domain shape. Theoretical studies are discussed later which attempt to provide an understanding of the inner structure and shape of the domain.

Both X-ray and neutron beam techniques have been applied to study the molecular organization of lipid monolayers at the air–water interface. Extensive literature of various techniques is available [28–32]. Specular reflectometry using X-rays or neutron beams probes the electron density or the neutron scattering length density profiles normal to the interface. The grazing incidence X-ray diffraction (GIXD) technique probes the molecular order in condensed phases of Langmuir monolayers (Fig. 8.6) [32–34]. X-ray beams obtained from synchrotron radiation facilities have been used for this purpose. The beam is incident on the surface at a very small angle ($\sim 0.1°$) and thus background scattering is avoided. The beam undergoes total reflection and only a part penetrates into the subphase. The scattered radiation is detected and the periodicity of molecular arrangements gives rise to the peaks in the scattered intensity.

The diffraction patterns obtained from the GIXD studies are always averaged over all domain orientations in the monolayer plane. As a result, while the vertical out-of-plane component, K_z, of the momentum transfer vector **K** is measurable separately, the two in-plane components (perpendicular to K_z), K_x and K_y, are not measurable separately. Only the combination $K_{xy} = (K_x + K_y)^{1/2}$ is measurable. Thus the orientation of the projection of the molecular segment, giving rise to the chirality with respect to any particular axis, is impossible to obtain from

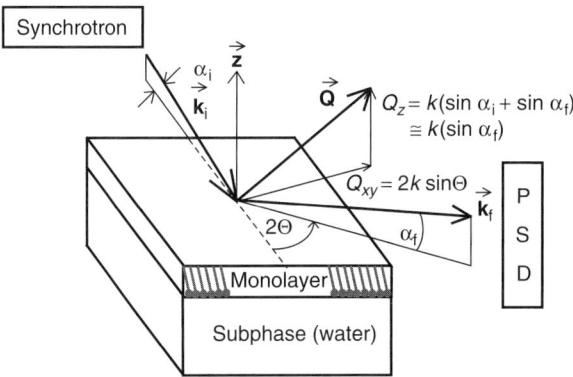

Figure 8.6 Principle of GIXD measurements for studying Langmuir monolayers. The diffracted beam is detected with a position sensitive detector (PSD).

GIXD data. The first-order peaks correspond to the distance between the neighboring molecules and are the most intense reflexes. When only one value of the in-plane component K_{xy} is available, it indicates equal separation between neighboring molecules and corresponds to hexagonal packing. Two and three distinct values of K_{xy} correspond to the rectangular and oblique unit cells, respectively. Thus important lattice information about the monolayer system can be obtained from GIXD studies.

It is important to note that an oblique structure in a monolayer system is chiral due to the lack of the plane of mirror symmetry. Also, achiral molecules can form chiral lattice structures when tilt or distortion happens in a direction intermediate between nearest neighbor (NN) and next nearest neighbor (NNN). On the other hand, a chiral molecule, which develops chiral domains in a condensed phase (oriented with a specific handedness), can have an achiral underlying lattice (e.g., centered rectangular) [35, 36]. The observation of an oblique structure is also possible for homochiral racemic mixtures when two enantiomers separate into the D- and L-isomers. This is an event of spontaneous chiral segregation. It is also to be noted that the local packing determined for a mesophase cannot be extended straightforwardly to the long-range scale because of finite positional correlations in such phases, unlike crystalline phases [32]. Explicitly, the symmetry of the molecule arrangement over a few intermolecular distances may not coincide with the long-range symmetry of the monolayer. Consequently, care must be taken to use the data obtained from GIXD studies for interpreting the long-range correlations present in monolayers. It is also important to note that the specific polar direction (normal to the interface) in the monolayer system couples molecular chirality and alkyl chain tilt. Thus the various values of the tilt azimuth observed in monolayers play a role in the formation of mesoscopic domain shapes [32]. Oblique lattices have been observed in both enantiomeric and racemic monolayers of DPPC [37]. It was suggested that the oblique lattice structure in DPPC is due to orientational ordering of the glycerol backbone, which links the two chains, at an oblique angle. On the other hand, only the enantiomeric monolayer of DPPE exhibits asymmetric lattice structure. It is assumed that the hydrogen bond network in the headgroup region in enantiomeric DPPE is one-dimensional (along the *b* axis), whereas in racemic DPPE the network is extended in two dimensions.

Phospholipids with more complicated chemical structure, such as triple-chain phospholipid monolayers of enantiomeric and racemic 3-*O*-hexadecyl-2-(2′-hexadecylstearoyl)glycerol-*sn*-1-phosphocholine and racemic 1-*O*-hexadecyl-2-(2′-hexadecylstearoyl)glycerol-3-phosphocholine have been investigated by the GIXD technique in the temperature range of 5–25 °C [4]. The contour plots of the X-ray intensities as a function of in-plane component and out-of-plane component of the scattering vector of the enantiomeric monolayer indicate that the alkyl chains are arranged in a nonsymmetric direction showing a lack of mirror symmetry in the lattice arrangement. The alkyl chains are tilted in NN direction for both the enantiomeric and racemic monolayers. The tilt angle also decreases with increasing surface pressure. Three chains require more cross-sectional area than that required by the headgroup and determine the packing at higher surface pressures. The spacing between the chiral carbon and the trimethyl ammonium group is larger than two lattice spacings. Consequently, for parallel orientation of the headgroup to the surface, local interaction between the headgroups is suggested to couple the headgroup with the lattice arrangement of the alkyl chains. It is suggested that the long-range interactions between the distributed charges in the headgroups of the chiral molecules give rise to an interaction that is chiral in nature and may drive the formation of the chiral shape. Short-range chiral interaction between the headgroups is possible provided the headgroups are regularly arranged [32]. Consequently, the GIXD studies in various amphiphilic systems indicate the presence of chirality at the lattice structure level. On the other hand, we observed chirality at the domain structural level, which is many times larger than the lattice dimension. Chiral interaction is operative in tightly packed arrangements between successive neighbors and the neighboring chiral interaction again prefers anisotropic molecular arrangements. Consequently, the manifestation of chirality at different length scale may be different but the fundamental nature of the interaction seems to be similar.

8.3 CHIRALITY IN MEMBRANE MIMETIC SYSTEMS: THEORETICAL STUDIES

Macroscopic as well as molecular theories have been proposed to explain the effects of molecular chirality on the aggregate morphology as discussed earlier. Molecular theories are more promising in this direction and provide a direct micro–meso correlation that is important to understanding membrane systems. Andelman and Orland [38] used a model in which the chiral molecule is assumed to be composed of a chiral carbon to which four groups are attached. This model is limited by modeling a three-dimensional molecule as a planar tripod and by consideration of only near neighbor distance dependence interactions. Subsequently, Nandi and Bagchi [39, 40] developed a more realistic theory of the helical structure in bilayers. In this theory, the effective pair potential between a pair of chiral molecules is calculated to find the mutual orientation at minimum energy. It is reasonable to infer that the subtle stereogenicity at the chiral center of a chiral molecule is responsible for driving the aggregate shape to a particular morphology.

It is possible to predict and understand the structure formation from an effective intermolecular pair potential between the chiral centers of the monomers of the aggregate [41]. Minimally, this potential should depend only on the distance and the orientation between the two participating chiral amphiphilic molecules. This effective pair potential can then be used to find out the relative arrangement of a pair of molecules. The minimal energy conformation of the aggregate can be studied by changing the orientation of the groups and by reducing the distance between the chiral centers. The molecules are in a tightly packed structure with a small separation between them. Thus one expects short-range, both repulsive and attractive, interactions to play an important role in these systems. It was found that, for mirror-image isomers in the racemic modification, the minimum energy conformation is a nearly parallel alignment of the molecules. On the other hand, the same for a pair of molecules of one kind of enantiomer favors a tilt angle between them, thus leading to the formation of a helical morphology of the aggregate. It was also shown that the sense of the helix can be predicted from the effective pair potential description. In all considered cases, complete agreement between theoretical prediction and experimental results [41] was observed. This surprising success of the simple and straightforward early molecular approach strongly indicates that the formation of mesoscopic or macroscopic chiral structures such as a helix can be predicted from the intermolecular energy profile when the molecular chiral structure is considered explicitly. However, the early studies [39, 40] made several approximations that were modified later [41].

In recent years, a detailed molecular approach has been attempted for monolayers [42–53] to understand chirality-related effects. The theory assumes that the intermolecular pair potential determines the mesoscopic chiral structure of the condensed phase domain. It investigates the orientation and distance dependencies of the pair potential between neighboring chiral amphiphilic molecules. The objective was to find the most favorable mutual azimuthal angle at various separations when a pair of molecules are in their minimal energy state with individual molecules in their most probable conformational state [41].

While entropic factors contribute to the aggregation of amphiphiles, study of the intermolecular energy profile suffices to explain the chirality-dependent structural features of amphiphilic aggregates for the following reason. Entropy changes during aggregation of pure enantiomers or racemic mixtures into respective domains are not significantly different because the changes in rotational, translational, and vibrational degrees of freedom of the molecules due to the transfer from a free state to an aggregated state should be similar for L-L (or D-D) and L-D pairs. Consequently, the entropy difference hardly contributes to the observed difference in the shape of the condensed phase domains such as the mirror-image relation of the opposite curvatures or the intersection angle of main growth directions. Thus the energy profile is important to determine chirality-related effects.

For molecules as large as amphiphiles, the potential energy profile is a function of a large number of degrees of freedom. In principle, the intermolecular energy profile should be calculated as a function of all intra- as well as intermolecular degrees of freedom, which are relevant to the interface. However, the intermolecular energy variation for many of them is redundant when chirality-induced effects are of interest. Systematic approximations have been made in numerical calculations to handle large molecular systems. To reduce the variables in the numerical calculation of pair potential, the molecular structure of the amphiphile has been approximated at various levels of molecular detail while retaining intact the spatial chirality around the stereogenic center. Initially, a tetrahedral model with equivalent sphere representation of the groups was used [39, 40]. In the spirit of the condensed state theories, the tetrahedral model assumes that equivalent spheres are effective representations of the groups or atoms of the molecular structure that are attached to the chiral center. Hence the calculated effective pair potential provides essential information about the chirality-induced orientation dependence between molecules. This model has been applied successfully to bilayers [39, 40] and monolayers [46, 47]. The handedness predicted theoretically agrees well with the experimental data. Subsequently, coarse-grained molecular models have been applied to DPPC and N-palmitoyl-aspartic acid (PAA) monolayers [48–51]. In these studies, the alkyl chains are assumed to be composed of spherical groups such as —CH_2—, CH_3—, and so on and are in all-*trans* configuration, which is a reasonable approximation of the molecular state in condensed phases where chirality effects are relevant. However, different headgroup conformations are possible, and it is necessary to locate the most probable conformational state of the headgroup, as considered in theoretical studies [48]. The atomic coordinates of PAA in its energy-minimized state, obtained from semiempirical quantum mechanical calculations, are used to calculate the intermolecular pair potential [52]. The effects of coarse-graining on the theoretical predictions are also evaluated [50].

The dependence of molecular interaction on chirality in DPPC monolayers [48] was calculated using the *EPP* theory. The morphology of DPPC monolayers has been extensively investigated experimentally and the domains of enantiomeric monolayers are triskelion shaped with arms curved in a specific direction. The domain curvature is specific for enantiomers and no curvature is observed for racemates. BAM studies suggested that the neighboring molecular directors are in a mutually oriented state along the width and the length of the arms of the triskelions [20–22]. The recent EPP study on DPPC monolayers is based on a coarse-grained description of the molecular structure [48]. Further atomistic studies have recently been carried out for palmitoyl aspartic acid [53], which confirm the conclusions based on the coarse-grained model. Computational studies on lipid systems are limited by the presence of multiple alkyl chains in lipids and enhanced computational costs in lipids compared to simpler model amphiphiles. However, studies in these directions seem promising in understanding the subtle intermolecular interaction in explaining aggregate morphology. This description is more realistic and an improvement over the equivalent sphere description of the groups used in previous theoretical studies [39, 40] and simulation studies, where the molecular structure was not considered [54]. As a result, the EPP profile obtained from the coarse-grained description is expected to depict the molecular interaction in a more detailed way than previous studies. As mentioned before, from the minima of the pair potential, the preferred orientation of a pair of aggregating molecules can be obtained. The mutual orientation between the pair of molecules provides information about the domain handedness of the aggregate in the condensed state, where the molecules are aligned in a next-to-next order. It is important to note that in the condensed state of the domains, the molecules are in close proximity and chiral interactions are dominant. The effect of the orientational distribution of the headgroup in the aqueous subphase on the chiral shape of the condensed phase domains has also been investigated [48].

In the coarse-grained description [46–51], the alkyl chains and the heads of the molecule are represented by an array of collinear spherical groups (e.g., CH_2, CH_3, CH, COO groups). While atomistic details are neglected in the calculation, the average orientations of the groups with respect to the chiral center are considered over a reasonably small length scale. The molecular segments are placed at the lattice positions. Information about the orientation of the reference molecule with respect to the normal and azimuthal tilt directions of the alkyl chain (projected on the perpendicular plane to the normal) is obtained from the grazing incidence X-ray diffraction (GIXD) data [55]. The calculation uses no adjustable parameter. Information about the orientation of the headgroup is not available from the GIXD studies. All possible orientations of the headgroup with respect to the interface normal and the corresponding possible azimuthal projections are considered in calculating the pair potential of a pair of headgroups of neighboring amphiphiles. The lowest value of all such minima is thus the most probable orientation of the headgroups in the aqueous subphase with respect to the interface. It is observed that the pair potential is favorable over a broad range of orientations but is most favorable when they are oriented at an angle with respect to the air–water interface.

The EPP profiles of different molecular segments such as alkyl chains and headgroups are considered. The results of the calculations show that, taking the orientation of an alkyl chain of any molecule to be the same as that obtained from GIXD data, the orientation of the other alkyl chain of the same molecule as well as the orientation of the alkyl chains of an adjacent molecule are oriented in a right-handed way. If one alkyl chain of the reference molecule is located closest to the observer and the other alkyl chain of the neighboring molecule is located farthest from the observer, the progress of such an arrangement can represent the growth process of the domain where the molecules are aligned in next-to-next fashion. Thus the interaction between a pair of alkyl chains and a pair of headgroups indicates that all pairs of molecular segments have a large favorable pair potential (measured pairwise with the reference) when the mutual azimuthal projection (counted anticlockwise) is between $\sim 350°$ and $\sim 235°$. This indicates that the molecule segments have a large favorable energy when they all are oriented in a right-handed way with respect to the alkyl chain, which is closest to the observer in the domains composed of D-enantiomer. This mutual orientation is cooperative in the sense that all segments favor the tendency to have a right-handed turn with respect to the reference at the minima of the pair potential. In all cases, a high-energy barrier separates the minimum of the EPP, which favors the opposite handedness (left-handedness). The favorable mutual azimuthal orientation gradually moves to a parallel arrangement with an increase in the molecular separation. Also, the EPP becomes increasingly shallow with increase in temperature. These facts corroborate that an increase in molecular separation (by lowering the surface pressure) or an increase in temperature destroys the effect of chirality. A favorable broad range of orientation of a molecular segment is also expected to diminish the effect of chirality due to effective sphericalization. The present calculation shows that the headgroup pair potential remained favorable over a broad range of mutual orientations of headgroups of the neighboring amphiphiles. The headgroup of the second molecule is found to be oriented in a right-handed way, keeping the reference molecule closest to the observer. This indicates that the effect of chirality is not destroyed even if a distribution of orientation of headgroups exists in the aqueous subphase.

Several questions relating to chirality effects in two dimensions can only be answered from a molecular consideration [56]. It is worth investigating why the mutual intermolecular orientation is present between a pair of molecules of a particular amphiphile in the condensed phase and for other amphiphiles such a mutual orientation is absent. Furthermore, the role of hydrogen bonding in the shape formation is also important to investigate, because an extensive hydrogen bonding network is present in several amphiphilic aggregates. Hydrogen bonds are also sensitive to the orientation of the donor and acceptor groups. Whether the hydrogen bonding interactions are acting in concert with chiral interaction is yet to be investigated in detail. With the advent of experimental techniques like GIXD and BAM, detailed structural features are revealed, and theoretical methods exploited the information to obtain a better understanding of the monolayer systems. Due to the wide variety of molecular structures of chiral amphiphiles, more theoretical studies based on molecular approaches are necessary to provide a clear understanding about the diverse chirality effects observed in monolayers. Also, theories that can combine the explicit molecular structure, electrostatic interaction, and line tension together are expected to provide a simple picture of the shape variations. As indicated before, these studies are expected to be helpful in understanding other relevant biological systems like membranes.

8.4 DIPOLAR INTERACTION IN MEMBRANE MIMETIC SYSTEMS: EXPERIMENTAL OBSERVATION

Little experimental information is available about the structure of headgroups or their dipolar nature for monolayers at the air–water interface. Drastic change in the domain morphology is noted in experiments, which is caused by subtle modification of the headgroup structure and suggests a dipolar origin [57]. In the following, we describe recent studies in monolayer systems, which show that small changes in the polarity by subtle variation in the molecular structure can affect shape and organization of the condensed phase domains.

To investigate how systematic alteration of the headgroup structure can affect the domain shape, four types of racemic amphiphilic monoglycerols have been compared in molecular detail using Brewster angle microscopy (BAM) (Fig. 8.7) and the grazing incidence X-ray diffraction (GIXD) technique [58–62]. The chemical structures of the four similar amphiphiles are monoglycerol amide, ether, ester, and amine (abbreviated as ADD, ETD, ESD, and AMD, respectively). The variation in the domain shape with the change in the molecular structure is rather dramatic. Out of the four compounds, the crystalline nature increases from amine (lowest) to amide (highest). The amide domains are thin and brittle and break up when interdomain collision occurs. Amide domains have the highest orientational correlation within the domain propagated up to a few

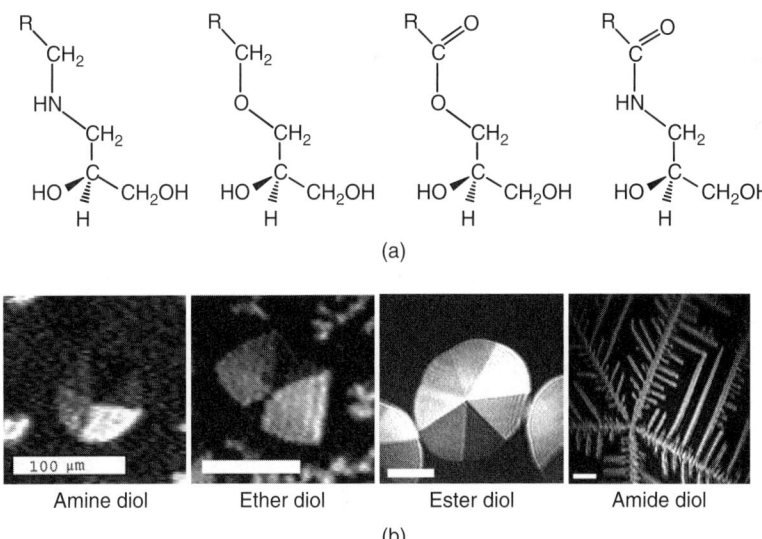

Figure 8.7 (a) Comparison of the chemical structure of the four monoglcerols. The variation in the chemical structure in between the monoglycerol group and the alkyl chain is shown and compared in detail in (b) the representative BAM images of four diols.

millimeters. On the other hand, the monoglycerol amine domains have a fractal nature with certain fluidity. Correspondingly, the orientational correlation in monoglycerol amine domains is several orders of magnitude smaller than that of the monoglycerol amide. The monoglycerol ester and ether domains have crystallinity and orientational correlation of the structure between the monoglycerol amide and amine. The molecular azimuthal projection changes within the domains in monoglycerol ester monolayers and such a change in the director direction is also observed for monoglycerol ether monolayers. However, the domains of the monoglycerol ester have a closed and round shape whereas the domains of the monoglycerol ether are branched with elongated arms. The studies of amide and amine glycerols are particularly interesting because the amide and amine groups are integral parts of the general structure of sphingolipids [63, 64]. The hydrogen bonding patterns of sphingolipids have been well investigated due to their enigmatic properties [65]. In all cases, racemic mixtures of the amphiphiles are used, as the effect of chirality on the domain shape is expected to be less significant than the two main factors controlling the domain shape, namely, line tension and electrostatic repulsion. Little knowledge is available about the line tension of monoglycerol monolayers at the molecular level. However, a comparison of the dipolar interaction can provide insight into the role of the dipolar repulsion in the present case. The results of the theoretical study of dipolar interactions are described in the next section.

Dipolar interactions in monolayers can have a subtle temperature dependence. It is largely accepted that the shape transition is governed by the competition between the shape-dependent dipolar repulsion, which favors elongated shapes, and the line tension between the condensed phase domain and the surrounding fluid phase, which favors compactness of the domain. Furthermore, the shapes could be anisotropic due to molecular chirality as well as the in-plane dipole moment, which scale as anisotropic line tension. It is commonly observed that the domain shapes become more compact with decreasing temperature. The electrostatic repulsion between headgroups is expected to be reduced by an increase in the static dielectric constant of the subphase with decreasing temperature [66]. The line tension is also expected to increase with decreasing temperature. Consequently, for the usual temperature dependence of the domain form, a compact shape would be favored with decreasing temperature whereas elongated shapes will be favored with increasing temperature. However, an anomalous temperature dependence is observed for the condensed phase domains of 1-hexadecyl-*rac*-glycerol monolayers as compact domain shapes are formed at higher temperature, but elongated arms or even long strips are developed as the temperature is lowered (Fig. 8.8). Such behavior indicates the increasing importance of the dipolar interaction at lower temperature.

Theoretical as well as experimental techniques have shown that the ratio of the difference in dipole density of the fluid and condensed phase of the monolayer at a particular temperature and the fluid/condensed phase line tension dictate the transition in the domain shape [19, 67–71]. Thus the observed anomaly is expected to be explained on the basis of the dipole density difference at various

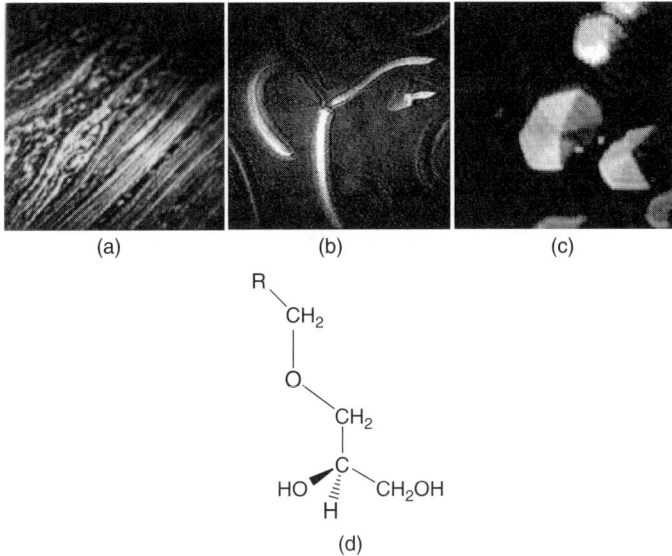

Figure 8.8 Domain shapes of rac-1-O-hexadecyl glycerol monolayers at (a) 278.15 K, $\pi = 0.1$ mN/m, (b) 288.15 K, $\pi = 2.2$ mN/m, and (c) 308.15 K, $\pi = 18$ mN/m. The chemical structure of an alkyl ether-diol molecule, where R = $(CH_2)_{14}(CH_3)$, is shown in (d).

temperatures. The dipole density difference at a particular temperature is dependent on the respective densities in the fluid and condensed phases. No significant average density difference is expected for the fluid phase at two different temperatures because the molecules are already randomly oriented at temperatures considered in the present case and the density is rather low compared to the condensed phase. The dipole density of the condensed phase should be responsible for the observed anomaly. Therefore we calculated the dipole moment of the monoglycerolether molecule with the variation in the temperature for different stable (energy-minimized) molecular conformations. Subsequently, we analyzed the in-plane and out-of-plane components of the dipole moment to understand the effect of the condensed phase dipole density.

The differences in the chemical structure of the polar headgroup in lipids are also noted as described for glycerolic amphiphiles. A comparison of the chemical structure of DPPC and DPPE shows that the two lipid structures are identical except for the headgroup region. While the DPPC has $—N^+(CH_3)_3$ present in the headgroup, the DPPE headgroup has $—N^+(CH_3)_3$ in the corresponding molecular segment. Despite this similarity, the domain shapes are different, revealing differences in the underlying molecular interaction (Fig. 8.4). Both equilibrium and nonequilibrium domain shapes of DPPC and DPPE have been investigated combining experimental techniques such as surface pressure isotherm measurement, BAM, GIXD, and theory, with a major focus on the effect of chirality, particularly on the shape of the DPPC [20–22, 24, 37, 48, 49, 71, 72]. Remarkable differences in the mesoscopic domain shapes can be observed by BAM. The equilibrium enantiomeric domain of DPPC develops elongated-arm triskelion structures with curvature in the arms, whereas the enantiomeric domains of DPPE are nearly round-shaped and subdivided by curved defect lines at which the orientation jumps into regions of varying molecular azimuthal projections. The curvature in the arm (as well as variation of the molecular azimuthal projection) in the case of DPPC was explained in terms of chirality of the constituent molecule [48] and the same reasoning can be used to explain the variation of the direction of molecular azimuthal projections in DPPE domains. However, the remarkable difference observed in the shapes, namely, elongated arms in DPPC and round shape in DPPE, indicates that the relative strengths of dipolar interaction and line tension play an important role in determining the drastically different shapes of the two lipids of closely similar chemical structure. It is expected that the observed differences in the overall shape (explicitly, the elongated arms of DPPC and the nearly round shape of DPPE) are due to the difference in the polarity of the headgroups. A comparison of the dipolar interaction can provide insight into the role of dipolar repulsion in determining the domain shape variation. We calculated the average dipole moment using quantum mechanical methods and its components using experimental information about the molecular orientation and lattice structure at the interface from the GIXD data [72]. The calculation of the *out-of-plane* and *in-plane* dipole moments and the area occupied by a molecule provided information about the dipolar repulsion in such systems.

Phospholipid Langmuir monolayers have widely been used as biomimetic model systems to investigate molecular-level interactions of biologically relevant molecules such as peptides and pharmaceutical drugs with cell membranes [73, 74]. Despite their simplicity compared to biological systems, Langmuir monolayers represent useful model systems to provide important insight into molecular interactions because the distance between molecules can easily be controlled by compression. In recent work, an interesting feature emerged, which is associated with the cooperative response of phospholipid monolayers to the presence of guest molecules, where a small concentration of the drug or peptide was already sufficient to cause measurable changes in monolayer properties [75–77]. An example was the interaction of dipyridamole (2,6-bis(diethanolamine)-4,8-dipiperidinopyrimido-[5,4-*d*]pyrimidine) (DIP), known as coronary vasodilator, antioxidant, or coactivator of antitumor activity and other drug effects [73], with DPPC [74, 77]. The properties of DPPC monolayers have been well documented in the literature [18, 19, 37, 48, 57, 58, 72] while the incorporation of DIP in DPPC monolayers was investigated using surface pressure isotherms, GIXD [76] surface potential measurements, fluorescence microscopy, and Fourier transform infrared reflection absorption spectroscopy (FT-IRRAS) [77]. These experimental studies revealed that the presence of DIP markedly changes the properties of a pure DPPC monolayer, due to the above-mentioned cooperative response. Moreover, the effects from DIP depend strongly on its relative concentration, c_{DIP}, when co-spread with DPPC. This is the case for the surface pressure at which the liquid-expanded (LE) to liquid-condensed (LC) phase transition starts. The ratio between the onset pressure for the mixed and pure DPPC, $\pi_{rel} = \pi_c(\text{DIP–DPPC})/\pi(\text{DPPC})$, gradually decreases from 1 to 0.75 for c_{DIP} in the range between 0 and 0.25 mol% and then increases back to unity. Similarly, the relative molecular area, $\Delta A_{rel} = \Delta A(\text{DIP–DPPC})/\Delta A(\text{DPPC})$, shows anomalous variation as a function of surface pressure for all c_{DIP} values ranging from 1 mol% to 15 mol%. ΔA_{rel} rises up to a maximum value of 1.08 for a surface pressure of 10 mN/m and then gradually decreases to 1.0 [76, 77]. The same applies to the monolayer surface potential: taking the surface potential at the onset of the LE–LC phase transition, $\Delta V_{rel} = \Delta V(\text{DIP–DPPC})/\Delta V(\text{DPPC})$ rises up to 1.075 for $c_{DIP} = 0.5$ mol% and sharply decreases to approximately 0.925 and retains this steady low value for higher c_{DIP}. More evidence coming from the FTIR data was that DIP interacts preferentially with the phosphate group of the zwitterion in the DPPC molecule, which was expected because DIP is protonated at the pH of the experiments [77]. It is clear that some sort of recognition process of DIP by DPPC at the molecular length scale is responsible for the observation of these macroscopic anomalies. DIP effects are also induced at the mesoscopic length scales, as demonstrated with fluorescence microscopy images indicating that in the domains of the DIP–DPPC system a larger number of arms grow from the nucleation center in comparison to the pure DPPC monolayer. In fact, at high surface pressures, triskelion-shaped domains are observed. The LE–LC boundary lines are more elongated in the DIP–DPPC monolayer than in the pure DPPC monolayer; that is related to the development of spikes with the increase in surface pressure. The incorporation of DIP induces changes in the monolayer properties at all length scales, from macro- to mesoscopic dimensions. The molecular recognition of the DIP–DPPC system raises several questions such as the concentration dependence of the surface pressure and surface potential isotherms of the mixed DIP–DPPC monolayers, and the appearance of spikes in the domains.

These studies indicate that polarity-dependent interactions can play an important role in modifying the structure as well as recognition in membrane mimetic systems. It is necessary to look into the difference in polarity of these amphiphiles and such studies are described next.

8.5 DIPOLAR INTERACTION IN MEMBRANE MIMETIC SYSTEMS: THEORETICAL STUDIES

Essentially, the dipolar nature of the headgroup is the major difference in the chemical structure of the amphiphiles described in the preceding section. It can be presumed that, other factors determining the morphology of the condensed phase domains being similar, an analysis of the dipolar nature of the headgroup can provide a clue to the molecular understanding of the observed phenomena. The large size of the molecule and the possibility of various conformations of the headgroup require that computation of the dipole moment is done at a moderate level of theory such as the semiempirical quantum chemical method. Since a reasonably large population of dipole moments needed to be calculated and only the relative strength of interactions would be studied, the use of the semiempirical method seems justified. The average dipole moment can be calculated using quantum mechanical methods and its components can be analyzed using experimental information about the molecular orientation and lattice structure at the interface from the GIXD data. We describe below quantum chemical studies aimed in this direction.

The shape variation of monoglycerol domains with small change in headgroup structure was described in the previous section [59–62]. The dipole moments of different energy-minimized conformations of four monoglycerol compounds have recently been calculated [43]. Subsequently, the contributions of the *in-plane* and *out-of-plane* dipole moments are analyzed using the GIXD data. Lattice structural information is used to determine the relative dipolar repulsion within neighboring molecules of the four types of monoglycerol monolayers. Due to the presence of the interface, the headgroup dipole is restricted within the aqueous subphase (generally insignificant polarity exists in the alkyl chain region), and it is useful to consider the in-plane (plane of interface) and out-of-plane components of the dipole moment. In the fluid phase of amphiphilic monolayers, the in-plane component cancels out on an average (due to free rotation of the molecule at the interface), whereas it is not canceled in the condensed monolayer phases due to restricted molecular rotation. It is possible to calculate the populations of the in-plane and out-of-plane components of the dipole moment assuming that the geometry of the molecule is rigid and the molecule is oriented on average according to the X-ray diffraction data. Accurate information about the tilt of the amphiphilic alkyl chain can be obtained from GIXD data, whereas such detailed experimental information about the headgroups is unavailable. Different conformations of headgroups are possible with different average molecular orientation and magnitude. As a result, the possibility of different tilts and magnitudes of headgroup dipoles exists. The dipole moments for different energy-minimized conformations of each of the four amphiphilic monoglycerol molecules are calculated. The calculated dipole moments are then used to compute the *in-plane* and *out-of-plane* (perpendicular) components according to the experimentally observed molecular tilt and azimuthal orientation from the GIXD data. As the GIXD data provide the alkyl chain orientation, the calculation assumes a rigid geometry of the molecule. The dipole moment was calculated using a standard molecular modeling technique with the help of the well-known quantum mechanical program package (CHEM 3D software) [78] and the theory used is at the semiempirical PM3 level [79]. The charges were obtained using the Mulliken population analysis (MPA) technique [80]. The *in-plane* and *out-of-plane* components of dipole moments of all energy-minimized conformations are calculated. The calculated components represent the distribution of the range of the magnitude of the dipole moment components for the corresponding molecule.

The magnitude of the dipolar interaction energy is computed as follows. The interaction energy between two dipoles, μ_1 and μ_2, separated by a distance r, oriented at angles θ_1 and θ_2 with respect to the line joining the center of the dipoles and a mutual angle φ between them, is given by

$$V_{\text{dipol}}(r, \theta_1 \theta_2 \phi) = -\frac{\mu_1 \mu_2}{4\pi \varepsilon_0 \varepsilon_s r^3}(2\cos\theta_1 \cos\theta_2 - \sin\theta_1 \sin\theta_2 \cos\phi) \quad (8.1)$$

where ε_0 is the vacuum permittivity and ε_s is the dielectric constant of the medium in which the dipoles are placed. Two extreme arrangements of *in-plane* dipole moments are possible (parallel and antiparallel). While tilt angle from the normal (to the interface) is the same in the two arrangements, the mutual arrangements of groups in the headgroup region are opposite. In the case of a racemic pair, the pair of molecules being mirror images of each other, the mutual orientation of the headgroups of neighboring molecules may remain in arrangements between two limiting possibilities. However, these two limiting group arrangements and the resulting dipolar orientation may result in both cooperativity as well as partial cancellation of the dipolar interaction due to the in-plane and out-of-plane dipole components depending on their mutual orientation. In one case, the μ_y and μ_z dipole directions give rise to repulsive interaction of magnitude $V_{\text{dipol}(1)} = \mu_{y/z}^2/4\pi\varepsilon_0\varepsilon_s r^3$; the μ_x components of neighboring dipoles are attracted by energy of magnitude $V_{\text{dipol}(2)} = -2\mu_x^2/4\pi\varepsilon_0\varepsilon_s r^3$. Similarly, for the other case, the μ_y and μ_z dipole directions give rise to repulsive interaction of magnitude $V_{\text{dipol}(1)} = \mu_{y/z}^2/4\pi\varepsilon_0\varepsilon_s r^3$; the μ_x components of neighboring dipoles are repelled by energy of magnitude $V_{\text{dipol}(2)} = 2\mu_x^2/4\pi\varepsilon_0\varepsilon_s r^3$. In the two cases, the net dipolar interaction energy should be different. It is impossible to say which of the two arrangements dominates the population of headgroup arrangements on the basis of knowledge of the available GIXD data. Both arrangements are considered as the two extreme limits of dipolar arrangements. The assignment of the dielectric constant at the aqueous interface is not straightforward [81–85]. For a microheterogeneous medium, like the headgroup region considered in the present case, the assignment of the dielectric constant is further complicated. It could be a function of different variables. One such variable could be the separation from the air–water dividing surface. Different empirical functions are used in the literature to represent the distance-dependent dielectric function [84, 85]. For example, recently, a hyperbolic tangent function was used to represent the dielectric function in the headgroup region [52]. However, such a choice of an empirical function could not be unique and caution must be exercised in representing the dielectric function of the headgroup region. A low value of dielectric constant of 2 is used in the calculation for all monoglycerols to compare the dipolar interactions of all compounds. The magnitude of the maximum dipolar repulsion increases in the order AMD < ETD < ESD < ADD.

Ab initio calculation of the intermolecular interaction between an enantiomeric pair of molecules of the four monoglycerols (ADD, ETD, ESD, AMD) is computationally expensive due to the large number of atoms present in the molecules. The discriminating factors in the intermolecular interactions between the pairs of molecules of each monoglycerol giving rise to the differences in the domain shape principally arise from the headgroup region. The intermolecular interaction is calculated by considering the complete molecular structure of the headgroup of each monoglycerol with a shorter alkyl chain (with $n = 1$ in the methylene group region and a CH_3—CH_2—group as a whole in the alkyl chain region). The geometry of the noncovalently bound pair of enantiomers of the four monoglycerol segments was optimized at the Hartree–Fock level of theory using the 6-31G** basis set. Ab initio calculations were performed using *Gaussian 03 W* software [86]. Due to the proximity of the hydrogen-bonded groups in the given geometry, a substantial part of the molecular interaction is due to hydrogen bond formation. The hydrogen bond properties are group properties and the binding energy of the hydrogen bond structure is greater than that of the sum of individual bonds. Consequently, the pair of molecules is expected to give a lower magnitude of binding energy compared to that in the aggregated state. However, the ab initio calculation of the energy of even a small part of the hydrogen bond cycle is unfeasible due to the high computational cost involved. The energy of interaction of a pair of enantiomers in the reoptimized geometry is denoted by E_{HB} and is given by

$$\Delta E_{HB} = E_{Dimer} - (E_{monomer1} + E_{monomer2}) \tag{8.2}$$

To understand the role of intra- as well as intermolecular hydrogen bonding interactions, the Coulombic attractions [87] between the donor and acceptor pairs of all monoglycerol headgroup segments were computed [43]. Isolated headgroup segments already have intramolecular interactions between the donor and acceptor groups in their individual energy-minimized structures. When two such molecules are brought together, intermolecular interactions develop and now the same hydrogen bond donor and acceptor groups are in competition to form intermolecular interactions [88]. As a result, the strength of the intermolecular hydrogen bond is reduced if the donor and acceptor groups are involved in strong intramolecular hydrogen bondings. The Coulombic interaction energies between the H-atom and the corresponding nearest atom of the acceptor group were calculated. The calculation is strongly dependent on the dielectric constant of the medium, which is effectively the local dielectric constant of the different monoglycerols. The local dielectric constant is difficult to estimate. The ratio of intermolecular versus intramolecular Coulombic interactions (E_c) is calculated assuming that, in the two cases, the local dielectric constant is not drastically different for the same monoglycerol and the effect of the dielectric constant is canceled out as follows:

$$E_c = \frac{\sum_{ij} \frac{q_i q_j}{4\pi \varepsilon_0 \varepsilon_s r_{ij}}}{\sum_{kl} \frac{q_k q_l}{4\pi \varepsilon_0 \varepsilon_s r_{kl}}} = \frac{\sum_{ij} \frac{q_i q_j}{r_{ij}}}{\sum_{kl} \frac{q_k q_l}{r_{kl}}} \tag{8.3}$$

where the sum *ij* runs over all intermolecular donor–acceptor pairs and *kl* runs over all intramolecular donor–acceptor pairs. The ratio gives the relative weightiness of the inter- or intramolecular hydrogen bonding interaction in each monoglycerol. The calculations clearly show that the magnitudes of both the *in-plane* and *out-of-plane* components of the dipole increase in the order AMD < ETD < ESD < ADD for the population of conformations within the surface pressure range studied. The observed increase in the dipolar nature of the monoglycerol headgroup in the order AMD < ETD < ESD < ADD is in perfect agreement with the fact that the ADD molecules with the largest dipole magnitude develop thin and straight-armed, highly crystalline domains with the largest orientational correlation length (about 5–10 mm), whereas the AMD monolayers form larger, fractal-like domains with the least orientational correlation length (<50 μm). The energy-minimized structures of the monoglycerol headgroup segments obtained on the basis of ab initio calculations and the corresponding ΔE_{HB} values indicate that the intermolecular interaction decreases in the order ETD < ESD < AMD < ADD.

For explaining the anomalous temperature dependence of the amphiphilic glycerolethers (Fig. 8.8), an analysis of the dipole moments of a population of conformers was carried out [44]. The dipole moments of the different conformations clearly indicated that the ratios of in-plane/out-of-plane dipole moment for all conformers are higher at low temperature relative to the corresponding ratios at high temperature. The GIXD data show that the tilt to the normal is gradually decreasing with increasing surface pressure. It is thus expected that the magnitude of the dipolar component to the normal (μ_z) should increase with increasing surface pressure and the in-plane component (μ_{xy}) should decrease. The lattice arrangement for a particular molecule has different temperature dependencies. This dependence is again controlled by intermolecular interactions as well as interaction with the aqueous interface, as shown by theoretical studies [89]. Thus the in-plane dipole moment is always dominant at lower temperatures in the pressure range investigated. The increase in the dominance of the in-plane dipole moment would lead to thinning of the domain in the corresponding tilt direction. The calculated effect of rotation of the amphiphile in a cone on the ratio of in-plane/out-of-plane dipole moment shows that the effect is not significant in general. McConnell and Moy [67] and Mayer and Vanderlick [90] draw attention to the fact that the in-plane dipole moment should tend to make condensed phase domains long and thin, provided they are all oriented in a particular direction. Although, measurements of the dipole density difference between fluid and condensed phase domains were carried out [91, 92], quantitative comparisons between theoretical predictions and experimental results were not made [90]. Experimental measurement of dipole moment is complicated for several reasons: for example, the complex nature of the interfacial dielectric function, the orientation and magnitude of the dipole moment, as well as assumptions involved in the theoretical models used to interpret the experimental data [93–95]. The role of the in-plane dipole moment in controlling the domain shape has not been demonstrated earlier due to these complications. However, theoretical study clearly demonstrates that the temperature dependence of molecular tilt and azimuthal orientation can have decisive influence on the contribution of the in-plane dipole moments aligned in a particular direction to the electrostatic repulsive interaction. A higher in-plane dipole moment at lower temperatures favors elongated shapes along the azimuthal tilt direction. The ordering of amphiphiles will be higher at lower temperature, which is also expected to favor the alignment of dipoles toward the azimuthal direction and thereby to increase the repulsive interaction in that direction. The study indicates the possibility of controlling the shapes of mesoscopic aggregates by designing the molecular structure and tuning dipolar interaction.

To explain the different shapes of the lipid domains of DPPC and DPPE [20–22, 24, 37, 48, 72], the dipole moments for different energy-minimized conformations of the DPPC and DPPE molecules were calculated [57]. The calculated dipole moments were used to compute the *in-plane* and *out-of-plane* (perpendicular) components according to molecular tilt and azimuthal orientation obtained from the GIXD data. DPPC has a higher population of the low energy conformation with large dipole moments than the low energy conformations of DPPE. The result can be understood from the fact that the size of the DPPC headgroup is larger than that of DPPE. The dipole moment of a species depends on its charge separation in the molecular charge distribution. Since the size of the DPPC headgroup is significantly larger than that of DPPE, due to the presence of three methyl groups, conformations with a larger charge separation are more probable for DPPC than DPPE. This leads to a greater population of conformations with larger dipole moments in DPPC than in DPPE, as shown in Fig. 8.4. As a result of the relatively higher population of conformations with large dipole moments in DPPC than DPPE, the domain of the former develops elongated arms whereas that of the latter has a round shape.

The *out-of-plane* dipole moment components for DPPC and DPPE obtained for the surface pressure region between 15 and 45 mN/m were compared and they were larger for DPPC than for DPPE. The out-of-plane component increases for both molecules with increasing surface pressure. This is expected as the long molecular axis is moved toward the normal to the interface and the dipolar component toward the normal increases with increasing surface pressure. On average, the in-plane dipole moment of DPPC is almost double that of DPPE at corresponding surface pressures within the complete surface pressure range investigated. Consequently, the in-plane dipole moment in DPPC is approximately double that in DPPE, which is due to the larger molecular tilt in the DPPC domain than in the DPPE domain. This correlates with the experimental finding that the enantiomeric DPPC domains develop elongated arms because of the larger dipolar repulsion, whereas the DPPE domains are compact. In the latter case, obviously, the dipolar repulsion cannot overcome the line tension. The larger in-plane component leads to growth in the direction toward the average tilt, whereas relatively low dipolar repulsion in the tilt direction allows growth in various directions favoring a compact shape of the DPPE domains.

To understand the recognition between DIP and DPPC molecules [75, 76] and the anomalous experimental features, conformations of the DPPC molecule and DIP molecule were energy minimized at the semiempirical PM3 level [45].

Charges were obtained using the Mulliken population analysis (MPA) technique. A total of 90 energy-minimized conformers of DPPC were studied, all of them with all-trans configuration for the alkyl chains but with different conformations in the headgroup region. Ideally, one should consider a number of possible geometric arrangements for DIP in relation to the DPPC headgroup, but the computational cost would be prohibitive. One can, however, systematically study how the DIP–DPPC monolayer differs from a pure DPPC monolayer by considering three limiting cases as follows. In the first arrangement, the plane containing the rings of the DIP molecule is collinear with the DPPC alkyl chains, which is referred to here as "vertical" arrangement. Second, the plane containing the rings of the DIP molecule is perpendicular to the DPPC alkyl chains, which is denoted as the "lateral" arrangement. Finally, the plane containing the rings of the DIP molecule could be in between the vertical and lateral arrangements with the DPPC alkyl chains, referred to as the "intermediate" arrangement. Obviously, this classification for the mutual arrangements between DIP and DPPC is arbitrary, but the final energy-minimized structures are unbiased by the classification and the conclusions reached are independent of the classification. The total dipole moment and its vertical component of 90 energy-minimized conformations for DIP–DPPC are calculated. The vertical component, $\mu_\perp$, is related with the surface potential, ΔV, as [94]

$$\Delta V = \frac{\mu_\perp}{\varepsilon_0 A}$$

where A is the area per molecule and ε_0 is the vacuum permittivity.

The experimental studies described earlier indicate that the surface potential (and therefore $\mu_\perp$) increases for low DIP concentrations and then decreases at larger concentrations. This change in the behavior was attributed to self-aggregation of DIP molecules as its relative concentration increased [75, 76]. A consistent description of the conformation of the DIP–DPPC system suggests that DIP positioning could change with the concentration, with DIP adopting a lateral or intermediate arrangement at higher concentrations. Because in all three arrangements the effect of DIP is to increase the dipole moment for the most conformers, the increased surface potential at very low DIP concentrations is readily explained. One may conclude that a vertical arrangement should be preferred at low DIP concentrations, although the intermediate or lateral arrangements cannot be discarded. The area per molecule is expected to increase due to the enhanced electrostatic repulsion in comparison with DPPC, as experimentally observed. On the other hand, at high DIP concentrations a vertical arrangement is unlikely because of the difficulties in accommodating the large vertical dipole moment component of DIP, and DIP appears to flip to a lateral (or intermediate) arrangement, probably in one of the conformers with a reduced normal component of dipole moment. The theoretical calculations presented point to an increased dipolar repulsion caused by incorporation of DIP, which is consistent with the development of spikes from the end of triskelions, as the growth of more elongated domains is a signature of enhanced electrostatic repulsion.

8.6 CONCLUSION

We have discussed how chirality and polarity of individual molecules influence the mesoscopic structure composed of lipids and glycerolic amphiphiles. The influence of both factors on the mesoscopic aggregate structure and recognition processes are discussed. Various experimental techniques, which can probe the molecular architecture of these mimetic systems at different length scales, have been used and macroscopic (thermodynamic) as well as microscopic (molecular) theories have been developed. These early studies focused on the role of chirality and polarity in the molecular interaction when the corresponding molecule is present in the condensed phase structure. The studies are still limited, considering the wide range of biological lipids and their complexity. However, molecular studies (experiment and theory) seem quite promising.

It may be noted that an obvious aim for such biomimetic studies is to transfer the knowledge gained to more complicated membrane systems, where the same fundamental physical principles are operative but the manifestation of the interaction is different due to the structural variation and complexity. Due to complexity of membranes where additional components such as proteins and carbohydrates are present, the manifestation is expected to be exotic and worth exploring due to the important biological role of membranes. More detailed and systematic studies are required to understand the role of chirality and polarity in membranes.

ACKNOWLEDGMENTS

The work described in this chapter has been supported in part by the Council of Scientific and Industrial Research, Government of India; Department of Science and Technology, Government of India; and the Alexander von Humboldt Stiftung, Germany.

REFERENCES

1. E. M. Arnett, N. G. Harvey, and P. L. Rose (1989). Stereochemistry and molecular recognition in two dimensions. *Acc. Chem. Res.* **22**: 131–138.
2. A. B. Harris, R. D. Kamien, and T. C. Lubensky (1999). Molecular chirality and chiral parameters. *Rev. Mod. Phys.* **71**: 1745–1757.
3. (a) C. M. Knobler (1990). Recent developments in the study of monolayers at the air–water interface. *Adv. Chem. Phys.* **77**: 397–449. (b) C. M. Knobler and R. C. Desai (1992). Phase transitions in monolayer. *Annu. Rev. Phys. Chem.* **43**: 207–236.
4. F. Bringezu, G. Brezesinski, P. Nuhn, and H. Möhwald (1996). Chiral discrimination in a monolayer of triple-chain phosphatidylcholine. *Biophys. J.* **70**: 1789–1795.

5. A. Dietrich, G. Brezesinski, H. Möhwald, B. Dobner, and P. Nuhn (1994). Domain shapes and monolayer structures of triple-chain phospholipids on water. *Nuovo Cimento D* **16**: 1537–1544.
6. A. Dietrich, H. Möhwald, W. Rettig, and G. Brezesinski (1991). Polymorphism of a triple-chain lecithin in two- and three-dimensional systems. *Langmuir* **7**: 539–546.
7. G. Brezesinski, C. Böhm, A. Dietrich, and H. Möhwald (1994). Condensed phases in monolayers of a triple-chain lecithin on water. *Phys. B* **198**: 146–149.
8. G. Brezesinski, A. Dietrich, B. Dobner, and H. Möhwald (1995). Morphology and structures in double-, triple- and quadruple-chain phospholipid monolayers at the air/water interface. *Prog. Colloid Polym Sci.* **98**: 255–262.
9. H. Gruniger, D. Mobius, and H. Meyer (1983). Enhanced light reflection by dye monolayers at the air–water interface. *J. Chem. Phys.* **79**: 3701–3710.
10. M. Orrit, D. Möbius, U. Lehmann, and H. Meyer (1986). Reflection and transmission of light by dye monolayers. *J. Chem. Phys.* **85**: 4966–4979.
11. R. Loschek and D. Möbius (1988). Metallation of porphyrins in lipid monolayers at the air–water interface. *Chem. Phys. Lett.* **151**: 176–182.
12. U. Lehmann (1988). Aggregation of cyanine dyes at Langmuir–Blodgett monolayers. *Thin Solid Films* **160**: 257–269.
13. D. Höenig and D. Möebius (1991). Direct visualization of monolayers at the air–water interface by Brewster angle microscopy. *J. Phys. Chem.* **95**: 4590–4592.
14. D. Höenig and D. Möebius (1992). Reflectrometry at the Brewster angle and Brewster angle microscopy at the air–water interface. *Thin Solid Films* **210**: 64–68.
15. S. Henon and J. Meunier (1991). Microscope at the Brewster angle: direct observation of first-order phase transitions in monolayers. *Rev. Sci. Instrum.* **62**: 936–939.
16. S. Henon and J. Meunier (1993). Phase transitions in Gibbs films: star textural defects in tilted mesophases. *J. Chem. Phys.* **98**: 9148–9154.
17. D. Vollhardt (1996). Morphology and phase behavior of monolayers. *Adv. Colloid Interface Sci.* **64**: 143–171.
18. H. Möhwald (1990). Phospholipid and phospholipid–protein monolayers at the air/water interface. *Annu. Rev. Phys. Chem.* **41**: 441–476.
19. H. M. McConnell (1991). Structures and transitions in lipid monolayers at the air–water interface. *Annu. Rev. Phys. Chem.* **42**: 171–195.
20. G. Weidemann and D. Vollhardt (1995). Long range tilt orientational order in phospholipid monolayers: a comparison of the order in the condensed phases of dimyristoylphosphatidylethanolamine and dipalmitoylphosphatidylcholine. *Colloid Surf. A* **100**: 187–202.
21. G. Weidemann and D. Vollhardt (1995). Long-range tilt orientational order in phospholipid monolayers: the inner structure of dimyristoyl-phosphatidyl-ethanolamine domains. *Thin Solid Films* **264**: 94–103.
22. (a) G. Weidemann and D. Vollhardt (1996). Long-range tilt orientational order in phospholipid monolayers: a comparative study. *Biophys. J.* **70**: 2758–2766. (b) R. M. Weis and H. M. McConnell (1984). Two-dimensional chiral crystals of phospholipid. *Nature* **310**: 47–49.
23. D. Vollhardt (2006). *Encyclopedia of Surface and Colloid Science*, 2nd ed. (P. Somasundaran, Ed.). Taylor & Francis, New York, Vol. 5, pp. 4104–4118.

24. W. M. Heckl and H. Möhwald (1986). A narrow window for observation of spiral lipid crystals. *Ber. Bunsenges. Phys. Chem.* **90**: 1159–1163.
25. W. M. Heckl, M. Lösche, D. A. Cadenhead, and H. Möhwald (1986). Electrostatically induced growth of spiral lipid domains in the presence of cholesterol. *Eur. Biophys. J.* **14**: 11–17.
26. L. A. Worthman, K. Nag, P. J. Davis, and K. M. Keough (1997). Cholesterol in condensed and fluid phosphatidylcholine monolayers studied by epifluorescence microscopy. *Biophys. J.* **72**: 2569–2580.
27. J. Als-Nielsen and H. Möhwald (1991). In *Handbook of Synchrotron Radiation* (S. Ebashi, M. Koch, and E. Rubenstein, Eds). Elsevier Sciene, Amsterdam, Vol. 4, pp. 1–2.
28. J. Als-Nielsen, D. Jacquemain, K. Kjaer, M. Lahav, F. Leveiller, and L. Leiserowitz (1994). Principles and applications of grazing incidence X-ray and neutron scattering from ordered molecular monolayers at the air–water interface. *Phys. Rep.* **246**: 251–313.
29. J. Als-Nielsen and K. Kjaer (1989). *Phase Transition in Soft Condensed Matter* T. Riste and D. Sherrington, Eds.), Plenum Press, New York.
30. J. Penfold and R. K. Thomas (1990). The application of the specular reflection of neutrons to the study of surfaces and interfaces. *J. Phys. Condensed Mat.* **2**: 1369–1412.
31. V. M. Kaganer, H. Möhwald, and P. Dutta (1999). Structure and phase transitions in Langmuir monolayers. *Rev. Mod. Phys.* **71**: 779–819.
32. I. Kuzmenko, H. Rapaport, K. Kjaer, J. Als-Nielsen, I. Weissbuch, M. Lahav, and L. Leiserowitz (2001). Design and characterization of crystalline thin film architectures at the air–liquid interface: simplicity to complexity. *Chem Rev.* **101**: 1659–1696.
33. B. Berge, P. F. Lenne, and A. Renault (1998). X-ray grazing incidence diffraction on monolayers at the surface of water. *Curr. Opin. Colloid Interface Sci.* **3**: 321–326.
34. R. Rietz, W. Rettig, G. Brezesinski, W. G. Bouwman, K. Kjaer, and H. Möhwald (1996). Monolayer behavior of chiral compounds at the air–water interface: 4-hexadecyloxy-butane-1,2-diol. *Thin Solid Films* **284–285**: 211–215.
35. R. Rietz, G. Brezesinski, and H. Möhwald (1993). Separation of enantiomers in a monolayer of racemic 3-hexadecyl-oxy-propane-1,2-diol. *Ber. Bunsenges. Phys. Chem.* **97**: 1394–1399.
36. G. Brezesinski, A. Dietrich, C. Struth, C. Böhm, G. Bowman, K. Kjaer, and H. Möhwald (1995). Influence of ether linkages on the structure of double-chain phospholipid monolayers. *Chem. Phys. Lipids* **76**: 145–157.
37. C. Böhm, H. Möhwald, L. Leiseowitz, J. Als-Nielsen, and K. Kjaer (1993). Influence of chirality on the structure of phospholipid monolayers. *Biophys. J.* **64**: 553–559.
38. D. Andelman and H. Orland (1993). Chiral discrimination in solutions and in Langmuir monolayers. *J. Am. Chem. Soc.* **115**: 12322–12329.
39. N. Nandi and B. Bagchi (1996). Molecular origin of the intrinsic bending force for helical morphology observed in chiral amphiphilic assemblies: concentration and size dependence. *J. Am. Chem. Soc.* **118**: 11208–11216.
40. N. Nandi and B. Bagchi (1997). Prediction of the senses of helical amphiphilic assemblies from effective intermolecular pair potential: studies on chiral monolayers and bilayers. *J. Phys. Chem. B.* **101**: 1343–1351.

41. N. Nandi and D. Vollhardt (2003). Effect of molecular chirality on the morphology of biomimetic Langmuir monolayers. *Chem. Rev.* **103**: 4033–4075.
42. N. Nandi and D. Vollhardt (2006). Chirality and molecular recognition in biomimetic organized films, in *Bottom Up Nanofabrication: Supramolecules, Self Assemblies and Organized Films* (K. Ariga, Ed.). American Scientific Publishers, Chap. 5, pp. 1–29.
43. K. Thirumoorthy, N. Nandi, and D. Vollhardt (2005). Role of electrostatic interactions for the domain shapes of Langmuir monolayers of monoglycerol amphiphiles. *J. Phys. Chem. B* **109**: 10820–10829.
44. N. Nandi and D. Vollhardt (2004). Anomalous temperature dependence of domain shape in Langmuir monolayers: role of dipolar interaction. *J. Phys. Chem. B* **108**: 18793–18795.
45. K. Thirumoorthy, N. Nandi, D. Vollhardt, and O. N. Oliveira, Jr. (2006). Semiempirical quantum mechanical calculations of dipolar interaction between dipyridamole and dipalmitoyl phosphatidyl choline in Langmuir monolayers. *Langmuir* **22**: 5398–5402.
46. N. Nandi and D. Vollhardt (2002). Prediction of the handedness of the chiral domains of amphiphilic monolayers: monolayers of amino acid amphiphiles. *Colloid Surf. A* **198–200**: 207–221.
47. N. Nandi and D. Vollhardt (2000). Microscopic study of chiral interactions in Langmuir monolayer: monolayers of N-palmitoyl aspartic acid and N-stearoyl serine methyl ester. *Colloid Surf. A* **183–185**: 67–83.
48. N. Nandi and D. Vollhardt (2002). Molecular origin of the chiral interaction in biomimetic systems: dipalmitoylphosphatidylcholine Langmuir monolayer. *J. Phys. Chem. B* **106**: 10144–10149.
49. N. Nandi, R. K. Roy, X. Anupriya, S. Upadhaya, and D. Vollhardt (2002). Chiral interaction in enantiomeric and racemic dipalmitoyl phosphatidylcholine Langmuir monolayer. *J. Surf. Sci. Tech.* **18**: 51–66.
50. N. Nandi and D. Vollhardt (2003). Chiral discrimination effects in Langmuir monolayers: monolayers of palmitoyl aspartic acid, N-stearoyl serine methyl ester, and N-tetradecyl-γ,δ-dihydroxypentanoic acid amide. *J. Phys. Chem. B* **107**: 3464–3475.
51. N. Nandi, D. Vollhardt, and G. Brezesinski (2004). Chiral discrimination effects in Langmuir monolayers of 1-O-hexadecyl glycerol. *J. Phys. Chem. B* **108**: 327–335.
52. N. Nandi, D. Vollhardt, and R. Rudert (2004). Molecular pair potential of chiral amino acid amphiphile in Langmuir monolayers on the basis of an atomistic model. *Colloid Surf. A* **250**: 279–287.
53. K. Thirumoorthy, N. Nandi, and D. Vollhardt (2006). Prediction of the handedness of the domains of monolayers of d-N-palmitoyl aspartic acid: integrated molecular orbital and molecular mechanics based calculation. *Colloid Surf. A* **282–283**: 222–226.
54. P. Krüger and M. Lösche (2000). Molecular chirality and domain shapes in lipid monolayers on aqueous surfaces. *Phys. Rev. E* **62**: 7031–7043.
55. (a) U. Dahmen-Levison, G. Brezesinski, and H. Möhwald (1998). Specific adsorption of PLA_2 at monolayers. *Thin Solid Films* **327–329**: 616–620. (b) G. Brezesinski (unpublished data recorded at 20 °C and 41 mN/m).
56. N. Nandi and D. Vollhardt (2007). Molecular interactions in amphiphilic assemblies: theoretical perspective. *Acc. Chem. Res.* **40**: 351–360.

57. K. Thirumoorthy, N. Nandi, and D. Vollhardt (2007). The role of dipolar interaction in the mesoscopic domains of phospholipid monolayers: dipalmitoyl phosphatidyl choline and dipalmitoyl phosphatidyl ethanolamine. *Langmuir* **23**: 6991–6996.
58. U. Gehlert and D. Vollhardt (1994). The phase behavior of an ether lipid monolayer compared with an ester lipid monolayer. *Prog. Colloid Polym. Sci.* **97**: 302–306.
59. U. Gehlert, G. Weidemann, and D. Vollhardt (1995). Morphological features in 1-monoglyceride monolayers. *J. Colloid Interface Sci.* **174**: 392–399.
60. G. Brezesinski, E. Scalas, B. Struth, F. Bringezu, H. Möhwald, U. Gehlert, G. Weidemann, and D. Vollhardt (1995). Relating lattice and domain structures of monoglyceride monolayers. *J. Phys. Chem.* **99**: 8758–8762.
61. U. Gehlert, G. Weidemann, D. Vollhardt, G. Brezesinski, R. Wagner, and H. Möhwald (1998). Relating domain morphology and lattice structure in monolayers of glycerol amide lipids. *Langmuir* **14**: 2112–2118.
62. U. Gehlert and D. Vollhardt (2002). Molecular packing and textures of 1-stearylamine-*rac*-glycerol monolayers. *Langmuir* **18**: 688–693.
63. J. M. Berg, J. L. Tymoczko, and L. Stryer (2003). *Biochemistry*, 5th ed. W. H. Freeman, New York.
64. D. L. Nelson and M. M. Cox (2004). *Lehninger Principles of Biochemistry*, 4th ed. W. H. Freeman, New York.
65. E. Mobelli, R. Morris, W. Taylor, and F. Fraternali (2003). Hydrogen-bonding propensities of sphingomyelin in solution and in a bilayer assembly: a molecular dynamics study. *Biophys. J.* **84**: 1507–1517.
66. *CRC Handbook of Chemistry and Physics*, 73rd ed. (D. R. Lide, Ed.). CRC Press, Boca Raton, FL.
67. H. M. McConnell and V. T. Moy (1988). Shapes of finite two-dimensional lipids. *J. Phys. Chem.* **92**: 4520–4525.
68. T. K. Vanderlick and H. Möhwald (1990). Mode selection and shape transitions of phospholipid monolayer domains. *J. Phys. Chem.* **94**: 886–890.
69. J. M. Deutch and F. E. Low (1992). Theory of shape transitions of two-dimensional domains. *J. Phys. Chem.* **96**: 7097–7101.
70. J. A. Miranda (1999). Closed form results for shape transitions in lipid monolayer domains. *J. Phys. Chem. B* **103**: 1303–1307.
71. R. de Koker and H. M. McConnell (1993). Circle to dogbone: shapes and shape transitions of lipid monolayer domains. *J. Phys. Chem.* **97**: 13419–13424.
72. E. Maltseva (2005). Doctoral thesis, University of Potsdam.
73. M. F. Nepomuceno, M. E. O. Mamede, D. V. Macedo, A. A. Alves, L. Pereira-da-Silva, and M. Tabak (1999). Antioxidant effect of dipyridamole and its derivative RA-25 in mitochondria: correlation of activity and location in the membrane. *Biochim. Biophys. Acta*. **1418**: 285–294.
74. P. M. Nassar, L. E. Almeida, and M. Tabak (1998). Binding of dipyridamole to DPPG and DPPC phospholipid vesicles: steady state fluorescence and fluorescence anisotropy decay studies. *Langmuir* **14**: 6811–6817.
75. H. Haas, W. Caetano, G. P. Borissevitch, M. Tabak, M. I. Mosquera Sanchez, O. N. Oliveira Jr., E. Scalas, and M. Goldman (2001). Interaction of dipyridamole with phospholipid monolayers at the air–water interface: surface pressure and grazing incidence X-ray diffraction studies. *Chem. Phys. Lett.* **335**: 510–516.

76. W. Caetano, M. Ferreira, M. Tabak, M. I. Mosquera Sanchez, O. N. Oliveira, Jr., P. Krüger, M. Schalke, and M. Lösche (2001). Cooperativity of phospholipid reorganization upon interaction of dipyridamole with surface monolayer on water. *Biophys. Chem.* **91**: 21–35.

77. O. N. Oliveira, Jr., A. Riul, Jr., and V. B. P. Leite (2004). Water at interfaces and its influence on the electrical properties of adsorbed films. *Brazilian J. Phys.* **34**: 73–83.

78. CHEM 3D Ultra, Version 8.0, © Cambridge Soft Corporation, USA.

79. J. P. Stewart (1989). Optimization of parameters for semiempirical methods I. Method. *J. Comp. Chem.* **10**: 209–220.

80. F. Jensen (1999). *Introduction to Computational Chemistry*. Wiley, Chichester, UK.

81. I. Benjamin (1995). Theory and computer simulations of solvation and chemical reactions at liquid interfaces. *Acc. Chem. Res.* **28**: 233–239.

82. I. Benjamin (1997). Molecular structure and dynamics at liquid–liquid interfaces. *Annu. Rev. Phys. Chem.* **48**: 407–451.

83. N. Nandi, K. Bhattacharyya, and B. Bagchi (2000). Dielectric relaxation and solvation dynamics of water in complex chemical and biological systems. *Chem. Rev*.**100**: 2013–2046.

84. M. A. Young, B. Jayaram, and D. L. Beveridge (1998). Local dielectric environment of B-DNA in solution: results from a 14 ns molecular dynamics trajectory. *J. Phys. Chem. B* **102**: 7666–7669.

85. E. W. Castner Jr., G. R. Fleming, B. Bagchi, and M. Maroncelli (1988). The dynamics of polar solvation: inhomogenous dielectric continuum models. *J. Chem. Phys.* **89**: 3519–3534.

86. M. J. Frisch et al (2004). *Gaussian 03*, Revision C.02, Gaussian, Inc., Wallingford, CT.

87. J. N. Israelachvili (1985). *Intermolecular and Surface Forces, with Applications to Colloidal and Biological Systems*. Academic Press, New York.

88. G. A. Jeffrey and W. Saenger (1991). *Hydrogen Bonding in Biological Structures*. Springer Verlag, Berlin.

89. Z. Cai and S. A. Rice (1990). Langmuir monolayers: structures and phase transitions. *Faraday Discuss. Chem. Soc.* **89**: 211–229.

90. M. A. Mayer and T. K. Vanderlick (1995). Calculation of shapes of dipolar domains in two-dimensional films: effect of dipole tilt. *J. Chem. Phys.* **103**: 9788–9794.

91. A. Miller, C. A. Helm, and H. Möhwald (1987). The colloidal nature of phospholipid monolayers. *J. Physique* **48**: 693–701.

92. H. M. McConnell (1993). Elementary theory of Brownian motion of trapped domains in lipid monolayers. *Biophys. J.* **64**: 577–580.

93. R. J. Demchak and T. Fort, Jr. (1974). Surface dipole moments of close-packed un-ionized monolayers at the air–water interface. *J. Colloid Interface Sci.* **46**: 191–202.

94. P. Dynarowicz-Latka, A. Cavalli, D. A. Silva Filho, M. C. dos Santos, and O. N. Oliveira, Jr. (2000). Dipole moments in Langmuir monolayers from aromatic carboxylic acids. *Chem. Phys. Lett.* **326**: 39–44.

95. C. N. Alves, M. Castilho, L. H. Mazo, M. Tabak, and A. B. F. da Silva (2001). Theoretical calculations on dipyridamole structure allow to explain experimental properties associated to electrochemical oxidation and protonation. *Chem. Phys. Lett.* **349:**146–152.

CHAPTER 9

Organic and Inorganic Osmolytes at Lipid Membrane Interfaces

PETER WESTH
MEMPHYS—Center for Biomembrane Physics, Odense, Denmark and NSM, Research Unit for Functional Biomaterials, Roskilde University, Roskilde, Denmark

GÜNTHER H. PETERS
MEMPHYS–Center for Biomembrane Physics, Odense, Denmark and Department of Chemistry, Technical University of Denmark, Lyngby, Denmark

CONTENTS

9.1	Introduction	228
9.2	Membrane Partitioning	230
9.3	Preferential Interactions	231
9.4	Free Energy of Membrane–Organic Osmolyte Interactions	234
9.5	Thermochemistry of Membrane–Organic Osmolyte Interactions	236
9.6	Driving Forces for Preferential Exclusion of Organic Osmolytes	241
9.7	Phase Behavior of Membranes in Organic Osmolyte Solutions	244
9.8	Organic Osmolytes and Membrane Stability	246
9.9	Inorganic Osmolytes at Membrane Interfaces	247
9.10	Thermodynamic Consideration	248
9.11	Inorganic Ions and Membrane Structure	253
9.12	Conclusion	256
	Acknowledgments	256
	References	257

Structure and Dynamics of Membranous Interfaces, edited by Kaushik Nag
Copyright © 2008 John Wiley & Sons, Inc.

9.1 INTRODUCTION

The word osmolytes generally refers to species that govern the intracellular osmotic pressure. Under normal conditions, the major part of the osmotic pressure in an animal cell is due to inorganic ions and Donnan-type colloid effects. Under water stress, however, osmotic regulation relies at least in part on *organic osmolytes*—a group of small uncharged or zwitterionic molecules, which includes sugars and sugar alcohols (e.g., trehalose, glycerol, and sorbitol), amino acids (e.g., proline, taurine, and glycine betaine), and methyl amines (e.g., urea and trimethyl amineoxide, TMAO). Many organic osmolytes exhibit a remarkable noneffect on cellular function, even when their concentration extends into the molar range. They are accordingly often denoted *compatible solutes* [1]. One noticeable exception is urea, an organic osmolyte found in some marine organisms [2]. This solute, a well-known denaturant, strongly perturbs the structure and function of proteins and is therefore a *noncompatible solute*. (*In vivo*, the perturbing effect of urea is counterbalanced through the coaccumulation of other organic osmolytes. In sharks, for instance, urea and TMAO always occurs in a 1:2 (urea:TMAO) molar ratio [3].) Many reports have pointed out that the effect of compatible solutes may include a compensation of the stress on, for example, protein structural stability, which arises from the increased concentration of ions and other effects coupled to water stress. It has therefore been suggested that *compensatory solutes* is a more appropriate name at least for some of these solutes [1]. Organic osmolytes (or subgroups of these) are also sometimes named according to a particular adaptation. *Cryoprotectants*, for example, are organic osmolytes accumulated during the adaptation to cold in temperate and polar invertebrates [4].

In this chapter, we discuss the interactions of organic osmolytes and membranous interfaces, and the effects of these interactions on the properties of the membrane. We also include a treatment of inorganic ions at the membrane interface since osmolyte effects involve a balance between organic and inorganic components. Before turning to the physicochemical discussion of interfacial interactions, we outline some central parts of the biology and biotechnology of organic osmolytes.

All animal cells have appropriate volumes, which are regulated through complex mechanisms including channels and transporters for inorganic ions. The importance of this regulation is obvious when the cell is exposed to variable salinities, but even at constant extracellular osmotic pressure, metabolism and other cellular activities generate changes, which require incessant osmotic adjustments [5]. Recent research has stressed that the importance of volume regulation is not limited to osmotic and mechanical problems of the cell, but directly influences an array of crucial processes including transepithelial transport, cell migration, cell growth, cell death, and metabolic regulation [6].

In addition to controlled fluxes of inorganic ions, the volume regulation relies on transport and *de novo* synthesis of organic osmolytes. One central role of this latter group is to act as osmoregulators in a nonspecific (colligative) fashion, but 25 years of research in the area has revealed a significant number of other biological functions of the organic osmolytes. Much of this research has been guided by the seminal review of Yancey et al. [7]—undoubtedly the most quoted contribution in the field. These workers discussed the compatibility of cellular function and osmolytes, based, for example, on *in vitro* experiments, which showed that the activity of some enzymes was essentially unaffected by organic osmolytes, whereas moderate concentrations of inorganic ions and other selected solutes strongly affected enzyme function [8]. Later work has confirmed these ideas [1, 9] and shaped the "compatibility paradigm" for organic osmolytes. This implies that evolution has selected a rather narrow group of organic osmolytes based on their noneffect on functional activity over a wide concentration range. Since the perturbing effect of inorganic ions precludes high intracellular concentrations, organic osmolytes (or compatible solutes) are evidently necessary for cellular volume regulation during exposure to moderate or high osmotic pressures. However, much evidence now suggests that organic osmolytes serve many other purposes than being just inert regulators of cell volume. These ideas of a broader role of organic osmolytes were also coined by Somero, Yancey, and co-workers [7, 10], who discussed the possible coevolution of biological macromolecules and their aqueous environment. It was suggested that regulation of the properties of biological solutions was at least in part brought about by changes in organic osmolyte concentrations. The evolutionary benefit of this strategy is genetic simplicity. Alternatively, a collection of isoenzymes would be required for the organism to cope with variable environmental conditions (e.g., salt concentrations). This is illustrated by a group of halophile Achaeans, which accumulate large amounts of inorganic ions and have developed enzymes that work under these conditions. However, many of these enzymes require >1 M salt for effective function and thus confines the organism to certain surroundings [7]. Plants, animals, bacteria, algae, and yeasts, on the other hand, all use organic osmolytes to regulate the cytosol properties and hence sustain the activity of one enzyme as conditions change [7, 10]. A large body of more recent work has developed this idea further by revealing that organic osmolytes are intimately involved in adaptations to abiotic stress vectors such as drought, heat, cold, salt, chemical toxicity, and oxidative stress in virtually all taxa [6, 11–13]. The molecular understanding of these compensatory effects of organic osmolytes primarily rests on the physicochemical characterization of interactions of water, ions, organic osmolytes, and macromolecules. We review some aspects of this but note here that organic osmolytes also induce cellular responses that appear to involve the specific binding to a receptor and a subsequent initiation of a regulatory cascade. Taurine, for example, is involved in the regulation of a surprising diversity of processes including heart rhythm, blood pressure, neuronal excitability, body temperature, and learning [14]. These specific effects may be related to the fact that many (amino acid) osmolytes are structurally related to neurotransmitters and hence are neuroactive [15]. The common neurotransmitter γ-amino butyric acid (GABA), for example, is an organic osmolyte in some bacteria [16].

Alongside the growing interest within physiology and biochemistry, there is an increasing focus on organic osmolytes in biotechnology. One example is the use of solutes to alleviate the physical instability that is inherent to most biotech products. Clearly, the natural tendency to protect against environmental stresses suggests a potential of osmolytes as so-called excipients in the formulation of pharmaceutical peptides, industrial enzymes, and other labile molecules. Today, lessons learned by *in vivo* studies of osmolytes are extensively used in formulation technologies [17]. While the use of osmolytes as excipients has successfully enhanced the shelf life of many products, there are also a few examples of technological problems relating to the stabilizing potential of organic osmolytes. Food preservation, for example, primarily relies on the application of physical stress, particularly cold. The adaptive response by pathogens and spoilage organisms includes osmolyte accumulation, and the control of such mechanisms is becoming relevant in the prevention of microbial growth in stored food [18]. A particularly interesting aspect of osmolyte technology comes from a number of attempts to genetically engineer plants to cope with suboptimal field conditions such as drought and increased salinity (e.g., see Ref. [11]). In many cases, the rationale behind such work has been to enhance osmolyte production in the modified plants.

9.2 MEMBRANE PARTITIONING

The vast majority of literature on membrane–solute interactions interprets the data along the lines of a partitioning approach, which assumes that the ratio of solute concentrations in, respectively, the membrane phase and the aqueous phase defines a conventional equilibrium constant, K_p. Hence $K_p = x_{mem}/x_{aq}$, where x is the concentration in some appropriate unit, and the subscripts specify, respectively, the membrane and the aqueous phase. In cases where this picture is indeed realistic, all thermodynamic parameters for the partitioning process can be defined by differentiating K_p with respect to pressure and temperature. This approach, however, carries a number of principal limitations. For example, the structured "solvent" in the membrane phase (i.e., the lipid chains) is likely to produce nonideal membrane–solute mixtures [19] and hence require some estimation of an activity coefficient for the calculation of K_p. Even more fundamental challenges come from the (extrathermodynamic) distinction between a partitioned and a free subpopulation of solute [20] and the specification of a concentration in an essentially two-dimensional system like a lipid bilayer. One of the problems defining a concentration is the hydration water, which is a part of the membrane phase but usually neglected in the calculations of solute concentration in the membrane. Since the number of intercalated water molecules in fluid PC is about 7–8 per lipid [21] and the total hydration probably involves over 30 water molecules [22], this shortcoming may severely skew the concentration calculation and hence the interpretation of partitioning coefficients. In spite of these (and other) limitations, the partitioning scheme has been effective in the description of membrane–solute interactions for highly hydrophobic solutes. For more polar solutes, on the other hand, the picture of a partitioning between ideal solutions in two separate phases does not provide a useful approach. The critical degree of hydrophobicity may be illustrated by comparing data obtained by methods, which either detect the concentration of solute in the membrane or rely on thermodynamic (model-independent) approaches. Doing so for the series of normal alcohols shows that thermodynamic and "contact detecting" methods provide identical results for alcohols larger than 1-propanol [23]. For shorter alcohols, on the other hand, the thermodynamic (vapor pressure) measurements consistently suggest lower affinity than, for example, radiotracer methods, which count the number of alcohol molecules in the membrane [23, 24]. This discrepancy most likely reflects an unfavorable solute interaction at the membrane interface, and for more polar solutes such as sugars, it becomes so strong that it corresponds to a negative partitioning coefficient, which is clearly not physically meaningful. We return to the origin of this effect later, but we note here that all osmolytes are less hydrophobic than 1-propanol, and hence that the analysis of osmolyte–membrane interactions cannot be rationalized on the basis of a partitioning scheme.

9.3 PREFERENTIAL INTERACTIONS

A number of approaches have been applied to overcome the shortcomings of the partitioning scheme for hydrophilic solutes. These include surface adsorption (or surface exchange) theories based on Langmuir isotherms [25] and modified partitioning schemes in which the hydration water at the membrane interface is implicitly described [26, 27]. A more general approach, however, comes from the theory of preferential interactions. This has its roots in the general thermodynamic treatment of three-component systems (e.g., water, solute, and biomolecule) [28]. In this section we briefly review a few of the central relationships in preferential interaction theory, which we use in subsequent paragraphs on membrane–osmolyte interactions. The theory has been developed and applied almost exclusively to the description of solute effects on protein and DNA transitions [29–32]. In a few cases, however, it has also been used (at least qualitatively) in discussions of membrane–solute interactions [23, 33–39].

The surface of a dissolved biomolecule interacts with the surrounding solvent, and these interactions contribute to the chemical potential, μ_2, of the biomolecule. Throughout this text we apply the so-called Scatchard notation in which subscripts 1, 2, and 3 refer, respectively, to water, biomolecule, and solute. If the biomolecule is transferred from pure water to a binary solvent (water + solute), its chemical potential changes by $\Delta\mu_2$. If the interaction of the solute and the biomolecule is more favorable than the water–biomolecule interaction, $\Delta\mu_2$ will be negative. Conversely, $\Delta\mu_2 > 0$ implies that water interacts more favorably than the solute with the biomolecule. The concentration dependence of μ_2 is known as the *preferential interaction parameter* (Eq. (9.1)) and the integral of this function from 0 to the concentration m_3 is the transfer free-energy change $\Delta\mu_2$ of moving the biomolecule from pure water to the mixed solvent [31]:

$$\left(\frac{\partial \mu_2}{\partial m_3}\right)_{T,P,m_2} = -\left(\frac{\partial m_3}{\partial m_2}\right)_{T,P,\mu_3}\left(\frac{\partial \mu_3}{\partial m_3}\right)_{T,P,m_2} \quad (9.1)$$

In Eq. (9.1), m denotes molal concentrations. For dilute biomolecule solutions ($m_2 \sim 0$) the first term on the right-hand side of the equation is the *preferential binding parameter*, $\Gamma_3 = (\partial m_3/\partial m_2)_{T,P,\mu_3}$. It signifies the number of solute molecules (subscript 3) that have to be added (or removed) to reestablish the solute chemical potential, μ_3, upon the addition of one biomolecule (subscript 2). As we see later, this is a useful parameter in discussions of membrane–solute interactions. It is also readily experimentally assessable since it is equal (to within a negligible approximation) to the binding number found by conventional dialysis equilibrium measurements [31]. For strong binding, such as the specific association with a receptor, Γ_3 is simply equal to the binding stoichiometry, but for weaker interactions structural (stoichiometric) interpretations and the thermodynamic binding concept expressed in Eq. (9.1) will diverge [40, 41]. In cases where $\Delta\mu_2 > 0$, the biomolecule interacts favorably with water and it follows that Γ_3 is negative; solute must be removed to reestablish its chemical potential upon addition of biomolecule. This condition is known as preferential hydration of the biomolecule or preferential exclusion of the solute. The most straightforward structural interpretation of this thermodynamic condition is that the interfacial layer has a lower concentration of solute than the bulk. This condition has been found for almost all investigated organic osmolytes at both protein and membrane interfaces.

Preferential interactions affect biomolecular equilibria. Thus if we consider an equilibrium between two states (A) and (B) (e.g., gel/fluid states of a membrane or native/denatured states of a globular protein), the effect of a solute will obviously depend on the difference $\Delta\mu_2(B) - \Delta\mu_2(A)$, that is, whether state (A) or (B) interacts more favorably with the solute. This has been stringently described in the linkage theory by Wyman [42]. One of the central results of this theory expresses the dependence of the equilibrium constant, K, on the solute activity, a_3 [43]:

$$\left(\frac{\partial \ln K}{\partial \ln a_3}\right) = \Gamma_3(2) - \Gamma_3(1) = \Delta\Gamma_3 \quad (9.2)$$

It follows from Eq. (9.2) that the change in the preferential binding parameter, $\Delta\Gamma_3$, directly specifies the solute-induced modulation of the equilibrium. In many cases, the effect of solutes on a biomolecular equilibrium is elucidated experimentally by measurements of a transition temperature, T_{trans}, as a function of the solute concentration. To analyze this type of data, the slope of the transition temperature may be derived from Eq. (9.2) [44]:

$$\frac{dT_{\text{trans}}}{dm_3} = \frac{RT_{\text{trans}}^2}{\Delta H} \Delta \Gamma_3 \frac{d \ln a_3}{dm_3} \qquad (9.3)$$

If we restrict our analysis to relatively dilute solutions, the nonideality of the aqueous solute may be neglected ($a_3 \sim m_3$), and Eq. (9.3) can be simplified to

$$\frac{dT_{\text{trans}}}{dm_3} \approx \frac{RT_{\text{trans}}^2}{\Delta H} \frac{\Delta \Gamma_3}{m_3} \qquad (9.4)$$

If applied to a phase transition such as the gel-to-fluid transition in a lipid bilayer, Eq. (9.4) describes the phase boundary analogously to the Clausius–Clapeyron equation. This analogy and its application to lipid phases were discussed by Koynova et al. [27], who also presented a derivation of a relationship corresponding to Eq. (9.4), which (in contrast to Eq. (9.4)) was based on phase equilibrium theory.

The preferential binding parameter may be interpreted on a molecular level through a so-called two-domain model [30]. In this approach there is a distinction between two zones, namely, the local (interfacial) domain, which is the aqueous solvent close to the biomolecule, and a bulk domain further away with thermodynamic properties corresponding to those outside the bag in a dialysis experiment. If $\Gamma_3 = 0$, the solute concentration will be equal in both domains, while negative values imply a diminished solute concentration in the local domain. If the numbers of solute and water molecules per lipid in the local domain are, respectively, B_3 and B_1, the preferential binding parameter may be written [30, 31]

$$\Gamma_3 = B_3 - B_1 \frac{m_3}{m_1} \qquad (9.5)$$

where m_1 is the number of moles of water in 1 kg H$_2$O (i.e., 55.5 mol/kg).

The degree of preferential interaction may be expressed as the partitioning coefficient [45], P, which is the ratio of concentrations in the two domains. Using molal concentration units, we have

$$P = \frac{^{\text{local}}m_3}{^{\text{bulk}}m_3} \approx \frac{\frac{B_3}{B_1 M_1}}{m_3} \qquad (9.6)$$

where M_1 is the molecular weight of water (in kg/mol). Since the bulk phase is in general much larger than the local domain, we neglect the difference between the average and the bulk concentrations ($m_3 \sim {}^{\text{bulk}}m_3$)

9.4 FREE ENERGY OF MEMBRANE–ORGANIC OSMOLYTE INTERACTIONS

The net affinity for PC membranes of some polyhydroxy osmolytes (sugars and sugar alcohols) has been investigated. In all cases, the osmolyte was found to be preferentially excluded from the interface ($\Gamma_3 < 0$) and the preferential binding parameter decreased linearly with the osmolyte concentration. (In these works the experimental observable was the so-called isoosmotic preferential binding parameter Γ_1. It was argued, however, that this function was numerically indistinguishable from Γ_3 for these systems [37, 38].) The slopes, Γ_3/m_3, were -0.14, -0.18, and -0.29 kg/mol, respectively, for glycerol [37], glucose (P. Westh, unpublished), and trehalose [46]. An even larger degree of preferential exclusion was found for dimethyl sulfoxide (DMSO), which may be considered a "synthetic osmolyte" inasmuch as it is an effective cryoprotectant [38]. For DMSO the slope Γ_3/m_3 at room temperature was about -0.35 kg/mol for both DMPC and POPC [38] (see Fig. 9.1). These results are in line with the much larger collection of data on protein–osmolyte interactions. These reports have collectively shown that organic osmolytes are preferentially excluded from the protein interface around room temperature (e.g., see Refs. 31 and 47). One exception to this is urea, which is weakly accumulated at the interface [31, 47] and this result is in accord with the noncompatibility of this solute discussed earlier.

The negative Γ_3 values show that PC membranes interact more favorably with water than with the osmolytes. This may be illustrated further by calculating the free-energy change, ΔG_{trans}, of transferring DMPC from pure water to a 1 molal osmolyte solution (integration of Eq. (9.1)). Results of this analysis at 30 °C show that ΔG_{trans} ranges from 0.3 kJ/mol (glycerol) to 0.9 kJ/mol (DMSO). The degree

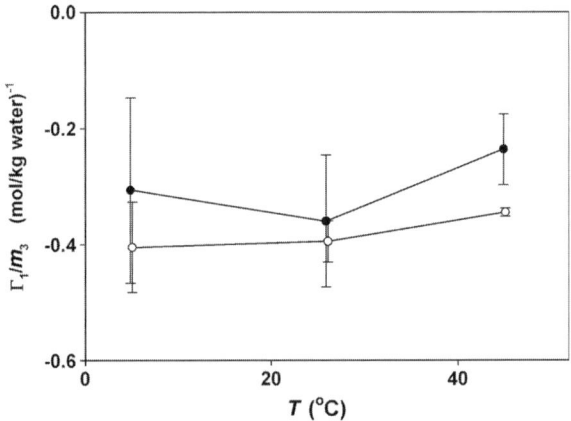

Figure 9.1 Preferential interactions of dimethyl sulfoxide (DMSO) and DMPC (filled symbols) and POPC (open symbols). The preferential binding parameter normalized with respect to the DMSO concentration, Γ_3/m_3, is plotted as a function of temperature. (Data from Ref. 38.)

of preferential exclusion of the osmolytes may be better illustrated through the application of the two-domain model. To do so, the boundary between the local and bulk domains must be specified on the basis of some structural information. If, for example, we define the local domain as a monolayer of water and use a per-lipid average lateral area of 60 Å^2, the maximal value of B_1 (number of water molecules in the local domain) will be about 6, given that one water molecule covers an area of about 10 Å^2 [48]. Insertion of this and the experimental Γ_3/m_3 given earlier in Eq. (9.5) gives B_3 values around or below zero. This implies the absence of osmolyte molecules in a local domain defined in this way. The projected lateral area, however, may be a poor measure of the hydrated membrane interface. Thus computer simulations of DPPC at 53°C showed a projected lateral area of ∼58 Å^2 but suggested that, due to the ruggedness of the interface, the water-accessible surface area was about twice as big (126 Å^2, cf. Fig. 9.7) [49]. In light of this, we suggest that monolayer coverage of DMPC involves about 13 water molecules. For glycerol, this translates into a B_3 value (Eq. (9.5)) of about 0.1 and hence $P = 0.4$ (Eq. (9.6)). It follows that the preferential exclusion of this osmolyte corresponds to a situation where its presence in the interfacial (mono-) layer is reduced to about 40% of the bulk concentration. For saccharides, B_3 is close to zero. This simply means that the preferential exclusion of these osmolytes corresponds to their absence from the first hydration layer. Finally, for the most excluded solute, DMSO, B_3 is still negative when $B_1 = 13$ is inserted in Eq. (9.5). In other words, the preferential exclusion of this solute extends beyond the average exclusion from the first hydration layer. The data for DMSO in Fig. 9.1 corresponds to a local domain of about 20 water molecules, which are free of DMSO, or to a partitioning coefficient (Eq. (9.5)) $P = 0.35$ when B_1 is set to 30, the amount of water that separates fully hydrated PC multilayers [50].

It may also be of interest to compare the degree of preferential exclusion of osmolytes for membranes and proteins. To this end, we apply the approach suggested by Courtenay et al. [47]. These workers compared the preferential exclusion of a number of protein–solute systems by normalizing the preferential binding Γ_3/m_3 with respect to the accessible surface area. For glycerol the normalized values ranged from -2.9×10^{-4} mol^{-1}(kg water)Å^{-2} for β-lactoglobulin, lysozyme, and RNase to -5.3×10^{-4} mol^{-1}(kg water)Å^{-2} for chymotrypsinogen A. If we again assume that the surface-accessible area of a hydrated PC membrane is about twice the lateral area [49], we arrive at a value of 120 Å^2 [21] for DMPC at 30 °C and a normalized preferential binding of about -11×10^{-4} mol^{-1}(kg water)Å^{-2}. For trehalose–DMPC, the normalized preferential binding is -24×10^{-4} mol^{-1}(kg water)Å^{-2}, while typical values for trehalose and globular proteins scale in the -7×10^{-4} to -11×10^{-4} mol^{-1}(kg water)Å^{-2} range [47]. It appears that the area-normalized preferential interaction is 2–3 times stronger for PC membranes, and thus that the interfacial preference for water over osmolytes is stronger for PC bilayers than for proteins. These conclusions only relate to polyhydroxy osmolytes—preferential interactions of amineosmolytes with membranes await investigation.

As already mentioned, the two-domain approach is a structural interpretation of the thermodynamic results. This interpretation is in accord with a number of other observations (discussed later), but other interpretations cannot be ruled out at present. One alternative possibility is that the solute enters the membrane phase, but copartitions with a larger number of water molecules. If, for example, there was a 2:1 (water:osmolyte) copartitioning, the bulk concentration of osmolyte, m_3, would increase in spite of the fact that osmolyte was removed. In other words, Γ_3 would be negative without necessarily reflecting depletion in the interfacial region. Some information has suggested that partitioning of peptides and small monoalcohols, such as ethanol, may increase the water content of membranes [51, 52], but it is unclear if this mechanism is important for small hydrophilic solutes.

9.5 THERMOCHEMISTRY OF MEMBRANE–ORGANIC OSMOLYTE INTERACTIONS

The appearance of highly sensitive isothermal titration calorimeters (ITCs) has opened an interesting avenue for studies of solute–membrane interactions [53, 54]. For solutes that are membrane permeable, the experimental design is straightforward (Fig. 9.2). Highly polar, impermeable solutes such as sugars, on the other hand, remain unexplored by this method since the contribution from the membrane–solute interaction cannot readily be singled out from the total heat, which also includes effects associated with osmotic shrinkage or expansion of the liposome [55]. Most organic (and all inorganic) osmolytes fall into the latter group, but glycerol and the "synthetic osmolyte" DMSO equilibrate over lipid membranes rapidly enough to use ITCs. One approach is to measure the enthalpy change, ΔH_{trans}, for the transfer of lipid membrane from water to an osmolyte solution. In practice, this function can be measured by applying the four-step calorimetric procedure illustrated in Fig. 9.2.

One of the advantages of ΔH_{trans} measured as illustrated in Fig. 9.2 is that the function is compatible with ΔG_{trans} values derived from integration of Eq. (9.1). Hence the transfer entropy may be calculated as $T\Delta S_{trans} = \Delta H_{trans} - \Delta G_{trans}$. An overview of the thermodynamic transfer functions for the glycerol–DMPC system at 30 °C is shown in Fig. 9.3.

Typical *in vivo* concentrations of glycerol (as well as other organic osmolytes) are in the 0.1–1 m range. It appears from Fig. 9.3 that the transfer enthalpy in this range is endothermic and rather small (e.g., about 0.2 kJ/mol DMPC for 1 mol(kg water)$^{-1}$ glycerol). The entropic contribution to the transfer is also small and unfavorable; $T\Delta S_{trans} \sim -0.1$ kJ/mol for 1 mol(kg water)$^{-1}$ glycerol. The slopes of the curves in Fig. 9.3 provide a measure of the strength of membrane–glycerol interactions [56–58]. For example, the positive slope of the solid curve shows that glycerol–DMPC interactions are endothermic or "enthalpically unfavorable." This behavior is parallel to the one found for both glycerol's interaction with a globular protein [58] and for DMPC's interaction with other small alcohols

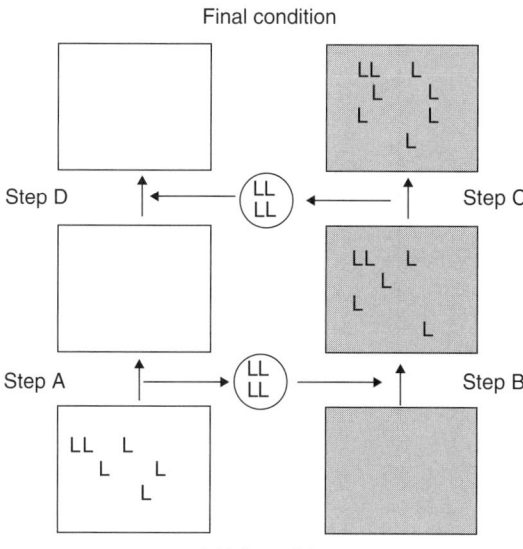

Figure 9.2 Four-step procedure to determine the enthalpy change, ΔH_{trans}, of transferring lipid from water (white) to a mixed solvent (shaded). Four separate measurements (A–D) are required. These are, injection of a concentrated liposome solution into, respectively, water (process A) and the solution (process B) and injection of pure water into, respectively, the lipid in mixed solvent (process C) and water (process D). The heat of transfer per mole of lipid can then be written $\Delta H_{trans} = (M/\rho w_L)[-q_A/V_A + q_B/V_B + v_w(-q_C/V_C + q_D/V_D)]$, where q_i and V_i are, respectively, the heat of mixing and the injected volume at process i (i = A, B, C, or D). M is the molar mass of the lipid, ρ is the density of the concentrated lipid suspension, and w_L and v_w are, respectively, the weight fraction of lipid and the volume fraction of water in the concentrated lipid suspension. Details can be found in Refs. 54 and 58.

54, 57–60. However, the magnitude of the slope for DMPC–glycerol is much smaller than for the other examples. This implies weak interactions and certainly the absence of any major molecular reorganization induced by glycerol.

To further discuss the thermodynamics of membrane–glycerol interactions illustrated in Fig. 9.3, we take a closer look at the structural interpretation of the interaction. As discussed earlier, the negative values for Γ_3 suggest a lowering of the glycerol concentration in the local domain. However, glycerol readily permeates both biological membranes and fluid lipid bilayers [61] and this permeability testifies to its presence (albeit at low concentrations) throughout the membrane. Highly sensitive isotope-label methods have suggested a water–bilayer partitioning coefficient $K_p = 0.06$ (in molal units) [24]. It follows that the interaction involves both the distribution between two phases (membrane and aqueous bulk) and the interfacial effect discussed earlier. This is illustrated in Fig. 9.4. Thermodynamic measurements, such as the vapor pressures underlying

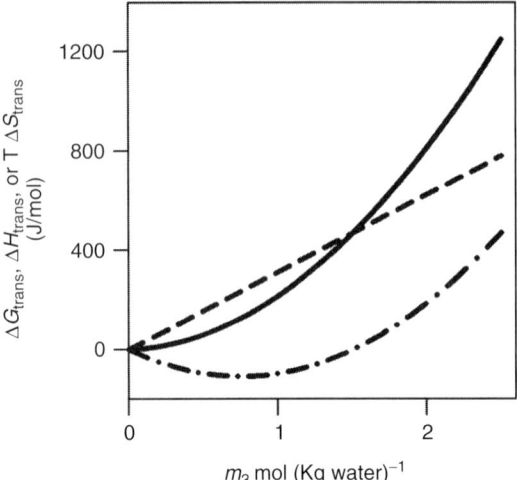

Figure 9.3 Thermodynamics for the transfer of unilamellar vesicles of DMPC from water to aqueous glycerol plotted as a function of the glycerol concentration. The dashed, solid, and dot-dash curves are, respectively, ΔG_{trans}, ΔH_{trans}, and $T\Delta S_{trans}$. (From Ref. 37.)

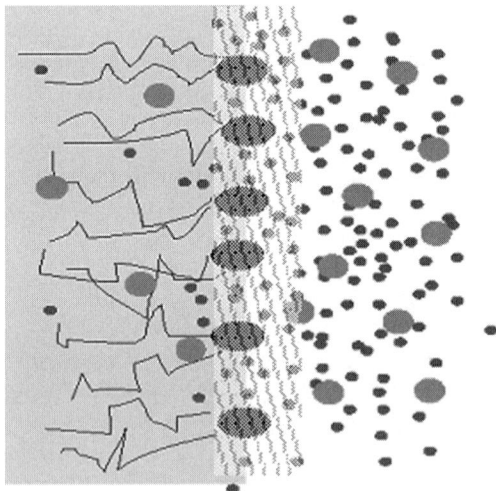

Figure 9.4 Illustration of the molecular picture for glycerol–membrane interactions discussed in the text. The lipid molecules are hydrated with water (small circles) and glycerol (large circles). Three different regions with separate glycerol concentration are considered. These are (1) the membrane interior (acyl chain region), which has a low concentration of both water and glycerol, (2) the membrane interface (hatched) from which the glycerol is preferentially excluded so that the concentration of water is increased, and (3) the bulk aqueous solution.

the Γ_3 data discussed previously, are thus governed by (at least) two effects. Partitioning of glycerol into the membrane will decrease the bulk concentration of the osmolyte and thus lower its chemical potential, μ_3. Conversely, the preferential exclusion of glycerol from the membrane interface will increase μ_3. The measured preferential interaction parameter, $\Gamma_3/m_3 = -0.14$ kg/mol, may therefore be written

$$\Gamma_3/m_3 = {}^{\text{intef}}\Gamma_3/m_3 + {}^{\text{part}}\Gamma_3/m_3 \qquad (9.7)$$

where the superscripts specify, respectively, the effect of partitioning and preferential interactions at the interface. Since K_p gives the amount of glycerol that partitions into the membrane, $^{\text{part}}\Gamma_3/m_3$ can readily be calculated [23, 37]. Using $K_p = 0.06$, we find $^{\text{part}}\Gamma_3/m_3 = 0.040$ kg/mol. It follows (Eq. (9.7)) that preferential exclusion at the interface is slightly stronger than the value derived directly from the experiments, namely, $^{\text{intef}}\Gamma_3/m_3 = -0.18$ kg/mol.

This analysis illustrates the important conclusion that the net effect of glycerol–membrane interactions is dominated by interfacial effects, whereas glycerol partitioning is 3–4 times less important. We suggest that the approach illustrated in Eq. (9.7) is generally valid for analysis of membrane–solute interactions. Clearly, the importance of the two terms on the right-hand side of the equation will depend strongly on the size and polarity of the solute. For 1-alkanols larger than propanol, the interfacial term becomes unimportant and for highly hydrophilic (membrane-impermeable) solutes such carbohydrates and inorganic ions, the partitioning term can safely be neglected. In such cases, the molecular picture can be simplified to either a surface (preferential) interaction or a bulk partitioning between two separate phases. For solutes of intermediate polarity such as methanol, ethanol, and DMSO, on the other hand, both contributions in Eq. (9.7) need to be considered to understand the nature of membrane–solute interactions.

Returning to glycerol, the thermodynamic data in Fig. 9.3 may be interpreted in the light of the picture suggested in Fig.9.4. If ΔH_{trans} is assumed to reflect the sum of the two effects, we may (in analogy to Eq. (9.7)) write

$$\Delta H_{\text{trans}} = {}^{\text{interf}}\Delta H_{\text{trans}} + {}^{\text{part}}\Delta H_{\text{trans}} \qquad (9.8)$$

We assume that the measured transfer enthalpy is governed by the hydration of glycerol. In other words, we examine the result of presuming that the measured heat changes solely reflect heats of dilution (or heats of concentration) of glycerol, while enthalpic effects of glycerol–membrane contacts can be neglected. This assumption has previously been demonstrated to be acceptable for a number of other small alcohols' interactions with lipid bilayers [54, 57, 60]. We have already discussed the relocation of glycerol molecules between the regions defined in Fig. 9.4. The enthalpic effects of these reorganizations of glycerol hydration can be calculated [37] on the basis on mixing enthalpy data for binary aqueous glycerol [62]. These enthalpy changes, which correspond to $^{\text{interf}}\Delta H_{\text{trans}}$

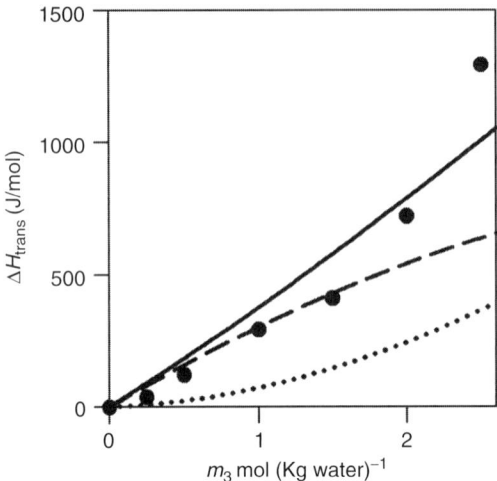

Figure 9.5 Total enthalpy change, ΔH_{trans}, for the transfer of unilamellar DMPC vesicles from water to aqueous glycerol at the concentration m_3. Experimental values are indicated by the points and the solid line represents ΔH_{trans} calculated as indicated in the main text. The figure also shows calculated values for the contribution of, respectively, preferential interactions at the interface ($^{interf}\Delta H_{trans}$, dotted line) and the membrane partitioning process ($^{part}\Delta H_{trans}$, dashed line) to the total transfer enthalpy, ΔH_{trans}.

and $^{part}\Delta H_{trans}$ defined in Eq. (9.8), are plotted in Fig. 9.5. It appears that the sum of the two processes (solid curve) is in good accordance with the experimental values (points), thus supporting the relevance of the approach. If we again focus on the physiologically most relevant composition range (0.1–1 mol (kg water)$^{-1}$ glycerol) the figure also shows that, although very few glycerol molecules partition into the membrane ($K_p = 0.06$), this process dominates the experimental transfer enthalpy at low concentrations. Preferential interactions at the interface (dotted curve) appear to be almost athermal for $m_3 < 1$ mol (kg water)$^{-1}$.

We now have estimates of the contributions from the two processes (Fig. 9.4) to the free energy and the enthalpy. The entropic contributions $T^{interf}\Delta S_{trans} = {}^{interf}\Delta H_{trans} - {}^{interf}\Delta G_{trans}$ and $T^{part}\Delta S_{trans}$ (defined analogously) were calculated and plotted together with the free-energy contributions in Fig. 9.6. The free-energy data in the upper panel again illustrates the comparably strong unfavorable effect of the preferential exclusion (dotted line) and the small favorable contribution to ΔG_{trans} from the partitioning (dashed line). More importantly (lower panel) it appears that the total entropic contribution to the transfer process is small (solid line) but composed of larger favorable and unfavorable effects from, respectively partitioning (dashed line) and interfacial interactions (dotted line). This provides a starting point for a general interpretation of the thermodynamics of glycerol–membrane interactions. It suggests that partitioning is driven

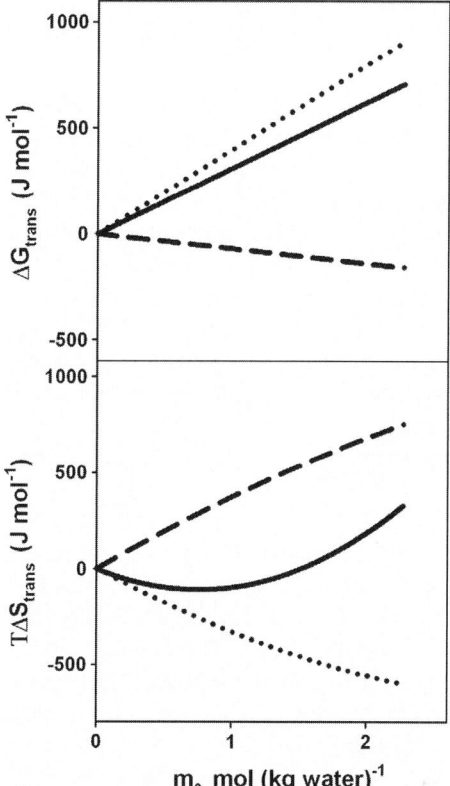

Figure 9.6 Estimated contributions to the transfer thermodynamics of DMPC vesicles into glycerol (ΔG_{trans}, upper panel and $T\Delta S_{trans}$, lower panel). Dotted lines specify contributions from the preferential interactions at the interface and dashed lines are the contributions from partitioning. The solid lines are the sums of the two contributions.

by the entropic gain of utilizing the volume available in the membrane phase. Conversely, the entropic penalty of establishing an uneven distribution of glycerol in the interfacial layer is the main contributor to the (unfavorable) transfer free-energy change associated with preferential interactions.

9.6 DRIVING FORCES FOR PREFERENTIAL EXCLUSION OF ORGANIC OSMOLYTES

Preferential interactions illustrate the effects of osmolytes on membrane equilibria (as exemplified in Eq. (9.2)), but this thermodynamic approach does not provide direct information on the origin of the effects. To this end it is necessary to elucidate the sources of the observed preferential interactions at the membrane interface. This is relevant, for example, for the technological application of osmolytes, since it paves the way for a rational selection of stabilizers in formulation procedures and thus limits the requirements of comprehensive empirical trials.

One source of preferential exclusion that has been discussed for proteins for over fifty years [63] is steric exclusion of the solute [29, 3164–66]. Basically, this effect reiterates that a small molecule (like water) has an easier access to an interface than larger molecules (such as osmolytes) [66]. This effect will be accentuated if the interface is rugged with imperfections of dimensions comparable to a water molecule. We have used results from computer simulations to investigate the surface accessibility of water and "osmolytes" to membranes (Hansen et al., in preparation). In this work we used a combination of two types of simulation. First, the algorithm of Lee and Richards [67] was used to define the water-accessible surface area of trajectories from an equilibrated DPPC membrane (contour in Fig. 9.7). Next, the contour was implemented as one wall in a simulation box containing two types of hard spheres with the size of, respectively, water and osmolyte (Fig. 9.7). The preferential interactions, which depend solely on steric accessibility, were found to be 30–40% of the experimental Γ_3/m_3 values discussed previously, thus suggesting a significant contribution from this effect on the preferential exclusion at the interface.

The main source of preferential exclusion appears to come from the "hydrophilicity" of the osmolyte, that is, from its favorable interaction with water. More specifically, this depends on the subtle balance between water–membrane and water–osmolyte interactions. Thus if water exhibits a particularly strong interaction with one or both of these components, its removal upon close contact of membrane and osmolyte will be disfavored. This relationship may be discussed in a number of ways. One interesting approach is the effect of the solute on the

Figure 9.7 Illustration of the steric accessibility to a membrane interface of spheres with the size of water (smallest), a "half-saccharide" like glycerol, a monosaccharide, and a disaccharide (largest). The contour covering the membrane interface is the water-accessible surface as defined by the Lee and Richards [67] algorithm.

surface tension. For protein–osmolyte interactions, many cases have been identified in which the preferential exclusion scales with the solute-induced increase in the surface tension of the binary (osmolyte–water) mixture (e.g., see Ref. [68]). One study has used similar arguments for osmolyte effects on membrane phase transitions [34]. A solute that increases the surface tension has a negative Gibbs surface excess, and this is tantamount to "preferential exclusion of solute" from the air–liquid interface. This, of course, cannot be applied directly to a biomolecule–water interface, but the experimental investigations mentioned earlier showed a good accordance and hence suggested that the interfacial behavior of the osmolytes was governed primarily by the hydration behavior in the aqueous bulk and less so by the specific nature of the interface. This nonspecific effect is related to the Hofmeister series, which ranks inorganic ions with respect to their relative effect on a number of biochemical equilibria and kinetically controlled processes [69]. The remarkable universality of the series clearly suggests that the underlying mechanism relies on the hydration of the ions, and this is reflected, for example, in the correlation between the ion-induced effect on surface tension and the position in the series [69]. Those ions that bring about the largest increases in surface tension (so-called strong kosmotropes) also tend to be strongly excluded from interfaces. Osmolytes are sometimes included in an extended Hofmeister series, where they are placed as kosmotropic ("structure making") solutes. The molecular rational of this, however, remains elusive. Strong kosmotropic inorganic ions such as fluoride, sulfate, and phosphate bind a hydration sphere tightly [69] (and thus tend to avoid interfacial locations where some of this strongly held water must be released). This has recently been illustrated calorimetrically in studies where the hydration of 1-propanol in aqueous "Hofmeister salts" was used as a model. It was found that kosmotropic ions showed large hydration numbers (e.g., 26 for $NaSO_4$ and 19 for NaF [70–72]) but that the properties of the bulk water outside the hydration shell were only lightly perturbed by the salt. Conversely, "structure breaking" (chaotropic) salts such as NaSCN and $NaClO_4$, which preferentially bind to proteins [31], strongly affected bulk water properties [70–72]. Analogous investigations on the hydration of 1-propanol in solutions of glycerol and fructose, however, did not show any strong analogy with the kosmotropic salts. While these osmolytes were indeed found to be "preferentially excluded" from propanol in the sense that $\left(\frac{\partial \mu_{propanol}}{\partial m_{osmolyte}}\right)_{m_{propanol}} > 0$ [70–72], analysis of the calorimetric data showed that both glycerol and fructose dampened the entropy fluctuations characteristic to pure water. This suggests that the mechanisms underlying the preferential exclusion of inorganic kosmotropes and osmolytes are different.

In addition to the two mechanism discussed previously, the preferential interaction depends on the chemical structure of the interface. For a PC bilayer, the majority of the water-accessible surface is the amine $(-N(CH_3)_3)^+$ and the phosphate group [49]. Since both tertiary amines and phosphate are kosmotropes [10], they are expected to show strong hydration and thus a propensity to expel solutes from the interface. The consistently kosmotropic nature of the interface may be at least one reason why polyhydroxy compounds are excluded 2–3 times stronger from membranes than from proteins (see above) since the latter has a more heterogeneous interface with charges and patches on the nonpolar area. In spite of the partial depletion of osmolytes from the interface, direct interactions such as hydrogen bonding with carbohydrates have recently been identified in computer simulations (see Ref. 73 and references therein). This interaction may become particularly important at very low water contents, and we return to this later.

9.7 PHASE BEHAVIOR OF MEMBRANES IN ORGANIC OSMOLYTE SOLUTIONS

Polyhydroxy osmolytes (sugars and sugar alcohols) consistently increase the temperature, T_m, of the main ($L_{\beta'} \rightarrow L_\alpha$) phase transition of lipid bilayers [27, 34, 74, 75]. For PE, they also decrease the high-temperature transition (T_h) from the fluid lamellar to the inverted hexagonal (H_{II}) phase [27]. The noncompatible (chaotropic) osmolyte, urea, on the other hand, has the opposite effect on both of these transitions [75]. Analogous solute effects are observed for inorganic ions for which kosmotropes (e.g., sulfate and fluoride) increase T_m and decrease T_h while chaotropes (e.g., thiocyanate and perchlorate) have the opposite effect [27, 75]. These systematic relationships are exemplified in Fig. 9.8 and may be

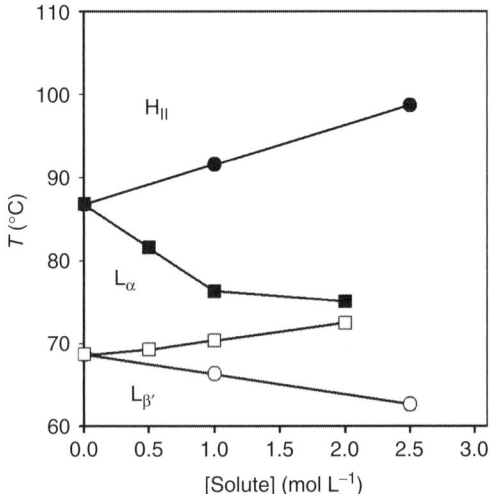

Figure 9.8 Phase behavior of DHPE in the osmolyte sucrose (squares) and the chaotropic salt NaSCN (circles). Filled symbols indicate the lamellar-to-hexagonal transition; open symbols the main (gel-to-fluid) transition. It appears that the preferentially excluded osmolyte favors the hexagonal and the gel phases over the fluid lamellar phase. Thus 2 M sucrose practically abolishes the L_α phase. Conversely, a chaotropic solute like NaSCN strongly expands the L_α phase space. These changes, discussed in the text, reflect the sign of $\Delta\Gamma_3$ (Eq. (9.2)). (Data from Ref. 27.)

rationalized through the preferential solute–membrane interactions. As pointed out by Cevc [33], "changing transition temperature T_m seems to reflect primarily the variations in the affinity of the lipid polar residues for different molecules in the solvent." This statement is contained in Eq. (9.4), which stipulates that negative $\Delta\Gamma_3$—that is, a situation where the affinity of the membrane for water is higher than for the solute—specifies a positive value of the slope dT_m/dm_3. This is because kosmotropes (such as osmolytes), which are preferentially excluded, generate an osmotic gradient in the solvent near the interface. This osmotic stress will displace equilibria toward conformations with a lower hydrated area, because that allows some dehydration and a concomitant movement of water down the osmotic gradient. Results like those in Fig. 9.8 imply that the L_α phase is the most hydrated, and hence disfavored by a preferentially excluded solute. Conversely, the most hydrated phase will be promoted by preferentially bound solutes. This is a general effect of preferential exclusion, and its application to membrane phase behavior has been discussed in various settings [27, 33, 35, 36, 38, 46, 75].

Amino osmolytes are less investigated for their effect on lipid phase behavior, but some have been shown to raise T_m [76] in accordance with their kosmotropic nature [10]. In other cases, amino acid concentrations up to a few hundred millimdar were too low to single out significant changes in T_m [77].

The simple coupling between preferential exclusion and phase behavior is only valid if the preferential interaction is independent of the chemical nature of the interface—in other words, if the degree of exclusion is proportional to the interfacial area. This assumption has proved acceptable in a number of cases of protein stabilization but is not generally valid [32]. Trehalose, for example, preferentially binds to RNase at 52 °C but nevertheless stabilizes the native (low surface area) state of the protein since it binds more to this conformation [78]. This implies a low affinity of trehalose for the specific moieties exposed during denaturation of RNase. Surface-specific effects are also evident for glycoproteins, which appear to preferentially bind osmolytes to the carbohydrates while excluding them from the peptide moieties [79]. The true measure of solute-stability linkage is the *change* in preferential interaction specified in Eq. (9.2). This function, $\Delta\Gamma_3$, has not been measured for membrane phase transitions but may be estimated by inserting calorimetric data into Eq. (9.4) (after solving for $\Delta\Gamma_3$). For the main transition in the DHPE–sucrose system, all required data (Eq. (9.4)) can be taken from Koynova et al. [27], and we find $\Delta\Gamma_3/m_3$ is about -0.05 kg water/mol throughout the available concentration interval. Similar comprehensive data is not available for PC membranes, but based on a number of data sets [27, 34, 74, 76, 80, 81] values of $\Delta\Gamma_3/m_3 \sim -0.06$ kg water/mol were found for sucrose and trehalose effects at the main transition of saturated PC. The experimental Γ_3/m_3 value for trehalose and fluid PC was about -0.29 kg water/mol, and it follows that the preferential exclusion of these osmolytes increase by about 20% at the main transition. A very similar relative change in lateral surface area occurs at the main phase transition of DPPC bilayers [82]. This suggests that Γ_3 is proportional to the interfacial area and thus that the sugar interacts equally (unfavorable) with the interface of the gel and the fluid phase.

9.8 ORGANIC OSMOLYTES AND MEMBRANE STABILITY

In the Introduction we mentioned the stabilizing or "compensatory" effects of osmolytes *in vivo* and in biotechnology. For globular proteins, this can readily be linked to the preferential exclusion. Thus breakdown of the native structure of a protein conformation will, in general, couple to an increase in the water-accessible area. In most cases, the degree of preferential exclusion will increase with the surface area and hence the osmolyte favors the native (low surface area) conformation according to Eq. (9.2). For membranes, the picture is more complex. The ability of many osmolytes to enhance the resistance of both lipid bilayers and biological membranes against physical stress is indeed well established [83–85]. The underlying mechanisms, however, are only partially understood. One example of membrane stabilization is the capacity of polyhydroxy compounds such as glycerol and sorbitol to alleviate injury associated with cold and freezing, which is utilized both in natural cold hardiness and in the technical cryopreservation of cells and tissue [4, 86]. The cryoprotective effect of these osmolytes has been suggested to rely on a number of mechanisms including the depression of the freezing point, limitation of cellular shrinkage during (extracellular) ice formation, promotion of glassy states, and the control of the shape, growth rate, and possible recrystallization of the ice crystals [86]. All these putative protective mechanisms are only related to the membrane–osmolyte interactions in the sense that the protective compound must be compatible with membrane structure and function at rather high concentrations, which results from "freeze-concentration." However, some reports have suggested that stabilization also relies on the specific interactions of the osmolyte and membrane [87–89]. A particularly clear example of osmolyte-induced stabilization of membranes, which is unrelated to the ice formation process, comes from the work of Fabrie et al. [90]. They found that the disaccharides trehalose and sucrose strongly reduced the leakage of a fluorescent probe, which otherwise occurs from DMPC liposomes around the main phase transition. As discussed by Crowe and co-workers [35], this type of protective mechanism may be rationalized along the lines of preferential osmolyte–membrane interactions. However, unlike for proteins, where the inactivation couples to an increased surface area, there is no straightforward structural fundament for discussions of interrelationships of stabilization and preferential interactions in membrane systems. In fact, simplistic evaluations of the effects of osmolytes on membranes might not point toward stabilization at all. Cryoinjury, for example, has often been linked to phase transitions, in which the fluid lamellar phase changes into condensed phases or nonlamellar arrangements such as the inverted hexagonal phase [86]. As illustrated in Fig. 9.8, a cryoprotectant such as sucrose destabilizes the fluid lamellar phase of a model membrane to the extent where it ceases to exist at a concentration of a few molar. In addition to its influence on the phase behavior, the osmotic stress at the interface exerted by a preferentially excluded solute will tend to change membrane properties in a number of ways. It may, for example, decrease interlamellar spacing, promote aggregation, induce tighter chain packing, reduce chain tilt and surface area, and increase thickness [33, 50]. The relationship between these effects on membrane structure and increased resistance against physical stress is only fragmentarily understood, and progress in this area relies particularly on a better understanding of the structural basis of membrane injury.

Perhaps the most remarkable stabilizing effect on membranous systems is seen for a number of sugars (particularly trehalose), which are essentially capable of making intact liposomes go through cycles of freeze-drying and rehydration [91]. A similar protection has been found for biological model membranes, which remained at least partially functional after drying and rehydration in the presence of trehalose [92]. Trehalose and sucrose are also prevalent protectants in anhydrobiotic (extremely drought tolerant) invertebrates and plants [92, 93]. The protective mechanism at these extremely low hydration levels, where any aqueous bulk is absent, appears to rely on the formation of hydrogen bonds between sugar—OH and polar groups on the membrane interface [94], although other explanations, such as the formation of a glassy sugar matrix, have also been proposed [39, 95]. The hydrogen bonding of the sugar is suggested to constitute a "water replacement" [96], which stabilizes the membrane conformation in the absence of water (and thus a hydrophobic driving force). This suggests rather different mechanisms of trehalose protection at high and low water contents. Thus, at full hydration, trehalose interacts less favorably than water with the membrane interface and is therefore preferentially excluded. During drying, at some critical water level, this situation appears to change and the sugar binds to the membrane interface and replaces the hydration water.

9.9 INORGANIC OSMOLYTES AT MEMBRANE INTERFACES

Many problems in interfacial chemistry and biophysics are related to the electrostatic structure of the interface [97–99], and the presence of charges and electrical fields at the interface between a dielectric and an electrolyte solution influences a variety of processes [100]. Therefore, many interfacial processes, such as adsorption of charged species and interaction of charged interfaces, require a detailed knowledge of the electrical properties of the dielectric–electrolyte solution interface. For instance, ionic interactions are critically important for the structure and function of many biological membranes [101]. Small changes in ionic environment can induce significant alterations not only in the surface potential but also in the membrane structure [102]. This provides a potential mechanism that can trigger by changes in the surface pH or concentration of monovalent and divalent ions [103, 104] for the membrane to adapt to environmental changes and to be an integral part of the signal transduction network. Furthermore, surface potential, which is affected by ionic strength, controls the activity of numerous ion channels including the Na^+/K^+-ATPase [105, 106]. Changes in the membrane structure induced by ions, on the other hand, result in lateral stress in the membrane, which may not only induce protein-to-protein interactions but may also influence protein conformation and thereby protein function. An idea that has been developed further by Robert Cantor and co-workers [107] is that the variation in lateral stress in biological membranes could serve as a mechanism for modulating protein function.

9.10 THERMODYNAMIC CONSIDERATION

At the main phase transition, membranes undergo several dynamical and conformational changes. These include (1) formation of rotational isomers within the lipid hydrocarbon chains [108–111], (2) the onset of rapid lateral diffusion of the lipid molecules [112–115] (ordered to fluid transition), and (3) an expansion of the bilayer area [116–118]. For a reversible phase transition at constant pressure, the molar free energy is given by

$$\Delta G^\circ = \Delta H^\circ - T\Delta S^\circ \tag{9.9}$$

where ΔH° and ΔS° denote the enthalpy and entropy change, respectively. At the main phase transition temperature, T_m, the molar free energies of the ordered and fluid phase are equal, and Eq. (9.9) reduces to

$$\Delta H^\circ = T_m \cdot \Delta S^\circ \tag{9.10}$$

The enthalpy change corresponds to the heat absorbed during the ordered to fluid phase transition. The enthalpy change may be written as a sum of nonelectrostatic and electrostatic contributions:

$$\Delta H^\circ = \Delta H^\circ_{nonelec} + \Delta H^\circ_{elec} \tag{9.11}$$

$$\Delta H^\circ = (H^\circ_{nonelec,fluid} - H^\circ_{nonelec,ordered}) + (H^\circ_{elec,fluid} - H^\circ_{elec,ordered}) \tag{9.12}$$

Hence, the change in the main phase transition caused by electrostatic interactions can be expressed as

$$\Delta T_m = T_m - T_{m,non-elec} = \Delta H^\circ_{elec}/\Delta S^\circ \tag{9.13}$$

T_m is the observed phase transition temperature, and $T_{m,nonelec}$ is the transition temperature in the absence of electrostatic interactions. To estimate the effect of the electrostatic contribution to the change in the phase transition temperature, one has to consider the surface potential governed by a balance between the membrane surface charge and ions distributed in solution. Cell membrane surfaces usually are negatively charged, leading to the formation of a double layer, since positive charged ions in solution tend to balance the negative surface charge [119–121]. This results in a diffuse layer of positive charges in the electrolyte opposing a centralized layer of negative charges at the membrane surface. Thus a double layer is formed in which the separation of each layer is on the order of a few angstroms [122]. To describe the motion of ions and the distribution of ions in an electrolyte close to a charged surface, several electric double-layer models have historically been proposed [123, 124], which will be briefly discussed in the following.

The first theory was proposed by Helmholtz, who developed the electric double-layer theory for metal–metal interfaces in the 1850s, and this theory was extended for electrolyte–metal interfaces in 1879 [125]. As shown in Fig. 9.9A, in the Helmholtz model, the solvated ions arrange themselves along the surface but are kept a distance away from the surface due to their hydration spheres. The positions of the ions is called the Outer Helmholtz Plane. Sometimes ions are able to escape their hydration spheres and are held closer to the surface. This location is called the inner Helmholtz plane. The potential as a function of distance from the surface changes linearly between the potential of the surface and the potential of the bulk solution. This simple model can account for the basic of an electric double layer at the interface of the metal electrode and electrolyte dielectric material [126], but it neglects the effect of thermal motion and the concentration gradient of ions in the electrolyte that occurs further away from the interface [127].

Gouy and Chapman independently further developed the electric double-layer theory in 1910 by incorporating thermal effects, which break up the rigid layer, leading consequently to a diffuse layer [128]. Ions of opposite charge (to the surface) cluster close to the surface, whereas ions of a similar charge (to the surface) are repelled from the surface. This leads to a concentration gradient of ions

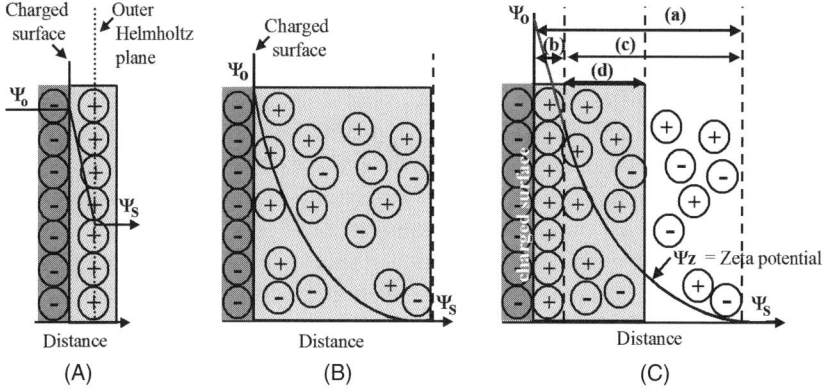

Figure 9.9 Illustrations of different models for the double layer discussed in the text. (A) Helmholtz model, (B) Gouy–Chapman model, and (C) Stern–Grahame model: (a) diffuse double layer, (b) Stern layer, (c) Gouy–Chapman layer, and (d) bound water.

at the interface (Fig. 9.9B), where the ion concentration decreases exponentially with distance from the surface.

A more refined model is the Stern model developed by Stern and Grahame [129]. Stern used both the Helmholtz model and the Gouy–Chapman model to define the electrical double layer, and the model is based on both an adsorbed layer of finite sized ions and a diffuse layer of ions in the electrolyte [130]. Grahame further developed Stern's model by defining two regions, which consist of a compact layer of finite sized ions adsorbed onto the surface referred to as the inner Helmholtz plane and a diffuse layer of ions in the electrolyte solution referred to as the outer Helmholtz plane (Fig. 9.9C).

Between the inner and outer Helmholtz planes there is a linear relationship between the potential difference and the distance from the surface. The original Stern and Grahame model describes the capacitance in great detail; however, it fails to account for a hydration shell of the ions in solution. In the 1960s, Bockris and co-workers defined a model that employed the impact of solvent on ions near the metal–electrolyte interface [131]. The model accounting for a hydration shell at the inner Helmholtz plane was later extended to allow for the fixed dipoles due to the charge in the adjacent surface. The hydration shell at the inner Helmholtz plane would be displaced due to the adsorption of ions at the charged surface, and the outer Helmholtz plane would consist of hydrated ions in a diffuse layer.

The models just outlined account for different levels of accurately describing the electrical double layer established at a charged surface. The Gouy–Chapman theory has proved remarkably successful in describing a number of biophysical processes [132]. For instance, the Gouy–Chapman theory has been used to describe the effect of salt on the surface potential above phospholipid monolayers [133], the electrostatic repulsion between charged bilayers [134], the surface potential as a function of distance from the membrane surface [135], and the zeta potential of phospholipid vesicles [136]. The Gouy–Chapman theory relates the charge density and electrical potential originated from a charged surface (e.g., charged membrane surface) to the concentrations of ions in the external solution [137], and as discussed earlier, the ion concentration at the interfacial plane decreases exponentially with the distance from the surface, implying that ion concentration can differ significantly from the concentrations in the contacting bulk medium [128, 137]. The Gouy–Chapman theory is based on a number of assumption including (1) the activity coefficient of each ion is independent of the distance from the membrane surface, (2) the external surface charge is uniformly distributed, and (3) the membrane is homogeneous [138]. The expression relating the double-layer potential, the external surface charge density, and the ionic concentrations in the external solution is then given by the Poisson–Boltzmann equation:

$$\frac{d^2\Psi_o}{dx^2} = \frac{-1}{\varepsilon_r \varepsilon_0} \sum_i^n C_{i,\infty} Z_i F \exp\left(\frac{-Z_i F \Psi_o}{RT}\right) \quad (9.14)$$

which with the assumption of constant activity coefficient (i.e., the activity coefficient of ions is independent of the distance from the surface [130]) can be integrated to yield the Grahame equation [128, 130]

$$\sigma^2 = 2\varepsilon_r\varepsilon_0 RT \sum_i^n C_{i,\infty}\left(\exp\left[\frac{-Z_i F \Psi_o}{RT}\right] - 1\right) \tag{9.15}$$

where σ is the charge density on the membrane surface expressed in coulombs per square meter (C/m^2); ε_r and ε_0 are the dielectric constant for water and the permittivity of a vacuum, respectively. R is the gas constant and T is temperature [129, 139]. The prefactor $2\varepsilon_r\varepsilon_0 RT$ has a value of 0.00345 at 25 °C for concentrations expressed in molarity. The sum is taken over ionic species, n. $C_{i,\infty}$ is the concentration of the ith ion in the bulk-phase medium. Z_i is the charge on the ith ion. F is the Faraday constant, and Ψ_o is the electrical potential at the membrane surface measured with respect to the bulk-phase medium (i.e., potential far from the surface). $-Z_i F \Psi_o / RT = -Z_i \Psi_o / 25.7$ at 25 °C for Ψ_o expressed in millivolts.

Equation (9.15) applies in the absence of any bound solute ions. However, under many experimental conditions, polyvalent cations are added to the external solution to neutralize surface charges, and Eq. (9.15) has to be modified. For polyvalent cations, the equilibrium for binding to the negatively charged groups at the surface is given by $M^{n+} + nS^- \Leftrightarrow MS_n$, and the equilibrium constant, K, becomes

$$K = \frac{[MS_n]}{[M^{n+}][S^-]^n} \tag{9.16}$$

where $[M^{n+}]$ denotes the concentration of the unbound ion at the membrane surface. $[S^-]$ is the free negative surface charge concentration, and $[MS_n]$ is the neutralized site concentration. $[M^{n+}]$ may be computed from bulk-phase concentrations, $[M^{n+}_\infty]$, by the Boltzmann equation:

$$[M^{n+}] = [M^{n+}_\infty]\exp\left(\frac{-Z_M F \Psi_o}{RT}\right) \tag{9.17}$$

where Z_M is the valence of the cation. R, T, F, and Ψ_o have the same meaning as defined in Eq. (9.15). Assuming that the valence of the polyvalent cations and the valence of the negative surface charge are equal, the total negative surface charge concentration is then $[S_{\text{tot}}] = [S^-] + [MS_n]$, and Eq. (9.16) becomes

$$K = \frac{[S_{\text{tot}}] - [S^-]}{[M^{n+}][S^-]^n} \tag{9.18}$$

Combining Eqs. (9.15), (9.16), and (9.17), the effective surface charges, σ_{eff}, can then be expressed by

$$\sigma_{\text{eff}} = \frac{1}{1 + K[M_\infty^{n+}]\exp\left(\frac{-Z_M F \Psi_o}{RT}\right)}$$

$$\times \left(2\varepsilon_r \varepsilon_0 RT \sum_i^n C_{i,\infty}\left[\exp\left(\frac{-Z_i F \Psi_o}{RT}\right) - 1\right]\right)^{1/2} \quad (9.19)$$

The first factor relates to binding (i.e., neutralization of the surface charge), whereas the second one refers to screening. The shortcoming of this theory comes from the assumptions that the activity coefficient of each ion is independent of the distance of the ions from the surface and that the surface is smooth. These assumptions have been challenged and extension of the Gouy–Chapman theory has been proposed [132, 140, 141]. Due to the mean field approximation, the theory fails when, for instance, charged groups on the surface are not univalent (e.g., phosphatidylglycerol or phosphatidylserine) but multivalent (e.g., trivalent phosphatidylinositol). In this case, the Gouy–Chapman formalism overestimates strongly the charge effect on the counterions [141]. Furthermore, many surfaces of biological interest are not smooth but are spatially structured. In these cases, the Gouy–Chapman formalism has been extended to include spatially distributed surface charges [142]. In particular, when localized processes are considered, and the distances between charges are comparable to or smaller than the characteristic screening distance, the discreteness of the charges has to be included in the model [143]. Here, the localized electric potential can significantly differ from the mean field calculation, and therefore adsorption at localized charge sites and the subsequent lateral interactions between groups require a more realistic treatment [144, 145]. Adsorption of basic peptides [146, 147] or basic regions of proteins such as proteins from the myosin family [148], protein kinase C [149, 150], and SH2/SH3 domains [151, 152] to acidic lipids in membranes are typical examples where refinement of the model is required, and a "rough" interfacial plane with spatially resolved charges has be considered [132]. Haage et al. [153] used the inner field compensation method to determine an effective potential that accounts for peptide-induced changes in membrane capacitance. Kinraide [130] developed a generalized Gouy–Chapman–Stern model that considers ion binding at the membrane surface and their relative activities at the membrane surface [154]. The binding then causes the surface charge to be variable and dependent on the concentration and type of ions. For mineral rhizotoxicity, Kinraide [130] showed that a generalized Gouy–Chapman–Stern model can predict the physiological effect of several inorganic toxicants.

The Gouy–Chapman formalism can also be used to describe the effect of electrostatic interaction on the phase transition temperature. Träuble and Eibl [155] solved the differential equation (9.14) to obtain the electrical surface potential of a membrane hydrated by an aqueous 1:1 electrolyte. The solution is given by

$$\Psi_o = 2 \frac{RT}{F} \frac{\sigma'}{A_{\text{lipid}}} \quad (9.20)$$

where σ' is the charge per polar group, which depends on the degree of dissociation of the polar group. A_{lipid} is the area of one lipid molecule. Defining $\Delta A = A_{\text{lipid},2} - A_{\text{lipid},1}$ as the change in the area per molecule from the ordered ($A_{\text{lipid},1}$) to fluid ($A_{\text{lipid},2}$) phase, the change in electrostatic potential becomes [155]

$$\Delta H_{\text{elec}}^\circ = -2RT \left(\frac{\Delta A}{A_{\text{lipid},1}} - \frac{(\Delta A)^2}{A_{\text{lipid},1} \cdot A_{\text{lipid},2}} \right) \frac{\sigma'}{F} \quad (9.21)$$

The change in phase transition temperature is then given by Eq. (9.13); that is, $\Delta T_m \propto \Delta H_{\text{elec}}^\circ \propto \sigma'$. Hence for highly charged bilayers, $|\Delta T_m|$ increases linearly with increasing surface charge, implying that T_m decreases linearly with increasing surface charge. According to Eq. (9.21), the electrostatic potential of the fluid bilayer is smaller than in the ordered state ($\Delta H_{\text{elec}}^\circ < 0$), and therefore an increase in the surface charge promotes the fluid state. An increase in the charge per polar group of dimyristoyl-sn-glycerol-3-phosphate by, for instance, changing the pH of the buffer can lower the transition temperature by up to 20°C [155]. Clearly, rather small changes in pH can be sufficient to induce phase transition at constant temperature, particularly, when the pH shift causes the protonation state of the lipid species to change (i.e., pK_a of the lipid is within the range of pH shift).

9.11 INORGANIC IONS AND MEMBRANE STRUCTURE

According to Eq. (9.19), counterions also affect the phase transition. Increased binding of counterions neutralizes the surface charge, thereby promoting the ordered phase [156] and increasing the barrier to pore growth [157]. Cevc and Marsh [158] showed that for charged 1,2-dimyristoyl-sn-glycerol-3-phosphoglycerol, Cs^+ induces a shift in pretransition temperature (T_p) with a gradient of $dT_p/d[\text{CsCl}]_{\text{bulk}} = 6$ K·L/mol. The authors explained the findings by increasing concentration of counterions near the charged lipid interface. For zwitterionic phosphocholine bilayers, the effect of counterions on the pretransition temperature was significantly smaller (gradient ≈ 1.25 K·L/mol) than observed for the negatively charged bilayer [158]. Dramatic increase in transition temperature can be induced by concentrations of calcium and magnesium ions in the millimolar range [159–161]. Papahadjopoulos and co-workers [160] suggested that the change in phase transition temperature is an important mechanism controlling vesicle fusion and membrane reorganization [162]. The presence of bivalent ions also improves the permeability properties of liposomes, indicating a tighter packing of the membrane in the presence of counterions [163, 164]. The effect of calcium ions on membrane structure is dependent on the character of the polar headgroup of the negatively charged phospholipid present [159]. Garidel and co-workers [165, 166] performed a detailed study to elucidate the effect of alkaline earth cations on the structure of 1,2-dimyristoyl-sn-glycerol-3-phosphoglycerol (DMPG). Using calorimetric and spectroscopic techniques, the authors demonstrated that Ca^{2+} affects DMPG much stronger than Mg^{2+} or Sr^{2+}. This could not be explained in terms of the hydrodynamic radii, which are similar for the three ions [166]. The authors suggested that in the case of calcium ions, the Ca^{2+}–lipid headgroup complex might have a perfect geometry, where the bivalent ion fits much better between the headgroups to interact favorably with the phosphate groups of the lipids in the membrane, resulting in the formation of an inner sphere complex [166]. Fragata and co-workers [167] conducted spectroscopic measurements and concluded that Mg^{2+} binds to the interfacial PG bilayer, which is accompanied with a loss of water of hydration, and strongly interacts with the phosphate group in the PG headgroup. The authors suggested that Mg^{2+} coordinates with eight phosphatidylglycerol lipids [168], thereby increasing chain packing/lipid order. Similar results were observed for negatively charged phosphatidylserine (PS) bilayers [162, 169–171]. Divalent ions promote ordered bilayer structures, where the Ca^{2+}–PS and Mg^{2+}–PS complexes show higher ordered chain packing than Sr^{2+}–PS and Ba^{2+}–PS complexes. Sodium ions, on the other hand, have no significant effect on phosphatidylglycerol or phosphatidylserine bilayers phase behavior [170]. However, Roux and Neumann [172] showed that the interaction of Li^+ with PS bilayers is distinct from that of other monovalent ions, which is probably caused by its ability to form strong highly dehydrated Li^+–PS complexes that resemble divalent cation–PS complexes. Based on fluorescence measurements using DMPG bilayers doped with the fluorescence probe LAURDAN, Pedersen et al. [170] assessed the effect of Ca^{2+} and Na^+ on DMPG bilayer properties. In Fig. 9.10, the generalized emission polarization (GP) [173] values for LAURDAN in DMPG vesicles are shown for temperatures below (15 °C) and above (37 °C and 57 °C) the gel/liquid-crystalline phase transition. The presence of 0.1 mM Ca^{2+} (black) induces an increase in the GP value compared to no Ca^{2+} (white); the higher the GP value, the lower the penetration of water [173, 174] caused by more ordered membrane packing [171, 175, 176]. The presence of an equivalent amount of Na^+ (grey) has no significant effect on the GP value, indicating that the effect of calcium on phosphatidylglycerol bilayers is more pronounced than that of sodium [170].

These experiments yield important information on structure-related properties but do not provide information on a single molecule level. Molecular dynamics simulations, on the other hand, can fill this gap and allow one to follow single molecules in space and time, thus providing insight on an atomic level [177, 178]. Simulations have been used to study membrane structure and dynamics [179] as well as distribution and dynamics of counterions on charged bilayer systems [170, 180, 181]. However, the number of stimulation studies

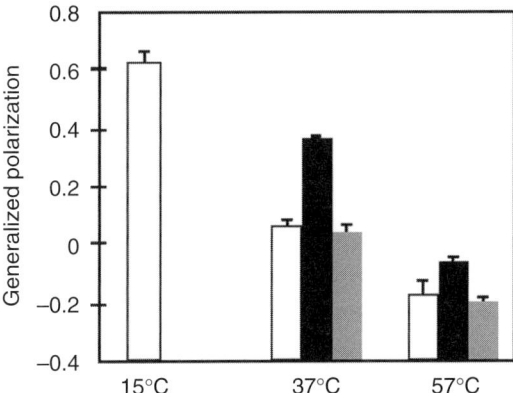

Figure 9.10 Generalized polarization (GP) values for Laurdan in DMPG multilamellar vesicles at temperatures below (15°C) and above (37 °C and 57 °C) the gel/liquid-crystalline phase transition of DMPG. White bar: 0.2 mM Na^+ solution; black bar: 0.1 mM Ca^{2+} + 0.2 mM Na^+ solution; grey bar: 0.3 mM Na^+ solution. $\lambda_{excitation}$ = 340nm; $\lambda_{emission}$ = 440 nm and 490 nm.

investigating the effect of counterions on negatively charged lipid bilayers is rather limited. The few simulations that appeared in the literature have mainly been carried out with calcium and sodium as counterions [170, 180, 181]. As suggested from experiments, simulation results showed that calcium ions are tightly bound to the negatively charged phospholipid, forming an integral part of the membrane interface [170]. Similar results have been observed for sodium ions, which intercalated into the interfacial region of phosphatidylserine bilayers [180, 181]. However, calcium ions appeared to be more strongly bound to the membrane surface than sodium ions, since a significant fraction of the monovalent ions were found to be dissolved in the aqueous bulk phase [180]. The distribution of sodium ions on the phosphatidylserine bilayer interface is about twice as broad as that observed for calcium ions [170, 181]. The sodium ions are able to penetrate deeply into the headgroup regions of the lipids and to coordinate to the carbonyl oxygens [180, 181], which is hardly observed for calcium ions [170]. Divalent ions bind predominantly to the phosphate group and to a lesser extent to the carboxylate group of the lipids. Typical examples of these interactions are illustrated in Fig. 9.11. The coordination of calcium ions by lipids is almost entirely due to interactions with the anionic groups, while interactions with the oxygen atoms of the ester groups (as observed for Na^+) are negligible. Hydration of bound calcium and sodium ions differs significantly. Calcium ions are on average coordinated by 3–4 water molecules, whereas hydration of sodium ions, which are intercalated into the region around the ester group of a phosphatidylserine membrane, involves only 2–3 water molecules [170, 181].

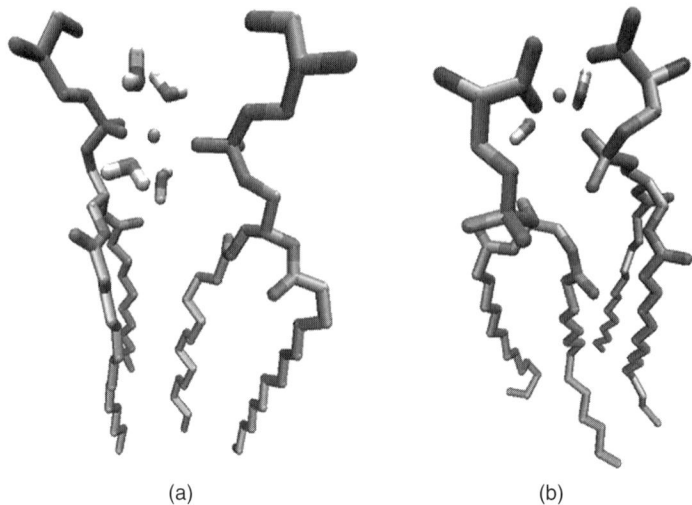

Figure 9.11 Representative snapshots of Ca^{2+} ion coordination in (A) DMPG bilayer and (B) DMPS bilayer. Coordination corresponds to a direct binding of Ca^{2+} to a lipid headgroup oxygen or water.

9.12 CONCLUSION

The physical properties of lipid membranes are strongly affected by the presence of "foreign molecules"—that is, components other than water and lipids—at the interface. In many cases, the solute-induced perturbations are rather complex and cannot be rationalized on the basis of traditional membrane–water partitioning schemes. This is because the perturbations reflect multifaceted distributions of the solute at the interfacial region. For organic osmolytes at uncharged (zwitterionic) lipid interfaces, there is a depletion of solute due to steric hindrance and hydration forces. Small nonionic osmolytes such as glycerol even require the analysis of three separate zones, namely, membrane interior, interface, and aqueous bulk, to account for experimental observations. The interfacial depletion generates an osmotic stress that favors conformations with small interfacial areas. For cations at anionic membranous interfaces, the interfacial distribution is a balance between electrostatic attraction and thermal movement.

ACKNOWLEDGMENTS

The authors acknowledge financial support from the Danish National Research Foundation via a grant to MEMPHYS—Center for Biomembrane Physics and the Danish Research Agency (grants 26-02-0160 and 21-04-0087). Simulations discussed here were performed at the Danish Center for Scientific Computing at the University of Southern Denmark.

REFERENCES

1. R. Gilles (1997). Compensatory organic osmolytes in high osmolarity and dehydration stresses: history and perspectives. *Comp. Biochem. Physiol. A* **117**: 279–290.
2. S. M. Bagnasco (2000). Urea: new questions about an ancient solute. *J. Nephrol.* **13**: 260–266.
3. E. A. Galinski, M. Stein, B. Amendt, and M. Kinder (1997). The kosmotropic (structure-forming) effect of compensatory solutes. *Comp. Biochem. Physiol. A* **117**: 357–365.
4. K. E. Zachariassen (1985). Physiology of cold tolerance in insects. *Physiol. Rev.* **65**: 799–832.
5. Y. Okada (2004). Ion channels and transporters involved in cell volume regulation and sensor mechanisms. *Cell Biochem. Biophys.* **41**: 233–258.
6. F. Wehner, H. Olsen, H. Tinel, E. Kinne-Saffran, and R. K. H. Kinne, (2004). Cell volume regulation: osmolytes, osmolyte transport, and signal transduction. *Rev. Physiol. Biochem. Pharmacol.* **148**: 1–80.
7. P. H. Yancey, M. E. Clark, S. C. Hand, R. D. Bowlus, and G. N. Somero (1982). Living with water-stress—evolution of osmolyte systems. *Science* **217**: 1214–1222.
8. R. D. Bowlus and G. N. Somero (1979). Solute compatibility with enzyme function and structure—rationales for the selection of osmotic agents and end-products of anaerobic metabolism in marine-invertebrates. *J. Exp. Zool.* **208**: 137–151.
9. A. J. Wang and D. W. Bolen (1996). Effect of proline on lactate dehydrogenase activity: testing the generality and scope of the compatibility paradigm. *Biophys. J.* **71**: 2117–2122.
10. G. N. Somero (1994). Adapting to water stress: convergence on common solutions, in *Water and Life* (G. N. Somero, C. B. Osmond, and C. L. Bolis, Eds.). Springer-Verlag, Berlin.
11. O. Borsani, V. Valpuesta, and M. A. Botella (2003). Developing salt tolerant plants in a new century: a molecular biology approach. *Plant Cell Tissue Org. Culture* **73**: 101–115.
12. E. A. Galinski (1993). Compatible solutes of halophilic eubacteria—molecular principles, water–solute interaction, stress protection. *Experientia* **49**: 487–496.
13. J. M. H. Stoop, J. D. Williamson, and D. M. Pharr (1996). Mannitol metabolism in plants: a method for coping with stress. *Trends Plant Sci.* **1**: 139–144.
14. S. Schaffer, K. Takahashi, and J. Azuma (2000). Role of osmoregulation in the actions of taurine. *Amino Acids* **19**: 527–546.
15. H. Pasantes-Morales, R. Franco, M. E. Torres-Marquez, K. Hernandez-Fonseca, and A. Ortega (2000). Amino acid osmolytes in regulatory volume decrease and isovolumetric regulation in brain cells: contribution and mechanisms. *Cell. Physiol. Biochem.* **10**: 361–370.
16. J. C. Measures (1975). Role of amino-acids in osmoregulation of non-halophilic bacteria. *Nature* **257**: 398–400.
17. J. C. Lee (2000). Biopharmaceutical formulation. *Curr. Opin. Biotech.* **11**: 81–84.
18. N. Beales (2004). Adaptation of microorganisms to cold temperatures, weak acid preservatives, low pH, and osmotic stress: a review. *Comp. Rev. Food Sci. Food Safety* **3**: 1–20.

19. L. R. DeYoung and K. A. Dill (1990). Partitioning of nonpolar solutes into bilayers and amorphous N-alkanes. *J. Phys. Chem.* **94**: 801–809.
20. T. L. Hill (1964). *Thermodynamics of Small Systems (Part 2)*. Dover Publications, New York.
21. P. Balgavy, M. Dubnickova, N. Kucerka, M. A. Kiselev, S. P. Yaradaikin, and D. Uhrikova (2001). Bilayer thickness and lipid interface area in unilamellar extruded 1,2-diacylphosphatidylcholine liposomes: a small-angle neutron scattering study. *Biochim. Biophys. Acta* **1512**: 40–52.
22. G. Cevc and D. Marsh (1987). *Phospholipid Bilayers: Physical Principles and Models*. Wiley, Hoboken, NJ.
23. P. Westh, C. Trandum, and Y. Koga (2001). Binding of small alcohols to a lipid bilayer membrane: Does the partitioning coefficient express the net affinity? *Biophys. Chem.* **89**: 53–63.
24. Y. Katz and J. M. Diamond (1974). Thermodynamic constants for nonelectrolyte partition between dimyristoyl lecithin and water. *J. Membr. Biol.* **17**: 101–120.
25. P. Westh and C. Trandum (1999). Thermodynamics of alcohol—lipid bilayer interactions: application of a binding model. *Biochim. Biophys. Acta* **1421**: 261–272.
26. Y. Katz and J. M. Diamond (1974). Nonsolvent water in liposomes. *J. Membr. Biol.* **17**: 87–100.
27. R. Koynova, J. Brankov, and B. Tenchov (1997). Modulation of lipid phase behavior by kosmotropic and chaotropic solutes—experiment and thermodynamic theory. *Eur. Biophys. J.* **25**: 261–274.
28. E. F. Casassa and H. Eisenberg (1964). Thermodynamic analysis of multicomponent solutions. *Adv. Protein Chem.* **19**: 287–395.
29. P. R. Davis-Searles, A. J. Saunders, D. A. Erie, D. J. Winzor, and G. J. Pielak (2001). Interpreting the effects of small uncharged solutes on protein-folding equilibria. *Annu. Rev. Biophys. Biomol. Struct.* **30**: 271–306.
30. M. T. Record, W. T. Zhang, and C. F. Anderson (1998). Analysis of effects of salts and uncharged solutes on protein and nucleic acid equilibria and processes: a practical guide to recognizing and interpreting polyelectrolyte effects, Hofmeister effects, and osmotic effects of salts. *Adv. Protein Chem.* **51**: 281–353.
31. S. N. Timasheff (1998). Control of protein stability and reactions by weakly interacting cosolvents: the simplicity of the complicated. *Adv. Protein Chem.* **51**: 355–432.
32. S. N. Timasheff (2002). Protein hydration, thermodynamic binding, and preferential hydration. *Biochemistry* **41**: 13473–13482.
33. G. Cevc (1988). Effect of lipid headgroups and (nonelectrolyte) solution on the structural and phase properties of bilayer-membranes. *Ber. Bunsenges. Phys. Chem.* **92**: 953–961.
34. L. M. Crowe and J. H. Crowe (1991). Solution effects on the thermotropic phase-transition of unilamellar liposomes. *Biochim. Biophys. Acta* **1064**: 267–274.
35. L. M. Crowe, C. A. Wistrom, and J. H. Crowe (1993). Does the preferential exclusion hypothesis apply to hydrated phospholipid bilayers? *Cryobiology* **30**: 224–225.
36. R. M. Epand and M. Bryszewska (1988). Modulation of the bilayer to hexagonal phase-transition and solvation of phosphatidylethanolamines in aqueous salt-solutions. *Biochemistry* **27**: 8776–8779.
37. P. Westh (2003). Unilamellar DMPC vesicles in aqueous glycerol: preferential interactions and thermochemistry. *Biophys. J.* **84**: 341–349.

38. P. Westh (2004). Preferential interaction of dimethyl sulfoxide and phosphatidyl choline membranes. *Biochim. Biophys. Acta* **1664**: 217–223.
39. J. Wolfe and G. Bryant (1999). Freezing, drying, and/or vitrification of membrane–solute–water systems. *Cryobiology* **39**: 103–129.
40. J. A. Schellman (1990). A simple-model for solvation in mixed-solvents—applications to the stabilization and destabilization of macromolecular structures. *Biophys. Chem.* **37**: 121–140.
41. J. A. Schellman (1993). The relation between the free-energy of interaction and binding. *Biophys. Chem.* **45**: 273–279.
42. J. Wyman (1948). Heme proteins. *Adv. Protein Chem.* **4**: 407–531.
43. J. Wyman and S. J. Gill (1990). *Binding and Linkage: Functional Chemistry of Biological Macromolecules.* University Science Books, Mill Valley.
44. E. L. Kovrigin and S. A. Potekhin (1997). Preferential solvation changes upon lysozyme heat denaturation in mixed solvents. *Biochemistry* **36**: 9195–9199.
45. D. J. Felitsky and M. T. Record (2004). Application of the local-bulk partitioning and competitive binding models to interpret preferential interactions of glycine betaine and urea with protein surface. *Biochemistry* **43**: 9276–9288.
46. P. Westh and C. Trandum (2002). Interactions of some cryoprotectives with lipid bilayers and biological membranes. *Biophys. J.* **82**:36A–36A.
47. E. S. Courtenay, M. W. Capp, C. F. Anderson, and M. T. Record (2000). Vapor pressure osmometry studies of osmolyte–protein interactions: implications for the action of osmoprotectants *in vivo* and for the interpretation of osmotic stress experiments *in vitro*. *Biochemistry* **39**: 4455–4471.
48. S. J. Gill, S. F. Dec, G. Olofsson, and I. Wadso (1985). Anomalous heat-capacity of hydrophobic solvation. *J. Phys. Chem.* **89**: 3758–3761.
49. E. Tuchsen, M. O. Jensen, and P. Westh (2003). Solvent accessible surface area (ASA) of simulated phospholipid membranes. *Chem. Phys. Lipids* **123**: 107–116.
50. R. P. Rand and V. A. Parsegian (1989). Hydration forces between phospholipid-bilayers. *Biochim. Biophys. Acta* **988**: 351–376.
51. C. J. Ho and C. D. Stubbs (1997). Effect of n-alkanols on lipid bilayer hydration. *Biochemistry* **36**: 10630–10637.
52. R. E. Jacobs and S. H. White (1989). The nature of the hydrophobic binding of small peptides at the bilayer interface—implications for the insertion of transbilayer helices. *Biochemistry* **28**: 3421–3437.
53. H. Heerklotz and J. Seelig (2000). Titration calorimetry of surfactant-membrane partitioning and membrane solubilization. *Biochim. Biophys. Acta* **1508**: 69–85.
54. C. Trandum, P. Westh, K. Jorgensen, and O. G. Mouritsen (1999). A calorimetric investigation of the interaction of short chain alcohols with unilamellar DMPC liposomes. *J. Phys. Chem. B* **103**: 4751–4756.
55. S. Nebel P. Ganz, and J. Seelig (1997). Heat changes in lipid membranes under sudden osmotic stress. *Biochemistry* **36**: 2853–2859.
56. Y. Koga (1996). Mixing schemes in aqueous solutions of nonelectrolytes: a thermodynamic approach. *J. Phys. Chem.* **100**: 5172–5181.
57. C. Trandum, P. Westh, K. Jorgensen, and O. G. Mouritsen (1999). Association of ethanol with lipid membranes containing cholesterol, sphingomyelin and ganglioside: a titration calorimetry study. *Biochim. Biophys. Acta* **1420**: 179–188.

58. P. Westh and Y. Koga (1997). Intermolecular interactions of lysozyme and small alcohols: a calorimetric investigation. *J. Phys. Chem. B* **101**: 5755–5758.
59. C. Trandum, P. Westh, K. Jorgensen, and O. G. Mouritsen (1999). Use of isothermal titration calorimetry to study the interaction of short-chain alcohols with lipid membranes. *Thermochim. Acta* **328**: 129–135.
60. C. Trandum, P. Westh, K. Jorgensen, and O. G. Mouritsen (2000). A thermodynamic study of the effects of cholesterol on the interaction between liposomes and ethanol. *Biophys. J.* **78**: 2486–2492.
61. S. Mitragotri, M. E. Johnson, D. Blankschtein, and R. Langer (1999). An analysis of the size selectivity of solute partitioning, diffusion, and permeation across lipid bilayers. *Biophys. J.* **77**: 1268–1283.
62. E. C. H. To, J. V. Davies, M. Tucker, P. Westh, C. Trandum, K. S. H. Suh, and Y. Koga (1999). Excess chemical potentials, excess partial molar enthalpies, entropies, volumes, and isobaric thermal expansivities of aqueous glycerol at 25 degrees C. *J. Sol. Chem.* **28**: 1137–1157.
63. H. K. Schachman and M. A. Lauffer (1949). The hydration, size and shape of tobacco mosaic virus. *J. Am. Chem. Soc.* **71**: 536–541.
64. D. W. Bolen (2004). Effects of naturally occurring osmolytes on protein stability and solubility: issues important in protein crystallization. *Methods* **34**: 312–322.
65. A. J. Saunders, P. R. Davis-Searles, D. L. Allen, G. J. Pielak, and D. A. Erie (2000). Osmolyte-induced changes in protein conformation equilibria. *Biopolymers* **53**: 293–307.
66. K. E. S. Tang and V. A. Bloomfield (2000). Excluded volume in solvation: sensitivity of scaled-particle theory to solvent size and density. *Biophys. J.* **79**: 2222–2234.
67. B. Lee and F. M. Richards (1971). Interpretation of protein structures—estimation of static accessibility. *J. Mol. Biol.* **55**: 379.
68. J. K. Kaushik and R. Bhat (2003). Why is trehalose an exceptional protein stabilizer? An analysis of the thermal stability of proteins in the presence of the compatible osmolyte trehalose. *J. Biol. Chem.* **278**: 26458–26465.
69. K. D. Collins and M. W. Washabaugh (1985). The Hofmeister effect and the behavior of water at interfaces. *Q. Rev. Biophys.* **18**: 323–422.
70. Y. Koga, P. Westh, J. V. Davies, K. Miki, K. Nishikawa, and H. Katayanagi (2004). Toward understanding the Hofmeister series. 1. Effects of sodium salts of some anions on the molecular organization of H_2O. *J. Phys. Chem. A* **108**: 8533–8541.
71. Y. Koga, P. Westh, and K. Nishikawa (2004). Effects of Na_2SO_4 and $NaClO_4$ on the molecular organization of H_2O. *J. Phys. Chem. A* **108**: 1635–1637.
72. P. Westh, H. Kato, K. Nishikawa, and Y. Koga (2006). Towards understanding the Hofmeister series. 3. Effects of sodium halides on the molecular organization of H_2O as probed by 1-propanol. *J. Phys. Chem. A* **110**: 2072–2078.
73. C. S. Pereira, and P. H. Hunenberger (2006). Interaction of the sugars trehalose, maltose and glucose with a phospholipid bilayer: a comparative molecular dynamics study. *J. Phys Chem. B.* **110**: 15572–15581.
74. R. Koynova and M. Caffrey (1998). Phases and phase transitions of the phosphatidylcholines. *Biochim. Biophys. Acta* **1376**: 91–145.
75. P. W. Sanderson, L. J. Lis, P. J. Quinn, and W. P. Williams (1991). The Hofmeister effect in relation to membrane lipid phase-stability. *Biochim. Biophys. Acta* **1067**: 43–50.

76. A. S. Rudolph and B. Goins (1991). The effect of hydration stress solutes on the phase-behavior of hydrated dipalmitoylphosphatidylcholine. *Biochim. Biophys. Acta* **1066**: 90–94.
77. M. Szogyi, T. Cserhati, and B. Bordas (1987). Phospholipid–amino-acid interactions. *Mol. Cryst. Liq. Cryst.* **152**: 267–278.
78. G. F. Xie and S. N. Timasheff (1997). The thermodynamic mechanism of protein stabilization by trehalose. *Biophys. Chem.* **64**: 25–43.
79. H. L. Bagger, C. C. Fuglsang, and P. Westh (2003). Preferential binding of two compatible solutes to the glycan moieties of *Peniophora lycii* phytase. *Biochemistry* **42**: 10295–10300.
80. J. Stumpel, W. L. C. Vaz, and D. Hallmann (1985). An X-Ray-diffraction and differential scanning calorimetric study on the effect of sucrose on the properties of phosphatidylcholine bilayers. *Biochim. Biophys. Acta* **821**: 165–168.
81. T. D. Tsvetkov, L. I. Tsonev, N. M. Tsvetkova, R. D. Koynova, and B. G. Tenchov (1989). Effect of trehalose on the phase properties of hydrated and lyophilized dipalmitoylphosphatidylcholine multilayers. *Cryobiology* **26**: 162–169.
82. J. F. Nagle and S. Tristram-Nagle (2000). Structure of lipid bilayers. *Biochim. Biophys. Acta* **1469**: 159–195.
83. J. H. Crowe, L. M. Crowe, J. F. Carpenter, A. S. Rudolph, C. A. Wistrom, B. J. Spargo, and T. J. Anchordoguy (1988). Interactions of sugars with membranes. *Biochim. Biophys. Acta* **947**: 367–384.
84. J. H. Crowe, L. M. Crowe, and S. A. Jackson (1983). Preservation of structural and functional-activity in lyophilized sarcoplasmic-reticulum. *Arch. Biochem. Biophys.* **220**: 477–484.
85. G. Strauss and H. Hauser (1986). Stabilization of lipid bilayer vesicles by sucrose during freezing. *Proc. Natl. Acad. Sci. USA* **83**: 2422–2426.
86. B. W. W. Grout and G. J. Morris (1987). *The Effect of Low Temperatures on Biological Systems*. Edward Arnold Publishers, London.
87. T. J. Anchordoguy, A. S. Rudolph, J. F. Carpenter, and J. H. Crowe (1987). Modes of interaction of cryoprotectants with membrane phospholipids during freezing. *Cryobiology* **24**: 324–331.
88. A. S. Rudolph and J. H. Crowe (1985). Membrane stabilization during freezing—the role of 2 natural cryoprotectants, trehalose and proline. *Cryobiology* **22**: 367–377.
89. G. Strauss, P. Schurtenberger, and H. Hauser (1986). The interaction of saccharides with lipid bilayer vesicles—stabilization during freeze-thawing and freeze-drying. *Biochim. Biophys. Acta* **858**: 169–180.
90. C. Fabrie, B. Dekruijff, and J. Degier (1990). Protection by sugars against phase transition-induced leak in hydrated dimyristoylphosphatidylcholine liposomes. *Biochim. Biophys. Acta* **1024**: 380–384.
91. T. D. Madden, M. B. Bally, M. J. Hope, P. R. Cullis, H. P. Schieren, and A. S. Janoff (1985). Protection of large unilamellar vesicles by trehalose during dehydration—retention of vesicle contents. *Biochim. Biophys. Acta* **817**: 67–74.
92. W. F. Wolkers, F. Tablin, and J. H. Crowe (2002). From anhydrobiosis to freeze-drying of eukaryotic cells. *Comp. Biochem. Physiol. A* **131**: 535–543.
93. F. A. Hoekstra, E. A. Golovina, and J. Buitink (2001). Mechanisms of plant desiccation tolerance. *Trends Plant Sci.* **6**: 431–438.

94. L. M. Crowe, D. S. Reid, and J. H. Crowe (1996). Is trehalose special for preserving dry biomaterials? *Biophys. J.* **71**: 2087–2093.
95. K. L. Koster, Y. P. Lei, M. Anderson, S. Martin, and G. Bryant (2000). Effects of vitrified and nonvitrified sugars on phosphatidylcholine fluid-to-gel phase transitions. *Biophys. J.* **78**: 1932–1946.
96. J. S. Clegg, P. Seitz, W. Seitz, and C. F. Hazlewood (1982). Cellular-responses to extreme water-loss—the water-replacement hypothesis. *Cryobiology* **19**: 306–316.
97. T. W. Cha, A. Guo, and X.-Y. Zhu (2006). Formation of supported phospholipid bilayers on molecular surfaces: role of surface charge density and electrostatic interaction. *Biophys. J.* **90**: 1270–1274.
98. M. Ø Jensen, O. G. Mouritsen, and G. H. Peters (2004). The hydrophobic effect: molecular dynamics simulations of water confined between extended hydrophobic and hydrophilic surfaces. *J. Chem. Phys.* **120**: 9729–9744.
99. T. R. Jensen, M. Ø. Jensen, N. Reitzel, K. Balashev, G. H. Peters, K. Kjaer, and T. Bjørnholm (2003). Water in contact with extended hydrophobic surfaces: direct evidence of weak dewetting. *Phys. Rev. Lett.* **90**:086101- 1–086101- 4.
100. E. A. Wasserman, and R. Felmy (1998). Computation of the electrical double layer properties of semipermeable membranes in multicomponent electrolytes. *Appl Environ. Microbiol.* **64**: 2295–2300.
101. G. Cevc (1990). Membrane electrostatics. *Biochim. Biophys. Acta* **1031**: 311–382.
102. S. Genet, R. Costalat, and J. Burger (2001). The influence of plasma membrane electrostatic properties on the stability of cell ionic composition. *Biophys. J.* **81**: 2442–2457.
103. F. C. Tsui, D. M. Ojcius, and W. L. Hubbell (1986). The intrinsic pK_a values for phosphatidylserine and phosphatidylethanolamine in phosphatidylcholine host bilayers *Biophys. J.* **49**: 459–468.
104. G. Cevc, A. Watts, and D. Marsh (1981). Titration of the phase transition of phophatidylserine bilayer membranes. Effects of pH, surface electrostatics, ion binding, and head-group hydration. *Biochemistry* **20**: 4955–4965.
105. M.-L. Ahrens (1983). Electrostatic control by lipds upon the membrane-bound (Na^+–K^+)-ATPase. II. The influence of surface potential upon the activating-ion equilibria. *Biochim. Biophys. Acta* **732**: 252–266.
106. C. M. Armstrong and G. Cota (1990). Modification of sodium channel gating by lanthanum. Some effects that cannot be explained by surface charge theory. *J. Gen. Physiol.* **96**: 1129–1140.
107. R. S. Cantor (1997). Lateral pressures in cell membranes: a mechanism for modulation of protein function *J. Phys. Chem. B* **101**: 1723–1725.
108. J. L. Lippert and W. L. Peticolas (1972). Raman active vibrations in long-chain fatty acids and phospholipid sonicates. *Biochim. Biophys. Acta* **282**: 2–17.
109. N. I. Yellin and W. Levin (1977). Hydrocarbon chain trans-gauche isomerization in phospholipid bilayer gel assemblies. *Biochemistry* **16**: 642–647.
110. Y. K. Levine, N. J. Birdsall, A. G. Lee, and J. C. Metcalfe (1972). ^{13}C nuclear magnetic resonance relaxation measurements of synthetic lecithins and the effect of spin-labeled lipids. *Biochemistry* **11**: 1416–1421.
111. H. D. Träuble and H. Haynes (1971). The volume change in lipid bilayer lamellae at the crystalline–liquid crystalline phase transition. *Chem. Phys. Lipids* **7**: 324–335.

112. O. G. Mouritsen and K. Jørgensen (1994). Dynamical order and disorder in lipid bilayers. *Chem. Phys. Lipids* **73**: 3–25.
113. P. H. Devaux and M. McConnell (1972). Lateral diffusion in spin-labeled phosphatidylcholine multilayers. *J. Am. Chem. Soc.* **94**: 4475–4481.
114. E. Sackmann and H. Träuble (1972). Studies of the crystalline phase transition of lipid model membranes. I. Use of spin labels and optical probes as indicators of the phase transition *J. Am. Chem. Soc.* **94**: 4482–4491.
115. H. J. C. Schweizer (1977). Single-molecule anisotropy measurements in lipid bilayers. Diplom Thesis, Technische Universität, Dresden, Germany, pp. 1–167.
116. J. Sonne, F. Y. Hansen, and G. H. Peters (2005). Methodological problems in pressure profile calculations for lipid bilayers. *J. Chem. Phys.* **122**: 1–9.
117. M. K. Chaudhury and S. Ohki (1981). Correlation between membrane expansion and temperature-induced membrane fusion. *Biochim. Biophys. Acta* **642**: 365–374.
118. M. C. Phillips and D. Chapman (1968). Monolayer characteristics of saturated 1,2,-diacyl phosphatidylcholines (lecithins) and phosphatidylethanolamines at the air–water interface. *Biochim. Biophys. Acta* **163**: 301–313.
119. A. H. Hainsworth and S. B. Hladky (1987). Effects of double-layer polarization on ion transport. *Biophys. J.* **51**: 27–36.
120. V. I. Zabolotskii, K. A. Lebedev, and E. G. Lovtsov (2003). Electric double layer at the membrane/solution interface in a three-layered membrane system. *Russ. J. Electrochem.* **39**: 1065–1072.
121. M. J. Bedzyk, G. M. Bommarito, M. Caffrey, and T. L. Penner (1990). Diffuse-double layer at a membrane–aqueous interface measured with X-ray standing waves. *Science* **248**: 52–56.
122. M. Endo, T. Takeda, Y. J. Kim, K. Koshiba, and K. Ishii (2001). High power electric double layer capacitors (EDLCs); from operating principle to pore size control in advanced activated carbons. *Carbon Sci.* **1**: 117–128.
123. L. A. Kibler and D. M. Kolb (2003). Physical electrochemistry: recent developments. *Z. Physikal. Chem.* **217**: 1265–1280.
124. P. Dynarowicz and M. Paluch (1992). Studies on the electrical double layer structure at the water/air interface. *Colloid Polym. Sci.* **270**: 349–352.
125. H. Helmholtz (1853). Über einige Gesetze der Verteilung elektrischer Ströme in körperlichen Leitern mit Anwendung auf die tierisch-elektrischen Versuche. *Ann. Phys.* **89**: 211–233.
126. M. Winter and R. J. Brodd (2004). What are batteries, fuel cells and supercapacitors? *J. Chem. Rev.* **104**: 4245–4269.
127. K. B. Wiles (2005). High performance disulfonated poly(arylene sulfone) co- and trepolymers for proton exchange membranes for fuel cell and transducer applications: synthesis, characterization and fabrication of ion conducting membranes. Virginia Polytechnic Institute and State University, Blackburg, Virginia, 1–360.
128. J. Barber (1980). Membrane surface charges and potentials in relation to photosynthesis. *Biochim. Biophys. Acta* **594**: 253–308.
129. D. C. Grahame (1947). The electrical double layer and the theory of electrocapillarity. *Chem. Rev.* **41**: 441–501.
130. T. B. Kinraide (1994). Use of a Gouy–Chapman–Stern model for membrane-surface electrical potential to interpret some features of mineral rhizotoxicity. *Plant Physiol.* **106**: 1583–1592.

131. J. O. Bockris, M. A. V. Devanathan, and K. Müller (2006) On the structure of charged interfaces. *Proc. R. Soc. London* **A274**: 55–79.
132. R. M. Peitzsch, M. Eisenberg, K. A. Sharp, and S. McLaughlin (1995). Calculations of the electrostatic potential adjacent to model phospholipd bilayers. *Biophys. J.* **68**: 729–738.
133. F. Lakhdar-Ghazal, J.-L. Tichadou, and J.-F. Tocanne (1983). Effect of pH and monovalent cations on the ionization state of phosphatidylglycerol in monolayers. *Eur. J. Biochem.* **134**: 531–537.
134. J. Marra (1986). Direct measurement of the interaction between phosphatidylglycerol bilayers in aqueous electrolyte solutions. *Biophys. J.* **50**: 815–825.
135. R. Kraayenhof, G. J. Sterk, and H. W. Sang (1993). Probing biomembrane interfacial potential and pH profiles with a new type of float-like fluorophores positioned at varying distances from the membrane surface. *Biochemistry* **32**: 10057–10066.
136. A. P. Winiski, A. C. McLaughlin, R. V. McDaniel, M. Eisenberg, and S. McLaughlin (1986). An experimental test of the discreteness-of-charge effect in positive and negative lipid bilayers. *Biochemistry* **25**: 8206–8214.
137. S. McLaughlin (1977). in *Current Topics in Membranes and Transport* (F. Bronner and A. Kleinzeller, Eds.). Academic Press, New York, pp. 71–144.
138. W. R. Fawcett and T. G. Smagala (2004). Ion size effects in a primitive level model of the diffuse double layer. *Cond. Matter Phys.* **7**: 709–718.
139. D. L. Gilbert and G. Ehrenstein (1969). Effect of divalent cations on potassium conductance of squid axons: determination of surface charge. *Biophys. J.* **9**: 447–463.
140. S. Carnie and S. McLaughlin (1983). Large divalent cations and electrostatic potentials adjacent to membranes. *Biophys. J.* **44**: 325–332.
141. S. Stankowski (1991). Surface charging by large multivalent molecules. *Biophys. J.* **60**: 341–351.
142. E. Donath and A. Voigt (1988). Effect of thick fixed-charge layers on electrostatic interaction—a theoretical approach. *Colloid Polym. Sci.* **266**: 1024–1030.
143. V. B. Arakelian, D. Walther, and E. Donath (1993). Electric potential distributions around discrete charges in a dielectric membrane—electrolyte solution system. *Colloid Polym. Sci.* **270**: 268–276.
144. B. E. Enos and D. A. McQuarrie (1981). The effect of discrete charges on the electrical properties of membranes. II. *J. Theor. Biol.* **93**: 499–522.
145. M. Ø Jensen, O. G. Mouritsen, and G. H. Peters (2004). Simulations of a membrane-anchored peptide: structure, dynamics, and influence on bilayer properties. *Biophys. J.* **86**: 3556–3575.
146. B. De Kruijff, A. Rietveld, N. Telders, and B. Vaandrager (1985). Molecular aspects of the bilayer stabilization induced by poly(L-lysines) of varying size in cardiolipin liposomes. *Biochim. Biophys. Acta* **820**: 295–304.
147. M. Mosior and S. McLaughlin (1992). Binding of basic peptides to acidic lipids in membranes: effects of inserting alanine(s) between basic residues. *Biochemistry* **31**: 1767–1773.
148. D. Li, M. Miller, and P. D. Chantler (1994). Association of a cellular myosin II with anionic phospholipids and the neuronal plasma membrane. *Proc. Natl. Acad. Sci. USA* **91**: 853–857.
149. A. C. Newton (1993). Interaction of proteins with lipid headgroups: lessons from protein kinase C. *Annu. Rev. Biophys. Biomol. Struct.* **22**: 1–25.

150. J. Kim, P. J. Blackshear, J. D. Johnson, and S. McLaughlin (1994). Phosphorylation reverses the membrane association of peptides that correspond to the basic domains of MARCKS and neuromodulin. *Biophys. J.* **67**: 227–237.
151. A. I. Magee, C. M. Newman, T. Giannakourous, J. F. Hancock, E. Fawell, and J. Armstrong (1992). Lipid modifications and function of the ras superfamily of proteins. *Biochem. Soc. Trans.* **20**: 497–499.
152. C. A. Buser, C. Sigal, M. D. Resh, and S. McLaughlin (1994), Membrane binding of myristoylated peptides corresponding to the NH_2-terminus of Src. *Biochemistry* **33**: 13093–13101.
153. S. O. Hagge, A. Wiese, U. Seydel, and T. Gutsmann (2004). Inner field compensation as a tool for the characterization of asymmetric membranes and peptide–membrane interactions. *Biophys. J.* **86**: 913–922.
154. S. Nir, C. Newton, and D. Papahadjopoulos (1978). Binding of cations to phosphatidylserine vesicles. *Bioelectrochem. Bioenerg.* **5**: 116–133.
155. H. Träuble and H. Eibl (1974). Electrostatic effects on lipid phase transitions: membrane structure and ionic environment. *Proc. Natl. Acad. Sci. USA* **71**: 214–219.
156. K. Kubica (1997). The effect of amphiphilic counterions on the gel–fluid phase transition of the lipid bilayer. *Appl. Math. Comp.* **87**: 261–270.
157. B. Y. Ha (2001). Stabilization and destabilization of cell membranes by multivalent ions. *Phys. Rev. E* **64**:051902- 1–051902- 5.
158. G. Cevc and D. Marsh (1983). Properties of the electrical double layer near the interface between a charged bilayer membrane and electrolyte solution: experiment vs. theory. *J. Phys. Chem.* **87**: 376–379.
159. P. W. van Dijck, B. De Kruijff, A. J. Verkleij, L. L. van Deenen, and J. de Gier (1978). Comparative studies on the effects of pH and Ca^{2+} on bilayers of various negatively charged phospholipids and their mixtures with phosphatidylcholine. *Biochim. Biophys. Acta* **512**: 84–96.
160. D. Papahadjopoulos, W. J. Vail, C. Newton, S. Nir, K. Jacobsen, G. Poste, and R. Lazo (1977). Studies on membrane fusion. III. The role of calcium-induced phase changes. *Biochim. Biophys. Acta* **465**: 579–598.
161. A. J. Verkleij, B. De Kruijff, P. H. Ververgaert, J.-F. Tocanne, and L. L. van Deenen (1974). The influence of pH, Ca^{2+} and protein on the thermotropic behaviour of the negatively charged phospholipid, phosphatidylglycerol. *Biochim. Biophys. Acta* **339**: 432–437.
162. H. Hauser and G. Shipley (1983). Interactions of monovalent cations with phosphatidylserine bilayer membranes. *Biochemistry* **22**: 2171–2178.
163. P. W. van Dijck, P. H. Ververgaert, A. J. Verkleij, L. L. van Deenen, and J. de Gier (1975). Influence of Ca^{2+} and Mg^{2+} on the thermotropic behaviour and permeability properties of liposomes prepared from dimyristol phosphatidylglycerol and mixtures of dimyristol phosphatidylglycerol and dimyristol phosphatidylcholine. *Biochim. Biophys. Acta* **406**: 465–478.
164. J. Bramhall (1984). Electostatic forces control the penetration of membranes by charged solutes. *Biochim. Biophys. Acta* **778**: 393–399.
165. P. Garidel, G. Förster, W. Richter, B. H. Kunst, G. Rapp, and A. Blume (2000). 1,2-Dimyristol- *sn* -glycero-3-phosphoglycerol (DMPG) divalent cation complexes: an X-ray scattering and freeze-fracture electron microscopy study. *Phys. Chem. Chem. Phys.* **2**: 4537–4544.

166. P. Garidel and A. Blume (1999). Interaction of alkaline earth cations with the negatively charged phospholipid 1,2-dimyristoyl-*sn*-glycero-3-phosphoglycerol: a differential scanning and isothermal titration calorimetric study. *Langmuir* **15**: 5526–5534.

167. M. Fragata, F. Bellemare, and E. K. Nenonene (1997). Mg(II) adsorption to a phosphatidylglycerol model membrane studied by atomic absorption and FT-IR spectroscopy. *J. Phys. Chem. B* **101**: 1916–1921.

168. D. L. Brautigan and F. M. Pinault (1993). Serine phosphorylation of protein tyrosine phosphatase (PTP1B) in HeLa cells in response to analogues of cAMP or diacylglycerol plus okadaic acid. *Mol. Cell Biochem.* 127–128: 121–129.

169. H. Hauser and G. Shipley (1984). Interactions of divalent cations with phosphatidylserine bilayer membranes. *Biochemistry* **23**: 34–41.

170. U. R. Pedersen, C. Leidy, P. Westh, and G. H. Peters (2006). The effect of calcium on the properties of charged phospholipid bilayers. *Biochim. Biophys. Acta* **1758**: 573–582.

171. J. R. Coorssen and R. P. Rand (1995). Structural effects of neutral lipids on divalent cation-induced interactions of phosphatidylserine-containing bilayers. *Biophys. J.* **68**: 1009–1018.

172. M. Roux and J.-M. Neumann (1986). Deuterium NMR study of head-group deuterated phosphatidylserine in pure and binary phospholipid bilayers: interactions with monovalent cations Na^+. and Li^+ *FEBS Lett.* **199**: 33–38.

173. T. Parasassi, G. De Stasio, G. Ravagnan, R. M. Rusch, and E. Gratton (1991). Quantitation of lipid phases in phospholipid vesicles by the generalized polarization of LAURDAN fluorescence. *Biophys. J.* **60**: 179–189.

174. T. Parasassi, E. K. Krasnowska, L. A. Bagatolli, and E. Gratton (1998). LAURDAN and PRODAN as polarity-sensitive fluorescent membrane probes. *J. Fluoresc.* **8**: 365–373.

175. T. Parasassi, M. Di Stefano, M. Loiero, G. Ravagnan, and E. Gratton (1994). Cholesterol modifies water concentration and dynamics in phospholipid bilayers: a fluorecence study using LAURDAN probe. *Biophys. J.* **66**: 763–768.

176. J. C. M. Lee, R. J. Law, and D. E. Discher (2001). Bending contributions to hydration of phospholipid and block copolymer membranes: unifying correlations between probe fluorescence and vesicle thermoelasticity. *Langmuir* **17**: 3592–3597.

177. T. Schlick, R. D. Skeel, A. T. Brunger, L. V. Kale, J. A. Board, J. Hermans, and K. Schulten (1999). Algorithmic challenges in computational molecular biophysics *J. Comput. Phys.* 151: 9–48.

178. G. H. Peters (2004). Computer Simulations: A tool for investigating the function of complex biological macromolecules in *Enzyme Functionality: Design, Engineering, and Screening* (A. Svendsen, Ed.). Marcel Dekker, New York, pp. 97–147.

179. D. P. Tieleman, S. J. Marrink, and H. J. C. Berendsen (1997). *Biochim. Biochim. Acta* 1331: 235–270.

180. S. A. Pandit and M. L. Berkowitz (2002). Molecular dynamics simulation of dipalmitoylphosphatidylserine bilayer with Na^+ counterions. *Biophys. J.* **82**: 1818–1827.

181. P. Mukhopadhyay, L. Monticelli, and D. P. Tieleman (2004). Molecular dynamics simulation of a palmitoyl-oleoyl phosphatidylserine bilayer with Na^+ counterions and NaCl. *Biophys. J.* **86**: 1601–1609.

CHAPTER 10

Protein Lipid Interactions from a Molecular Dynamics Simulation Point of View

CHRISTIAN KANDT
Department of Biological Sciences, University of Calgary, Calgary, Alberta, Canada

EDIT MÁTYUS
Department of Biological Sciences, University of Calgary, Calgary, Alberta, Canada and Semmelweis University, Institute of Biophysics and Radiation Biology, Budapest, Hungary

D. PETER TIELEMAN
Department of Biological Sciences, University of Calgary, Calgary, Alberta, Canada

CONTENTS

10.1	Introduction	268
10.2	Lipids as Solvent	270
	10.2.1 Bacteriorhodopsin	270
	10.2.2 OmpF Porin	272
10.3	Lipids as Cofactors	273
	10.3.1 Rhodopsin	273
	10.3.2 KcsA	275
10.4	Lipids as Substrate	276
	10.4.1 Magainins	276
	10.4.2 $cPLA_2$-C2 Proteins	278
10.5	Outlook	279
Acknowledgments		280
References		280

Structure and Dynamics of Membranous Interfaces, edited by Kaushik Nag
Copyright © 2008 John Wiley & Sons, Inc.

10.1 INTRODUCTION

A fundamental precondition for life is the compartmentalization of cells and organelles from their environment. These confined and self-contained reaction spaces are provided by biological membranes all sharing the general structure of a thin film of lipid and protein molecules held together by mainly noncovalent interactions. Biomembranes are essentially two-dimensional (2D) fluids most frequently organized as a continuous double layer of lipids and embedded membrane proteins. Even though the lipid bilayer is the most common form, lipid monolayers do also occur, for instance, in lung surfactant [1], where they are crucial for the breathing process.

Lipids form a highly diverse class of amphipathic molecules. One way of classification is based on a lipid's central building block—the lipid backbone: *glycerolipids* are based on glycerol whereas the foundation of *sphingolipids* is the long-chain amine sphingosine. Both glycero- and sphingolipids can have sugars and/or phosphates incorporated into their headgroups resulting in phospholipids, glycolipids, or phosphoglycolipids. A third major lipid class is based on *sterols*, with cholesterol being a prominent example. Sterols do not contain fatty acid components. For more detailed information about lipids and lipid classification see, for example, Refs. 2 and 3.

Beyond their basic role as a barrier, biomembranes facilitate a number of other functions that are mainly determined by the type of proteins associated with or embedded in the bilayer. Located at the interface between cell or organelle interior and exterior, membrane proteins are of high biological relevance as key players in fundamental processes such as energy conversion, transport, signal recognition, and signal transduction. Despite their significance, membrane proteins are still structurally underrepresented in the Protein Data Bank. To date, the three-dimensional (3D) structures of 120 different membrane proteins have been solved, which is more than two orders of magnitudes lower than the number of structures available for soluble proteins [4]. Gaining deeper insight and understanding into the architecture of membrane proteins is thus a major goal in modern structural biology.

Apart from further development and improvement of structure resolving experimental techniques, computational methods have become increasingly important, utilizing a palette of techniques ranging from bioinformatics and homology modeling to quantum mechanical calculations and molecular dynamics (MD) simulations. The main method we use, and one of the most powerful computational methods to study matter at the level of atoms and molecules, is MD simulation. In MD, the forces between atoms are described by a simplified empirical potential that tries to mimic the "real" underlying interaction potential. The key terms in this simplified potential function are shown in Fig. 10.1 and consist of harmonic springs for bonds and angle, a cosine expansion for rotations around a central bond, and Lennard-Jones and Coulomb interactions. Modern quantum chemistry can reproduce almost exactly the real potential for simple cases involving on the

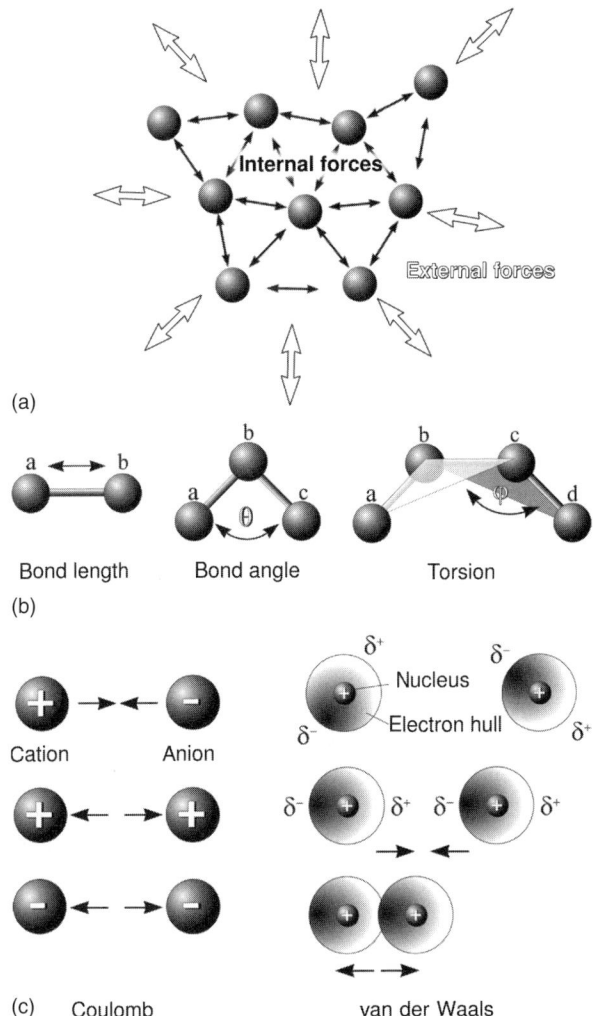

Figure 10.1 (a) Molecular dynamics (MD) simulation investigates the motions of a system of particles under the influence of internal and external forces. In the MD simulation of biomolecules, external forces comprise temperature and pressure coupling, as well as potentially electric fields or biasing forces that mimic experiments, whereas the bonded and nonbonded interactions contribute to the internal forces. (b) Bonded interactions are mediated by covalent bonds and include stretching and bending vibrations as well as torsions. (c) Nonbonded interactions comprise long-range electrostatics and short-range van der Waals forces. The latter are based on transient fluctuating dipoles that are due to small asymmetries in the electron distributions. Depending on the distance of the atoms, both attractive or repulsive effects occur.

order of tens or a hundred atoms, but it is currently too expensive computationally for most of the problems we are interested in and contains details that are not likely to be essential for these problems. Using this potential function, the forces (the derivative of the potential with respect to position) on all atoms in the system of interest are calculated and used to solve classical equations of motions to generate a trajectory of all atoms in time. The potential function can be modified to include additional energy terms, such as an externally applied electric field.

The primary result of an MD simulation is a trajectory of all atoms in time, from which dynamic, structural, and thermodynamic properties of the simulated system can be calculated. The main strength of MD simulations is the amount of detail: all or nearly all atoms are incorporated in the model and few assumptions are necessary. Its main weaknesses are the computational cost and the associated limited time (realistically, at this time, approximately 100–500 ns) and limited length scale (approximately $10 \times 10 \times 10$ nm) and the approximations inherent in the potential function. The use of fixed partial charges centered on the atoms is a particularly significant limitation, but there are other areas for improvement (e.g., additional terms that describe coupling between bonded interactions). There is a substantial ongoing effort to improve the accuracy of the potential functions, but this does not change the fundamental method. Review articles about recent MD simulations of membrane proteins and how to set up such simulations can be found in Refs. 5–7.

In this chapter we discuss recent examples of MD simulation studies of membrane proteins to illustrate three different levels of lipid–protein interactions. The first level encompasses interactions where lipids function merely as a solvent not affecting the protein structure and function. At the second level lipids act as essential cofactors whose presence or absence is crucial for the protein to function properly. In the third level of interaction lipids are the actual target or substrate of a (membrane) protein or peptide. Giving a complete overview of the rapidly growing field of MD simulations of membrane proteins is not possible in the framework of this book chapter. We therefore confine ourselves to present only two or three examples for each of the three levels of lipid–protein interaction.

10.2 LIPIDS AS SOLVENT

10.2.1 Bacteriorhodopsin

Bacteriorhodopsin (bR) is a small seven-helix transmembrane protein found in some halophilic archaebacteria living in shallow salt lakes, for example, Owens Lake in California. As the water temperature rises, the oxygen concentration decreases and it is only under these increasingly anaerobic conditions that bR is synthesized to provide a simple (if not the simplest) form of photosynthesis. The protein acts as a light-driven proton pump using the energy of photons absorbed by its retinal chromophore to build up a proton gradient over the plasma membrane powering the synthesis of ATP. Since its discovery 35 years ago [8], bacteriorhodopsin has been the subject of intensive research that made the protein a model for the architecture of membrane proteins and the functional mechanism of ion-transporting systems. It was also the first integral membrane protein whose 3D structure was determined [9].

A remarkable property of bacteriorhodopsin is the protein's apparent insensitivity to its lipid environment. It is routinely reconstituted in, for example, palmitoyloleoyl phosphatidylcholine (POPC), wherein bR remains fully functional, even though this lipid does not occur in the protein's native environment—the purple membrane—which is additionally characterized by a high protein/lipid ratio of 3:1. Over the years many theoretical studies have been performed on bR using both molecular mechanical and quantum mechanical approaches [10–20], and again POPC has been a popular choice when a bilayer was represented explicitly. In part this is also due to simulation parameters for lipids available at the time a given MD investigation is carried out. In one of these MD studies, bR

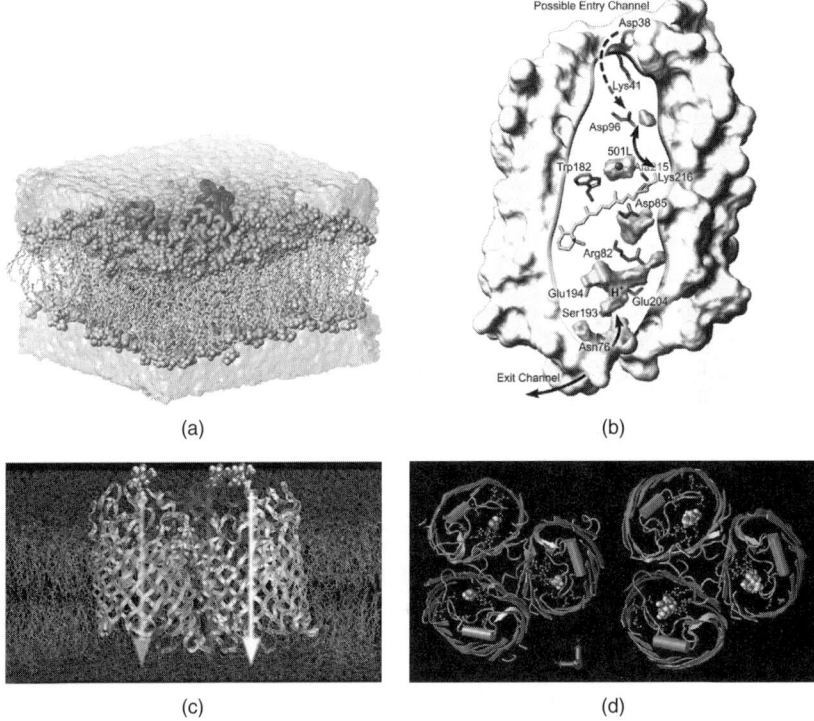

Figure 10.2 Lipids as solvent. (a) In a simulation of the bacteriorhodopsin trimer, the dynamics of protein-internal water molecules was used to determine a complete proton conduction pathway that connects all key residues involved in proton transfer. (b) The cloud-like objects represent areas of high water residence probability. (c) Alanine and α-methylglucose were pulled through the OmpF trimer. (d) Both molecules display a preferred orientation during the permeation process, aligning strongly in the electric field of the bottleneck region of the pore. (Part (b) is from Ref. 18; parts (c) and (d) were adapted from Ref. 31.)

was simulated for 5 ns in its trimeric quaternary structure embedded in a POPC bilayer (Fig. 10.2a) to investigate unanswered detail questions about the bacteriorhodopsin proton transfer mechanism [18, 19]. It is generally known that proton conduction in bR (and other proton-conducting systems) is based on hydrogen bonded networks formed by protein-internal water molecules and adjacent polar residues and occurs in a Grotthuss-like mechanism. Going back to the works of C. J. T. de Grotthuss on electrolysis in 1806, the observed proton conduction in bacteriorhodopsin is not a single proton particle traveling from one end of the protein to the other but rather the net result of a domino effect of many hydrogen bonds forming and breaking between adjacent water molecules and polar residues and several local protons shifting from one end of their hydrogen bridge to the other. Although some key residues of this hydrogen bonded network in bR have been identified, it is still unclear if this list is complete or how proton transfer between certain groups can occur—especially when the distance between them is too large to be bridged by hydrogen bonds. Since a molecular mechanics approach was used in this study, the question of proton conduction could not be treated directly as this requires a quantum mechanical description—which itself is limited to system sizes and simulation times much shorter than classical MD. Instead, the dynamics of protein-internal water molecules were monitored and their hydrogen bonding pattern to adjacent residues in the protein's ground state structure was analyzed. Besides model accuracy, another classical issue with MD simulations is that of gaining an adequate sampling of conformational space. This was in part dealt with by simulating bR in its trimeric quaternary structure, thus tripling the achieved total sampling time of the protein behavior from 5 to 15 ns. Using this approach, a complete proton pathway could be obtained that connects all known relevant key residues by water diffusion (Fig. 10.2b). New relevant residues could also be identified and successful predictions could be made regarding the effect of certain mutations on specific absorption bands in the IR arising from water dynamics in the proton release channel.

10.2.2 OmpF Porin

Porins form aqueous channels that act as the passive diffusion pores for small hydrophilic molecules. They play major roles in ion and nutrient uptake but are also involved in cell death [21, 22] and multidrug resistance [23]. Porins belong to the β-barrel type of integral membrane proteins, which to date have only been found in the outer membrane of Gram-negative bacteria, where interestingly integral membrane proteins of the α-helical type seemed not to occur. Recently, however, the crystal structure of the integral outer membrane Wza showed an α-helical barrel in the transmembrane region [24]. Following the endosymbiont hypothesis [25], β-barrel proteins are also expected in the outer membranes of mitochondria and plastids [26]. The voltage-dependent ion channel VDAC appears to be such an example [27].

OmpF porin from *Escherichia coli* is a nonspecific pore that permits transport of ions and small molecules up to a mass of about 650 kDa [28]. OmpF has been widely studied experimentally and theoretically [5, 29] and was the first membrane protein forming X-ray grade crystals [30]; unfortunately, the phasing problem could not be solved and the first OmpF crystal structure was determined 12 years later [28].

The protein was also the subject of a molecular dynamics simulation study in which the OmpF trimer was simulated in a dimyristoyl phosphatidylcholine (DMPC) bilayer. Over the course of 1 ns, two OmpF substrates—alanine and α-methylglucose each in a separate MD run—were pulled through the porin to the other side of the membrane [31] (Fig. 10.2c). It was found that both dipolar molecules do not diffuse through the pore in a randomly tumbling manner but instead exhibit a preferred orientation during the permeation process. They align strongly in the electric field of the bottleneck region of the pore (Fig. 10.2d), where they experience the maximum adhesion force. Binding to this part of OmpF was not observed.

OmpF is another example of a membrane protein that is fairly unaffected by its particular environment. The simulation study presented earlier is somewhat limited by the simulation length and thus by the speed at which the two substrates are pulled across the bilayer, which currently is much faster than reasonable biological rates. As computational power increases, longer simulation times become available and concomitantly slower and thus more realistic pulling velocities. On the other hand, the main strength of the simulation approach here is obviously the actual possibility of pulling substrates through a membrane protein in a highly controlled manner, gaining highly detailed structural and energy information that could not be obtained with experimental techniques. A combination of detailed atomistic simulations and experimental methods is a very powerful tool if details of interactions are important. One interesting example for OmpF is its interaction with certain antibiotics. Single-channel electrophysiology measurements can follow binding and unbinding of antibiotics to OmpF [32]. OmpF has also been engineered to be selective for calcium ions [33]. Because of its stability and relatively easy expression, it forms an attractive basis for nanotechnology applications, where rational redesign of the protein can benefit significantly from molecular modeling and simulation.

10.3 LIPIDS AS COFACTORS

10.3.1 Rhodopsin

Rhodopsin is the primary light receptor in the animal visual system and—like bacteriorhodopsin—uses retinal as a molecular light sensor (Fig. 10.3a). But contrary to bR, in rhodopsin the retinal chromophore leaves the protein during the photocycle, in the course of which large-scale conformational changes of the protein structure occur [34]. As a G-protein coupled receptor (GPCR), rhodopsin belongs to the largest known protein superfamily whose members are of critical importance in a wide range of biological signaling processes [35]. Rhodopsin is also the only GPCR whose structure is known at atomic resolution [36].

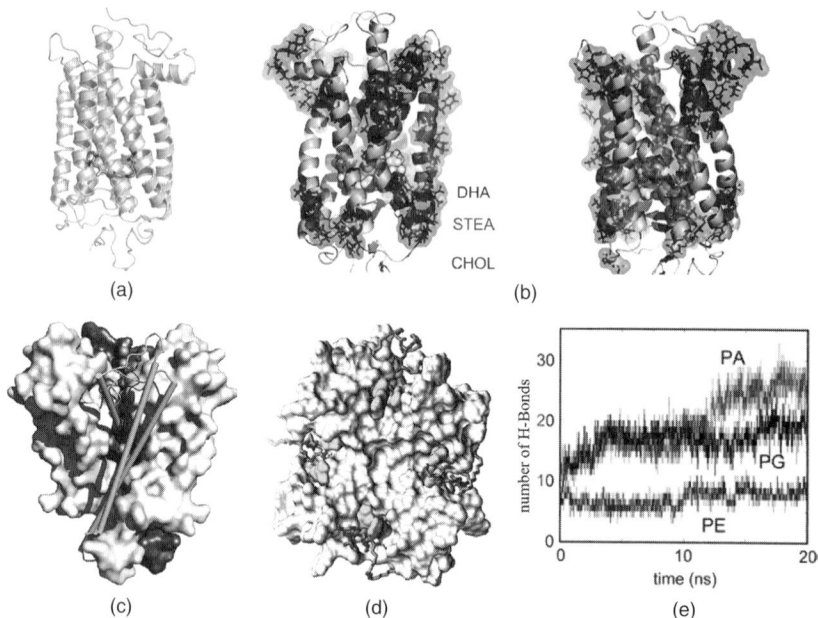

Figure 10.3 Lipids as cofactors. (a) Rhodopsin is the primary light receptor in the animal visual system and—like bacteriorhodopsin—uses retinal as a molecular light sensor. Rhodopsin was simulated in a bilayer containing a mixture of cholesterol (CHOL) and lipids with a saturated stearic acid chain (STEA) and a polyunsaturated docosahexaenoic acid chain (DHA). (b) The protein appears to have specific regions that preferentially interact with one membrane component. DHA is known to destabilize the protein and accelerate the photocycle, whereas cholesterol has the opposite effect. (c) The KcsA potassium channel has a tetrameric architecture with fourfold symmetry about a central pore. The spheres represent K^+ ions as seen in the crystal structure. (d) The channel requires the presence of anionic lipids to function and contains four lipid binding sites. These sites were tested with anionic POPA and POPG and neutral POPE lipids. Hydrogen bond analysis showed that POPA has the highest level of interaction with KcsA. (Part (b) was adapted from Ref. 44; parts (d) and (e) were adapted from Ref. 50.)

The protein's native environment is characterized by a high content of ω-3 polyunsaturated fatty acids [37, 38] and a high concentration of cholesterol in newly formed disk membranes that eventually decreases as these membranes mature [39]. Absorption and fluorescence spectroscopy experiments showed that the presence of lipids with polyunsaturated chains destabilizes the native state of rhodopsin and accelerates the kinetics of the photocycle, whereas cholesterol has the opposite effect [40–42]. NMR investigations indicated that rhodopsin is preferentially solvated by lipids containing ω-3 polyunsaturated docosahexaenoic acid (DHA) [43]. The molecular image of these interactions could be elucidated by long-time multicopy molecular dynamics simulations of rhodopsin in a heterogeneous membrane environment rich in ω-3 fatty acid DHA [44]. A series of 26 independent 100-ns simulations were performed with the protein embedded in a 99-lipid bilayer containing a 2:2:1 mixture of 1-stearoyl-2-docosahexaenoyl phosphatidylethanolamine, 1-stearoyl-2-docosahexaenoyl phosphatidylcholine, and cholesterol. The former two contained an 18-carbon saturated chain (STEA) and a 22-carbon ω-3 polyunsaturated chain (DHA). The packing between rhodopsin and each lipid chain was computed. DHA chains were found to form tight associations with the protein in a small number of specific binding sites different from the nonspecific interactions made by unsaturated chains and cholesterol (Fig. 10.3b).

Furthermore, since DHA binding primarily occurs in grooves between helices and thus weakens the interhelical packing of the protein, this suggests a possible means by which DHA modulates rhodopsin stability, kinetics, and function.

Experimental findings indicated that one lipid component has a specific effect on the stability and function of a membrane protein. Using MD it was possible to distinguish and quantify the interaction of the rhodopsin with different lipids and lipid chains in terms of preferred interaction sites. Evidence was also provided for the molecular scenario and details that cause lipids with ω-3 polyunsaturated chains to destabilize protein structure, which is of special physiological relevance as rhodopsin undergoes large-scale conformational changes during its photocycle [34]. On the other hand, even though 26×100 ns have been simulated, this is still too short to appropriately simulate the entire photocycle. Beyond that, a quantum mechanical treatment would be desirable to adequately describe the trigger event of the light-induced retinal isomerization.

10.3.2 KcsA

Ion channels act as extremely selective gateways, allowing specific ions to pass in and out of the cell in response to various signals. They are fundamental for physiological processes such as the formation and transduction of nerve impulses, muscle contraction, or osmoregulation. The first X-ray crystal structure of an ion channel was determined for the potassium channel KcsA in 1998 [45].

Potassium channels control the electric potential across cell membranes by catalyzing the rapid, selective diffusion of K^+ ions down their electrochemical gradient. KcsA is a tetramer with fourfold symmetry about a central pore (Fig. 10.3c) and requires the presence of negatively charged phospholipids for function [46], with no apparent specificity for a particular anionic lipid [47, 48]. A recent KcsA crystal structure revealed the presence of a lipid binding site involving two arginine residues, but only a lipid fragment could be identified due to a distorted electron density at this position [48]. Mass spectroscopy experiments indicated that KcsA preferentially binds to phosphatidylglycerol (PG) rather than to phosphatidylcholine (PC) [49].

In a molecular dynamics study [50], lipid–protein interactions of the KcsA channel were investigated by constructing three models with different lipids present at the binding site defined by the lipid fragment in the X-ray structure (Fig. 10.3d). Palmitoyloleoyl phosphatidic acid (POPA), palmitoyloleoyl phosphatidylethanolamine (POPE), and palmitoyloleoyl phosphatidylglycerol (POPG) were tested to explore the headgroup specificity of the binding site. All three models were inserted into a POPC bilayer and simulated for 20 ns. Hydrogen bond analysis revealed the anionic POPA and POPG show the highest level of interaction with the KcsA arginine binding site, whereas for the neutral POPE much less hydrogen bonding was observed (Fig. 10.3e). Another 27-ns simulation, where the channel was embedded in a POPE/POPG bilayer without any lipids initially placed at the binding sites, showed a clear preference of KcsA to interact with POPG over POPE.

KcsA appears to require the presence of negatively charged phospholipids to function and the X-ray crystal structure revealed lipid fragments closely associated with the protein but could not identify which particular lipids were present. Using the simulation approach, possible candidates could be tested and ranked based on their KcsA interaction in terms of hydrogen bond frequency. Again the quality of the employed molecule descriptions (topologies) is crucial for this and any MD study. Also, the small sampling rate of conformational space is a major issue here, as only single MD runs were performed. This should be kept in mind when interpreting MD results even when—as in this case—experimental results like the KcsA preference for POPG over POPC could be reproduced well.

10.4 LIPIDS AS SUBSTRATE

Because of the importance of membrane organization to different biological activity, it is of interest to understand how the membrane adsorption of proteins affects the structural and electrostatic properties of the membrane. Lipids reorganize themselves easily in the vicinity of embedded proteins to optimize the protein–membrane interactions. Antimicrobial peptides (AMPs) are short peptides that differ widely in structure and amino acid composition, although they often are amphipathic and cationic [51]. Their accumulation might lead to a localized collapse of the lipid bilayer to form a transient hole, leading to cell death [52]. AMPs may undergo substantial conformational changes to interact and insert into the target membrane. The strength of the interactions between AMPs and bacterial membranes depends on the size, charge, and structural properties of the peptides [53, 54]. In this section, two examples are discussed from the large diversity of peptide–lipid interaction mechanisms that may occur [51, 55–58]. One is an AMP, the other a phospholipase enzyme. The antimicrobial peptides magainins interact directly with the membrane without the involvement of a specific receptor and form transient pores. The phospholipase $cPLA_2$-C2 protein docks to phosphatidylcholine-rich intracellular membranes, which brings the 600-residue catalytic domain in close proximity to the membrane. This domain catalyzes the hydrolysis of zwitterionic phospholipids, to form lysophospholipids and free fatty acids.

10.4.1 Magainins

Magainins are secreted by the skin of the African clawed frog *Xenopus laevis* and exhibit a wide range of antimicrobial and antifungal activity. At high concentrations they permeabilize the lipid, likely by forming toroidal pores that involve lipid headgroups lining the porewall, stabilized by the presence of the peptides. Experimentally, it is observed that the peptides have a random coil structure in aqueous solution at neutral pH [59]. Interacting with the lipid bilayer, they fold into the α-helical secondary structure upon binding to acidic phospholipid bilayers [60, 61]. As the concentration of bound peptides increases, they self-aggregate and insert perpendicular to the membrane surface, permeabilizing the lipid bilayer [62]. The polar region of the bilayer is expanded [63] and an increase in lipid flip-flop is observed [64]. Due to the positive curvature strain imposed by the peptides, pore formation is triggered.

Based on those experimental observations and the detection of toroidal structure using neutron scattering [65], it is accepted that magainins form toroidal pores. Despite the wide scale experimental studies, the peptide–lipid interactions, the structure of the pore, and the mechanism of pore formation are not known in detail.

Atomistic molecular dynamics simulations of magainin-2 with POPC bilayers have identified the key peptide–lipid interactions responsible for its antimicrobial action [66]. Strong interactions of lysine residues with the oxygens of the lipid headgroup regions were detected (Fig. 10.4a) and the hydrogen bonds formed were persistent for the duration of the 20-ns simulations. Simulations also suggest that the α-helical stability of magainin-2 appears related to the abundance of lysine residues, which are capable of shielding the backbone to protect intrahelical hydrogen bonds.

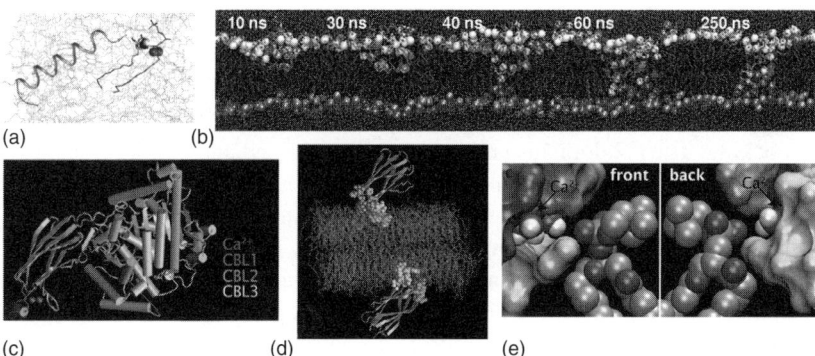

Figure 10.4 Lipids as substrate. (a) Snapshot highlighting the hydrogen bonding between lysine and lipid oxygens. (b) Snapshots of spontaneous pore formation by magainins in the DPPC lipid bilayer: the peptides aggregate while binding to the membrane interface (10 ns); after binding one peptide moves toward the bilayer interior (30 ns); and rapidly pore formation occurs (40 ns); the pore relaxes to a toroidal shape, which remains stable until the end of the simulation (250 ns). (c) The complete cPLA$_2$-C2 molecule: the C2 domain is shown in blue, the catalytic domain in green, and the Ca^{2+} ions in purple. The three Ca^{2+} binding loops: CBL1 in red, CBL2 in orange, and CBL3 in yellow. (d) Snapshot of the simulated system: two isolated cPLA$_2$-C2 domains in a POPC bilayer. (e) Front and back view of CBLs and the three water molecules shielding the Ca^{2+} ions from lipids. (Parts (a) and (b) are from Ref. 66; parts (c)–(e) are from Ref. 67.)

The mechanism of pore formation by magainin and the structure of the pore are explored in a landmark 0.25-μs molecular dynamics simulations [67] (Fig. 10.4b). The pore-forming peptides maintain a predominantly parallel orientation with respect to the membrane interface and aggregate. The lysine and phenylalanine residues play a dominant role in dragging the peptides deeply into the lipid–water interface, thereby imposing a large curvature stress. Although the peptide is embedded deeply within the membrane, the charged lysine residues remain hydrogen bonded. The system is metastable for ∼10 ns, until one solvent molecule from the peptide-free interface interacts with hydrophilic groups of the embedded peptide (Fig. 10.4b, $t = 30$ ns). This contact leads to the formation of a contiguous pore. During this process, the peptide together with some lipid molecules moves across the membrane (Fig. 10.4b, $t = 40–60$ ns). In the remaining part of the simulation, the pore relaxes toward a toroidal shape, which is stable for the duration of the simulation.

Surprisingly, typically only one peptide is found near the pore center, while other peptides remain close to the pore edge. For spontaneous pore formation peptide aggregation and a peptide to lipid ratio of $>1/64$ ratio are required. The presence of full α-helicity is not a prerequisite and the increased flexibility of the peptide might facilitate pore formation.

10.4.2 cPLA$_2$-C2 Proteins

Cytosolic phospholipase A$_2$ (cPLA2) is an 85-kDa enzyme that initiates the synthesis of leukotrienes and prostaglandin, important mediators of inflammation [68]. Its 138-residue C2 domain (cPLA$_2$-C2) (Fig. 10.4c) docks to phosphatidylcholine-rich intracellular membranes in response to a second messenger Ca-signal to establish membrane proximity of the catalytic domain of enzyme, which hydrolytically liberates arachidonic acid from zwitterionic phospholipids after docking. The C2 domains consist of an eight-strand antiparallel β-sandwich, and three negatively charged calcium-binding loops (CBLs) that typically bind two or three Ca^{2+} ions and drive membrane docking [69].

Mechanic studies have shown that the docking mechanisms of C2 domains are dominated by both electrostatic and hydrophobic interactions [70]. Fluorescence [71], NMR [72] and EPR [73] studies proposed that the two calcium ions bound by the CLBs are directly coordinated by phospholipid headgroups in the membrane-docked state. Those measurements defined the depth and the orientation of the membrane-inserted domain relative to the bilayer surface.

Understanding the cPLA$_2$-C2–membrane molecular interactions is essential for elucidating which C2 domains stably associate with specific target membrane. Molecular dynamics simulations based on EPR membrane depth measurements and the crystal structure of cPLA$_2$-C2 were carried out to investigate the cPLA$_2$-C2–membrane interactions in atomic detail [74] (Fig. 10.4e). During equilibration, the POPC bilayer generated a cavity with a hydrophobic basin formed by the lipid alkyl chains and a hydrophilic rim formed from lipid phosphate, choline, and carbonyl groups. Thus the bilayer adopts a shape complementary to that of the protein [74]. Each CBL induced its own docking site, because the bilayer reorganized itself to maximize energetically favorable interactions with protein. The polar collar of the membrane-docked protein interacts with polar lipid headgroup components, and apolar side chains promote interactions in the basin of the cavity. The lipid hydrocarbon chains interact solely with CBL1 and CBL3, while avoiding close interactions with charged residues. Carbonyl contacts are very similar to those of alkyl chains, and glycerols have the fewest contacts. Because of the CLB2 lacking the hydrophobic residues, most of the lipids interact with them.

This simulation suggests that instead of direct calcium coordination by lipids an indirect coordination mediated by water molecules occurs [75] (Fig. 10.4d). The headgroup phosphates are unable to coordinate calcium directly, because the lipids are too bulky to infiltrate the CBL calcium-binding cage. Calcium is directly coordinated only by protein and water oxygens. This model shows that the calcium ion neutralizes the CBLs, thereby allowing membrane docking and penetration into the bilayer core.

Molecular dynamics simulations show important details of the targeted docking process of the C2 domain of cytosolic phospholipase A_2 enzyme.

10.5 OUTLOOK

In the past few years, growth in the number of simulation studies and the complexity of the biological systems simulated have been phenomenal. The papers described in this chapter give a good overview of the type of problems that can be investigated by computer simulations. In general, such simulations provide a detailed view of lipid–protein interactions. The main limitations are the accuracy of the parameters that describe such interactions and the limited time that can be simulated, but in both areas rapid improvements continue to be made. Simulations of pure lipids are now investigating phenomena on length scales of 10–30 nm and time scales of hundreds of nanoseconds, and can be compared more directly to experimental and theoretical work at mesoscopic scales. The availability of fast software and relatively inexpensive computers will no doubt lead to a further increase in the complexity of the systems that can be simulated. It should also lead to increased methodological development and more extensive control simulations to improve the statistical accuracy of phenomena observed. Simulations are generating questions that can be tested experimentally, which will no doubt lead to increasingly stronger links with experiments. The recent growth in the number of high-resolution membrane protein structures is an exciting development. Simulations can provide additional dynamic details of the static snapshots given by membrane protein structures and can be used to investigate and build models of other conformational states.

It is worth reflecting on the time and effort spent on a particular type of simulation. Eight years ago, it took 6 months to create a starting model of the bacterial porin OmpF in a lipid bilayer and another 6 months on a very expensive computer to simulate its motions for 1 ns [76]. Recently, we repeated a very similar calculation in a day or two, using desktop computers. We believe that computational studies can provide a valuable contribution to many problems in biochemistry, biophysics, and related areas, often as a small part of an integrated approach to solving a specific problem that combines experiment and simulation.

ACKNOWLEDGMENTS

This research was supported by the Natural Science and Engineering Research Council (NSERC). DPT is an AHFMR Senior Scholar and CIHR New Investigator. CK is an AHFMR Postdoctoral Fellow. EM was supported by the Hungarian Scientific Research Fund, grants OTKA F043192 and OTKA D048670.

REFERENCES

1. E. J. Veldhuizen and H. P. Haagsman (2000). *Biochim. Biophys. Acta* **1467**: 255–270.
2. O. G. Mouritsen (2005). Life—as a matter of fat. Springer, Heidelberg.
3. E. Fahy, S. Subramaniam, H. A. Brown, C. K. Glass, A. H. Merrill, Jr., R. C. Murphy, C. R. Raetz, D. W. Russell, Y. Seyama, W. Shaw, T. Shimizu, F. Spener, G. van Meer, M. S. VanNieuwenhze, S. H. White, J. L. Witztum, and E. A. Dennis (2005). *J Lipid Res*. **46**: 839–861.
4. S. H. White (2004). *Protein Sci*. **13**: 1948–1949.
5. W. L. Ash, M. R. Zlomislic, E. O. Oloo, and D. P. Tieleman (2004). *Biochim. Biophys. Acta Biomembr*. **1666**: 158–189.
6. J. Gumbart, Y. Wang, A. Aksimentiev, E. Tajkhorshid, and K. Schulten (2005). *Curr. Opin. Structural Biol*. **15**: 423–431.
7. C. Kandt, W. L. Ash, and D. P. Tieleman (2007). Setting Up and Running Membrane Protein Simulations Methods 41(4): 475–488.
8. D. Oesterhelt and W. Stoeckenius (1971). *Nature New Biol*. **233**: 149–152.
9. R. Henderson and P. N. Unwin (1975). *Nature* **257**: 28–32.
10. J. Baudry, E. Tajkhorshid, F. Molnar, J. Phillips, and K. Schulten (2001). *J. Phys. Chem. B* **105**: 905–918.
11. M. Ben-Nun, F. Molnar, H. Lu, J. C. Phillips, T. J. Martinez, and K. Schulten (1998). *Faraday Discuss*. 447–462.
12. O. Edholm, O. Berger, and F. Jahnig (1995). *J. Mol. Biol*. **250**: 94–111.
13. S. Grudinin, G. Buldt, V. Gordeliy, and A. Baumgaertner (2005). *Biophys. J*. **88**: 3252–3261.
14. S. Hayashi and I. Ohmine (2000). *J. Phys. Chem. B* **104**: 10678–10691.
15. S. Hayashi, E. Tajkhorshid, and K. Schulten (2002). *Biophys. J*. **83**: 1281–1297.
16. W. Humphrey, D. Xu, M. Sheves, and K. Schulten (1995). *J. Phys. Chem*. **99**: 14549–14560.
17. F. Jahnig and O. Edholm (1992). *J. Mol. Biol*. **226**: 837–850.
18. C. Kandt, K. Gerwert, and J. Schlitter (2005). *Proteins* **58**: 528–537.
19. C. Kandt, J. Schlitter, and K. Gerwert (2004). *Biophys. J*. **86**: 705–717.

20. S. Tanizaki and M. Feig (2006). *J. Phys. Chem. B* **110**: 548–556.
21. R. Olson, H. Nariya, K. Yokota, Y. Kamio, and E. Gouaux (1999). *Nature Structural Biol.* **6**: 134–140.
22. L. Z. Song, M. R. Hobaugh, C. Shustak, S. Cheley, H. Bayley, and J. E. Gouaux (1996). *Science* **274**: 1859–1866.
23. V. Koronakis, A. Sharff, E. Koronakis, B. Luisi, and C. Hughes (2000). *Nature* **405**: 914–919.
24. C. Dong, K. Beis, J. Nesper, A. L. Brunkan-Lamontagne, B. R. Clarke, C. Whitfield, and J. H. Naismith (2006). *Nature* **444**: 226–229.
25. S. D. Dyall, M. T. Brown, and P. J. Johnson (2004). *Science* **304**: 253–257.
26. G. E. Schulz (2002). *Biochim. Biophys. Acta. Biomembr.* **1565**: 308–317.
27. C. A. Mannella (1997). *J Bioenerg. Biomembr.* **29**: 525–531.
28. S. W. Cowan, T. Schirmer, G. Rummel, M. Steiert, R. Ghosh, R. A. Pauptit, J. N. Jansonius, and J. P. Rosenbusch (1992). *Nature* **358**: 727–733.
29. R. Koebnik, K. P. Locher, and P. Van Gelder, (2000). *Mol. Microbiol.* **37**: 239–253.
30. R. M. Garavito and J. P. Rosenbusch (1980). *J. Cell Biol.* **86**: 327–329.
31. K. M. Robertson and D. P. Tieleman (2002). *FEBS Lett.* **528**: 53–57.
32. E. M. Nestorovich, C. Danelon, M. Winterhalter, and S. M. Bezrukov (2002). *Proc. Natl. Acad. Sci. USA* **99**: 9789–9794.
33. H. Miedema, M. Vrouenraets, J. Wierenga, D. Gillespie, B. Eisenberg, W. Meijberg, and W. Nonner (2006). *Biophys. J.* **91**: 4392–4400.
34. W. L. Hubbell, C. Altenbach, C. M. Hubbell, and H. G. Khorana (2003). *Adv. Protein Chem.* **63**: 243–290.
35. U. Gether (2000). *Endocr. Rev.* **21**: 90–113.
36. T. Okada, M. Sugihara, A. N. Bondar, M. Elstner, P. Entel, and V. Buss (2004). *J. Mol. Biol.* **342**: 571–583.
37. K. Boesze-Battaglia and A. D. Albert (1989). *Exp. Eye Res.* **49**: 699–701.
38. W. L. Stone, C. C. Farnsworth, and E. A. Dratz (1979). *Exp. Eye Res.* **28**: 387–397.
39. K. Boesze-Battaglia, T. Hennessey, and A. D. Albert (1989). *J. Biol. Chem.* **264**: 8151–8155.
40. D. C. Mitchell, S. L. Niu, and B. J. Litman (2001). *J. Biol. Chem.* **276**: 42801–42806.
41. S. L. Niu, D. C. Mitchell, and B. J. Litman (2001). *J. Biol. Chem.* **276**: 42807–42811.
42. S. L. Niu, D. C. Mitchell, and B. J. Litman (2002). *J. Biol. Chem.* **277**: 20139–20145.
43. O. Soubias and K. Gawrisch (2005). *J. Am. Chem. Soc.* **127**: 13110–13111.
44. A. Grossfield, S. E. Feller, and M. C. Pitman (2006). *Proc. Natl. Acad. Sci. USA* **103**: 4888–4893.
45. D. A. Doyle, J. Morais Cabral, R. A. Pfuetzner, A. Kuo, J. M. Gulbis, S. L. Cohen, B. T. Chait, and R. MacKinnon (1998). *Science* **280**: 69–77.
46. I. M. Williamson, S. J. Alvis, J. M. East, and A. G. Lee (2003). *Cell Mol. Life Sci.* **60**: 1581–1590.
47. L. Heginbotham, L. Kolmakova-Partensky, and C. Miller (1998). *J. Gen. Physiol.* **111**: 741–749.
48. F. I. Valiyaveetil, Y. Zhou, and R. MacKinnon (2002). *Biochemistry* **41**: 10771–10777.

49. J. A. Demmers, A. van Dalen, B. de Kruijff, A. J. Heck, and J. A. Killian (2003). *FEBS Lett.* **541**: 28–32.
50. S. S. Deol, C. Domene, P. J. Bond, and M. S. Sansom (2006). *Biophys. J.* **90**: 822–830.
51. R. M. Epand and H. J. Vogel (1999). *Biochim. Biophys. Acta* **1462**: 11–28.
52. R. G. Efremov, D. E. Nolde, A. G. Konshina, N. P. Syrtcev, and A. S. Arseniev (2004). *Curr. Med. Chem.* **11**: 2421–2442.
53. I. Zelezetsky, S. Pacor, U. Pag, N. Papo, Y. Shai, H. G. Sahl, and A. Tossi (2005). *Biochem. J.* **390**: 177–188.
54. K. Matsuzaki, K. Sugishita, M. Harada, N. Fujii, and K. Miyajima (1997). *Biochim. Biophys. Acta* **1327**: 119–130.
55. Y. Shai (2002). *Biopolymers* **66**: 236–248.
56. Z. Oren, J. Ramesh, D. Avrahami, N. Suryaprakash, Y. Shai, and R. Jelinek (2002). *Eur. J. Biochem.* **269**: 3869–3880.
57. Y. Shai and Z. Oren (2001). *Peptides* **22**: 1629–1641.
58. Z. Oren and Y. Shai (1998). *Biopolymers* **47**: 451–463.
59. K. Matsuzaki, Y. Mitani, K. Y. Akada, O. Murase, S. Yoneyama, M. Zasloff, and K. Miyajima (1998). *Biochemistry* **37**: 15144–15153.
60. D. J. Hirsh, J. Hammer, W. L. Maloy, J. Blazyk, and J. Schaefer (1996). *Biochemistry* **35**: 12733–12741.
61. B. Bechinger, M. Zasloff, and S. J. Opella (1993). *Protein Sci.* **2**: 2077–2084.
62. K. Matsuzaki, O. Murase, H. Tokuda, S. Funakoshi, N. Fujii, and K. Miyajima (1994). *Biochemistry* **33**: 3342–3349.
63. K. Matsuzaki (1999). *Biochim. Biophys. Acta* **1462**: 1–10.
64. K. Matsuzaki, O. Murase, N. Fujii, and K. Miyajima (1996). *Biochemistry* **35**: 11361–11368.
65. S. J. Ludtke, K. He, W. T. Heller, T. A. Harroun, L. Yang, and H. W. Huang (1996). *Biochemistry* **35**: 13723–13728.
66. S. K. Kandasamy and R. G. Larson (2004). *Chem. Phys. Lipids* **132**: 113–132.
67. H. Leontiadou, A. E. Mark, and S. J. Marrink (2006). *J. Am. Chem. Soc.* **128**: 12156–12161.
68. C. C. Leslie (2004). *Prostaglandins Leukot. Essent. Fatty Acids* **70**: 373–376.
69. J. Bai and E. R. Chapman (2004). *Trends Biochem. Sci.* **29**: 143–151.
70. E. A. Nalefski, M. A. Wisner, J. Z. Chen, S. R. Sprang, M. Fukuda, K. Mikoshiba, and J. J. Falke (2001). *Biochemistry* **40**: 3089–3100.
71. E. A. Nalefski and J. J. Falke (1998). *Biochemistry* **37**: 17642–17650.
72. G. Y. Xu, T. McDonagh, H. A. Yu, E. A. Nalefski, J. D. Clark, and D. A. Cumming (1998). *J. Mol. Biol.* **280**: 485–500.
73. N. J. Malmberg, D. R. Van Buskirk, and J. J. Falke (2003). *Biochemistry* **42**: 13227–13240.
74. S. Jaud, D.J. Tobias, J.J. Falke, and S.H. White (2007). *Biophys. J.* 92(2):517–24.
75. S. Malkova, F. Long, R. V. Stahelin, S. V. Pingali, D. Murray, W. Cho, and M. L. Schlossman (2005). *Biophys. J.* **89**: 1861–1873.
76. D. P. Tieleman and H. J. Berendsen (1998). *Biophys. J.* **74**: 2786–2801.

PART III
COMPLEX MEMBRANOUS SYSTEMS

■■■■■■■ CHAPTER 11

Molecular Analysis of Bacterial Membranous Systems

SALIM SIOUD

IUT de Béthune, Département Chimie, Université d'Artois, Béthune, France

NICOLAS JOLY and PATRICK MARTIN

Unité de Catalyse et de Chrimie du Solide, site de l'Artois - UMR CNRS 8181 - I.U.T. de Béthune, Département Chimie, 1230 rue de l'Université, 62408 Béthune cedex, France

JOSEPH BANOUB

Department of Chemistry, Memorial University of Newfoundland, St. John's Newfoundland, Canada and Fisheries and Oceans Canada, Science Branch, Special Projects, St. John's, Newfoundland, Canada

CONTENTS

11.1	Introduction	286
11.2	Separation Methods	288
	11.2.1 Chromatographic Separation	289
	11.2.2 Capillary Electrophoresis	295
11.3	Analysis Methods	297
	11.3.1 Mass Spectrometry	297
	11.3.2 Fast Atom Bombardment	298
	11.3.3 MALDI-ToF-MS	299
	11.3.4 MALDI-IM-ToF-MS	300
	11.3.5 High Pressure Liquid Chromatography Coupled with Electrospray Ionization–Mass Spectrometry (HPLC-ESI-MS)	300
	11.3.6 ESI–Tandem Mass Spectrometry (ESI-MS/MS)	302
11.4	Conclusion	306
References		307

Structure and Dynamics of Membranous Interfaces, edited by Kaushik Nag
Copyright © 2008 John Wiley & Sons, Inc.

11.1 INTRODUCTION

The existence of transmembrane phospholipid asymmetry was first reported thirty years ago. Since that time, an assortment of phospholipid distributions has been elucidated in a variety of membranes, often revealing a high degree of asymmetry [1]. There has been a considerable uncertainty in the data for certain membrane types such as intracellular and bacterial membranes, which have reinforced the need to rediscover the distribution values, in order to lower the uncertainties [1]. The asymmetric arrangement appears to control an array of cellular functions. Chemicals that alter the phospholipid distribution are likely to have a variety of biological and pharmaceutical applications [1].

Phospholipids are amphiphilic molecules that are fundamental components of the cellular membrane and as such confer to the membrane particular properties that are studied in this book. It is well known that phospholipids are the major part of the lipid components of the biological membranes and, as such, have the same structure as the triglycerides. Phospholipids, in addition, carry one carboxylic terminal and a phosphorylated group. These phosphorylated groups (polar end and/or end absorbent) can ionize and be charged and, for this reason, increase the water solubility, whereas the nonpolar lipid chains are insoluble in water (nonpolar end and/or hydrophobic end) [2]. If one deposits oil in water, the two bodies do not mix, and consequently the molecules of lipid will form micelles, with their hydrophobic end pointing in the central position [2]. Phospholipids occur in nature as a combination of two fatty chains. Since these fatty chains can vary in length and degree of unsaturation, each natural phospholipid contains numerous molecular species. These species differ greatly in their chemical and biological properties and their identification and quantification are of great interest.

Phospholipids are classified as a mixture of different diagnostic molecules, such as phosphatidylcholine (PC), phosphatidylethanolamine (PE), phosphatidylinositol (PI), phosphatidic acid (PA), and phosphatidylserine (PS). The occurrence of higher concentrations of other lipid families tends to be more organ specific [3].

The chemical structure of a typical phospholipid, namely, phosphatidylethanolamine, is shown in Fig. 11.1. The top polar region is composed of the NH_3 terminal of the phophorylated polar group, which is connected to the glycerol moiety, which in turn is attached to two fatty acid tails. One of the tails is a straight chain fatty acid (saturated). The other has a twist in the tail because of a *cis* double bond (unsaturated). This bend influences the packing and the movement of the lateral plane of the membrane [3].

Existing as many different kinds of compounds, phospholipids play an important role in the signal transduction process. Amphiphilic phospholipids self-associate in aqueous solutions, at concentrations above their critical micelle concentration, and accordingly form micelles [4]. An examination of chemical composition of mammalian cell membranes reveals that over 90% of their mass (dry weight) is comprised of proteins and phospholipids [5].

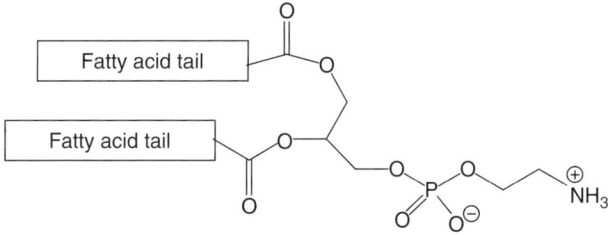

Figure 11.1 Example of the phospholipid phosphatidylethanolamine.

All of the phospholipids consist of an assortment of long-chain fatty acids and phosphoryl constituents; and given that the fatty chains can vary in length and in the degree of unsaturation, each phospholipid class has numerous molecular species with different chemical and biological properties. Accordingly, the identification and quantification of phospholipids in biological samples have shown great promise in revealing the associated biological activities [5].

There are two major classes of phospholipids: the phosphoglyceride (or glycerophospholipid) that has a glycerol backbone and the sphingosine-based phospholipid, known as sphingomyelin. Usually, the phosphoglyceride consists of glycerol-3-phosphate, which is esterified at its C1 and C2 positions with fatty acids; the phosphoryl group is attached to another group X, to form the sphingo class of compounds (Fig. 11.2a) [6]. The X group can typically be choline, ethanolamine, serine, or inositol.

The chemical structures of the corresponding phosphoglycerides are named phosphatidylcholine (PC) and phosphatidylethanolamine (PE), phosphatidylserine (PS), and phosphatidylinositol (PI) and are shown in Fig. 11.2. Please note that

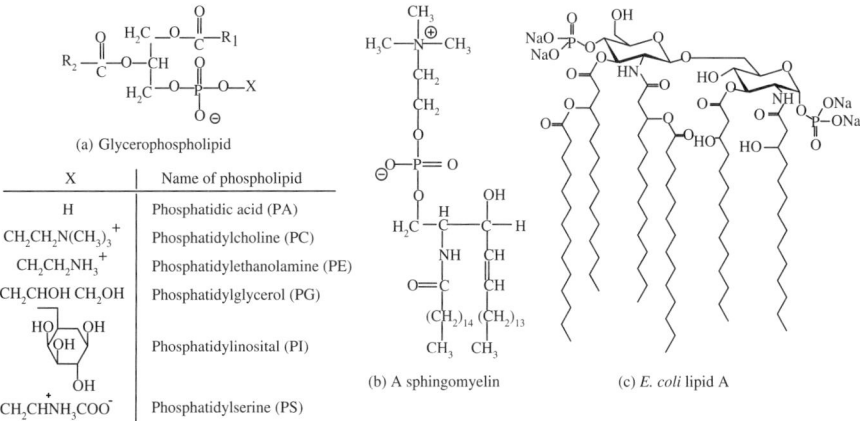

Figure 11.2 Structures of phospholipids: (a) glycerophospholipid, (b) a sphingomyelin, and (c) *E. coli* lipid A.

some other phospholipids that are present in very low concentrations are also called minor phospholipids, such as platelet activating factor (PAF), phosphatidic acid, and sphingosine-1-phosphate. PAF denotes a unique class of 2-O-acetylated phospholipids (1-O-alkyl-2-O-acetyl-*glycero*-3-phosphocholine), which has the property of platelet activation [5]. Sphingomyelins are ceramides bearing either a phosphocholine or a phosphoethanolamine group (Fig. 11.2b).

Lipid A (Fig. 11.2c), the hydrophobic anchor of the lipopolysaccharide (LPS; endotoxin) component of the cell wall of the gram-negative bacteria, is an important class of phospholipid that causes sepsis and endotoxic shock. Lipid A is comprised of a β-D-GlcNp-(1→6)-D-GlcNp disaccharide backbone, in which the free sugar hydroxyl and amino groups are acylated with several fatty acid chains. Lipid A triggers the biosynthesis of diverse mediators of inflammation and is suspected to be the causative agent of septic shock syndrome [7–9]. Lipid A exists in two forms: diphosphoryl lipid A and monophosphoryl lipid A [10].

Figure 11.2a classifies various polar molecules and in addition introduces a new class of lysophosphatic acids (LPAs), which are more polar than the phosphatic acid (PA) class. In a given phospholipid class (PC, PS, PE, PI, or PG; LPA or PA) the polarity of the molecule is connected to the constituent fatty acids' polarity. It was noted that the phopholipids' polarity increases as follows:

- For glycerophosphatids, the polarity increases in the direction PC > PS > PE.
- For phosphatidylglycolipides, PI > PG.

As mentioned previously, phospholipids are not only major constituents of membranes but are also important molecules in a variety of biological events. In recent years, lysophosphatidylcholine (LPC) has been reported for a variety of diseases, for example, inflammation [10, 11], atherosclerosis [12], and diabetes [13]. In addition to LPC, other phospholipids may play a functional role in the pathogenesis of various diseases; thus simultaneous analysis of phospholipid classes is important.

11.2 SEPARATION METHODS

Detailed elucidation of lipid composition has been the subject of research investigations for several decades [14, 15]. In 1958, Inouye and Noda [16] reacted egg phospholipids with mercuric acetate before submitting them to paper chromatography. Chromatographic resolution was a function of the differences in mercuration depending on the degree of unsaturation of the phospholipid fatty acyl chains.

In the 1960s, Collins [17, 18] published a series of papers in which a method for the countercurrent distribution of rat liver PC molecular species was described. Separation was incomplete but provided some information on species composition.

Renkonen [19, 20] utilized acetolysis and enzymatic hydrolysis with phospholipase C followed by acetylation of the diglyceride residues and fractionation into subclasses based on degree of unsaturation by silver nitrate thin layer chromatography (TLC) [21–24]. The different subclasses were then submitted to gas chromatographic analysis (GC) of the fatty acids. This method provided useful information about the relative abundance of fatty acids in different subfractions of chicken egg yolk, bovine brain, and human serum phospholipids. It also provided a conceptual framework for future method development in subsequent decades.

With the advent of lipophilic derivatives of Sephadex [25, 26], a new solid phase for liquid chromatography was available which was used with some success for molecular species separations [27–29]. This partition mode of separation was much further improved when solid phases for high pressure liquid chromatography (HPLC) composed of alkylated (usually C8 chains) silica particles became available. This has become one of the most utilized separation methods for molecular species characterization in liquid chromatography. It was found that when trying to separate and/or purify components of the phosphatidylglycolipid and glycerophosphatid families by HPLC, the polarity of these compounds interfered differently, according to the mobile phase used.

Even though reverse phase on TLC plates has been commercially available for a length of time, it has found little use in phospholipid molecular species separation. The utilization of TLC for this category of separations is mostly confined to separation according to degree of acyl chain unsaturation (i.e., silver ion chromatography).

More recently, hyphenated methods like liquid chromatography–electrospray ionization–mass spectrometry (LC-ESI-MS) methods have provided highly selective analysis of phospholipid species [30] and have also been used in the analysis of intact molecular species of bovine milk sphingomyelin (N-acylsphingosine-l-phosphocholine) [31]. These novel ESI-MS methods are discussed further in this rationale.

11.2.1 Chromatographic Separation

Thin Layer Chromatography Thin layer chromatography (TLC) is a widely used chromatography technique employed to separate chemical compounds. It involves a stationary phase consisting of a thin layer of adsorbent material, usually silica gel, aluminum oxide, or cellulose immobilized onto a flat, inert carrier sheet. A liquid phase consisting of the solution to be separated and dissolved in an appropriate solvent is drawn through the plate via capillary action, separating the experimental solution. It can be used to determine the pigments contained in a plant, to detect pesticides or insecticides in food, in forensics to analyze the dye composition of fibers, or to identify compounds present in a given substance, among other uses. It is a quick and generic method for organic reaction monitoring [32, 33].

A growing number of methods have been developed to attempt one-dimensional phospholipid separations. Some of these methods use TLC plates prepared with various buffer systems [34]. Others specify the use of standard silica gel plates [35].

Dugan [36] carried out the analysis of phospholipids, using one-dimensionnal high performance thin layer chromatography (HPTLC) plates. Biological samples were supplied as whole bovine brain, gerbil liver, and soybean extracts; separations were performed on preadsorbent high efficiency silica gel, type HLF plates [37]. In this study, a large number of solvent combinations were examined for use as possible mobile phases to produce a complete separation of the major phospholipid groups. One advantage of this mobile phase is the insensitivity of the separation to small changes in mobile phase composition. Good separation of major phospholipid groups results when the amounts of either chloroform or ammonium hydroxide were varied between 5 and 7 mL. Figure 11.3 demonstrates the stability of R_f values in these regions [37].

This method has demonstrated an ability to handle natural mixtures of phospholipids in biological extracts. Small variations in the mobile phase do not adversely affect the separation. The mobile phase of denatured ethanol/chloroform/ammonium hydroxide used in conjunction with preadsorbent silica gel HPTLC plates provides a good separation of the major phospholipid groups in a single development.

Another example that combines TLC and HPLC was developed after the administration of eicosapentaenoic acid ethyl ester, for the assay of fatty acid compositions of individual phospholipids, in platelets from non-insulin-dependent diabetes mellitus (NIDDM) patients [38]. The author described the effect of ethyl cis-5,8,11,14,17-eicosapentaenoate (EPA-E) on the fatty acid composition of individual phospholipids in platelets obtained from patients with NIDDM [38]. Miwa et al. [38] described the one-dimensional TLC system for the separation

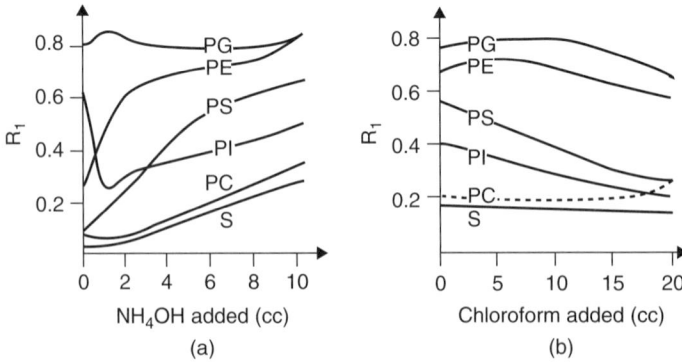

Figure 11.3 (a) The influence of 5-mL additions of ammonium hydroxide on the R_f values of the phospholipids. (b) The influence of chloroform on the R_f values of the phospholipids when 50 mL of ethanol chloroform is added to 50:5 ethanol/ammonium hydroxide. (From Ref. 37.)

on a commercially available TLC plate without any pretreatment of the plate (Fig. 11.4). This method showed a satisfactory precision in analyzing the fatty acids composition of individual platelet phospholipids.

To confirm the practical utility of the above method in the assay of biological materials, Miwa et al. [38] applied it to the determination of fatty acids incorporated into each major phospholipid class (PC, PE, PS, and PI) of platelets obtained from ten fasting NIDDM patients. In the TLC, glycolipids and neutral lipids including free fatty acids, which have an R substantially greater than the farthest migrating class of phospholipid, such as PG, did not interfere with the phospholipid separation (Fig. 11.4).

Complete separation of the six major phospholipid groups was found to require two-dimensional TLC development. Although the two-dimensional technique is quite useful for complex mixtures, it suffers from the disadvantage of being able to handle only one sample per plate [34].

An improved HPLC method has been developed for the analysis of 28 saturated and mono- and polyunsaturated fatty acids (C8:0–C22:6) including *cis/trans* isomers and positional isomers of double bonds in the fatty acid chain [39].

It is well known that the phospholipid concentration can vary widely between subclasses and the different organs; in addition, it is well accepted that these

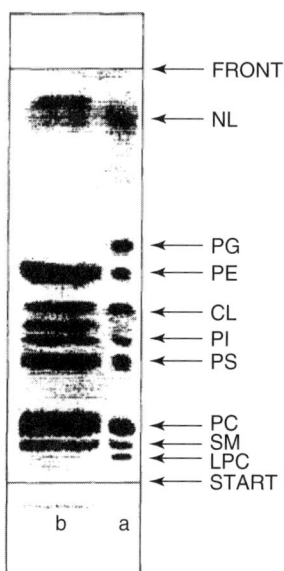

Figure 11.4 Thin layer chromatogram of (a) eight phospholipid standards and (b) whole lipids of platelet from a NIDDM patient on an LK-5 plate (20 cm × 5 cm × 0.25 mm). (NL = neutral lipids; PG = L-α-phosphatidyl-DL-glycerol; PE = L-α-phosphatidylethanolamine; CL = cardiolipin; PI = L-α-phosphatidylinositol; PS = c-α-phosphatidyl-L-serine; PC = L-α-phosphatidylcholine; SM = sphingomyelin; LPC = L-c-lysophosphatidylcholine) Staining done with iodine vapor. (Adapted from Ref. 38.)

variations can also be induced under certain pathological conditions [40–42]. Accordingly, novel accurate methods for the simultaneous separation of different phospholipids continue to be reported for the study of fatty acid composition [43, 44].

Several one-dimensional TLC methods, which are simpler and more rapid than two-dimensional TLC methods, have been developed [45–50]. One of the major weaknesses of some of these methods is that they need a lengthy and cumbersome pretreatment of the TLC plates before application of the lipids to the plates. Another drawback is that most methods cannot successfully separate phospholipids on a preparative scale owing to the poor resolution of the different phospholipids.

High Performance Liquid Chromatography High performance liquid chromatography (HPLC) has emerged as an exceedingly valuable separation protocol that can be adapted for normal or reverse-phase separation strategies. The separation of lipids using normal phase silica is analogous to TLC separation with the advantage that the extracted phospholipids remain in solution. It has been found that the analysis of an individual species of glycerophosphocholine lipids can reveal an excellent chromatographic separation. One of the difficulties in carrying out normal phase HPLC is the lack of reproducibility in the retention time since the mobile phase often contains some amount of water. Water alters the hydration state of silica and therefore alters the affinity of the phospholipid for the silica. One can regenerate normal phase columns by injecting 1,2-dimethoxypropane to react with any tightly bound water on the normal phase HPLC column, restoring activity to the column [51].

A change in HPLC retention time nonetheless can be quite disconcerting, but an understanding of the basic principles involved in the separation process as well as the dynamic alteration of the column partition characteristics can increase confidence in the use of this technique for phospholipid analysis.

Reverse-phase HPLC, in contrast to normal phase HPLC, is an exceedingly reproducible technique in terms of retention time for individual phospholipid components [52, 53]. The major determinant of the retention time is the lipophilicity of the fatty acyl chains, rather than the hydrophilicity of the polar headgroup. Thus care should be taken while separating classes of phospholipids, as individual molecular species of widely divergent phospholipid classes may elute with a very similar retention time. However, the combination of both normal phase chromatography to separate individual molecular species of widely divergent phospholipid classes (by polar headgroup), and a reverse-phase chromatography to separate by lipophilicity presents a powerful approach to isolate individual molecular species within a complex mixture of phospholipids. Such a strategy can be particularly valuable if one is looking for a minor component in a mixture of highly abundant normal phospholipid components. A review by Peterson and Cummings [52] discusses in detail the different HPLC methods for the assessment of phospholipids in biological samples.

The application of HPLC to separate phospholipid components was first reported by Jungalwala et al. [53]. HPLC utilizes both a solid- or liquid-coated solid stationary phase and a liquid mobile phase. HPLC also uses a lower temperature than the GC methods, hence reducing the risk of isomerization of unsaturated fatty acids. HPLC has a direct advantage over TLC because it decreases the exposure of the sample to atmospheric oxygen, decreasing the risk of autooxidation of phospholipids [54]. In addition, separated fractions can be collected for further analysis [55]. HPLC is used mainly for lipid class separations; also, some methods for separating the molecular species of a specific class have been developed using reverse-phase chromatography [56]. However, these reverse-phase HPLC methods are time intensive and offer poor resolution, owing to the wide range of polarities. This inadequacy results from the different carbon chain lengths and degrees of saturation of the fatty acid chains of phospholipids.

Liquid Chromatography Coupled with ELSD and UV Christie and Hunter [57] used the evaporative light-scattering detection (ELSD) principle for the reverse-phase separation of rat liver PC [58, 59]. Sotirhos et al. [60] compared the ELSD with ultraviolet (UV) detection in the analysis of intact egg yolk PC and PE and also of rat liver PC. They found that the ELSD had several advantages over UV detection. The most important observation was that the response factor was largely independent of the number and configuration of double bonds in the fatty acid chains, thus making quantification more straightforward. Since ELSD was much less sensitive to baseline shift during gradient elution than the UV detector, greater freedom in the selection of solvents was permitted. This greatly expanded the systems available for improving chromatographic resolution.

The light-scattering detector has become increasingly used in the field of lipid analysis but not many workers have utilized this detector for molecular species analysis of intact molecules [61]. One reason for this might be that UV detectable lipid derivatives are perceived as somewhat more amenable to reverse-phase chromatography than intact phospholipids. Kaufmann and Olsson [61] used this detector for the separation of molecular species of intact PC and PE from bovine milk. Partial resolution of phospholipid molecular species was achieved by Van der Meeren et al. [62] on a normal phase HPLC system with light-scattering detection. In this case, quantification was best when a calibration standard was employed that had a fatty acid composition that was similar to the samples.

For example, the separation and quantification of cholesterol and major phospholipid classes in human semen by high performance liquid chromatography and light-scattering detection were studied by Grizard et al. [63]. In this study, a HPLC method coupled to a light-scattering detector was used to separate and accurately quantify cholesterol and the main phospholipid classes of human spermatozoa and seminal plasma (SP). Lipids were separated with a good resolution and high reproducibility and it was found that 25 mL of seminal plasma was sufficient for the accurate quantitative analysis of phosphatidylethanolamine (PE), phosphatidylserine (PS), phosphatidylinositol (PI), phosphatidylcholine

(PC), sphingomyelin (SM), and cholesterol. In semen, the composition of sperm membrane lipids is dependent on the exchanges of lipids between spermatozoa and seminal plasma [64]. Recently, it was reported that the major phospholipid classes of human semen have been separated by thin layer chromatography (TLC) in most studies. The different classes of phospholipids were quantified either by the determination of inorganic phosphorus [65, 66] or, after staining, by a scanning method with a spectrodensitometer [67, 68].

High performance liquid chromatography coupled with light-scattering detection has been described to separate and accurately quantify major lipid classes in biological samples [69–72]. The evaporative light-scattering detector measures the intensity of the light scattered by the fine droplets formed by the solute upon evaporation of the column effluent.

Separation shows a typical HPLC chromatogram obtained with a standard mixture of pure lipid classes (Fig. 11.5). The cholesterol and phospholipid

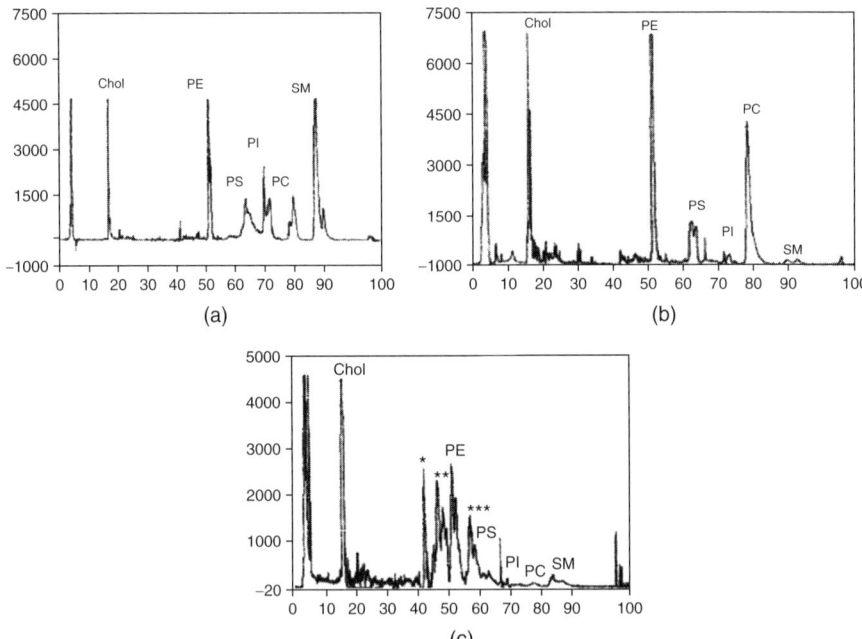

Figure 11.5 HPLC separation of cholesterol and phospholipid classes using an Inerstil column and light-scattering detection. Peaks refer to the following components: Chol, cholesterol; PE, phosphatidylethanolamine; PS, phosphatidylserine; PI, phosphatidylinositol; PC, phosphatidylcholine; SM, sphingomyelin. Experimental conditions are given in Experimental. (a) Chromatogram of a mixture of lipid standards: (50 nmol of Chol, 3.6 nmol of PE, 6.7 nmol of PS, 7.5 nmol of PI, 7.0 nmol of PC, and 17 nmol of SM). (b) Chromatogram of a lipid extract of 6×10^6 spermatozoa. (c) Chromatogram of a lipid extract of 25 mL seminal plasma. The marked peaks have the same retention times as cerebroside, cardiolipin, and cholesterol sulfate. (Adapted from Ref. 64.)

classes were separated with a good resolution. The peak appearance and the polarity-dependent retention time were highly reproducible. Peaks of cholesterol and PE were high and very narrow. PI, PC, and PS eluted as broad peaks, probably because the time retention for molecular species belonging to the same phospholipid class slightly varies with the acyl group of the molecule. SM was eluted as a double peak presumably because of partial separation of molecular species; such a result had been described previously by several authors [71–73]. The results of this study have demonstrated that HPLC coupled with light-scattering detection can be applied to the separation and quantification of a wide range of lipid classes in semen.

As mentioned before, light-scattering detection (also denoted as mass detection) is more and more used in lipid analysis because in contrast to the ultraviolet (UV) detection it is possible to perform gradient elution even with solvents of low UV transparency, such as chloroform, and it is not dependent on the degree of unsaturation of fatty acids [65, 74]. However, each phospholipid class is composed of individual molecular species with different saturated and unsaturated fatty acids in their molecules, which cannot be separated from each other using this normal phase chromatographic method. Recently, however, a new method using reverse-phase HPLC on a RP 18 column and a light-scattering detector has allowed, with success, a quantitative analysis of PC molecular species from a variety of biological samples including boar sperm [66]. Differential analysis of phospholipids reveals that PC and PE are the dominant phospholipids in spermatozoa, whereas the predominant phospholipid in seminal plasma is SM. Such data are in accordance with previous studies using other evaluation techniques [67, 75, 76]. HPLC and light-scattering detection can detect nanomolar quantities of cholesterol and different classes of phospholipids that are commonly present in human semen in less time than other techniques.

Gas Chromatography It is well known that phospholipid analysis using gas chromatography (GC) requires a derivatization step to render the phospholipids "gas phase chromatographable." It was also found that the physical characteristic of the phospholipids changes by decreasing polarity and increasing the sensitivity of detection. In order to render phospholipids GC compatible, these are usually hydrolyzed by phospholipase and derivatized with either *tert*-butyldimethylsilyl (TBDMS) [82, 83], trimethylsilyl (TMS) [77, 78], or pentafluorobenzoyl (PFB) [79]. The major steps of the various derivatization procedures are shown in Fig. 11.6. The reaction mixture usually is dried under nitrogen and then partitioned between hexane and water. The glycerol portion resides in the hexane layer, while the polar headgroup was recovered from the water layer and treated with TMS. Comparably, phospholipase C treatment of phospholipids makes the detection of the polar headgroup difficult, due to the purification problem.

11.2.2 Capillary Electrophoresis

Traditionally, thin layer chromatography (TLC) has been the simplest technique in lipid analyses, but quantification by scanning of the stained chromatograms

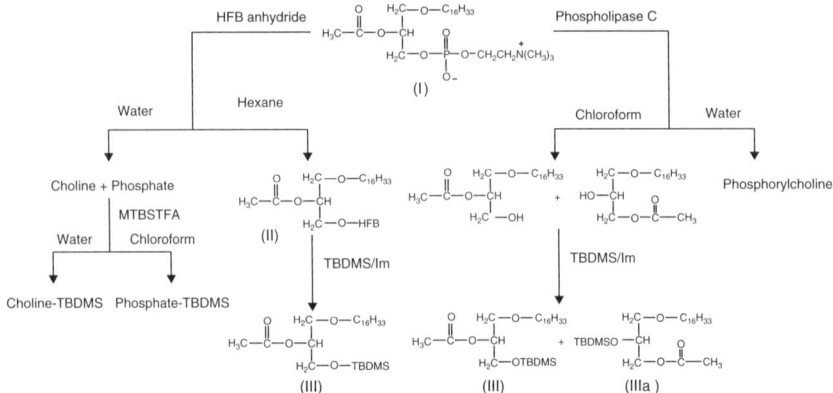

Figure 11.6 Derivatization procedures for analysis of phospholipid [80].

lacked precision because of problems with nonuniformity of staining of both the phospholipids and the background. Normal phase high performance liquid chromatography (NP-HPLC) has been used for the separation of phospholipid classes in recent years. The obvious disadvantage of most HPLC methods in the analysis of phospholipids is the high cost of the mobile phase and the time consumed during the analysis, which would not meet the needs for high throughput analysis.

Separation of phospholipids by capillary zone electrophoresis with indirect ultraviolet detection has been studied by Gao et al. [81]. Capillary electrophoresis (CE) provides an alternative to HPLC in the analysis of phospholipid classes for its high separation efficiency, application versatility, and instrument simplicity. Generally, the phospholipids are divided into several classes based on differences in their polar headgroups and their distributions in the different organs [82].

While the methodologies of CE are well established for analyzing a variety of substances, its application for phospholipid analysis has been limited because of the poor aqueous solubility and the low UV absorbance of the phospholipids. Raith et al. [83] separated phospholipids by CE online coupled to electrospray ionization–mass spectrometry (CE-ESI-MS) using no aqueous buffer. It was found that the separation for most of phospholipids was acceptable, but an extra pressure had to be applied to the inlet of the capillary column to guarantee a sufficient liquid flow between CE capillary and the electrospray ionization (ESI) needle. Guo et al. [84] separated and determined the different phospholipids in plant seeds, using the nonaqueous capillary electrophoresis (NACE) with UV detector, but the detection sensitivity was not satisfactory.

On the other hand, indirect UV detection exhibits greater sensitivity for phospholipids using an appropriate chromophore additive. The optimal CE conditions were investigated and the method was also validated. This method has been used to separate phospholipids in human blood samples. The CE was applied to human blood serum, which was analyzed by this capillary zone electrophoresis (CZE)

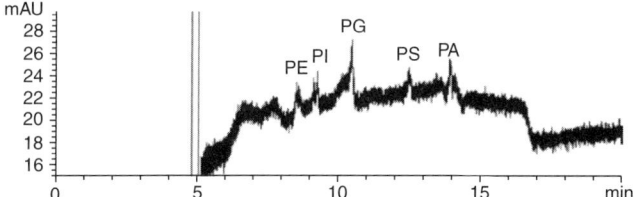

Figure 11.7 The electropherogram of human serum sample, 5 mmol/L AMP in methanol/water (9:1, v/v), pH 9.5; applied voltage, 30 kV; temperature, 25 °C; indirect UV at 259 nm; injection, 50 mbar, 5 s. (Adapted from Ref. 85.)

(with indirect ultraviolet detection method); the results are shown in Fig. 11.7 [82]. Even though the baseline slightly drifts, due to the matrix effect, a separation of the main phospholipids was obtained. Phospholipids in human blood serum were profiled and tested. The method showed great potential for determination of phospholipids in biological samples [85–87].

11.3 ANALYSIS METHODS

11.3.1 Mass Spectrometry

Several strategies are quite effective in the analysis of phospholipids. Each strategy often emphasizes a particular type of information that is desired from the mass spectrometric experiment. Mass spectrometry determines the masses of molecules in which an electrical charge is placed on the molecule and the resulting ions are separated by their mass to charge ratio (m/z). There are numerous types of mass spectrometers available that vary in the type of ionization source and mass analyzer that are used. This section overviews the various ionization methods used for phospholipid analysis.

Electron ionization (EI) was the first method of ionization used for phospholipid analysis. The ion source consists of a filament that is heated to give off electrons. These electrons are accelerated and collide with the sample, imparting energy. This method creates a large degree of fragmentation, and often the molecular ion is not detected. EI also requires the sample to be in the gas phase, so the sample must be volatile. Unfortunately, phospholipids are not volatile; for this reason, EI is only used in conjunction with GC (GC-EI-MS) for the analysis of derivatized fatty acid residues. Volatility restrictions for phospholipid analysis have been circumvented by the use of desorption techniques, which have allowed the production of quasimolecular ions (protonated molecules, sodiated molecules, etc.) of thermally fragile biomolecules. The hugely increased mass range made accessible by new ionization techniques has marked much of the progress in the field of mass spectrometry over the last two decades. The key ionization techniques include field desorption

(FD), ^{252}Cf-plasma desorption (PD), desorption chemical ionization (DCI), liquid secondary ion mass spectrometry (LSIMS), and fast atom bombardment (FAB), with significant contributions being made by each of those methods [88, 89]. Two further new soft ionization techniques—electrospray ionization (ESI) and matrix-assisted laser desorption ionization (MALDI)—have made such an important contribution to biological mass spectrometry that they have resulted in the award of the Nobel Prize in Chemistry (2002) to John Fenn (for ESI) and Koichi Tanaka (for MALDI) [90, 91].

Soft ionization techniques such as FAB ionization, MALDI, and ESI allow conventional single-stage direct MS analysis and characterization of the phospholipid molecular species. The FAB and MALDI methods involve the dissolution of the phospholipid sample in a solvent, termed the matrix. The matrix will absorb some of the internal energy of ionization produced by either the FAB canon or the MALDI laser and convert it into thermal motion energy, permitting sample ionization and desorption [92].

In addition to single-stage mass spectrometry, tandem mass spectrometry or mass spectrometry/mass spectrometry (MS/MS) has become a particularly important analytical application in the structural characterization of phospholipids. These fragmentations are primarily induced in collisions of a tandem mass spectrometer with neutral gas molecules. A multitude of reports have been published concerning the use of tandem mass spectrometry to unravel complete phospholipid sequencing information. These tandem mass spectrometric methods used several different kinetic MS/MS instruments, which involved metastable kinetic ion decomposition (MIKE) as well as low energy and high energy collision-induced dissociation (CID) analyses of the deprotonated phospholipid anions, including sector instruments, and Fourier transform ion cyclotron mass spectrometry [92–96].

11.3.2 Fast Atom Bombardment

Fast atom bombardment (FAB) as a mechanism for ionization was first described in the scientific literature by Barber et al. [87] in 1982 and rapidly became a very successful technique in mass spectrometry [87–90]. The ionizing beam consists of atoms that are typically obtained from an inert gas such as argon or xenon. This technique is similar to liquid secondary ion mass spectrometry (LSIMS), in which cesium ions are used as the ionizing beam. FAB was found to be extremely useful for the analysis of polar and natural biomolecules.

The resurgence of interest in the mass spectrometry of phospholipids was due primarily to the development of FAB and MS/MS [91, 92]. FAB allows nonvolatile phospholipids to be analyzed without prior derivatization, and MS/MS allows phospholipid mixtures to be fully characterized without prior separation [93–95].

The usefulness of mass spectrometry in phospholipid analysis depends on the ability to detect and identify the phospholipids in complex matrices and mixtures. This requires a high degree of selectivity. For a method to be useful in obtaining a profile of phospholipids present in bacteria in natural samples, it must have high sensitivity as well as high selectivity. Most of the phospholipid information is contained in the conventional FAB mass spectrum [93–95].

FAB tandem MS/MS analysis of selected molecular ion precursors produced by either low energy or high energy CID has proved to be useful in phospholipid structure analysis. FAB-CID-MS/MS analysis is extremely useful in obtaining information on the content, structures, and relative positions of the fatty acyls present on individual phospholipids [95–100]. Constant neutral loss scanning for polar head functional groups has been shown to be very useful in the differentiation of phospholipid classes and the analysis of phospholipids in complex matrices. However, in low energy CID, some classes of phospholipids do not readily lose their polar headgroup as a neutral fragment, whereas others may share a neutral loss with an interfering species. Therefore low energy CID may not always provide an unambiguous differentiation of the phospholipid classes contained in a complex mixture. Several researchers have taken advantage of the ability of either the triple-quadrupole mass spectrometer or hybrid quadrupole-quadrupole time-of-flight tandem mass spectrometry (QqToF-MS/MS) instruments to perform ion-molecule reactions to differentiate among closely related compounds [101–117].

11.3.3 MALDI-ToF-MS

MALDI is a soft ionization technique in which the energy from the laser is spent in volatilizing the matrix rather than in degrading the polymer. The MALDI technique is based on an ultraviolet absorbing matrix pioneered by Karas and Hillenkamp [118].

With the advent of MALDI in 1992, the challenge has been to discover appropriate matrix materials for use with synthetic polymers, since previous efforts were centered on biopolymers. Synthetic water-soluble polymers have been shown to be capable of analysis using similar conditions to those of biopolymers. Synthetic, organic-soluble polymers, however, have exhibited analysis complications due to their seeming incompatibility with the matrix materials. Because of this fact, only structurally simplistic synthetic water-soluble and organic-soluble polymers have been investigated to date using MALDI analysis. MALDI-ToF-MS is an emerging technique offering promise for the fast and accurate determination of numerous polymer characteristics.

Traditional methods for the detection of phospholipids in tissue by mass spectrometry involve extraction and possible purification prior to mass analysis [119]. These sample preparation steps are time consuming and have to be altered according to the specific phospholipid to be detected.

Actually, matrix-assisted laser desorption/ionization time-of-flight mass spectrometry (MALDI-ToF-MS) is being developed for the direct analysis of large biomolecules, mainly peptides and proteins, from tissue [117–122]. MALDI mass spectral profiles have been generated to map the location of biomolecules or drugs in tissue. MALDI is well suited for the direct analysis of biomolecules in tissue because of its high sensitivity, high tolerance for salts and other contaminants, and a wide mass range with little fragmentation.

Despite the success of MALDI for the direct analysis of peptides and proteins in tissue, very little work has been undertaken for the direct tissue analysis of lipids, probably because most of the lipids in tissue have a molecular weight below 1000 Da, which is a spectral region that is complicated by chemical noise and mass interferences from matrix molecular ions. In addition, identification of lipid analytes can be difficult because of background interference ions from the preparation of tissue sections, for example, stains or optimal cutting temperature compound. Conversely, MALDI was recently used for the direct analysis of phospholipids in slices of fresh and fixed lens tissues [122, 123].

11.3.4 MALDI-IM-ToF-MS

Ion mobility (IM) spectrometry is a robust method that allows for the rapid separation and analysis of a wide range of compounds [124]. The combining of a MALDI source with IM has allowed for a wide range of samples to be analyzed by MALDI-IM-ToF-MS [125, 126]. In a recent study, MALDI-IM-ToF-MS has been used to analyze complex mixtures of test biomolecules of known structure, including sphingomyelin and other lipids [127]. In that work, isobaric lipids, peptides, and oligonucleotides were preseparated prior to mass analysis by differences of up to 30% in IM drift time.

For example, the direct tissue analysis of phospholipids in rat brain using MALDI-ToF-MS and MALDI-IM-ToF-MS was carried out by Jackson et al. [128]. In particular, MALDI-IM-ToF-MS will be used because of its ability to separate molecules by class, which should make the identification of low molecular weight drug molecules easier when compared to conventional MALDI-ToF-MS.

11.3.5 High Pressure Liquid Chromatography Coupled with Electrospray Ionization–Mass Spectrometry (HPLC-ESI-MS)

With the recent coupling of HPLC to mass spectrometry (MS), new detection opportunities have been opened for the structural analysis of phospholipid species [30].

In the past, several ionization techniques have been used for the purpose of linking HPLC to mass spectrometers such as particle beam [129, 130], thermospray [131, 132], and plasmaspray [131].

Online mass spectral analysis of the HPLC effluent is possible by electrospray ionization (ESI) even though highly volatile solvents, including chlorinated solvents, are present in the mobile phase. The only important feature is that the mobile phase must be conductive to transfer the electrical charge from the electrospray needle to the droplets as they are forming.

ESI is a soft atmospheric ionization technique that can accommodate flow rates up to 1000 mL/min and is increasingly becoming the most popular ionization technique for LC-MS [132]. The use of HPLC coupled online with a mass spectrometer is a very useful approach to improve the specificity of phospholipid analysis. It is possible to combine the separation power of HPLC with the selective mass spectrometry (MS) detection and enables the structural analysis of intact PLs in a relatively short time. The analysis of intact molecular species of PLs without the need of derivatization decreases the risk of the formation of artifacts; thus the information on the fatty acid composition of the lipid class is preserved [133].

The lipids have also been introduced directly into the ESI source of triple-quadrupole instruments, without a previous chromatographic separation loop injection [133–138]. ESI-triple-quadrupole-MS/MS instruments facilitate the structural determination of the individual species in one single analysis. The molecular ion of interest is selected by MS^1 and fragmented by CID. The resulting product ion spectrum in MS^2 shows fragments that can identify the phospholipid species.

HPLC-ESI-MS has been applied successfully for the analysis of individual molecular species in human blood [139, 140], human gastric juice [131], mice synaptic plasma membrane [141], rat glial tumor C-6 cells [142], and Atlantic salmon head kidney [143]. In addition, it was recently reported by Hayakawa and Okabayashi [144] that the simultaneous analysis of eight phospholipid classes in human high density lipoprotein (HDL) was measured by HPLC-ESI-MS. The HPLC gradient conditions in the composition of the mobile phase were examined to obtain adequate separation of eight phospholipid classes and shorter run time. The improvement resulted in continuous analysis of 40-min cycles, as shown in Fig. 11.8.

During the quantification and validation of phospholipids in human HDL, it was found that such biological samples may contain more than one phospholipid molecular species. Therefore it is very difficult to obtain a phospholipid standard with the same composition as a biological sample, which can be used for the MS calibration curve. For this reason, it is advisable to use a series of selected standards of the dipalmitoyl type of PC, PG, PS, PA, and PE, the palmitoyl type of lysophosphatidylcholine (LPC), and a mixture of molecular species for SM and PI (for which the dipalmitoyl type was not available). It is reasonable to expect that the ion efficiency of the individual phospholipids will be different for each molecular species [145]. Good separation of PE, PI, PC, SM, and LPC was obtained, but PG, PS, and PA were not detected (Fig. 11.8). Identification of individual phospholipid classes was confirmed using the mass spectra data and the retention time. The total ion chromatogram (TIC) profile is shown in Fig. 11.8. The chromatograms for quantification and each peak area were obtained from the total ion in the m/z range set for each phospholipid class. In Fig. 11.8, we can understand that it is possible to simultaneously and continuously quantify eight phospholipid classes in biological samples within 40-min cycles as demonstrated for human HDL. Although PS and PA were not detected in human HDL, it was possible to quantify them in other biological samples [145]. The single-stage ESI-MS of phosphatidylcholine (PC) is shown in Fig. 11.9.

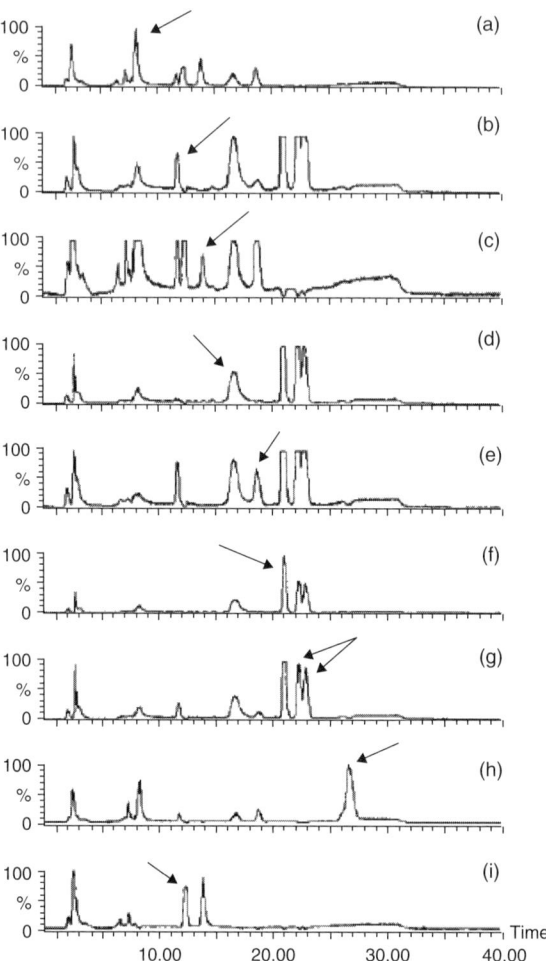

Figure 11.8 Total ion chromatograms of authentic phospholipid standards at selected mass range in positive ion mode: (a) phosphatidylglycerol (PG) (m/z range: 545–610); (b) phosphatidylethanolamine (PE) (m/z range: 690–800); (c) phosphatidylinositol (PI) (m/z range: 550–660); (d) phosphatidylserine (PS) (m/z range: 730–860); (e) phosphatidic acid (PA) (m/z range: 660–760); (f) phosphatidylcholine (PC); (g) sphingomyelin (SM) (m/z range: 675–850); (h) lysophosphatidylcholine (LPC) (m/z range: 490–570); (i) 12:0–12:0 phosphatidylethanolamine (PE) (m/z range: 580–620). (Adapted from Ref. 112.)

11.3.6 ESI–Tandem Mass Spectrometry (ESI-MS/MS)

ESI was developed in 1989 by John Fenn and co-workers [90] and involves dissolving the sample in a solvent that passes through a narrow orifice in the ion source, creating a spray. An electric potential is applied to the spray nozzle, creating an ionized solution. Eventually, the solvent evaporates and charge is

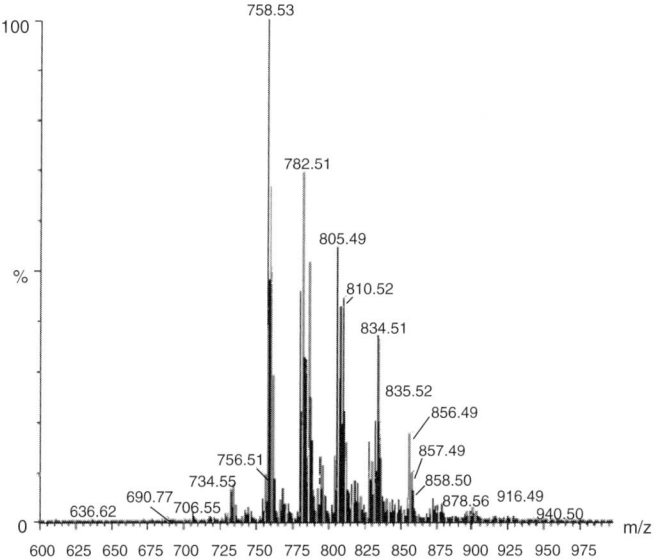

Figure 11.9 Positive ion mode ESI-MS of phosphatidylcholine (PC).

applied to the sample. The electric potential can vary in polarity and creates negative ions, molecules that have lost a proton, or positive ions, molecules that have gained a proton. Kim et al. [135] first described HPLC coupled to ESI-MS for the analysis and quantification of phospholipid molecular species, giving 50 times the sensitivity of thermospray techniques.

Taguchi et al. [146] first introduced the technique of coupling capillary columns, which have lower flow rates and less background suppression, to ESI-MS. This method allowed the low levels of phospholipids that are present from cell extracts to be quantitified as individual molecular species. Using a normal phase silica column (150 mm × 0.3 mm), phospholipids were separated by class and then introduced by direct injection to the ESI-MS. Over 500 molecular species of 10 different phospholipid classes were tentatively identified, using sample concentrations of only 5 pmol.

The biggest advantage of ESI over MALDI or FAB is that it has much higher signal-to-noise ratios, resulting in higher sensitivity [147]. The ability to couple ESI-MS directly to HPLC provides for a more efficient and faster analysis overall. Instead of running two separate instruments or experiments, the steps can be automated to be performed together. However, DeLong et al. [148] demonstrated that chromatographic steps to separate phospholipids cause changes in the molecular species, specifically a loss of molecular species with zero to three double bonds. ESI-MS is the most frequently used method today for the analysis of phospholipids. A comprehensive review of ESI-MS of phospholipids was published by Pulfer and Murphy [54].

Tandem mass spectrometry of phospholipids after electrospray ionization was performed by passage of the mass-selected precursor ion of individual phospholipids from the first quadrupole into the third quadrupole, through the second quadrupole which contains the collision cell, where dissociation was induced through collisional activation with a specific collision energy, which was adjusted through variation of the DC offset voltage and the collision gas pressures. In analysis of phospholipids, tandem mass spectrometry analyses have been shown to yield great specificity and sensitivity in detection [149–155]. Examination of the positive ions generated by the ESI technique can yield relevant information as to molecular weight. Nonetheless, it was shown that all phospholipids generate $[M + H]^+$ ions; except for the acidic phospholipids (phosphatic acid, lysophosphatidic acid, and phosphatidylinositol), which do not generate positive ions as abundantly as their corresponding negative ions $[M - H]^-$ [155].

ESI tandem mass spectrometric analysis of phospholipids in positive ion mode, in most cases, yields valuable information, revealing the diagnostic polar headgroup present. As seen with the following examples of low energy ESI-CID-MS/MS analyses of the common phospholipid classes (+ion mode), one typically can see the loss of the polar headgroup, as a neutral species, leading to abundant product ion, which is a signature of the polar headgroup. For example, the loss of 141 Da from $[M + H]^+$ is very strong evidence to support the identification of the component as a glycerophosphoethanolamine lipid [156].

It is important to point out that phospholipid classes can also generate more abundant negative ions, rather than positive ions. In fact, it has been noted that glycerophosphocholine lipids and sphingomyelin generate mainly negative ions [156]. Therefore, as indicated in Scheme 11.1,we can note the formation of the resulting demethylated product ion $[M - 15]^-$ obtained from the deprotonated molecule of phosphatidylcholine. In particular, with electrospray ionization, it is possible to detect even the presence of the cluster ion as an adduct $[M + Cl]^-$ or $[M + \text{acetate}]^-$ as ion species [156].

One of the most powerful aspects of tandem mass spectrometry applied to phospholipid analysis is the ability to characterize the fatty acyl substituent esterified to the glycerophospholipid backbone, due to the simplistic formation of the carboxylate anions following low energy CID-MS/MS analysis of the precursor deprotonated molecule $[M - H]^-$ of glycerophosphocholine [157]. Likewise, it is also possible to gain information on the exact site for each esterified fatty acyl group, either at the sn-1 position or the sn-2 position. For most fatty acyl groups of similar chain length, for example, 16–20 carbon atoms with 1–4 double bonds, it was found as a general rule that the most abundant carboxylate anion observed is usually derived from the sn-2 position [157]. A major exemption to this rule is for phosphatidic acid, where the most abundant carboxylate anion arises from the fatty acid esterified at the sn-1 position. There are exceptions to this rule, for example, the short-chained fatty acyl groups or very long-chained and polyunsatured fatty acyl groups [157].

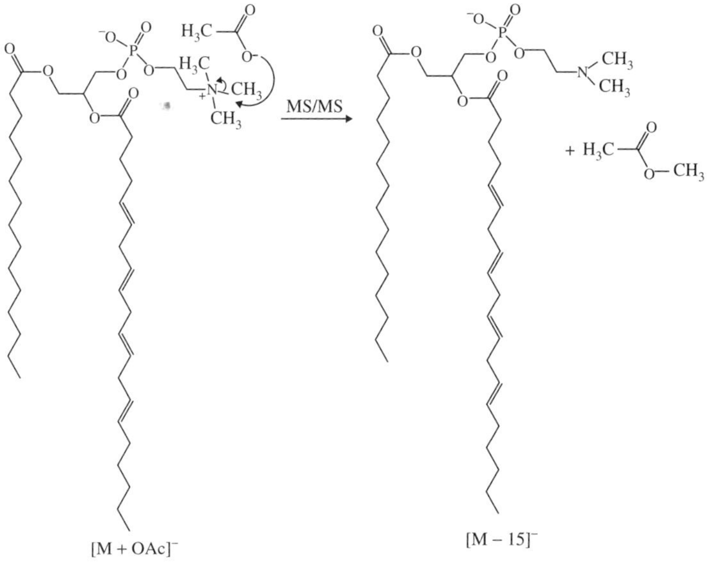

Scheme 11.1 Low energy CID-MS/MS formation of the demethylated product ion [M − 15]⁻ obtained from the deprotonated molecule [M − H]⁻ of phosphatidylcholine.

There are additional minor product ions formed during tandem mass spectrometric analyses, such as the loss of ketene neutral species from the *sn*-2 fatty acyl substituent and the loss of neutral carboxylic acid (R-COOH) from predominantly the *sn*-2 position.

It is possible to determine the position of double bonds present in a fatty acyl chain using tandem mass spectrometric techniques, in particular, if collisions take place at high energies such as those obtainable in a hybrid magnetic sector instrument (high energy CID-MS/MS) or a ToF/ToF tandem mass spectrometric instrument. Conversely, modification of the fatty acyl chains, which have occurred through free radical events or chemical modification, can significantly alter the behavior of phospholipids. This is valid also for their extraction and HPLC characteristics and, consequently, for their mass spectrometric decomposition behavior [157]. Another discrepancy is that additional product ions can also be formed by collisional activation during CID-MS/MS processes, which is not observed with normal fatty acyl substituents. These are most likely a result of alternative sites of proton attachment such as with an aldehydic or ozonide structure elsewhere in the molecular structure. These groups are often quite distant from the polar headgroup site of ionization, so that multiple charge, two sites of protonation, and one phosphate anion can result in a net single-charge state on the phospholipid molecule.

The structure of the *E. coli* type lipid A is composed of a diphosphorylated disaccharide backbone substituted, for the major species, by six fatty acid chains following the degree of acylation of this expressed heterogenic molecule. The hexaacyl component bears two chains in amide linkage at the C2 and C2′ positions: a (R)-3-hydroxymyristic acid [C14:0(3-OH)] substituting the GlcN I ring at the C2 position and a (R)-3-hydroxymyristic acid O-acylated by either a lauric acid [C14:0(3-O(12:0))] or an hydroxylauric acid [C14:0(3-O(12:0(3-OH)))] substituting the GlcN II ring at the C2′ position. In addition, the lipid A is substituted by ester linkages: a (R)-3-hydroxymyristic acid [14:0(3-OH)] chain at the C3 position and a (R)-3-hydroxymyristic acid that is O-acylated by a myristic acid [14:0(3-O)(14:0)] at the C3′ position. The gram-negative bacterial lipid A was analyzed by negative electrospray ion trap tandem mass spectrometry (ESI-QIT-MS^n) instrument. Stepwise dissociations of deprotonated species, under low energy CID conditions, were studied [158]. Sequential MS^n experiments of various precursors obtained from the underivatized lipid permitted the characterization of the substituting side chain. Mono- and diphosphorylated singly or doubly deprotonated lipid A molecules have been studied and provided complementary information [158].

Recently, the molecular structure of lipid A isolated from a rough mutant lipopolysaccharide *Aeromonas salmonicida*, using electrospray ionization quadrupole time-of-flight tandem mass spectrometry (ESI-QqToF-MS/MS) (negative ion mode), was elucidated [158]. The structural analysis of this lipid A indicated the existence of molecular heterogeneities in the fatty acid chains and the phosphate groups.

11.4 CONCLUSION

We have established that the phospholipids are an important class of compounds. Therefore the exact molecular characterization and quantification of these phospholipids in biomolecules in biological samples becomes a significant research and development area. High efficiency and high sensitivity analytical methods, such as TLC, HPLC, GC, CE, FAB-MS, MALDI-MS, ESI-MS, and MS/MS analyses, were found to excel for the exact molecular characterization and quantification of the phospholipid biomolecules. The usefulness of these techniques for the assessment of physiological processes of mammals has not yet been appreciated.

Furthermore, for all the methods we discussed, mass spectrometry holds promise for helping to elucidate the role of lipids in many disease states, such as cancer, and can also be useful for the determination of enzyme activities and kinetic studies of new drugs [159]. Knowledge gained in mass spectrometry analysis, in addition to the other analytical techniques in lipidomics, possibly will lead to further advances in proteomics and metabolomics. Additionally, many of these analytical techniques need to be continued and optimized for biological tissues. Such optimizations would increase the ability to accurately differentiate and quantitate the multitude of remarkably similar phospholipids in biological samples [52].

REFERENCES

1. S. S. Davies, A. V. Pontsler, G. K. Marathe, K. A. Harrison, R. C. Murphy, J. C. Hinshaw, G. D. Prestwich, A. St. Hilaire, S. M. Prescott, G. A. Zimmerman, and T. M. McIntyre (2001). Oxidized alkyl p-phospholipids are specific, high affinity peroxisome proliferator-activated receptor γ-ligands and agonists. *J. Biol. Chem.* **276**:16015.
2. S. L. Wolfe (1993). *Molecular and Cellular Biology*. Wadsworth Publishing Company, Belmont, CA, p. 155.
3. Y. Wang, I. S. Krull, C. Liu, and J. D. Orr (2003). Derivatization of phospholipids. *J. Chromatogr. B* **793**: 3–14.
4. D. M. Small (1986). In *Handbook of Lipid Research*, Vol. 4. Plenum Press, New York, Chap. 12.
5. D. J. Hanahan (1997). *A Guide to Phospholipid Chemistry*. Oxford University Press, New York, Chap. 1.
6. D. Voet and J. G. Voet (1995). *Biochemistry*. Wiley, Hoboken, NJ, Chap. 11.
7. C. Galanos, E. T. Rietschel, O. Luederitz, and O. Westphal (1971). Interaction of lipopolysaccharides and lipid A with complement. *Eur. J. Biochem.* **19**: 143–152.
8. K. H. Ali, T. W. Feeley, M. Bieber, B. McGrath, and N. N. Teng (1987). Cardiovascular effect of intravenous lipid A in rabbits. *Circulatory Shock* **23**: 285–293.
9. C. Galanos, O. Luederitz, E. T. Rietschel, O. Westphal, H. Brade, L. Brade, M. Freudenberg, U. Schade, M. Imoto, H. Yoshimura, S. Kusumoto, and T. Shiba (1985). Synthetic and natural *Escherichia coli* free lipid A express identical endotoxic activities. *Eur. J. Biochem.* **148**: 1–5.
10. A. Kaltoft Ryborg, B. Deleuran, H. Søgaard, and K. Kragballe (1998) Intracutaneous injection of lysophosphatidylcholine induces skin inflammation and accumulation of T-cells *J. Dermatol. Sci*, **16**(1): S168.
11. E. Lissauer, B. Johnson, S. Shi, T. Gentle, and M. Scalea (2007) Decreased lysophosphatidylcholine levels are associated with sepsis compared to uninfected inflammation prior to onset of sepsis *J. Surg. Res*., **137**(2): 206.
12. N. Zurgil, E. Afrimzon, Y. Shafran, O. Shovman, B. Gilburd, H. Brikman, Y. Shoenfeld, and M. Deutsch (2007) Lymphocyte resistance to lysophosphatidylcholine mediated apoptosis in atherosclerosis, *Atheroscl*., **190**(1): 73–83.
13. M.E. Dunlop, E. Muggli, S. Clark (1997) Differential disposition of lysophosphatidylcholine in diabetes compared with raised glucose: implications for prostaglandin production in the diabetic kidney glomerulus in vivo, *Biochim. Biophys. Acta Lipids and Lipid Metabolism*, **1345**(3): 306–316.
14. W. W. Christie (1982). *Lipid Analysis*. Pergamon Press, Oxford, UK, pp. 148–151.
15. W. W. Christie (1987). *HPLC and Lipids*. Pergamon Press, Oxford, UK, pp. 169–210.
16. Y. Inouye and M. Noda (1958). Paper chromatography of egg lecithins. *Arch. Biochem. Biophys.* **76**: 271–285.
17. F. D. Collins (1963). Studies on phospholipids. 9. The composition of rat-liver lecithins, *Biochem. J.*, **88**: 319–324.
18. F. D. Collins (1967). Counter-current distribution of lecithins. *Chem. Phys. Lipids* **1**: 91–99.

19. O. Renkonen (1964). *Acta Chem. Scand.* **18**: 271.
20. O. Renkonen (1966). Individual molecular species of phospholipids: III. Molecular species of ox-brain lecithins. *Biochim. Biophys. Acta* **125**: 288–309.
21. T. H. Bevan, D. A. Brown, G. I. Gregory, and T. Malkin (1953). Acyl migration during dephosphorylation and a suggested mechanism. *J. Chem. Soc.* **35**: 127–129.
22. M. J. MacFarlane and B. C. J. G. Knight (1941). The biochemistry of bacterial toxins. *Biochem. J.* **35**: 884–902.
23. S. Winstein and H. J. Lucas (1938). The coordination of silver ion with unsaturated compounds. *J. Am. Chem. Soc.* **60**: 836–847.
24. H. P. Kaufmann, H. Wessels, and C. Bondopadhyaya (1963). Dünnschicht-Chromatographie auf dem Fettgebiet XI: Die Analyse der Lecithine und der hydrolytischen Spaltprodukte der Phosphatide. *Fette Seifen Anstrichmittel* **65**: 543–547.
25. J. Porath and P. Flodin (1959). Gel filtration: a method for desalting and group separation. *Nature* **183**: 1657–1658.
26. T. C. Laurent (1993). History of a theory. *J. Chromatogr. A* **633**: 1–9.
27. J. Ellingboe, E. Nyström, and J. Sjrvall (1968). A versatile lipophilic Sephadex derivative for "reversed-phase" chromatography. *Biochim. Biophys. Acta* **152**: 803–805.
28. J. Ellingboe, B. Alme, and J. Sjövall (1970). Introduction of specific groups into polysaccharide supports for liquid chromatography. *Acta Chem. Scand.* **24**: 463–467.
29. J. Ellingboe, E. Nyström, and J. Sjövall (1970). Liquid-gel chromatography on lipophilic-hydrophobic Sephadex derivatives. *J. Lipid Res.* **11**: 266–273.
30. H.-Y. Kim and N. Salem, Jr. (1993). Liquid chromatography–mass spectrometry of lipids. *Prog. Lipid Res.* **32**: 221–245.
31. A. Valeur, N. U. Olsson, P. Kaufmann, S. Wada, C.-G Kroon, G. Westerdahl, and G. Odham (1994). Quantification and comparison of small natural sphingomyelins by on-line high-performance liquid chromatography/discharge-assisted thermospray mass spectrometry. *Biol. Mass Spectrom.* **23**: 313–319.
32. E. A. Dugan (2003). Analtech, Inc. EBSCO Publishing.
33. A. Kuksis (1977). Routine chromatography of simple lipids and their constituents. *J. Chromatogr.* **143**: 3–30.
34. P. Gentner, M. Bauer, and I. Dietrich (1981). Thin-layer chromatography of phospholipids: separation of major phospholipid classes of milk without previous isolation from total lipid extracts. *J Chromatogr. A* **206**: 200–204.
35. T. R. Watkins (1982). HRC8CC5, 104.
36. E. A. Dugan (2003). Analtech, Inc. Biological 13.
37. J. C. Touchstone, S. S. Levin, M. F. Dobbins, and P. J. Carter (1981). HRCVCC4, 423.
38. H. Miwa, M. Yamamoto, T. Futata, and K. Kan (1996). Thin-layer chromatography and high-performance liquid chromatography for the assay of fatty acid compositions of individual phospholipids in platelets from non-insulin-dependent diabetes mellitus patients: effect of eicosapentaenoic acid ethyl ester administration. *J. Chromatogr. B* **677**: 217–223.
39. L. Svennerholm (1968). Distribution and fatty acid composition of phosphoglycerides in normal human brain. *J. Lipid Res.* **9**: 570–579.

40. L. A. Morson and M. T. Clandinin (1985). Dietary linoleic acid modulates liver plasma membrane unsaturated fatty acid composition, phosphatidylcholine and cholesterol content, as well as glucagon stimulated adenylate cyclase activity. *Nutr. Res.* **5**: 1113–1120.

41. C. Ailing, L. Gustavsson, A. Kristensson-Aas, and S. Wallerstedt (1984). Changes in fatty acid composition of major glycerophospholipids in erythrocyte membranes from chronic alcoholics during withdrawal. *Scand. J. Clin. Lab. Invest.* **44**: 283–289.

42. D. N. Palmer, D. R. Husbands, and R. D. Jolly (1985). Phospholipid fatty acids in brains of normal sheep and sheep with ceroid-lipofuscinosis. *Biochim. Biophys. Acta* **834**: 159–163.

43. V. Skipski, R. F. Peterson, and M. Barcly (1963). Quantitative analysis of phospholipids by thin-layer chromatography. *Biochem. J.* **90**: 374–378.

44. E. Vitiello and J. P. Zanetta (1978). Thin-layer chromatography of phospholipids. *J. Chromatogr. A* **166**: 637–640.

45. J. C. Touchstone, J. C. Chen, and K. M. Beaver (1980). Improved separation of phospholipids in thin layer chromatography. *Lipids* **15**: 61–62.

46. J. B. Fine and H. Sprecher (1982). Unidimensional thin-layer chromatography of phospholipids on boric acid- impregnated plates. *J. Lipid Res.* **23**: 660–663.

47. M. Goppelt and K. Resch (1984). Densitometric quantitation of individual phospholipids from natural sources separated by one-dimensional thin-layer chromatography. *Anal. Biochem.* **140**: 152–156.

48. L. Gustavsson (1986). Densitometric quantification of individual phospholipids: improvement and evaluation of a method using molybdenum blue reagent for detection. *J. Chromatogr. B* **375**: 255–266.

49. K. Harrison, K. L. Clay, and R. C. Murphy (1999). Negative ion electrospray and tandem mass spectrometric analysis of platelet activating factor (PAF) (1-hexadecyl-2-acetyl-glycerophosphocholine). *J. Mass Spectrom.* **34**: 330–335.

50. G. M. Patton and S. J. Robins (1998). Separation and quantitation of phospholipid classes by HPLC. *Methods Mol. Biol.* **110**:193–215.

51. F. H. Chilton and R. C. Murphy (1986). Remodeling of arachidonate-containing phosphoglycerides within the human neutrophil. *J. Biol. Chem.* **261**: 7771–7777.

52. B. L. Peterson and B. S. Cummings (2006). A review of chromatographic methods for the assessment of phospholipids in biological samples. *Biomed. Chromatogr.* **20**: 227–243.

53. F. B. Jungalwala, R. J. Turel, J. E. Evans, and R. H. McCluer (1975). Sensitive analysis of ethanolamine- and serine-containing phosphoglycerides by high-performance liquid chromatography. *Biochem. J.* **145**: 517–526.

54. M. Pulfer and R. C. Murphy (2003). Electrospray mass spectrometry of phospholipids. *Mass Spectrom. Rev.* **22**: 332–364.

55. S. J. Robins and G. M. Patton (1986). Separation of phospholipid molecular species by high performance liquid chromatography: potentials for use in metabolic studies. *J. Lipid Res.* **27**: 131–139.

56. N. U. Olsson and N. Salem, Jr. (1997). Molecular species analysis of phospholipids. *J. Chromatogr. B* **692**: 245–256.

57. W. W. Christie and M. L. Hunter (1985). Separation of molecular species of phosphatidylcholine by high-performance liquid chromatography on a PLRP-S column. *J. Chromatogr. A* **325**: 473–476.
58. W. W. Christie (1992). In *Advances in Lipid Methodology-One* (W. W. Christie, Ed.). Oily Press, Ayr, pp. 239–271.
59. J. M. Charlesworth (1978). Evaporative analyzer as a mass detector for liquid chromatography. *Anal. Chem.* **50**: 1414–1420.
60. N. Sotirhos, C. Thrmgren, and B. Herslrf (1985). Reversed-phase high-performance liquid chromatographic separation and mass detection of individual phospholipid classes. *J. Chromatogr. A* **331**: 313–320.
61. P. Kaufmann and N. U. Olsson (1993). Determination of intact molecular species of bovine milk 1,2-diacyl->sn-glycero-3-phosphocholine and 1,2-diacyl-sn-glycero-3-phosphoethanolamine by reversed phase HPLC, a multivariate optimization. *Chromatographia* **35**: 517–523.
62. P. Van der Meeren, J. Vanderdeelen, G. Huyghebaert, and L. Baert (1992). Partial resolution of molecular species during liquid chromatography of soybean phospholipids and effect on quantitation by light-scattering. *Chromatographia* **34**: 557–562.
63. G. Grizard, B. Sion, D. Bauchart, and D. Boucher (2000). Separation and quantification of cholesterol and major phospholipid classes in human semen by high-performance liquid chromatography and light-scattering detection. *J. Chromatogr. B* **740**: 101–107.
64. L. Huacuja, N. M. Delgado, L. Calzada, A. Wens, R. Reyes, N. Pedron, and A. Rosado (1981). Exchange of lipids between spermatozoa and seminal plasma in normal and pathological human semen. *Arch. Androl.* **7**: 343–349.
65. A. Poulos and I. G. White (1973). The phospholipid composition of human spermatozoa and seminal plasma. *J. Reprod. Fertil.* **35**: 265–272.
66. S. M. Sebastian, S. Seharaj, M. M. Arnldhas, and P. Govindara Julu (1987). Pattern of neutral and phospholipids in the semen of normospermic, oligospermic and azoospermic men. *J. Reprod. Fertil.* **79**: 373–378.
67. J. G. Alvarez and B. T. Storey (1992). Evidence for increased lipid peroxidative damage and loss of superoxide dismutase activity as a mode of sublethal cryodamage to human sperm during cryopreservation. *J. Androl.* **13**: 232–241.
68. G. Haidl and C. Opper (1997). Changes in lipids and membrane: an isotropy in human spermatozoa during epididymal maturation. *Hum. Reprod.* **12**: 2720–2723.
69. H. Bunger and U. Piso (1995). Quantitative analysis of pulmonary surfactant phospholipids by high-performance liquid chromatography and light-scattering detection. *J. Chromatogr. B* **672**: 25–31.
70. C. Silversand and C. Haux (1997). Improved high-performance liquid chromatographic method for the separation and quantification of lipid classes: application to fish lipids. *J. Chromatogr. B* **703**: 7–14.
71. L. Landi, M. C. Galli, L. Cabrini, G. Hakim, C. Carru, and D. Fiorentini (1998). HPLC and light scattering detection allow the determination of phospholipids in biological samples and the assay of phospholipase A_2. *Biochem. Mol. Biol. Int.* **44**: 1157–1166.
72. P. Juaneda, G. Rocquelin, and P. O. Astorg (1990). Separation and quantification of heart and liver phospholipid classes by high-performance liquid chromatography using a new light-scattering detector. *Lipids* **25**: 756–759.

73. W. W. Christie (1985). Rapid separation and quantification of lipid classes by high performance liquid chromatography and mass (light-scattering) detection. *J. Lipid Res.* **26**: 507–512.

74. J. F. Brouwers, B. M. Gadella, L. M. Van Golde, and A. G. Tielens (1998). Quantitative analysis of phosphatidylcholine molecular species using HPLC and light scattering detection. *J. Lipid Res.* **39**: 344–353.

75. M. Oda, K. Satouchi, K. Yasunaga, and K. Saito (1985). Molecular species of platelet-activating factor generated by human neutrophils challenged with ionophore A23187. *J. Immunol.* **134**: 1090–1093.

76. K. Satouchi and K. Saito (1979). Use of t-butyldimethylchlorosilane/imidazole reagent for identification of molecular species of phospholipids by gas–liquid chromatography mass spectrometry. *Biomed. Mass Spectrom.* **6**: 396–402.

77. K. Satouchi and K. Saito (1976). Studies on trimethylsilyl derivatives of 1-alkyl-2-acylglycerols by gas–liquid chromatography mass spectrometry. *Biomed. Mass Spectrom.* **3**: 122–126.

78. C. Yon and J.-S. Han (2000). Analysis of trimethylsilyl derivatization products of phosphatidylethanol by gas chromatography–mass spectrometry. *Exp. Mol. Med.* **32**: 243–245.

79. C. S. Ramesha and W. C. Pickett (1986). Measurement of sub-picogram quantities of platelet activating factor (AGEPC) by gas chromatography/negative ion chemical ionization mass spectrometry. *Biomed. Mass Spectrom.* **13**: 107–111.

80. R. K. Satsangi, J. C. Ludwig, S. T. Weintraub, and R. N. Pinckard (1989). A novel method for the analysis of platelet-activating factor: direct derivatization of glycerophospholipids. *J. Lipid Res.* **30**: 929–937.

81. F. Gao, J. Dong, W. Li, T. Wang, J. Liao, Y. Liao, and H. Liu (2006). Separation of phospholipids by capillary zone electrophoresis with indirect ultraviolet detection. *J. Chromatogr. A* **1130**: 259–264.

82. G. Ceve (Ed.) (1993). *Phospholipids Handbook*. Marcel Dekker, New York, p. 745.

83. K. Raith, R. Wolf, J. Wagner, and R. H. H. Neubert (1998). Separation of phospholipids by nonaqueous capillary electrophoresis with electrospray ionisation mass spectrometry. *J. Chromatogr. A* **802**: 185–188.

84. B. Y. Guo, B. Wen, X. Q. Shan, S. Z. Zhang, and J. M. Lin (2005). Separation and determination of phospholipids in plant seeds by nonaqueous capillary electrophoresis. *J. Chromatogr. A* **1074**: 205–213.

85. P. Britz-Mckibbin and S. Terabe (2003). On-line preconcentration strategies for trace analysis of metabolites by capillary electrophoresis. *J. Chromatogr. A* **1000**: 917–934.

86. R. D. Plattner, R. J. Stack, J. M. Al-Hassan, B. Summers, and R. S. Griddle (1988). Identification of platelet activating factor by tandem mass spectrometry. *Org. Mass Spectrom.* **23**: 834–840.

87. M. Barber, R. S. Bordoli, G. J. Elliot, R. D. Sedgwick, and A. N. Tyler (1982). Fast atom bombardment mass spectrometry. *Anal. Chem.* **54**: 645–657.

88. D.J. Surman and J.C. Vickerman (1981) Fast atom bombardment quadrupole mass spectrometry, *J. Chem. Soc. Chem. Commun.* **7**: 324–325.

89. W. Aberth, K. M. Straub, and A. L. Burlingame (1982). Secondary ion mass spectrometry with cesium ion primary beam and liquid target matrix for analysis of bioorganic compounds. *Anal. Chem.* **54**: 2029–2034.

90. J. B. Fenn, M. Mann, C. K. Meng, S. F. Wong, and C. M. Whitehouse (1989). Electrospray ionization for mass spectrometry of large biomolecules. *Science* **246**: 64–74.
91. K. Tanaka (2003). The origin of macromolecule ionization by laser irradiation. *Angew. Chem. Int. Ed.* **42**: 3860–2870.
92. S. A. McLuckey and M. Wells (2001). Mass analysis at the advent of the 21st century. *Chem. Rev.* **101**: 571–606.
93. M. Karas, D. Bachmann, U. Bahr, and F. Hillenkamp (1987). Matrix-assisted ultraviolet laser desorption of non-volatile compounds. *Int. J. Mass Spectrom. Ion Processes* **78**: 53–68.
94. F. W. McLafferty (1983). *Tandem Mass Spectrometry*. Wiley, Hoboken, NJ.
95. K. L. Busch, G. L. Glish, and S. A. McLuckey (1988). *Mass Spectrometry/Mass Spectrometry*. VCH Publishers, New York.
96. J. H. Banoub, R. P. Newton, E. Esmans, D. F. Ewing, and G. Mackenzie (2005). Recent developments in mass spectrometry for the characterization of nucleosides, nucleotides, oligonucleotides, and nucleic acids. *Chem. Rev.* **105**: 1869–1916.
97. S. T. Weintraub, C. S. Lear, and R. N. Pinckard (1990). Analysis of platelet-activating factor by GC-MS after direct derivatization with pentafluorobenzoyl chloride and heptafluorobutyric anhydride. *J. Lipid Res.* **31**: 719–725.
98. D. N. Heller, R. J. Cotter, C. Fenselau, and O. M. Uy (1987). Profiling of bacteria by fast atom bombardment mass spectrometry. *Anal. Chem.* **59**: 2806–2809.
99. D. L. Balkwill, F. R. Leach, J. T. Wilson, J. F. McNabb, and D. C. White (1988). Equivalence of microbial biomass measures based on membrane lipid and cell wall components, adenosine triphosphate, and direct counts in subsurface aquifer sediments. *Microb. Ecol.* **26**: 73–84.
100. J. A. Zirrolli, K. L. Clay, and R. C. Murphy (1991). Tandem mass spectrometry of negative ions from choline phospholipid molecular species related to platelet activating factor. *Lipids* **26**: 1112–1116.
101. B. N. Pramanik, J. M. Zechman, P. R. Das, and P. I. Bartner (1990). Bacterial phospholipid analysis by fast atom bombardment mass spectrometry. *Biomed. Environ. Mass Spectrom.* **19**: 164–170.
102. M. M. Ross, R. A. Neihof, and J. E. Campana (1986). Direct fatty acid profiling of complex lipids in intact algae by fast-atom-bombardment mass spectrometry. *Anal. Chim. Acta* **181**: 149–157.
103. W. R. Sherman, K. E. Ackerman, R. H. Bateman, B. N. Green, and I. Lewis (1985). Mass-analysed ion kinetic energy spectra and B1E-B2 triple sector mass spectrometric analysis of phosphoinositides by fast atom bombardment. *Biomed. Mass Spectrom.* **12**: 409–413.
104. N. J. Jensen, K. B. Tomer, and M. L. Gross (1986). Fast atom bombardment and tandem mass spectrometry of phosphatidylserine and phosphatidylcholine. *Lipids* **21**: 580–588.
105. N. J. Jensen, K. B. Tomer, and M. L. Gross (1987). FAB MS/MS for phosphatidylinostitol, -glycerol, -ethanolamine and other complex phospholipids. *Lipids* **22**: 480–489.
106. M. J. Cole and C. G. Enke (1989). Presented at the 37th Annual Conference on Mass Spectrometry and Allied Topics, Miami Beach, FL, May 21–26.

107. A. Hayashi, T. Matsubara, M. Masanori, T. Kinoshita, and T. J. Nakamura (1989). Structural analysis of choline phospholipids by fast atom bombardment mass spectrometry and tandem mass spectrometry. *J. Biochem. (Tokyo)* **106**: 264–269.
108. B. J. Sweetman, M. Tamura, K. Higashimori, T. Inagami, and I. A. Blair (1987). Presented at the 35th Annual Conference on Mass Spectrometry and Allied Topics, Denver, CO, May 24–29.
109. H. Mtinster and H. Budzikiewicz (1988). Structural and mixture analysis of glycerophosphoric acid derivatives by fast atom bombardment tandem mass spectrometry. *Biol. Chem. Hoppe-Seyler* **369**: 303–308.
110. K. A. Kayganich and R. C. Murphy (1990). Presented at the 38th Annual Conference on Mass Spectrometry and Allied Topics, Tucson, AZ, June 4–9.
111. C. Easton, D. W. Johnson, and D. W. Poulos (1988). Determination of phospholipid base structure by CA MIKES mass spectrometry. *J. Lipid Res.* **29**: 109–112.
112. D. N. Heller, C. M. Murphy, R. J. Cotter, C. Fenselau, and O. M. Uy (1988). Constant neutral loss scanning for the characterization of bacterial phospholipids desorbed by fast atom bombardment. *Anal. Chem.* **60**: 2787–2791.
113. R. R. Pachuta, H. I. Kenttӕmaa, R. G. Cooks, T. M. Zennie, C. Ping, C. J. Chang, and J. M. Cassady (1988). Analysis of natural products by tandem mass spectrometry employing reactive collisions with ethyl vinyl ether. *Org. Mass Spectrom.* **23**: 10–15.
114. E. L. White and M. M. Bursey (1989). Distinguishing positional isomers of hexachlorinated biphenyls by ion-molecule reactions in a triple-quadrupole instrument. *Biomed. Environ. Mass Spectra* **18**: 413–415.
115. W. J. Meyerhoffer and M.M Bursey (1989). Differentiation of the isomeric 1,2-cyclopentanediols by ion-molecule reactions in a triple-quadrupole mass spectrometer. *Org. Mass Spectrom.* **24**: 169–175.
116. R. Kostiainen and S. Auriola (1988). Isomer-specific determination of tetrachlorodibenzo-p-dioxins by tandem mass spectrometry using low energy reactive collisions between oxygen and negative molecular ions. *Rapid Commun. Mass Spectrom.* **2**: 135–137.
117. Z. H. Huang, D. A. Gage, and C. C. Sweeley (1992). Characterization of diacylglycerylphosphocholine molecular species by FAB-CAD-MS/MS: a general method not sensitive to the nature of the fatty acyl groups. *J. Am. Soc. Mass Spectrom.* **3**: 71–78.
118. M. Karas and F. Hillencamp (1988). Laser desorption ionization of proteins with molecular masses exceeding 10,000 daltons. *Anal. Chem.* **60**: 2299–2301.
119. R. C. Murphy (2002). *Mass Spectrometry of Phospholipids*. Illuminati Press, Denver, CO, Chap. II.
120. S. A. Schwartz, M. L. Reyzer, and R. M. Caprioli (2003). Direct tissue analysis using matrix-assisted laser desorption/ionization mass spectrometry: practical aspects of sample preparation. *J. Mass Spectrom.* **38**: 699–708.
121. P. Chaurand and R. M. Caprioli (2002). Direct profiling and imaging of peptides and proteins from mammalian cells and tissue sections by mass spectrometry. *Electrophoresis* **23**: 3125–3135.
122. P. J. Todd, T. G. Schaaff, P. Chaurand, and R. M. Caprioli (2001). Organic ion imaging of biological tissue with secondary ion mass spectrometry and matrix-assisted laser desorption/ionization. *J. Mass Spectrom.* **36**: 355–369.

123. M. Rujoi, R. Estrada, and M. C. Yappert (2004). In situ MALDI-TOF MS regional analysis of neutral phospholipids in lens tissue. *Anal. Chem.* **76**: 1657–1663.
124. J. I. Baumbach and G. A. Eiceman (1999). Ion mobility spectrometry: arriving on site and moving beyond a low profile. *Appl. Spectrosc.* **53**: 338–355.
125. K. J. Gillig, B. Ruotolo, E. G. Stone, D. H. Russell, K. Fuhrer, M. Gonin, and A. J. Schultz (2000). Coupling high-pressure MALDI with ion mobility/orthogonal time-of-flight mass spectrometry. *Anal. Chem.* **72**: 3965–3971.
126. G. Von Helden, T. Wyttenbach, and M. T. Bowers (1995). Conformation of macromolecules in the gas phase: use of matrix-assisted laser desorption methods in ion chromatography. *Science* **267**: 1483–1485.
127. A. S. Woods, M. Ugarov, T. Egan, J. Koomen, K. J. Gillig, K. Fuhrer, M. Gonin, and J. A. Schultz (2004). Lipid/peptide/nucleotide separation with MALDI–ion mobility–TOF MS. *Anal. Chem.* **76**: 2187–2195.
128. S. N. Jackson, H. Y. Wang, A. S. Woods, M. Ugarov, T. Egan, and J. A. Schultz (2005). Direct tissue analysis of phospholipids in rat brain using MALDI-TOFMS and MALDI–ion mobility–TOFMS. *J. Am. Soc. Mass Spectrom.* **16**: 133–138.
129. M. Careri, M. Dieci, A. Mangia, P. Manini, and A. Raffaelli (1996). Liquid chromatography/mass spectrometry of phospholipids in soybean products using particle beam and ionspray interfaces. *Rapid Commun. Mass Spectrom.* **10**: 707–714.
130. R. Nilsson and C. Liljenberg (1996). Separation and identification of plant glycerolipid molecular species by particle beam–high-performance liquid chromatography–mass spectrometry. *Phytochem. Anal*. **7**: 228–232.
131. H. Y. Kim and N. Salem (1986). Phospholipid molecular species analysis by thermospray liquid chromatography/mass spectrometry. *Anal. Chem.* **58**: 9–14.
132. H.-Y. Kim and N. Salem (1987). Application of thermospray high-performance liquid chromatography/mass spectrometry for the determination of phospholipids and related compounds. *Anal. Chem.* **59**: 722–726.
133. F. A. Kuypers, P. Bütikofer, and C. H. L. Shackleton (1991). Application of liquid chromatography–thermospray mass spectrometry in the analysis of glycerophospholipid molecular species. *J. Chromatogr. B* **562**: 191–206.
134. Y.-C. Ma and H.-Y. Kim (1995). Development of the on-line high-performance liquid-chromatography thermospray mass spectrometry method for the analysis of phospholipid molecular species in rat brain. *Anal. Biochem.* **226**: 293–301.
135. H. Y. Kim, T. C. L. Wang, and Y. C. Ma (1994). Liquid chromatography/mass spectrometry of phospholipids using electrospray ionization. *Anal. Chem.* **66**: 3977–3982.
136. A. A. Karlsson, P. Michelsen, A. Larsen, and G. Odham (1996). Normal-phase liquid chromatography class separation and species determination of phospholipids utilizing electrospray mass spectrometry/tandem mass spectrometry. *Rapid Commun. Mass Spectrom.* **10**: 775–780.
137. A. Ravandi, A. Kuksis, L. Marai, J. J. Myher, G. Steiner, G. Lewisa, and H. Kamido (1996). Isolation and identification of glycated aminophospholipids from red cells and plasma of diabetic blood. *FEBS Lett.* **381**: 77–81.
138. E. A. A. M. Vernooij, J. J. Kettenes-Van den Bosch, and D. J. A. Crommelin (1998). Rapid determination of acyl chain position in egg phosphatidylcholine by high performance liquid chromatography/electrospray mass spectrometry. *Rapid Commun. Mass Spectrom.* **12**: 83–86.

139. X. Han and R. W. Gross (1994). Electrospray ionization mass spectroscopic analysis of human erythrocyte plasma membrane phospholipids. *Proc. Natl. Acad. Sci. USA* **91**: 10635–10639.

140. S. Uran, A. Larsen, P. B. Jacobsen, and T. Skotland (2001). Analysis of phospholipid species in human blood using normal-phase liquid chromatography coupled with electrospray ionization ion-trap tandem mass spectrometry. *J. Chromatogr. B.* **758**: 265–275.

141. U. Igbavboa, J. Hamilton, H. Y. Kim, G. Y. Sun, and W. G. Wood (2002). A new role for apolipoprotein E: modulating transport of polyunsaturated phospholipid molecular species in synaptic plasma membranes. *J. Neurochem.* **80**: 255–261.

142. P. B. W. Smith, A. P. Snyder, and C. S. Harden (1995). Characterization of bacterial phospholipids by electrospray ionization tandem mass spectrometry. *Anal. Chem.* **67**: 1824–1830.

143. E. Hvattum, C. Rosjo, T. Gjoen, G. Rosenlund, and B. Ruyter (2000). Effect of soybean oil and fish oil on individual molecular species of Atlantic salmon head kidney phospholipids determined by normal-phase liquid chromatography coupled to negative ion electrospray tandem mass spectrometry. *J. Chromatogr. B* **748**: 137–149.

144. J. Hayakawa and Y. Okabayashi (2005). Simultaneous analysis of eight phospholipid classes by liquid chromatography/mass spectrometry: application to human HDL. *J. Liquid Chromatogr. Technol.* **28**: 1473–1485.

145. J. Hayakawa and Y. Okabayashi (2004). Simultaneous analysis of phospholipid in rabbit bronchoalveolar lavage fluid by liquid chromatography/mass spectrometry. *J. Pharm. Biomed. Anal.* **35**: 583–592.

146. R. Taguchi, J. Hayakawa, Y. Takeuchi, and M. Ishida (2000). Two-dimensional analysis of phospholipids by capillary liquid chromatography/electrospray ionization mass spectrometry. *J. Mass Spectrom.* **35**: 953–966.

147. X. Han and R. W. Gross (2004). Shotgun lipidomics: electrospray ionization mass spectrometric analysis and quantitation of cellular lipidomes directly from crude extracts of biological samples. *Mass Spectrom. Rev.* **24**: 367–412.

148. C. J. DeLong, P. R. Baker, M. Samuel, Z. Cui, and M. J. Thomas (2001). Molecular species composition of rat liver phospholipids by ESI-MS/MS: the effect of chromatography. *J. Lipid Res.* **42**: 1959–1968.

149. K. Ekroos, I. V. Chernushevich, K. Simons, and A. Shevchenko (2002). Quantitative profiling of phospholipids by multiple precursor ion scanning on a hybrid quadrupole time-of-flight mass spectrometer. *Anal. Chem.* **74**: 941–949.

150. G. Liebisch, W. Drobnik, B. Lieser, and G. Schmitz (2002). High-throughput quantification of lysophosphatidylcholine by electrospray ionization tandem mass spectrometry. *Clin. Chem.* **48**: 2217–2224.

151. B. Brugger, G. Erben, R. Sandhoff, F. T. Wieland, and W. D. Lehmann (1997). Quantitative analysis of biological membrane lipids at the low picomole level by nano-electrospray ionization tandem mass spectrometry. *Proc. Natl. Acad. Sci. USA* **94**: 2339–2344.

152. R. C. Murphy. Library of Congress control number 2002110793 (Denver Colorado, USA).

153. M. J. Cole and C. G. Enke (1991). Direct determination of phospholipid structures in microorganisms by fast atom bombardment triple quadrupole mass spectrometry. *Anal. Chem.* **63**: 1032–1038.

154. K. A. Harrison and R. C. Murphy (1995). Negative electrospray ionization of glycerophosphocholine lipids: formation of [M − 15]$^-$ ions occurs via collisional decomposition of adduct anions. *J. Mass Spectrom.* **30**: 1772–1773.
155. J. Adam and M. L. Gross (1987). Tandem mass spectrometry for collisional activation of alkali metal-cationized fatty acids: a method for determining double bond location. *Anal. Chem.* **59**: 1576–1582.
156. G. Madalinski, F. Fournier, F.-L. Wind, C. Afonso, and J.-C. Tabet (2006). Gram-negative bacterial lipid A analysis by negative electrospray ion trap mass spectrometry: stepwise dissociations of deprotonated species under low energy CID conditions. *Int. J. Mass Spectrom.* **249/250**: 77–92.
157. J. L. Kerwin, A. R. Tuininga, and L. H. Ericsson (1994). Identification of molecular species of glycerophospholipids and sphingomyelin using electrospray mass spectrometry. *J. Lipid Res.* **35**: 1102–1114.
158. A. El-Aneed and J. Banoub (2005). Elucidation of the molecular structure of lipid A isolated from both a rough mutant and a wild strain of *Aeromonas salmonicida* lipopolysaccharides using electrospray ionization quadrupole time-of-flight tandem mass spectrometry. *Rapid Commun. Mass Spectrom.* **19**: 1683–1695.
159. M. Petkovic, J. Schiller, M. Muller, S. Benard, S. Reichl, K. Arnold, and J. Arnold (2001). Detection of individual phospholipids in lipid mixtures by matrix-assisted laser desorption/ionization time-of-flight mass spectrometry: phosphatidylcholine prevents the detection of further species. *Anal. Biochem.* **289**: 202–216.

CHAPTER 12

Thermodynamics of the Nervous Impulse

THOMAS HEIMBURG and ANDREW D. JACKSON

Niels Bohr Institute, University of Copenhagen, Copenhagen, Denmark

CONTENTS

12.1	Introduction	318
	12.1.1 Propagation of Pulses Along Nerve Membranes	318
	12.1.2 Thermodynamics of Biological Systems	318
	12.1.3 Macroscopic Versus Microscopic Models in Biology	320
	12.1.4 The Hodgkin–Huxley Model	321
12.2	Thermal and Mechanical Properties of Nerve Membranes	324
	12.2.1 Heat Release During the Action Potential	324
	12.2.2 Origin of the Reversible Heat Release	326
	12.2.3 Mechanical and Structural Changes During the Action Potential	327
12.3	Propagating Pulses in Cylindrical Membranes	328
	12.3.1 Melting in Biological Membranes and Changes in the Elastic Constants	329
	12.3.2 Mechanical Pulses	331
12.4	Excitation of The Nerve Pulse	334
	12.4.1 Pulse Generation	334
	12.4.2 Free Energy of a Pulse	334
12.5	Conclusion	335
Acknowledgments		337
References		337

Structure and Dynamics of Membranous Interfaces, edited by Kaushik Nag
Copyright © 2008 John Wiley & Sons, Inc.

12.1 INTRODUCTION

12.1.1 Propagation of Pulses Along Nerve Membranes

Since the classical paper of Hodgkin and Huxley [1], nerve pulses have been explained as voltage pulses generated by the transient opening of ion channel proteins, and the resulting flux of ions across the nerve membrane. Even though this picture is popular and described in numerous textbooks, there exists quite compelling evidence that it cannot be correct. Several authors have found reversible temperature changes and reversible heat release during the action potential of both myelinated and nonmyelinated nerves. This finding indicates that the underlying basis of nerve pulse transmission must be reversible physics while the Hodgkin–Huxley model is exclusively based on irreversible phenomena. Furthermore, various changes in mechanical properties and changes in membrane state have been observed. We introduce here the thermodynamics of nerves and biological membranes. We show that the physiological conditions imply the possibility of localized density pulse propagation along the nerve membrane that is in agreement with the thermodynamics findings in nerves.

12.1.2 Thermodynamics of Biological Systems

Thermodynamics is one of the foundations of physics. It is based on two fundamental postulates: the conservation of energy and the maximum entropy principle. Thermodynamics is strictly true on all scales of physics from atomic scales to cosmology. Thus thermodynamics is also the theory underlying biological processes.

The first law assumes the form

$$dE = T\,dS - p\,dV - \Pi\,dA - f\,dl + \psi\,dq + \cdots + \sum_i \mu_i\,dn_i \quad (12.1)$$

Each term on the right-hand side is the product of an intensive variable (independent of system size) and the differential of an extensive variable (dependent on system size). The second law can be expressed as

$$dS = dS_r + dS_i \quad (12.2)$$

where S is the entropy. S_r is the reversible part of the entropy related to the reversible adsorption or release of heat given by

$$dS_r = \frac{dQ}{T} \quad (12.3)$$

The irreversible part S_i is related to spontaneous changes within the system that are not coupled to heat exchange with the environment and the performance of work on the exterior. This part is always larger than zero:

$$dS_i \geq 0 \quad (12.4)$$

The irreversible part of the entropy is typically responsible for the progress of chemical reactions and other equilibration processes.

The laws of thermodynamics also hold when the system under consideration is not in equilibrium. The second law states that each system approaches the state of maximum entropy, that is, the most probably state. Around this state one finds fluctuations due to thermal motion. In the field of nonequilibrium thermodynamics it is typically assumed that on small scales detailed balance (i.e., microscopic reversibility) is obeyed. Thus, nonequilibrated systems also rely on the laws of equilibrium thermodynamics.

The maximum entropy law applies to the complete system under consideration but not to arbitrarily chosen subsystems. One can easily construct examples where the second law holds for the total system but not for individual parts. For instance, two gas containers with different pressure that are coupled by a piston are not in equilibrium (Fig. 12.1). After equilibration, the pressure in both containers is the same. (In fact, in each equilibrated system the intensive quantities T, p, Π, ψ, μ_i ... are homogeneous.) During the equilibration process, the volume of one container increases and the volume of the other one decreases. Thus the entropy of one container increases while that of the other container decreases. The total entropy, however, approaches a maximum. Another example would be the cold unfolding of proteins. It is well known that proteins denature (or unfold) upon temperature increase. During this process the configurational entropy of the protein backbone increases. Many proteins, however, also denature upon cooling to temperatures far below room temperature [2]. This is an unexpected (and not very well-known) phenomenon that is related to the strong temperature dependence of the interactions of proteins with water. During cold unfolding, the entropy of the amino acid chains increases upon cooling. The laws of thermodynamics state, however, that entropy always decreases with decreasing temperature. How then is cold unfolding possible? In fact, cold unfolding does not violate the laws of thermodynamics because the entropy of the proteins including the entropy of the associated water shell decreases upon cold unfolding even though the entropy of

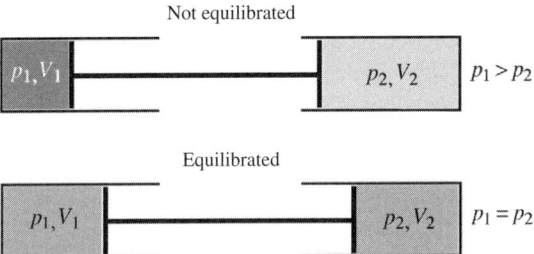

Figure 12.1 Two gas containers with different pressures that are coupled by a piston. After equilibration, the pressure in both containers is the same. The entropy of the total system has increased. The entropy of the right-hand gas container, however, has decreased. Thus the maximum entropy law does not hold for arbitrary subsystems if they are coupled to other systems.

the chains alone increases. The entropy increase of the chains is compensated by the decrease in the entropy of the water associated to the protein. We learn from this that the laws of thermodynamics apply to the whole system but not to subsystems. This statement is especially important when considering biological systems. Much of the research on biological systems focuses on individual molecules and their function. The proteins are considered separately from the membrane lipids. Typically, experiments on these molecules are done in test tubes at large dilution, where the molecule under consideration may be considered individually. It would, however, be a mistake to assume that one can understand the behavior of biological organisms by adding the functions of the individual molecules. In the biological environment molecules may act in a cooperative manner. The thermodynamic laws apply to the cell as a whole rather than to the isolated individual molecules. Thus the understanding of individual molecules may not necessarily lead to an understanding of complex biological systems.

If a system is not in equilibrium, the thermodynamic forces drive the system back to equilibrium. The progress of chemical reactions, diffusion processes, electrical currents, and soon are consequences of these forces. Typically, the thermodynamic forces consist in the gradients of the intensive quantities, for example, $-\text{grad } T$, $-\text{grad } \psi$, and $-\text{grad } \mu_i$. An exception is chemical reactions of nature $\nu_A \cdot A + \nu_B \cdot B + \cdots \longleftrightarrow \nu_C \cdot C + \nu_D \cdot D + \cdots$ (with the stoichiometries ν_i), where the thermodynamic force is given by the affinity A of a reaction defined by $A = -\sum \nu_i \mu_i$. Thermodynamic forces determine the progress of all reactions. All molecular species enter the thermodynamic equations via their chemical potentials. Thus thermodynamics is also applicable to nonequilibrium systems.

12.1.3 Macroscopic Versus Microscopic Models in Biology

Considering biological membranes, the previous statements on the entropy of large systems imply that the laws of thermodynamics apply to the membrane as a whole but not necessarily to each isolated molecule in this membrane when they interact with other molecules. Therefore it is dangerous to base theories of large systems on the behavior of single molecules.

In this chapter we discuss the propagation of nerve pulses. The textbook pictures for these phenomena are based on the action of single molecules and, in particular, on ion channel proteins. Nerve pulse propagation has commonly been explained by the voltage- and time-dependent opening and closing of such ion channels.

Models based on microscopic detail contain some further risks. A well-accepted strategy in physics is to search explanations of phenomena by using models of similar length scale. For instance, the propagation of sound is well described by the differential equation

$$\frac{\partial^2 \rho}{\partial t^2} = c_0^2 \frac{\partial^2 \rho}{\partial x^2} \tag{12.5}$$

where c_0 is the speed of sound and ρ is the density. It is a function of the isentropic compressibility, κ_S, and is given by $c_0 = \sqrt{1/\kappa_S \rho}$. This equation holds for all gases independent of their molecular composition. It would be impossible to understand the propagation of sound on the basis of molecular mechanisms, for example, by applying molecular dynamics simulations. This is because sound propagation is basically an entropic phenomenon involving ensemble properties on large scales, that is, on the order of their wavelength. It is therefore more meaningful to determine the compressibility on a macroscopic scale.

A similar case is the propagation of the nerve pulse in its presently accepted picture. Hodgkin and Huxley proposed a model in 1952 in which they explained the features of a nerve pulse on the basis of individual ion channel proteins that open and close in a voltage- and time-dependent manner [1]. The typical diameter of an ion channel is about 5 nm. The nerve pulse in a myelinated nerve propagates at about 100 m/s and lasts about 1 ms, resulting in a typical pulse length of about 10 cm. A nonmyelinated squid axon displays a velocity of about 25 m/s and a length of 3 ms, corresponding to 6 cm. Thus the length scale of the nerve pulse is about 2.5×10^7 times larger than that of the ion channels. This is a similar difference in scale as, for instance, a coffee cup and the size of Europe. Nobody, however, would tend to explain macroscopic phenomena of the length scale of Europe (e.g., earthquakes, growth of mountain chains, or large storms) on the basis of objects the size of a coffee cup. Similar differences in scale, however, form the basis for the biological models of nerve pulses.

In this chapter we wish to apply the above concepts to the propagation of nerve pulses. During nerve pulses, in phase with voltage changes, one finds a reversible release of heat. We outline next that this implies that the nervous impulse is a consequence of reversible physics. To be able to describe the underlying thermodynamic theory, we describe in the next section the basics of the textbook model for nerve pulses, in particular, the Hodgkin–Huxley model that explains the propagating pulse as a consequence of specific resistors called ion channels. We show that this model predicts changes in heat release that are inconsistent with experimental findings.

12.1.4 The Hodgkin–Huxley Model

Nerve cells possess extended axons along which voltage pulses called "action potentials" can propagate. In 1952, Hodgkin and Huxley proposed a model for the action potential that was based on the conductance properties of nerve membranes obtained by voltage clamp experiments [1]. In such experiments an electrode is inserted into a nerve, and a voltage difference between the inside and outside of the nerve is applied. Since the voltage is kept constant along the whole nerve, no pulses propagate in such experiments. Schematically, this experiment is shown in Fig. 12.2.

The transmembrane current is described by

$$I_m = C_m \frac{dU}{dt} + g_K(U - E_K) + g_{Na}(U - E_{Na}) + g_L(U - E_L) \qquad (12.6)$$

where $g_K(V, t)$ and $g_{Na}(V, t)$ are voltage- and time-dependent conductivities of potassium and sodium channels, and E_K and E_{Na} are the corresponding Nernst

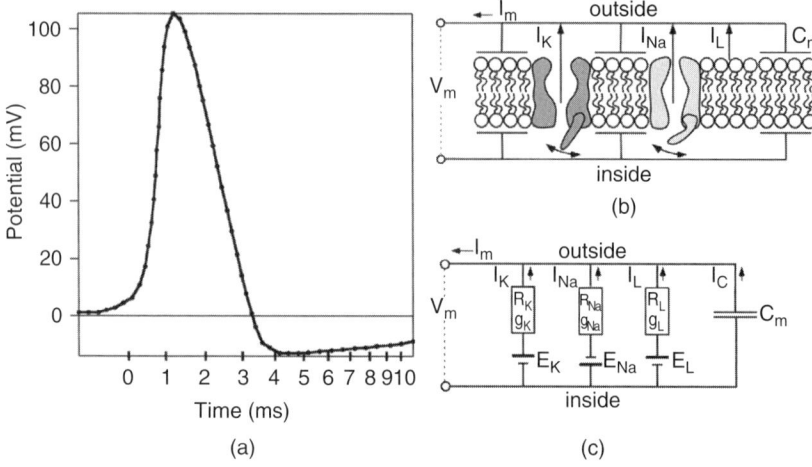

Figure 12.2 Molecular basis of the Hodgkin–Huxley model. (a) The action potential as given by Hodgkin and Huxley [1]. (b) Schematic drawing of the membrane including sodium and potassium channel proteins that are selective conductors for sodium and potassium. (c) Equivalent circuit picture of the same membrane (from Ref. 3).

potentials that depend on the ion concentration differences inside and outside the nerve cell. For the squid axon they are $E_K = -80$ mV and $E_{Na} = +60$ mV. The terms with index L correspond to leakage currents of other ions and will not be considered in the following. Typical maximum conductances of squid axon membranes are on the order of 4×10^{-3} S/cm² (=A/V·cm²) for both potassium and sodium channels. The capacitance of the membrane is on the order of 1 µF/cm² (which is calculated for a capacitor of capacitance $C_m = \varepsilon \cdot \varepsilon_0 \cdot A/d$ with plate separation of $d = 3.5$ nm (thickness of the hydrophobic core of the membrane) and a dielectric constant within the membrane of $\varepsilon = 4$).

Equation (12.6) contains currents through resistors and capacitive currents. The charge on a capacitor is given by $q = C_m U$. Thus the capacitive current is given by

$$I_C = \frac{dq}{dt} = C_m \frac{dU}{dt} + U \frac{dC_m}{dt} \qquad (12.7)$$

Thus in the Hodgkin–Huxley description, it was assumed that the capacitance stays constant. We show later that this is probably incorrect. Note also that the neglected term dC_m/dt has the units of a conductance [38].

The differential equation for the propagation of the nerve pulse is

$$\frac{a}{2R_i} \frac{\partial^2 U}{dt^2} = C_m \frac{\partial U}{dt} + g_K(U - E_K) + g_{Na}(U - E_{Na}) \qquad (12.8)$$

where a is the radius of the nerve and R_i is the internal resistance of the aqueous medium along the nerve interior. According to Hodgkin and Huxley [1], the maximum amplitude of the potassium current is on the order of 800 µA/cm², while the maximum sodium current amplitude is about +800 µA/cm².

The Hodgkin–Huxley model relies on equivalent circuits (Kirchhoff circles). Since the model contains resistors one expects a heat production described as

$$\frac{dQ}{dt} = P = U \cdot I \cong \sum_i g_i (U - E_i)^2 \qquad (12.9)$$

In the following we want to estimate the heat observed by a heat sensor placed (e.g., a thermocouple) on the nerve surface.

Using Eq. (12.9), one obtains a heat production for the potassium channels at maximum voltage of about

$$\frac{dQ}{dt} \approx 4 \times 10^{-3} \frac{A}{V \cdot cm^2} \cdot (0.1 V)^2 = 40 \times 10^{-6} \frac{J}{s \cdot cm^2} \qquad (12.10)$$

This is the heat per cm² of heat sensor surface released per second. The heat produced by the sodium currents should be of similar order. The data underlying this calculation have been taken from the original paper of Hodgkin and Huxley [1] on the squid axon. Since heat release is independent of the direction of the current, heat release should always be positive.

The energy of the membrane capacitor at maximum voltage is

$$E_C = \frac{1}{2} C U^2 = \frac{1}{2} \times 10^{-6} \frac{F}{cm^2} \cdot (0.1 V)^2 = 5 \times 10^{-9} \frac{J}{cm^2} \qquad (12.11)$$

This corresponds to the heat recorded by 1 cm² of heat sensor if the membrane is charged from 0 to 100 mV. Simultaneously, this is the energy of the nerve pulse per unit area.

The squid axon has a propagation velocity of about 25 m/s, and the action potential lasts about 3 ms when measured at one point on the nerve. This results in a pulse length of about 6–9 cm, which is quite extended.

This means that the heat dissipated during the squid action potential by Na⁺ and K⁺ currents (that would be measured at one point of the nerve) is on the order of (cf. Eq. (12.10))

$$Q = 200 \times 10^{-9} \text{ J/cm}^2 \qquad (12.12)$$

The heat from the currents that would be recorded by a heat sensor is therefore more than one order of magnitude larger than the energy necessary to charge the capacitor. The latter is the energy of the action potential, because the information transported during the action potential lies in the propagating segment of charged capacitor. Most of the chemical free energy stored in the ion concentration gradients is therefore dissipated. Assuming a membrane density of about 1 g/cm³ and a membrane thickness of 5 nm, one obtains a heat of about 0.4 J/g from the ion currents of membrane while the energy of the pulse as calculated from the capacitive energy is on the order of 0.01 J/g of membrane. All these calculations are estimates intended to indicate the order of magnitude of the respective effects.

The action potential is a dissipative process due to the flux of ions through resistors from the high concentration to the low concentration side. Thus in the electrical picture of Hodgkin and Huxley, there should always be heat dissipation, and the dissipation of heat during the pulse is much larger than the pulse energy. If a nerve had 1 meter of length, the dissipated heat over the whole length would be about 500 times larger than the capacitive energy of the pulse that is the actual signal. Thus the propagation of the nerve pulse in the electrical picture is a very "expensive" process.

12.2 THERMAL AND MECHANICAL PROPERTIES OF NERVE MEMBRANES

Studies on heat release, as considered in the previous section, can in fact be found in the literature. The thermodynamics of the nerve pulse is described next.

12.2.1 Heat Release During the Action Potential

The heat release during action potentials has been measured by a number of authors 4–8. We show some examples from different nerves, both myelinated and nonmyelinated (Figs. 12.3–12.5).

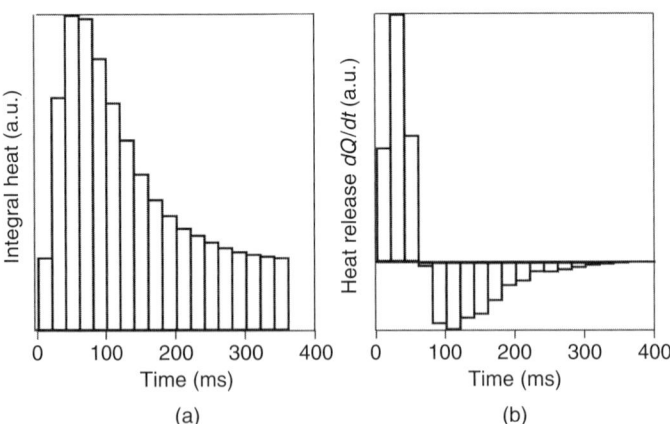

Figure 12.3 (a) Integrated heat and (b) heat release per time during the action potential of nonmyelinated maja nerve bundles (adapted from Ref. 4). A period of heat release is followed by a phase of heat reabsorption. Thus the heat is nearly completely reabsorbed by the nerves, indicating that the physical processes underlying the action potential are mostly reversible processes.

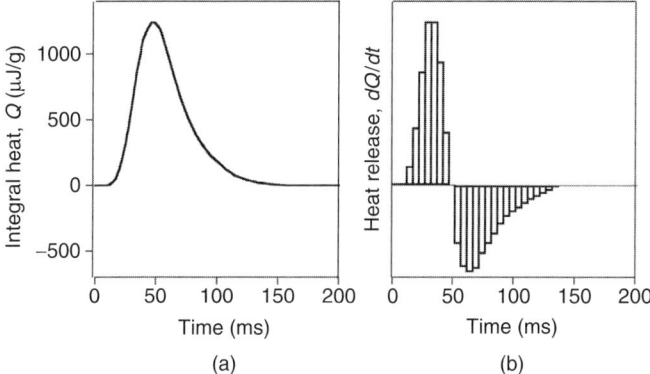

Figure 12.4 Heat production in nonmyelinated fibers of garfish olfactory nerve. (a) The integrated heat (given per gram of nerve) and (b) the differential heat release of the same experiment from a heat block analysis. (Data adapted from Ref. 6.)

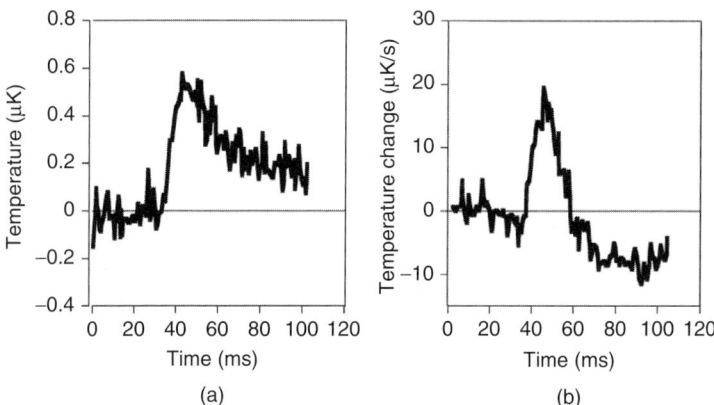

Figure 12.5 Temperature response in myelinated bullfrog sciatic nerve. (a) Absolute temperature change and (b) rate of temperature change indicating the transient release and reabsorption of heat. (Adapted from Ref. 8.)

All authors found the same interesting phenomenon: during the action potential an initial heat liberation is found that is followed by a subsequent heat reabsorption. The total heat of the pulse is zero within experimental error, meaning that the magnitudes of the heat liberation and of the heat absorption are about equal. This is a striking finding that seems to be in conflict with the statements of the Hodgkin–Huxley model that predicts significant positive heat liberation during the nerve pulse due to fluxes of currents (Section 12.4). Interestingly, poisons like tetrodotoxin (that are thought to block ion channels) inhibit the reversible heat pulses [6].

12.2.2 Origin of the Reversible Heat Release

What might be the origin of the reversible heat release? Isn't the presence of a positive and a negative phase of the heat surprising?

The Hodgkin–Huxley picture of the action potential uses equivalent circuits that are an electrical analogy of a biological situation of quite different nature. Instead of electrons flowing along a metal wire, one rather has a flux of hydrated ions through aqueous pores. As already pointed out by Abbott et al. [4], the situation in a nerve resembles much more the expansion of an ideal gas through semipermeable walls. Van't Hoff argued in the 19th century that the concentration of molecules in a solution is closely related to the partial pressure of a gas. Therefore an ion concentration difference across the membrane corresponds to a pressure difference. The concentration of potassium inside the nerve is high (about 400 mM) inside giant squid axons and it is low outside (around 20 mM). For sodium the situation is reversed. According to the Hodgkin–Huxley model, during the action potential the membrane becomes transiently permeable to sodium and potassium. The ions move along their gradients from the high temperature to the low temperature side. During this process, work can be performed by each of the ions. This resembles the motion of a piston during the expansion of an ideal gas. If no work is performed, no heat is absorbed by the system. If work is performed, for instance, by charging the membrane capacitor, heat is absorbed and results in a transient cooling of the nerve. Thus the picture of Hodgkin and Huxley (flux of ions through pores) contains the possibility of cooling even though the electrical analogy is clearly insufficient to describe such effects. Ritchie and Keynes [6] showed that the reversible heat release during the action potential is proportional to the energy of the membrane capacitor given by $E_C = 0.5 C_m U^2$ (Fig. 12.6). They indicated, however, that the reversible heat release is significantly higher than the calculated capacitive energy (see previous section) and thus ruled out that the charging of the capacitor is responsible for the heat changes. For a thorough evaluation of the possible origin of the reversible heat in nerves see Refs. 4 and 5.

In Section 12.1 we showed that the laws of thermodynamics apply to the complete system under consideration but not to arbitrarily separated subsystems. In our present case, the total system is the nerve as a whole, and not just the sum of capacitor, resistors, and other components. Therefore one has to consider the thermal response globally. The fact that the heat changes found in real nerves are within error completely reversible indicates that no net heat is generated and that the entropy of the system remains constant. This is so because $dQ = TdS$ and $\oint dQ/T = 0$, but $\oint dQ = 0$ only holds if one reverses the path when moving back to the original state. This statement must be true for the total nerve even if one can imagine individual processes that release and the absorb heat. If one makes the analogy between the expansion of a gas and the performance of work on the membrane capacitor, a reversible heat change can only be generated by a reversible expansion of a gas. During expansion of the gas the capacitor is charged and heat is absorbed, while discharging of the capacitor compresses the gas, leading to the same concentration differences as before the pulse. This is

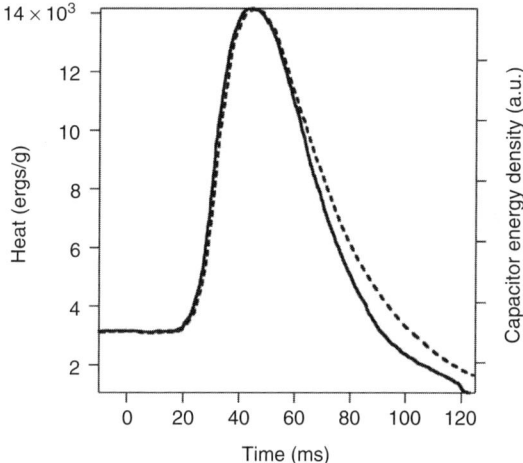

Figure 12.6 Integrated heat release (solid line) and capacitor energy density (dashed line) for garfish olfactory nerve. The two functions are proportional but of different magnitude. (Adapted from Ref. 6.)

definitely not the situation in the Hodgkin–Huxley model in which the pulse generation is driven by permanent fluxes of the ions along their gradients, but never against their gradients.

12.2.3 Mechanical and Structural Changes During the Action Potential

It has been shown by the group of I. Tasaki that the action potential is also accompanied by mechanical and structural changes.

If a thin piston is placed on a squid axon it is displaced by about 1 nm during the nerve pulse (Fig. 12.7b). Simultaneously, a force of up to 2 nN acts on the piston. Both force and dislocation are, within experimental error, proportional to the action potential. Similar results were reported in Refs. 7, 9 and 11–14. These findings indicate that the Hodgkin–Huxley model cannot be complete because it does not contain any mechanical element. If the change in the dimensions of the nerve is due to changes in the membrane thickness, then it is likely that the capacitance is in fact not constant during the action potential. This, however, was assumed by Hodgkin and Huxley [1].

Another interesting finding is that some fluorescence markers alter their fluorescence properties during the action potential [15–19]. In Fig. 12.8 we show the fluorescence changes of the membrane marker pyronine B during the action potential in crab nerves. The fluorescence displays a difference in the two polarization planes. This indicates changes in fluorescence anisotropy and related changes in membrane viscosity during the action potential. For this reason, it has been proposed that the action potential is accompanied by structural changes

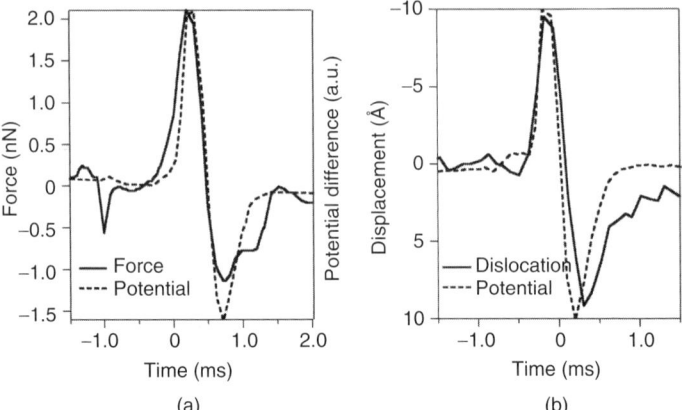

Figure 12.7 Mechanical changes during the action potential. (a) Force on a piston during the action potential in a squid axon. The solid line represents the voltage changes, the dotted curve the force. (b) During the nerve pulse in a squid axon the thickness of the nerve changes proportional to the voltage. (Data adapted from Ref. 10.)

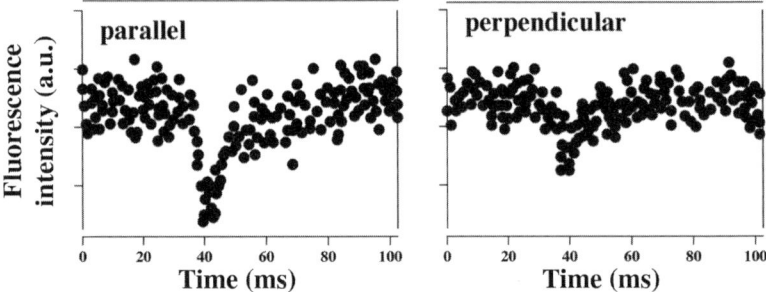

Figure 12.8 Changes of the fluorescence of pyronine B during the action potential of crab nerves (Adapted from Ref. 19). The differences in the fluorescence changes between parallel and perpendicular polarization indicate changes in the rotational anisotropy of the membrane components. This indicates changes in the membrane viscosity. Kobatake et al. [19] concluded from these data that the macromolecules of the nerve membranes undergo structural changes during the action potential.

in the macromolecules and that it is related to transitions in either macromolecules or membranes [14, 19–21].

12.3 PROPAGATING PULSES IN CYLINDRICAL MEMBRANES

The possibility of transitions in the nerve membranes is explored in this section. We show that the thermodynamic findings described earlier can be explained if one assumes that the action potential consists of propagating density pulses. Heimburg and Jackson [21] demonstrated that one could obtain stable propagating density pulses in cylindrical lipid membranes provided that the membrane exists in a physical state slightly above a melting transition. In the following we outline the underlying basis of this model.

12.3.1 Melting in Biological Membranes and Changes in the Elastic Constants

Many biological membranes display melting transitions slightly below body temperature [3, 21, 22, 39]. It is known that the lipid compositions of bacteria and also of fish cells change as a function of growth temperature and hydrostatic pressure [23–26]. In Fig. 12.9 the melting transitions of native *Escherichia coli* membranes (including all their proteins) are shown. One finds a pronounced lipid melting peak slightly below body temperature that is affected by growth temperature of the bacteria, by hydrostatic pressure, and by pH [3]. Furthermore, one finds several protein unfolding peaks slightly above body temperature. We assume in the following that lipid melting processes slightly below physiological temperature are common and that they exist in nerve membranes.

The melting transitions of such membranes display a melting temperature, T_m, a melting enthalpy, ΔH, and a melting entropy, ΔS, given by $\Delta S = \Delta H / T_m$.

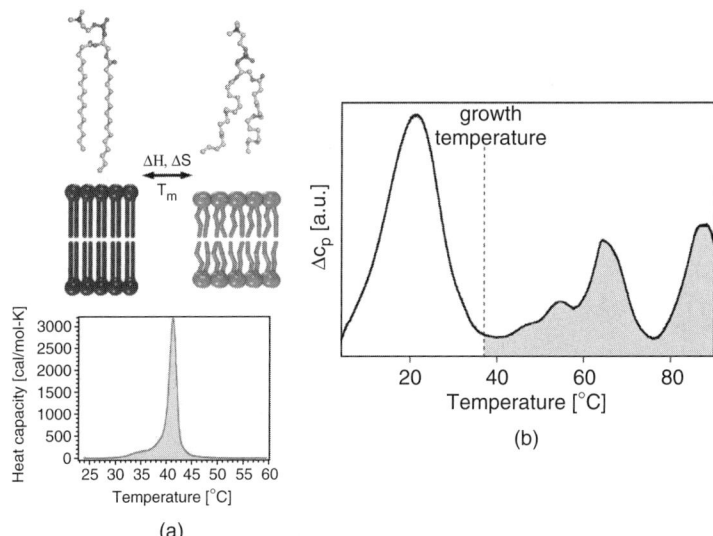

Figure 12.9 (a) Schematic picture of the melting process in lipid membranes and the associated change in the specific heat capacity. (b) Melting profile of the native membranes of *E. coli* grown at 37 °C. The growth temperature is indicated as a dashed line. The peaks below growth temperature belong to the melting of lipid membranes; the peaks shaded in grey above the growth temperature are attributed to protein unfolding. (From Ref. 3.)

Furthermore, volume and area of the membrane change during the melting process. For the model lipid DPPC (dipalmitoyl phosphatidylcholine), a major lipid component of lung surfactant, one finds $T_m = 314.2$ K, $\Delta H = 35$ kJ/mol, $\Delta S = 111.4$ J/mol·K, $\Delta V/V = 0.04$, and $\Delta A/A = 0.246$. These values give the order of magnitude but vary between different lipid species.

The heat capacity, shown for the *E. coli* membranes in Fig. 12.9, is related to the fluctuations in enthalpy through the relation

$$c_P = \frac{\langle H^2 \rangle - \langle H \rangle^2}{RT^2} \tag{12.13}$$

The isothermal and the area compressibility are related to the fluctuations in volume and area as

$$\kappa_T^V = \frac{\langle V^2 \rangle - \langle V \rangle^2}{\langle V \rangle \cdot RT} \quad \text{and} \quad \kappa_T^A = \frac{\langle A^2 \rangle - \langle A \rangle^2}{\langle A \rangle \cdot RT} \tag{12.14}$$

It has been shown [27, 28] that in the chain melting transition volume and area changes are proportional to enthalpy changes. Therefore, in the lipid melting transition, the heat capacity is directly related to the excess isothermal volume and area compressibility, such that one obtains the following relation for the area compressibility:

$$\kappa_T^A(T) = \kappa_{T,0}^A(T) + \Delta\kappa_T^A(T) = \kappa_{T,0}^A(T) + \frac{\gamma_A^2 T}{A}\Delta c_P(T) \tag{12.15}$$

The function $\kappa_{T,0}^A$ is the temperature-dependent area compressibility of the pure phases, which has to be taken from the literature. The factor γ_A assumes the values $\gamma_A = 0.89$ m²/J. One can see that the area compressibility assumes a maximum at the temperature where the heat capacity is at maximum.

The adiabatic compressibility, important for the determination of the lateral sound velocity, can be determined once the isothermal compressibility is known. It assumes the form [27]

$$\kappa_S^A = \kappa_T^A - \frac{T}{A \cdot c_P}\left(\frac{dA}{dT}\right)_\Pi^2 \tag{12.16}$$

where the heat capacity c_P is that of the membrane plus the aqueous environment that transiently absorbs heat from the membrane upon compression. For a periodic density variation, it is therefore a function of frequency (for details see Ref. 27). It is obviously smaller than the isothermal compressibility. Therefore one concludes that the adiabatic compressibility is in general frequency dependent, meaning that dispersion is found. The frequency dependence of relaxation phenomena in the lipid melting transition has also been documented in experiments [29, 30]. While moving a membrane through the lipid melting transition, the lateral density, ρ^A, of the membrane decreases and the specific area, A, increases. It has been shown by Heimburg [27] that the change in membrane area, ΔA, is proportional to the change in enthalpy. The heat capacity is defined by $c_P = (dH/dT)_P$. Thus the lateral compressibility becomes a nonlinear function of density that can be calculated from experimental heat capacity data.

The lateral sound velocity within the membrane plane is given by

$$c = \sqrt{\frac{1}{\kappa_S^A \rho^A}} \qquad (12.17)$$

The adiabatic compressibility is a nonlinear function of the area density of the membrane, and it follows that the sound velocity is a nonlinear function of the density that close to the lipid melting transition can be expanded into a power series, such that

$$c^2 = c_0^2 + p(\Delta \rho^A) + q(\Delta \rho^A)^2 + \cdots \qquad (12.18)$$

where c_0 is the sound velocity in the fluid phase of the membrane. p and q are parameters that have to be determined from the known dependence of the sound velocity on the density. For unilamellar DPPC membranes slightly above the transition, one finds experimentally that $c_0 = 176.6$ m/s (the lateral sound velocity in the fluid phase at low frequencies), $p = -16.6 c_0^2/\rho_0^A$, and $q = 79.5 c_0^2/(\rho_0^A)^2$ (for details see Ref. 21). Here, $\rho_0^A = 4.035 \times 10^{-3}$ g/m^2 is the lateral area density in the fluid phase of the membrane slightly above the melting point. Similar values were found for lung surfactant or native *E. coli* membranes.

12.3.2 Mechanical Pulses

In the following we explore propagation phenomena in cylindrical membranes. It has been shown by Heimburg and Jackson [21] that the propagation of density changes can be described by the following hydrodynamic relation:

$$\frac{\partial^2}{\partial t^2} \Delta \rho^A = \frac{\partial}{\partial x} \left[c^2 \frac{\partial}{\partial x} \Delta \rho^A \right] - h \frac{\partial^4}{\partial x^4} \Delta \rho^A \qquad (12.19)$$

The second term is chosen ad hoc to describe the frequency dependence of the sound velocity in a linear way using a parameter h (for details see Ref. 21). It implies that the frequency dependence of the sound velocity is linearly related to the frequency, which is just a first-order approximation. Nerve pulses typically last several milliseconds. Therefore the frequency regime of interest is the kilohertz regime. Unfortunately, the frequency dependence of lipid membranes in the melting transition has not been investigated in this range. In other frequency regimes, however, the frequency dependence of the sound velocity is well known [29, 30]. The parameter h is the only one that has not yet been determined in an experiment. We will see later that the only role of the parameter h is to set the linear scale of the propagating pulse.

We have shown earlier that the sound velocity is a function of the area density, ρ^A. Introducing Eq. (12.18) into Eq. (12.19), we obtain

$$\frac{\partial^2}{\partial t^2}\Delta\rho^A = \frac{\partial}{\partial x}\left[(c_0^2 + p\Delta\rho^A + q(\Delta\rho^A)^2 + \cdots)\frac{\partial}{\partial x}\Delta\rho^A\right] - h\frac{\partial^4}{\partial x^4}\Delta\rho^A \quad (12.20)$$

and after the coordinate transformation $z = x - v \cdot t$ (introducing the propagation velocity, v) we arrive at the time-independent form describing the shape of a propagating density excitation:

$$v^2\frac{\partial^2}{\partial z^2}\Delta\rho^A = \frac{\partial}{\partial z}\left[(c_0^2 + p\Delta\rho^A + q(\Delta\rho^A)^2 + \cdots)\frac{\partial}{\partial z}\Delta\rho^A\right] - h\frac{\partial^4}{\partial z^4}\Delta\rho^A \quad (12.21)$$

This equation has an analytical localized solution [31]:

$$\Delta\rho^A(z) = \frac{p}{q} \cdot \frac{1 - \left(\frac{v^2 - v_{min}^2}{c_0^2 - v_{min}^2}\right)}{1 + \left(1 + 2\sqrt{\frac{v^2 - v_{min}^2}{c_0^2 - v_{min}^2}}\cosh\left(\frac{c_0}{h}z\sqrt{1 - \frac{v^2}{c_0^2}}\right)\right)} \quad (12.22)$$

Such localized solutions are known as solitary waves or solitons. Here we see that the linear scale is proportional to z/h. Therefore the only influence of the parameter h on the solution of the differential equation in Eq. (12.21) is to introduce a scaling factor for the pulse length. A representative soliton profile is shown in Fig. 12.10. The minimum velocity v_{min} in Eq. (12.22) is given by

$$v_{min} = \sqrt{c_0^2 - p^2/6q} \quad (12.23)$$

v_{min} of a soliton in DPPC membranes is found to be $v_{min} = 115$ m/s, which is very close to the velocity of the action potential found in myelinated nerves. The minimum velocity is the velocity of the soliton when its amplitude reaches the maximum value

$$\Delta\rho_{max}^A = |p|/q \quad (12.24)$$

corresponding to an overall density change of $\Delta\rho_{max}^A/\rho_0^A = 0.21$ (Fig. 12.10). Solitons with larger density change do not exist.

The total area change when going through a melting transition is $\Delta\rho_{max}^A/\rho_0^A = 0.246$ (for DPPC). Thus at maximum amplitude, the soliton forces the lipid membrane by about 85% through the melting transition. This will cause a transient heat release corresponding to 85% of the melting enthalpy (which is on the order of 35 kJ/mol or ~13 kT per lipid).

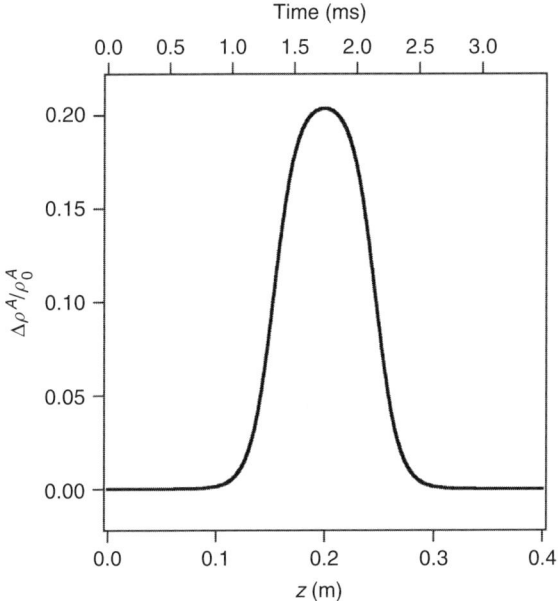

Figure 12.10 Soliton profile for a soliton velocity of $v = 0.651c_0$ calculated for $h = 2m^4/s^2$. This soliton has a maximum amplitude of $\Delta\rho^A/\rho_0^A = 0.21$. Its width is approximately 10 cm and it lasts about 1 ms.

During the pulse the membrane is reversibly moved by about 85% through the melting transition. Simultaneously, the thickness of the membrane will change by 85% of the thickness change in the transition from fluid to gel (7.4 Å for DPPC). Since the soliton is linked to changes in lipid state, the fluorescence anisotropy will also change in an experiment. It is well known that the anisotropy (related to the rotational mobility) is higher in the gel phase than in the fluid phase. Such changes have all been found in real nerves under the influence of the action potential (see Section 12.2.3 and reference therein). The order of magnitude of these changes matches the data found for such nerves.

The heat of melting of DPPC is about 35 kJ/mol or 48 J/g of lipid. If the membrane is pushed by 85% through the transition, this corresponds to 40 J/g of lipid. This heat is two orders of magnitude higher than that hypothetically dissipated by the ion currents during a pulse (0.4 J/g, see Section 12.1.3) and more than three orders of magnitude higher than the energy of the membrane capacitor during the action potential (0.01 J/g, see Section 12.1.3). It has been pointed out by Abbott et al. [4] that the positive and negative phases of heat liberation during a pulse are so close in time that in real experiments they are partially averaged out due to the response time of the heat sensor and by its physical size. Therefore the experimentally observed heat is just a low-end estimate of the real heat that must be significantly larger.

The mechanical energy of the solitons can also be calculated by using Lagrangian formalism [21]. The result is that the combined kinetic and potential energy of the soliton is about 40 times larger than the capacitor energy. The heat in the soliton, however, seems to be about 100 times larger than the mechanical energy of the pulse. This is possible close to transitions where the work to compress the membranes is small, and where entropy and internal energy (or enthalpy) changes compensate. This effect is often called entropy–enthalpy compensation. It seems as if the potential energy obtained by Lagrangian formalism is in fact a free-energy change. In transitions, free-energy changes can be very small even if the associated heat and enthalpy changes are large.

12.4 EXCITATION OF THE NERVE PULSE

12.4.1 Pulse Generation

How are the density pulses initiated? As mentioned earlier, the membrane is pushed by 85% through its phase transition during the soliton. Therefore everything that is able to move a lipid membrane through its transition should display the potential to start a pulse. The nature of the solitons that are the solutions to Eq. (12.21) has carefully been investigated [31]. It was found that if one locally generates a gel-like membrane region (domain) in a membrane above the melting transition regime, it should fall apart into two solitons propagating in opposite directions. One possibility to generate local gel domains is a local sudden lowering of temperature. It has been described by Kobatake et al. [19] that in fact nerves start firing when they are cooled locally, while firing is inhibited upon heating. Another possibility to shift membranes through transitions is the local lowering of pH [32]. Biological lipid membranes are typically negatively charged. Lowering of pH leads to protonation of the negatively charged groups. Simultaneously, the melting points of charged membranes increase by up to 20 K. Thus protons may induce a transition in membranes. It is known that action potentials can be triggered by local acidification, but the mechanism is so far not understood [33]. Furthermore, a pulse may be induced by local increase in calcium concentration because it is known that membrane melting transitions shift to higher temperatures upon calcium binding. An increase of calcium concentration from 1.8 to 20 mM has been shown to increase the capability of neuroblastoma cells to fire repetitively [34]. Another possibility to generate pulses is a change in membrane potential due to the electrostatic potential of a negatively charged lipid membrane. This is the most common way to evoke action potentials.

12.4.2 Free Energy of a Pulse

The free energy required to generate a local gel domain is closely related to the distance from the transition midpoint to physiological temperature [35]. At

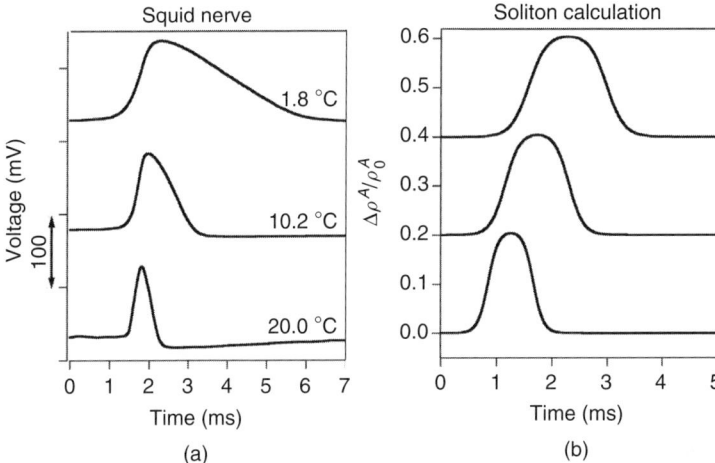

Figure 12.11 Width of the action potential after constant trigger as a function of temperature. (a) Squid action potentials recorded at different temperatures (adapted from Ref. 36). (b) Solitons generated with the same amount of free energy at increasing ambient temperature. As in the squid experiment, the broader signals stem from lower temperatures.

constant pressure it is given by

$$\Delta G = \Delta H \cdot \left(\frac{T_m - T}{T_m} \right) \tag{12.25}$$

If a constant amount of free energy is provided to a nerve (e.g., by a local change in voltage or a fixed number of protons or calcium ions), the pulse will possess smaller energy content when the temperature is higher (because more free energy is required to start the pulse at higher temperatures). This is shown in Fig. 12.11. The solitons display a smaller width when the energy content is smaller. Eventually, the amplitude will also decrease. In experiments on squid axons, it has been found that the width of the action potential becomes smaller when temperature is raised [36].

It has been shown recently [35] that the effect of general anesthetics is also related to the free-energy change necessary to move a membrane through a transition. Anesthetics are known to lower lipid membrane melting points, and thus increase the amount of free energy to start a pulse. This picture is able to quantitatively explain anesthesia. Thus the soliton model automatically implies a mechanism for anesthesia [3, 35].

12.5 CONCLUSION

Thermodynamics is a theory that can be applied to complex systems even if the molecular details of this system are unknown. The second law states that in equilibrium the entropy assumes a maximum. We have shown that during the action potential of nerves no net heat is dissipated. A phase of heat release is immediately followed by a phase of heat reabsorption. The finding of zero net heat exchange implies a conservation of the entropy of the system (no entropy is dissipated). This means that the entropy after completion of the action potential is the same as before the action potential. The consequence is that the physical processes underlying the action potential must be of reversible nature. The Hodgkin–Huxley model, which is the textbook picture of the nervous impulse, however, is exclusively based on irreversible processes (consisting of fluxes of ions through ion channels from the side of high concentration to that of low concentration) in which entropy is not maintained. The model can therefore not be correct. It may be possible to think about processes that release heat and other processes that absorb heat. Nevertheless, thermodynamics applies to the system as a whole and not to individual subsystems. Therefore the fact that no entropy is dissipated during the nerve pulse has to be taken seriously. It will be impossible to divide the total nerve signal into subprocesses (with some of them releasing and others absorbing heat) that result in an overall heat dissipation of zero and still be consistent with a model based on dissipative processes. For this reason, we described here an alternative approach that assumes that the nervous impulse is an entropic pulse (with zero entropy change after completion of the pulse). Such a pulse would show some features that have been described for nerves: reversible heat release, reversible changes in thickness, and reversible changes in membrane state. Since proteins are parts of biological membranes, they may also contribute to these changes as long as the processes are reversible. A reversible pulse involving proteins is favored by Tasaki [14].

The thermodynamic behavior of artificial and biological membranes gives rise to a number of interesting mechanical features. In particular, slightly below physiological temperature, one finds a cooperative lipid melting transition in which the heat capacity and also the elastic constants assume maxima. Under these conditions, localized mechanical (entropic) pulses called solitons can propagate with a velocity of about 100 m/s, close to that of myelinated nerves. One many loosely refer to them as sound pulses. However, like all mechanical changes, they are linked to changes in other variables of the system including changes in charge density, in the electrostatic potential, in heat content, in nerve thickness, and in lipid chain order parameter. Such changes have been found in real nerves. The classical studies on nerves have focused mainly on measuring voltage changes since they are large (100 mV) and therefore easy to measure. However, many other features change simultaneously and have to be measured to obtain a complete picture of the underlying physical processes. This implies thickness changes, density changes, and heat release, but also transient pH or calcium concentration changes. Thermodynamics clearly helps to reinterpret classical findings on nerve pulse propagation and provide a different viewpoint on processes that are seemingly in conflict with each other.

ACKNOWLEDGMENTS

Thanks to Benny Lautrup at the Niels Bohr Institute, who accompanied the development of these thoughts with major contributions, encouragement, and criticism. We are very grateful to Konrad Kaufmann, who discussed these issues with one of us for many years. Some of these issues have already been pointed out in his earlier manuscripts [37, 38]. We thank Dr. I. Tasaki for his interest and important comments and suggestions.

REFERENCES

1. A. L. Hodgkin and A. F. Huxley (1952). A quantitative description of membrane current and its application to conduction and excitation in nerve. *J. Physiol. (London)* **117**: 500–544.
2. P. L. Privalov (1990). Cold denaturation of proteins. *Crit. Rev. Biochem. Mol. Biol.* **25**: 281–305.
3. T. Heimburg and A. D. Jackson (2007). On the action potential as a propagating density pulse and the role of anesthetics. *Biophys. Rev. Lett.* **2**: 57–78.
4. B. C. Abbott, A. V. Hill, and J. V. Howarth (1958). The positive and negative heat associated with a nerve pulse. *Proc. R. Soc. London B* **148**: 149–187.
5. J. V. Howarth, R. D. Keynes, and J. M. Ritchie (1968). The origin of the initial heat associated with a single impulse in mammalian non-myelinated nerve fibres. *J. Physiol.* **194**: 745–793.
6. J. M. Ritchie and R. D. Keynes (1985). The production and absorption of heat associated with electrical activity in nerve and electric organ. *Q. Rev. Biophys.* **392**: 451–476.
7. I. Tasaki, K. Kusano, and M. Byrne (1989). Rapid mechanical and thermal changes in the garfish olfactory nerve associated with a propagated impulse. *Biophys. J.* **55**: 1033–1040.
8. I. Tasaki and P. M. Byrne (1992). Heat production associated with a propagated impulse in bullfrog myelinated nerve fibers. *Jpn. J. Physiol.* **42**: 805–813.
9. K. Iwasa and I. Tasaki (1980). Mechanical changes in squid giant-axons associated with production of action potentials. *Biochem. Biophys. Res. Commun.* **95**: 1328–1331.
10. K. Iwasa, I. Tasaki, and R. C. Gibbons (1980). Swelling of nerve fibers associated with action potentials. *Science* **210**: 338–339.
11. I. Tasaki, K. Iwasa, and R. C. Gibbons (1980). Mechanical changes in crab nerve fibers during action potentials. *Jpn. J. Physiol.* **30**: 897–905.
12. I. Tasaki and K. Iwasa (1982). Further studies of rapid mechanical changes in squid giant axon associated with action potential production. *Jpn. J. Physiol.* **32**: 505–518.
13. I. Tasaki and M. Byrne (1990). Volume expansion of nonmyelinated nerve fibers during impulse conduction. *Biophys. J.* **57**: 633–635.
14. I. Tasaki (1999). Evidence for phase transition in nerve fibers, cells and synapses. *Ferroelectrics* **220**: 305–316.

15. I. Tasaki, A. Watanabe, R. Sandlin, and L. Carnay (1968). Changes in fluorescence, turbidity, and birefringence associated with nerve excitation. *Proc. Natl. Acad. Sci. USA* **61**: 883–888.

16. I. Tasaki, L. Carnay, and A. Watanabe (1969). Transient changes in extrinsic fluorescence of nerve produced by electric stimulation. *Proc. Natl. Acad. Sci. USA* **64**: 1362–1368.

17. I. Tasaki, L. Carnay, R. Sandlin, and A. Watanabe (1969). Fluorescence changes during conduction in nerves stained with Acridine Orange. *Science* **163**: 683–685.

18. F. Conti and I. Tasaki (1970). Changes in extrinsic fluorescence in squid axons during voltage-clamp. *Science* **169**: 1322–1324.

19. Y. Kobatake, I. Tasaki, and A. Watanabe (1971). Phase transition in membrane with reference to nerve excitation. *Adv. Biophys.* **2**: 1–31.

20. P. K. J. Kinnunen and J. A. Virtanen (1986). A qualitative, molecular model of the nerve pulse. Conductive properties of unsaturated lyotropic liquid crystals, in *Modern Bioelectrochemistry* (F. Gutmann and H. Keyzer, Eds.). Plenum Press, New York, pp. 457–479.

21. T. Heimburg and A. D. Jackson (2005). On soliton propagation in biomembranes and nerves. *Proc. Natl. Acad. Sci. USA* **102**: 9790–9795.

22. J. L. C. M. van de Vossenberg, A. J. M. Driessen, M. S. da Costa, and W. N. Konings (1999). Homeostasis of the membrane proton permeability in *Bacillus subtilis* grown at different temperatures. *Biochim. Biophys. Acta* **1419**: 97–104.

23. S. V. Avery, D. Lloyd, and J. L. Harwood (1995). Temperature-dependent changes in plasma membrane lipid order and the phagocytic activity of the *Amoeba acanthamoeba castellanii* are closely correlated. *Biochem. J.* **312**: 811–816.

24. E. F. DeLong and A. A. Yayanos (1985). Adaptation of the membrane lipids of a deep-sea bacterium to changes in hydrostatic pressure. *Science* **228**: 1101–1103.

25. J. R. Hazel (1979). Influence of thermal acclimation on membrane lipid composition of rainbow trout liver. *Am. J. Physiol. Regul. Integrative Comp.Physiol.* **287**: R633–R641.

26. J. R. Hazel (1995). Thermal adaptation in biological membranes: Is homeoviscous adaptation the explanation? *Annu. Rev. Physiol.* **57**: 19–42.

27. T. Heimburg (1998). Mechanical aspects of membrane thermodynamics. Estimation of the mechanical properties of lipid membranes close to the chain melting transition from calorimetry. *Biochim. Biophys. Acta* **1415**: 147–162.

28. H. Ebel, P. Grabitz, and T. Heimburg (2001). Enthalpy and volume changes in lipid membranes. I. The proportionality of heat and volume changes in the lipid melting transition and its implication for the elastic constants. *J. Phys. Chem. B* **105**: 7353–7360.

29. S. Mitaku and T. Date (1982). Anomalies of nanosecond ultrasonic relaxation in the lipid bilayer transition. *Biochim. Biophys. Acta* **688**: 411–421.

30. S. Halstenberg, W. Schrader, P. Das, J. K. Bhattacharjee, and U. Kaatze (2003). Critical fluctuations in the domain structure of lipid membranes. *J. Chem. Phys.* **118**: 5683–5691.

31. B. Lautrup, A. D. Jackson, and T. Heimburg (2005). The stability of solitons in biomembranes and nerves. (http://www.arXiv.org/biophysics/0510106).

32. H. Träuble, M. Teubner, P. Woolley, and H. Eibl (1976). Electrostatic interactions at charged lipid membranes. 1. Effects of pH and univalent cations on membrane structure. *Biophys. Chem.* **148**: 319–342.
33. M. Vukicevic and S. Kellenberger (2004). Modulatory effects of acid-sensing ion channels on action potential generation in hippocampal neurons. *Am. J. Physiol. Cell Physiol.* **287**: C682–C690.
34. W. H. Moolenaar and I. Spector (1979). The calcium action potential and a prolonged calcium dependent after-hyperpolarization in mouse neuroblastoma cells. *J. Physiol.* **292**: 297–306.
35. T. Heimburg and A. D. Jackson (2007). The thermodynamics of general anesthesia. *Biophys. J.* **92**: 3159–3165.
36. J. J. C. Rosenthal and F. Bezanilla (2000). Seasonal variation in conduction velocity of action potentials in squid giant axons. *Biol. Bull.* **199**: 135–143.
37. K. Kaufmann (1989). Action potential, Caruaru (http://membranes.nbi.dk/Kaufmann/pdf/Kaufmann_book4_org.pdf).
38. K. Kaufmann (1989). Lipid membrane Caruaru (http://membranes.nbi.dk/Kaufmann/pdf/Kaufmann_book5_org.pdf).
39. M. Larsson, T. Nylander, K. M. W. Keough, and K. Nag (2006). An X-ray diffraction study of alterations in bovine lung surfactant bilayer structures induced by albumin. *Chem. Phys. Lipids* **144**: 137–145.

CHAPTER 13

Relationships Between Surface Viscosity, Monolayer Phase Behavior, and the Stability of Lung Surfactant Monolayers

JOSEPH A. ZASADZINSKI, CORALIE ALONSO, JUNQI DING, FRANK BRINGEZU, HEIDI WARRINER, TIM ALIG, and SIEGFRIED STELTENKAMP

Departments of Chemical Engineering and Materials, University of California, Santa Barbara, California

ALAN J. WARING

Department of Pediatrics, Harbor UCLA Medical Center, Torrance, California and Department of Medicine, UCLA, Los Angeles, California

CONTENTS

13.1	Introduction	342
	13.1.1 Monolayer Collapse Determines the Minimum Surface Tension	345
	13.1.2 Surface Viscosity and Surfactant Distribution in the Airways	349
13.2	From Surface Viscosity to Molecular Structure in "Simple" Monolayers: A History of Surface Viscosity	350
13.3	Surface Viscosity of Phospholipids and Lipid Mixtures	358
13.4	Effect of Lung Surfactant Proteins on Surface Viscosity	364
	13.4.1 Surface Viscosity of Clinical Surfactants	372
	13.4.2 Adjusting the Surface Viscosity of Replacement Surfactants	373
13.5	Physiological Implications for Surface Viscosity	375
	References	377

Structure and Dynamics of Membranous Interfaces, edited by Kaushik Nag
Copyright © 2008 John Wiley & Sons, Inc.

13.1 INTRODUCTION

Lung surfactant is a complex mixture of lipids and proteins originating from the type II cells that line the alveolar epithelial surfaces of humans and other mammals [1, 2]. Lung surfactant spreads as an interfacial film at the alveolar air–liquid interface; its function is to reduce the surface tension in the alveolar spaces, minimize the work of breathing, promote uniform lung inflation, and prevent alveolar collapse during the cycles of lung expansion and compression. A lack of functional surfactant due to prematurity leads to neonatal respiratory distress syndrome (nRDS) [1, 3]. A lack of functional surfactant is also implicated in the development of acute RDS in infants and adults [2, 4–12].

Lung surfactant (LS) has to achieve several seemingly contradictory results for proper lung function to occur. First, LS must form a stable interfacial film with sufficient integrity to sustain high surface pressures (low surface tensions) on compression of the interface during exhalation [13]. At the same time, LS must spread quickly to recover the interface during inspiration. This requires a subtle balance of lipids and proteins in the interfacial films and may require different interfacial compositions during lung expansion and compression [2, 14–22]. A low minimum surface tension on compression is generally ensured by a significant fraction of dipalmitoyl phosphatidylcholine (DPPC) and other disaturated phospholipids in lung surfactant, although this ranges from ~40% for Infasurf, to ~60% for Curosurf, to ~70% for Survanta, to ~80% for Exosurf, the four commonly used clinical replacement lung surfactants in the United States [23–25].

Respreading is associated with unsaturated lipids, although surfactant proteins SP-B and SP-C also play a significant role [15, 17]. Both proteins are likely involved in maintaining a reservoir of material available for respreading in the vicinity of the interface [14, 18–22, 26, 27]. The role of fatty acids and alcohols (such as the palmitic acid (PA) added to Survanta and the hexadecanol added to Exosurf) is less well known, although recent work has shown that PA and DPPC or hexadecanol and DPPC form highly ordered, crystalline monolayer phases at higher temperatures than DPPC alone [16, 28–30], which make the monolayer significantly more rigid [15–17, 31]. Infasurf has the highest content of unsaturated lipids and is the only clinical surfactant to include cholesterol [25, 32].

One of the difficulties in establishing the optimal lipid/protein composition for replacement lung surfactant arises from the inherent variability of sampling a nanometer-thick film within the deep lung *in vivo* [23–25]. While there is general consensus that the dominant lipid in all mammalian lung surfactant is DPPC, there is little agreement as to the fraction of DPPC or other lipids in native surfactant. There is still controversy whether cholesterol is present in lung surfactant, or if the cholesterol is a contaminant from the cell debris in the lung lavage needed to obtain surfactant [32, 33]. Cholesterol is removed from Survanta and Curosurf but is left in Infasurf. The compositions of replacement surfactants vary even more due to variations in the extraction procedures and in the number and type of additives used. At present, there is no universally accepted native or replacement lung surfactant composition [2, 24, 25, 32].

The first two replacement surfactants approved by the U.S. Food and Drug Administration (FDA) for the treatment of nRDS were the purely synthetic Exosurf, which contains DPPC (about 80%) combined with hexadecanol and tyloxapol, and Survanta, which is an organic extract of bovine lung mince supplemented with synthetic DPPC (for a final DPPC concentration of about 60–70%), palmitic acid (PA), and tripalmitin. While chemically quite distinct, administration of Survanta or Exosurf to premature infants proved to be very effective treatments for nRDS [1, 3, 34–37]. Benefits of therapy include improvement in systemic oxygenation, reduced need for ventilation, more uniformly inflated lungs, increased lung compliance, increased stability during deflation, and an increased functional residual capacity [1, 3, 34–39]. The initiation of replacement therapies coincided with a significant decline in the mortality rate of nRDS infants. SRT has been shown to reduce mortality rates by 30–50% for nRDS infants, and 80% of the decline in the infant mortality rate of the United States between 1989 and 1990 could be attributed to surfactant therapy [40, 41].

More recent replacement LS include Curosurf, an extract of whole mince of porcine lung tissue purified by column chromatography, and Infasurf, a chloroform–methanol extract of neonatal calf lung lavage. All clinical surfactants are primarily lipid (>98%), but the animal extracts contain small fractions of the amphiphilic surfactant proteins SP-B and SP-C, but not the hydrophilic SP-A and SP-D [25]. While Survanta, Curosurf, and Infasurf differ in both lipid and protein constituents, the effects of these differences are not clear. All are effective in clinical trials, and comparison trials do not conclusively show advantages of one over the other [36, 37, 42]. This suggests either that the detailed lipid and protein composition of a replacement lung surfactant is unimportant, or, more likely, that the currently used clinical surfactants are still far from optimal.

One approach toward optimizing replacement surfactant composition is to determine the effects of various lipid and protein species on a limited number of physiologically relevant and measurable physical parameters. The most commonly measured parameters deal with surface tension reduction, the primary function of LS. These include the equilibrium spreading pressure and the minimum surface tension measured in Langmuir troughs [14–17, 20–22, 31, 43–48] or by captive [18, 49, 50] or pulsating bubble [51, 52] techniques. However, all of the currently used replacement lung surfactants are capable of lowering the surface tension to near zero under optimized conditions, so it is necessary to examine in more detail the physical phenomena of monolayer collapse (which determines the minimum surface tension) and respreading (which determines the reproducibility of the surface tension on expansion and compression cycling) to elucidate the biochemical parameters on which these phenomena depend.

This has led us to develop the magnetic needle surface viscometer (Fig. 13.1), a new method for studying the viscous properties of monolayer and multilayer films at the air–water interface [31, 47, 48, 53]. Surface viscosity depends on

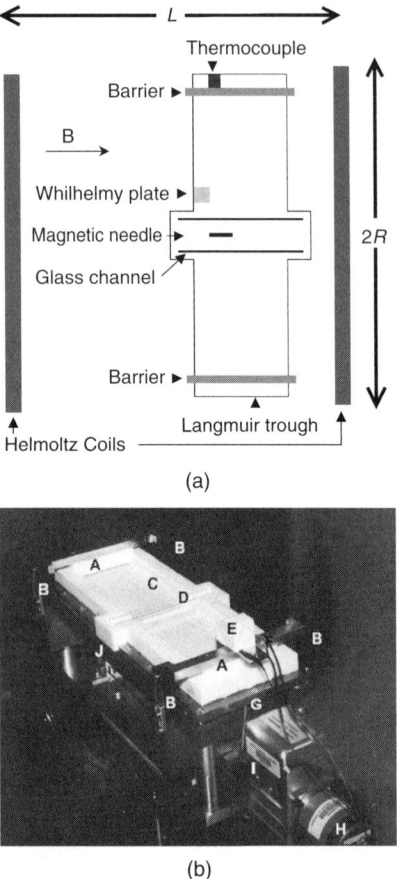

Figure 13.1 (a) Schematic drawing of the MNSV. The Helmholtz coils are circular coils of copper wire of diameter equal to the spacing between the coils ($L = 2R$). A current run through the coils provides a constant magnetic field gradient that pulls the magnetic needle with a constant force along the surface of the trough. The magnetic field changes sign in the middle of the gap between the coils, so the trough must be located offset from the center. The startup and terminal velocities are measured to determine the surface viscosity. The barriers can set a constant surface pressure and the thermocouple is used to help maintain a constant temperature [47]. (b) Photograph of Langmuir trough/surface viscometer. (A) Dual Teflon barriers are used, instead of the flat barrier in most commercial troughs. The barriers are notched to minimize leakage at high surface pressure. (B) Springs keep the Teflon barrier in position and provide tight contact with the trough. (C) Teflon trough with effective area $260\,\text{cm}^2$. (D) Two hydrophilic glass plates positioned in two Teflon slots to make a channel. The plates are parallel with an adjustable width. (E) Wilhelmy-plate surface pressure sensor. (F) Themistor for temperature measurement. (G) Thermoelectric coolers (TECs) sandwiched between a copper plate attached to the Teflon trough and a polysulfone water cell (J) for recirculating cooling water. (H) Stepper motor connected to a left–right lead screw. (I) Hard-limit switch for motor control. The whole trough is mounted on an adjustable lab-jack to align the trough along the center line between the Helmholtz coils (not shown) and is fixed to a vibration-free table [48].

monolayer composition, phase behavior, temperature, monolayer history, method of deposition, and other variables [15–17, 31, 47, 48]. While the importance of the LS surface viscosity had been recognized early on [54], there is little experimental data on the surface viscosity of any of the lipids common to lung surfactant [55], less on the effects of proteins on lipid monolayer rheology, and no real understanding or even hypothesis on what the ideal surface viscosity of a lung surfactant might be. Here we review how lipid chemistry and lateral organization in monolayers affect surface viscosity, how phase separation and more macroscopic organization influence the flow properties of monolayers, and the effects of lung surfactant specific proteins on the surface viscosity. The surface rheology of the clinical lung surfactants Survanta, Curosurf, and Infasurf is presented and correlated to the lipid and protein composition and phase behavior of these mixtures, and a new hypothesis on the optimal surface viscosity of a lung surfactant film is proposed.

13.1.1 Monolayer Collapse Determines the Minimum Surface Tension

One of the essential features of a good lung surfactant is to reduce the surface tension at the alveolus air–water interface to near zero. However, there is no simple relationship between the chemical or physical properties of a monolayer and its ability to lower surface tension. The surface tension properties of a monolayer on dynamic expansion and compression are linked not only to equilibrium properties of the film but to kinetic properties such as the surface viscosity.

The surface pressure of a monolayer in contact with a bulk phase of surfactant in the subphase under static conditions is known as the equilibrium spreading pressure, π_{ESP}. Monolayer films of single-hydrocarbon chain lipids and detergents (such as sodium dodecyl sulfate, SDS) maintain π_{ESP} during expansion and compression of the interface via rapid molecular adsorption and desorption from the interface into the subphase. Such molecules have micro- to millimolar solubility in water. If there is a reservoir of molecular surfactant in the bulk, the interface maintains a roughly constant surface concentration, or area per molecule during changes in interfacial area. The equilibrium surface pressure exerted by such a surface active, soluble detergent or soluble protein (such as albumin) is generally a logarithmic function of the bulk concentration up to a point at which the surface becomes saturated, that is, reaches a limiting area per molecule [11, 56, 57]. Higher concentrations have a minimal additional effect on the surface pressure but, for detergents like SDS, result in the formation of micelles or other aggregates in solution.

Such soluble surfactants could not control surface tension in the alveolus so as to allow normal breathing. The maximum surface pressure needs to be substantially greater than π_{ESP}, and the mechanical stability of the lungs requires that the surface pressure change on expansion and compression of the interface. For example, Curosurf, Infasurf, and Survanta all have π_{ESP} in the range of 40–45 mN/m [10]. This is well below the >60-mN/m surface pressure (<10-mN/m minimum surface tension) that characterizes an effective lung surfactant.

However, the lipids and proteins in lung surfactant are not soluble in saline, so the number of surfactant molecules at the interface does not change during expansion and compression of the interface. On compression, the surface pressure of the film quickly exceeds π_{ESP}, and the monolayer becomes metastable [58] as the area per molecule decreases below the equilibrium area per molecule. After sufficient compression, the molecular area (and surface pressure) reach a limiting value beyond which the film cannot be compressed further, and the monolayer "collapses" by folding, buckling, expelling material above or below the monolayer, and so on [22, 26, 58–60], to try to adjust the surface concentration back to the equilibrium value.

The surface pressure at which the monolayer collapses determines the maximum surface pressure that the film can produce, and hence the ability of the surfactant to reduce surface tension. This is generally dictated by the mechanism of monolayer collapse [26], which in turn depends on the mechanical properties of the interfacial film [60]. Current theories propose that monolayer collapse is similar to classical nucleation and growth, as of a crystal from a melt or a liquid droplet from a vapor [60].

Whenever the surface pressure exceeds π_{ESP}, the monolayer is unstable relative to the formation of a three-dimensional phase. As in crystal nucleation, the transition between the unstable monolayer with $\pi > \pi_{ESP}$ and the collapsed film occurs via the formation of some form of "collapse nuclei," that is, structures intermediate between the monolayer and the collapsed film [14, 20, 21, 26, 27, 58, 61–64]. The formation of collapse nuclei generally requires overcoming an activation energy barrier that is a function of monolayer chemical and physical properties, compression speed, surface pressure, temperature, and so on [60–62, 64]. As the surface pressure increases, the barrier to nucleating collapse decreases, and the formation of critical nuclei of the three-dimensional (3D) phase becomes more probable [58]. Once formed, the critical nuclei can rapidly grow into the bulk 3D phase and the monolayer collapses and the monolayer eventually returns to the equilibrium surface pressure. Smith and Berg [64] showed that this picture could describe the slow decrease in area of a monolayer with time when held at a constant pressure above π_{ESP}.

A common feature (Fig. 13.2) among collapsed lung surfactant monolayers are multilayer patches formed from buckled or folded areas of the monolayer [14, 20, 21, 26, 27, 58, 60, 62–65]. Hence some type of elemental fold or defect in the monolayer is believed to be the critical nuclei for collapse [60, 62]. The LS monolayer can be idealized as a deformable thin plate under compression [60]. As the pressure is increased past some limiting value, the plate can (1) fracture and break, (2) buckle at constant total area, or (3) lose material (and hence interfacial area) depending on the properties of the monolayer [14–17, 26, 27, 59, 62, 63, 66, 67]. While the limiting surface pressure at collapse, π_c, determines the minimum surface tension for a given monolayer, the collapse mechanism also partially determines the reversibility—that is, what fraction of the monolayer remains at the interface and how well the monolayer respreads to cover the interface as surface pressure is decreased.

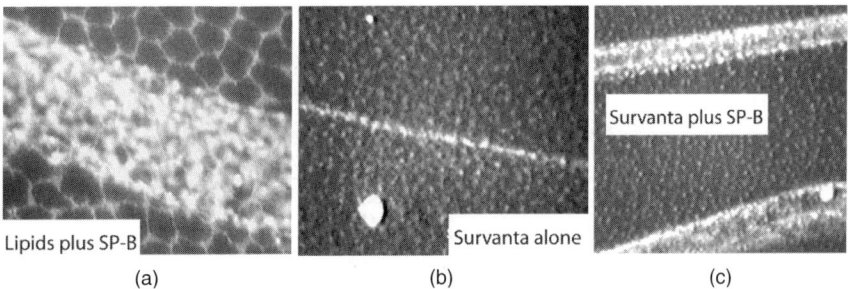

Figure 13.2 Fluorescence images of collapse structures in (a) DPPC/POPG/PA plus SP-B lipid films, in (b) Survanta, and in (c) Survanta with added SP-B [20]. The large variation in the collapse morphology likely depends on the composition, phase behavior, and mechanical properties of the films. (Part (a) from Ref. 20.)

In general, disordered, liquid-expanded (LE) phase [68–70] monolayers, such as those formed by phospholipids above their critical temperatures, collapse at relatively low π_c ($\pi_{ESP} \sim \pi_c$) via the ejection of material to the subphase [14–17, 26, 27, 59, 62, 63, 66, 67]. More ordered and rigid liquid-condensed (LC) or solid (S) phase [68–70] monolayers collapse at higher π_c, often by fracturing, followed by loss of portions of the monolayer to the subphase or formation of multilayered crystalline aggregates at the air side of the interface [27]. Collapse via fracture or solubilization is often irreversible; the collapsed phase material generally does not reincorporate into the monolayer as π is decreased [27] below π_{ESP}. Two-phase monolayers with a continuous LE phase separating islands of LC or S phase at π_c can undergo large amplitude buckling into the subphase (Fig. 13.2) [26]. The folded regions coexist with the flat monolayer; further compression changes the fraction of material in the folds relative to the flat monolayer at constant π_c, indicative of a first-order phase transition. The morphology of a continuous network of fluid LE phase separating LC phase domains alters the elasticity of the monolayer, which allows the monolayer to bend rather than break. The folds remain attached to the monolayer at the interface and are reincorporated into the monolayer upon expansion with little loss of material [14–17, 26, 27, 59, 62, 63, 67]. However, there is a significant difference in the surface pressure at which the folds form (π_c) and the surface pressure at which the folds reincorporate back into the monolayer. This is likely the cause of the hysteresis observed in compression/expansion isotherms of lung surfactants.

A recent theoretical description of the structure and activation energy of the critical nuclei for folding monolayer collapse [60] is a modification of the Euler-type buckling of a flat sheet to include the hydrophobic–hydrophilic asymmetry of the monolayer. At equilibrium, the fold pictured in Fig. 13.3 would be stable if the net surface tension force pulling on the monolayer (2γ) was less than the self-adhesion energy, W, of the hydrophobic part of the monolayer for itself. The situation is similar to pulling on both ends of a piece of adhesive tape folded

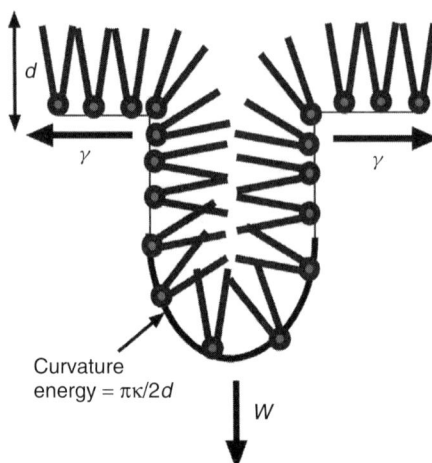

Figure 13.3 Schematic diagram of collapse structure in monolayer of surface tension γ. The hydrocarbon portion of the monolayer prefers to stick to itself, rather than be exposed to air, with energy of W/area. To form the fold requires creating highly curved, high energy regions at the tip and shoulders of the fold. These high energy regions provide an activation energy that leads to higher surface pressures on compression before the fold can nucleate. This results in a higher collapse pressure than the equilibrium spreading pressure that depends on the surface viscosity of the film. (Adapted from Ref. 60.)

onto itself in the center; as long as the pulling force is less than the adhesion, the tape remains stuck to itself. Similarly, a self-adhering monolayer maintains folds at equilibrium to shield the hydrocarbon chains from air if the surface tension of the interface drops below $W/2$. From the drawing (Fig. 13.3), $W/2$ is the free-energy cost of creating a unit area of hydrocarbon surface exposed to air, or approximately the surface tension of a hydrocarbon fluid in contact with air. For most hydrocarbon fluids, the range of surface tensions is rather narrow, $\gamma_H = 25 \pm 4$ mN/m (mJ/m^2 in terms of energy/area) [60]. This corresponds to equilibrium between the folded, essentially bilayer collapse structure and the monolayer occurring at a surface pressure of $\pi_{ESP} \sim \gamma_W - \gamma_H = 72 - (25 \pm 4) = 47 \pm 4$ mN/m, which is close to the equilibrium spreading pressure of replacement lung surfactants and many saturated and unsaturated phospholipid monolayers [10].

However, to nucleate such a fold can require a substantial activation energy, which implies that the collapse pressure of a monolayer under compression, π_c, can be much greater than π_{ESP}. Nucleating a fold from a flat monolayer requires creating a highly curved region at the edges of the fold with energy per length of $\pi\kappa/2d$; κ is the bending modulus of the monolayer and d is the monolayer thickness [60]. This opposes the decrease in energy due to the elimination of the hydrocarbon–air interface and determines the critical fold length, L^*; for $L > L^*$, folds grow spontaneously, while for $L < L^*$ a fold is absorbed into the monolayer (in vapor–liquid droplet nucleation, the increase in the surface energy of the droplet is balanced against the latent heat of condensation to determine the critical droplet radius). The critical fold length, L^*, and the nucleation energy barrier, $\Delta E = \frac{\pi \kappa L^*}{4d}$, are proportional to the Young's modulus of the film, $Y = \frac{4KG}{(K+G)}$, in which K is the area-expansion modulus of the monolayer (related to the slope of the isotherm near the collapse pressure) and G is the shear modulus of the monolayer. G is related to the surface shear viscosity: $\eta_s = \frac{G}{\tau_s}$ in which η_s is the surface shear viscosity and τ_s is the stress relaxation time [60]. In terms of the experimental parameters: $L^* = \kappa Y/d(\pi - \pi_{ESP})$. This model suggests that fold formation should occur only for surface pressures above $\pi_{ESP} \sim 47 \pm 4$ mN/m on dynamic compression. How far π_c is above π_{ESP} depends on the surface viscosity and compressibility of the surfactant films. The greater the surface viscosity, the larger the activation energy for nucleating collapse. Hence the surface viscosity and shear modulus of lung surfactant films could provide important information on the nature of monolayer collapse, and hence minimum surface tension. Viscous films (large G) should collapse differently than less viscous films [60]. We see long, narrow cracks (Fig. 13.2) in viscous films of *Survanta*, broader and more three-dimensional folds in more fluid *Curosurf*, and small round defects in low viscosity *Infasurf*. SP-B also appears to influence the formation and shape of folds (Fig. 13.2) [20]. Systematic measurements of the surface viscosity and surface shear modulus of these films is necessary to understand the relationships between chemical composition and ability to reduce surface tension and respread after collapse.

13.1.2 Surface Viscosity and Surfactant Distribution in the Airways

Lung surfactant also is at least partially responsible for maintaining a surface tension gradient from the alveolus to the trachea. For proper lung function, the surface tension in the alveoli must drop to nearly zero on exhalation [2, 71, 72]. However, in the trachea, the surface tension is ~ 30 mN/m [73] and in the connecting airways the surface tension is ~ 15 mN/m [74] and is roughly constant as there is no mechanism for expansion or compression of these surfaces. Hence a surface tension gradient of varying magnitude exists between the trachea, airways, and alveoli during breathing. Such surface tension gradients induce flow of surfactant and the associated epithelial fluids, along with other interfacial materials in the direction of the higher surface tension. This flow is likely necessary to promote removal of particulates and other debris from the deep lung [2] where no cilia are available to actively promote fluid flow.

For surface tension driven flows, the rate of transport is proportional to $\Delta \gamma / R_{\text{eff}}$. $\Delta \gamma$ is the difference in surface tension between the alveoli (where surfactant is generated and expressed) and the airways (where no surfactant is expressed) and R_{eff} is an effective flow resistance that depends on the surface viscosity, η, of the surfactant film [75], the bulk viscosity of the epithelial lining fluid, η_B [76–78], the thickness of the epithelial lining fluid [76–78], the branching and geometry of the airways, surface interactions, particulates, and so on [2]. Instillation of replacement surfactants in premature infants (in which the higher surface tension is in the alveolus) occurs by surface tension driven flows in the smaller airways and alveoli and takes only minutes to occur in practice [76–78]. Reversing the direction of the surface tension gradient in normal lungs would suggest that the flow of surfactant from the lungs must be just as fast in the absence of a sufficiently high flow resistance. Moreover, the total internal area of the terminal bronchioles is about 3000 cm^2, and the total internal area of the tracheobronchial tree from larynx to terminal bronchioles is about 4000 cm^2 [79]. By comparison, alveolar surface area is 80–150 m^2, or about 200–300 times larger. Hence only a fraction of the alveolar surfactant could line the whole airway. For surface tension gradients to be maintained, there must be a strong resistance to surfactant flow at low surface tension. How surfactant regulates its surface viscosity is likely very important to the distribution and maintenance of surfactant throughout the lungs.

Monolayer viscosity is also likely to be an important parameter in the stability and structure of many cell membranes and is especially likely to play a role in the organization of lipids and proteins at the cell surface. For example, cholesterol is hypothesized to promote micro- to nanoscale phase separation in lipid membranes, which can lead to the formation of "membrane rafts." Such rafts are believed to have significantly different physical properties, including surface viscosity, than the surrounding membrane. A local increase or decrease in the surface viscosity would influence the rates of lateral diffusion and rearrangement of membrane proteins. Such rafts are under increasing scrutiny for their roles in protein localization and function. A better understanding of the relationship between lipid and protein composition and surface viscosity might shed light on how such rafts form and how they function in cell membranes [80, 81].

13.2 FROM SURFACE VISCOSITY TO MOLECULAR STRUCTURE IN "SIMPLE" MONOLAYERS: A HISTORY OF SURFACE VISCOSITY

Even single-component Langmuir monolayers of fatty acids exhibit a rich structural polymorphism that is reflected in the surface viscosity [68–70]. The carbon backbone of the fatty acid molecules can adopt several orientations with respect to the interface as well as with neighboring molecules, which leads to a wide range of crystalline, semicrystalline, hexatic (orientationally ordered but positionally disordered), and disordered phases [82], each with its own rich flow behavior. The historical and conceptual development of the relationships between local molecular packing in monolayers and properties such as surface viscosity began with fatty acid and fatty alcohol monolayers. The basic relationships between molecular order and organization found in these simple films can help explain much of the behavior observed in more complex, multicomponent lung surfactant films presented in the following section.

The first systematic attempts to measure surface viscosity began to appear in the 1930s and 1940s. This data [83–90] contributed to the discovery of new two-dimensional phase transitions in both fatty acids and alcohols. One of the leaders in this initial charge was William D. Harkins [84, 85, 87], who with his students and co-workers discovered that many tilted monolayer phases (traditionally known as "liquid-expanded" or LE phases) showed an exponential increase in surface viscosity with surface pressure [86] (or decrease in area per molecule [47, 55, 91, 92]). A second set of observations showed that large, discontinuous increases in the surface viscosity accompanied higher surface pressure phase transitions, leading to the idea of a transition from a "liquid-condensed" to a "solid" phase on increasing surface pressure. This discontinuous increase in surface viscosity was what first led researchers to the idea of disordered to ordered phase transitions in monolayers. Subsequent work showed that not every fatty acid or alcohol obeyed the simple liquid-expanded–liquid condensed–solid rules with increasing surface pressure; transitions to "superliquid" films actually showed a drop in surface viscosity with increasing surface pressure that were difficult to explain given the available experimental tools at the time [93].

The development of grazing incidence X-ray diffraction methods using synchrotron sources over the past twenty years has shown that the surface viscosity of fatty acid monolayers can be correlated with (1) the extent of positional and orientational molecular correlations, and (2) the molecular tilt direction and amplitude, all of which are functions of the surface pressure at a given temperature [68]. These more detailed phase descriptions have superseded the LE–LC–S descriptions of the monolayer structure of fatty acid films. Most fatty acids can be represented by the generic phase diagram shown in Fig. 13.4, provided that the temperature is properly scaled. As first pointed out by Langmuir more than seventy years ago [94], varying the temperature is equivalent to varying the fatty acid chain length, with the simple rule that each additional carbon atom is equivalent to increasing the temperature by about 5 °C. The prediction of the exact position of phase boundaries is given by a linear relationship between pressure, temperature, and number of carbons [95]. Fatty acids with chain lengths from about 14 to 26 or more exhibit roughly the same sequences of phases as a function of surface pressure and temperature [68, 94, 96, 97].

The molecular packing in monolayers [68] of fatty acids is dictated primarily by the packing of the hydrocarbon portion of the molecules and can be understood in analogy to the three-dimensional rules for alkane packing given by Kitaigorodskii [98]. For a given temperature, at low surface pressure, fatty acid molecules pack in a herringbone (or pseudoherringbone) motif with the main axis tilted with respect to the vertical (L_2, L'_2, L''_2). In the herringbone packing, the elliptically shaped carbon backbone of an all-*trans* fatty acid chain alternates orientation with its neighbors by 90 ° to give it the closest possible packing, given the constraints imposed by the headgroup packing [28, 29, 67, 68, 82, 99, 100]. (In Fig. 13.4, the herringbone motif is shown for the CS or crystalline solid phase.) There are four generally accepted low surface pressure tilted phases with distinct symmetries as shown in Fig. 13.4. The low temperature L''_2 phase is a two-dimensional crystal with the molecules tilted toward their nearest neighbor (NN). In all the tilted phases, increasing the surface pressure decreases the tilt,

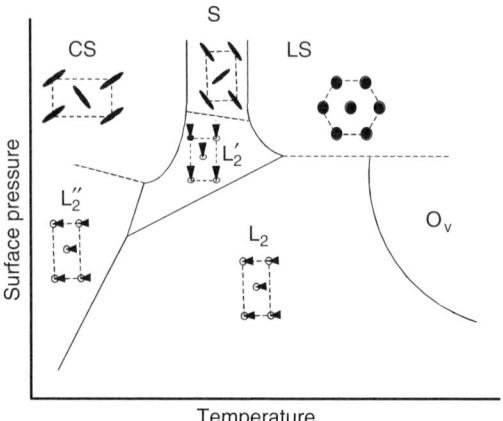

Figure 13.4 Universal phase diagram of fatty acid monolayers and schematic diagrams showing transformations of the unit cell according to X-ray diffraction data [68, 97]. With the exception of the LS and O_v phases, the molecular packing of the alkane chains are in a herringbone motif. The all-*trans* alkane chains of the fatty acids form an elliptical shape, which changes orientation by 90° relative to its nearest neighbors. In the LS and O_v phases, the molecules rotate freely around their long axes and the effective shape of the molecule is a cylinder. The lower surface pressure monolayers are tilted; in both the L_2'' and L_2 phases, the molecules are tilted toward the nearest neighbor as indicated by the dark triangles. In the L_2' phase, the molecules are tilted toward the next nearest neighbor. The high surface pressure CS, S, and LS phases are untilted, with crystalline, semicrystalline, and hexatic order.

up to the point where the molecules are oriented normal to the interface in the high surface pressure CS, S, and LS phases.

As the temperature is increased at low surface pressure, the long-range crystalline ordering of the L_2'' phase gives way to a more short-range semicrystalline positional order in the L_2 phase, which also has NN tilt. Increasing the surface pressure in the L_2 phase also decreases the molecular tilt until the transition either to the L_2' phase, which has a next-nearest neighbor (NNN) tilt, or the LS phase, which is untilted. The positional order in the L_2' phase is also short ranged and semicrystalline, and the tilt decreases with increasing surface pressure up to the transition to the untilted S or LS phase [68, 101].

At the highest surface pressures, the molecules adopt a close-packed arrangement with the chain axis perpendicular to the interface (CS, S, LS). For a given surface pressure, the lattice expands with temperature, allowing more degrees of freedom to the molecules, which eventually leads to the LS and O_v rotator phases. In the LS and O_v phases, the alkane chains rotate freely about their axes, which effectively changes the elliptical shape of the alkane chain to a more cylindrical shape. This eliminates the need for the herringbone packing common at lower temperatures. The LS phase is an untilted, hexagonal packing of the freely

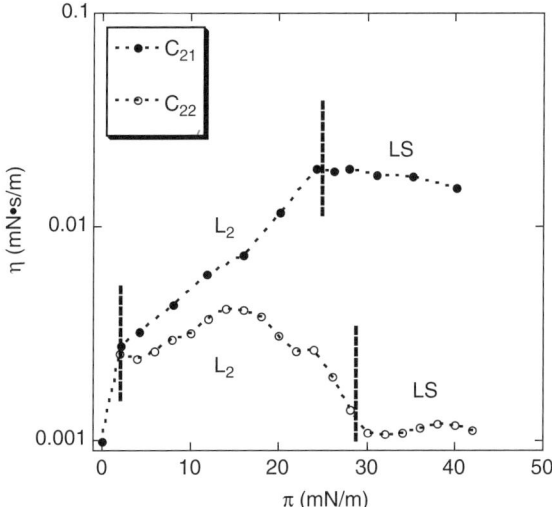

Figure 13.5 Surface shear viscosity η as a function of surface pressure π for monolayers of heneicosanoic acid (C_{21}) at 30 °C and behenic acid (C_{22}) at 25 °C spread on water [119]. The monolayers undergo a disordered to L_2 transition at about 2 mN/m and then L_2 to LS rotator phase transition at 29 and 25 mN/m, respectively (both transitions are marked with dashed lines). For C_{21} in the L_2 phase at 30 °C, the viscosity increases exponentially as the molecular tilt decreases, which is what is found for most fatty acids in these phases [86, 93, 103, 104]. However, for C_{22}, the surface viscosity in the L_2 phase shows a peak, which is correlated to the symmetry of the in-plane crystalline lattice. The molecular lattice undergoes a change from distorted hexagonal along one axis to hexagonal to distorted hexagonal along the perpendicular axis, with the viscosity maximum corresponding to the most highly symmetric hexagonal packing [103, 104] At the transition to the untilted, hexagonal rotator phase LS, the surface viscosity does not change with surface pressure and remains low, also showing that once the tilt is fixed, the surface viscosity does not change significantly [93]. Despite the high symmetry and untilted packing, the LS phase has a low viscosity due to the poor cohesion between molecules induced by the molecular rotation of the alkane portion of the molecule. The surface viscosity in the LS phase of C_{21} is more than an order of magnitude greater than the longer chain length C_{22} even at a lower temperature, although this is also observed in other untilted phases at high surface pressure [86].

rotating fatty acid molecules with short-ranged positional (semicrystalline) order; the O_v phase is tilted with short-ranged positional order.

Figure 13.5 shows some of the complex ways that the surface shear viscosity, η varies as a function of surface pressure, π, for monolayers of heneicosanoic acid (C_{21}) at 30 °C and behenic acid (C_{22}) at 25 °C. Both monolayers undergo the L_2 to LS phase transition, depicted by a dashed line, at ~28 mN/m. However, two regimes can easily be distinguished on the $\eta(\pi)$ curves, which indicate a transition pressure of 25 mN/m for C_{21} and 29 mN/m for C_{22}. This discrepancy may be due to the compression speed, as many of the phase transitions in monolayers are above the equilibrium spreading pressure and hence are metastable. The barriers move continuously for the isotherm measurements, whereas the pressure is held at a constant value for the viscosity measurements. However, surface viscosity may be a more sensitive measure of phase change than breaks or kinks in the isotherm [84, 102].

For C_{21} at zero surface pressure, η is low, ~ 0.001 mN·s/m, typical of a structureless, fluid monolayer. As compression is initiated, η increases fivefold on entry into the L_2 phase, which is accompanied by the formation of semicrystalline domains. On further compression, but still below the L_2–LS transition, η increases exponentially with surface pressure, which is similar to many other fatty acid and alcohol films in the L_2 phase [86, 93]. This increase in the monolayer rigidity correlates with the reduction of the tilt and η is continuous going into the LS phase. After the transition to the LS phase, η remains constant at the last value reached for the L_2 phase, about 0.02 mN·s/m.

However, for C_{22} in the L_2 phase, η gradually increases with surface pressure, but reaches a maximum at 14 mN/m and then decreases until the transition pressure to the LS phase. The $\pi-A$ isotherm does not show any discontinuity around 14 mN/m; hence this maximum in η is not due to a phase transition. A comparison of surface viscosity and X-ray diffraction shows that the surface viscosity increases along with the symmetry of the chain lattice and to a lesser extent with the headgroup symmetry [103, 104]. In the L_2 phase at low pressure, the molecular lattice is distorted hexagonal [68, 104]. Upon compression, the distortion of the lattice changes from being elongated along one direction to being elongated along the perpendicular crystallographic direction [101]. The maximum of η corresponds to the crossover at which the lattice is most nearly hexagonal. For C_{22} at 25 °C, the crossover point was 14 mN/m by X-ray diffraction [101], which is in good agreement with the measured maximum in surface viscosity. This shows that the surface viscosity measured with a macroscopic probe is extremely sensitive to subtle changes in molecular packing.

As the pressure is increased further into the LS phase, η remains constant at about 0.001 mN·s/m. As for C_{21} in the LS phase, η is independent of surface pressure when the molecules are untilted. Surprisingly, the surface viscosity in the LS phase of C_{21} is more than an order of magnitude greater than in the longer chain length C_{22}. In general, for the untilted phases, the surface viscosity decreases with increasing chain length for both fatty acids and alcohols [86, 93, 102]. For C_{22} and C_{21}, the odd–even effect also appears to play a role in the surface viscosity in the LS phase; the orientation of the last C_{n-1}—C_n bond determines the tilt of the whole molecule for alkanes: a chain with an even C_{2n} number of carbons has a larger tilt than the consecutive odd C_{2n+1} chain in rotator phases [105–107]; this greater tilt may explain the significantly lower surface viscosity in the LS phase of C_{22}. These relatively low viscosities at high surface pressures gave the LS phase its historical name of "superliquid."

From Fig. 13.4, the L_2 phase can evolve into the L_2' or LS phase with increasing surface pressure. In Fig. 13.6, η is shown in the vicinity of the L_2–L_2'

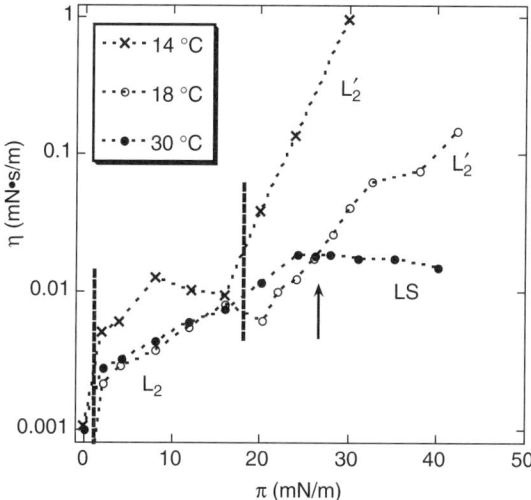

Figure 13.6 Surface shear viscosity η as a function of surface pressure π for a monolayer of heneicosanoic acid (C_{21}) spread on water at three different temperatures: 14, 18, and 30 °C [119]. The disordered to L_2 and L_2 to L_2' phase transitions at 14 °C and 18 °C are marked by dashed lines and the L_2 to LS transition at 30 °C is marked by an arrow. At low pressure at 18 °C and 30 °C in the L_2 phase, the two curves roughly superimpose and show an exponential increase in surface viscosity with surface pressure. At 14 °C ($\pi = 8$ mN/m) and 18 °C ($\pi = 16$ mN/m), the surface viscosity goes through a local maximum similar to C_{22} at 25 °C that correlates with the changes in molecular symmetry (see Fig. 13.4). At 14 and 18 °C the monolayer undergoes a transition from a L_2 to a L_2' phase marked by a change of slope in the surface viscosity. At 30 °C, the high pressure phase is LS or superliquid phase, which shows a constant, low viscosity.

transition for C_{21} at 14 °C and 18 °C and the L_2–LS transition at 30 °C. For the three curves, at zero surface pressure, η is low as expected for fluid monolayers [103, 104]. In the L_2 phase at 18 and 30 °C, the measured surface viscosities are in the same range and increase exponentially with surface pressure, consistent with measurements of Boyd and Harkins in the late 1930s [86]. However, at 14 °C, after the initial linear increase with surface pressure, a maximum in η occurs at ~8 mN/m, the same surface pressure that showed hexagonal symmetry by X-ray diffraction [101], and the same surface pressure at which Ghaskadvi et al. [104] observed a maximum in G'' in oscillatory measurements. The surface viscosity at 18 °C also has a local maximum at about 16 mN/m just before the transition to the L_2' phase; this is also consistent with the Ghaskadvi et al. [104] results that showed the surface pressure of the peak in G'' increased with increasing temperature. X-ray measurements also showed the surface pressure at which hexagonal symmetry occurred increased with increasing temperature [101]. The absence of a surface viscosity maximum at 30 °C is consistent with the transition to hexagonal symmetry occurring at the LS phase transition rather than within the L_2 phase.

In early work, L_2 phases that showed this hexagonal symmetry transition and have a maxima in the surface viscosity were called L_{2h} phases and were distinguished from those that did not, which were called L_{2d} phases [101]. An important distinction in these earlier phase diagrams was that the L_{2h} phase always transformed into the L'_2 phase on increasing surface pressure, while the L_{2d} phase transformed into the LS phase [101]. Ghaskadvi et al. [103, 104] showed a maximum in the surface viscosity for C_{21} at temperatures at which there was an L_2–L'_2 transition. However, from Fig. 13.5, for C_{22}, the appearance of a maximum in surface viscosity that correlates with the hexagonal symmetry transition occurs at a temperature at which there is a transformation from the L_2 (nominally the L_{2d} phase described in Ref. 101) to the LS phase. Hence this characteristic surface viscosity maximum does not require that the higher surface pressure phase be L'_2. The distinction between L_{2h} and L_{2d} phases has not been retained in the more recent phase diagrams on which Fig. 13.4 is based [68]. This is consistent with Figs. 13.5 and 13.6; there is no distinction between L_{2h} and L_{2d} phases based on these surface viscosity measurements.

The surface viscosity at the L_2–L'_2 phase transition for heneicosanoic acid (26 mN/m at 14 °C, 27 mN/m at 18 °C) shows a discontinuous change of slope of η with π. In the L'_2 phase, η also increases exponentially with surface pressure; however, the slope is larger than for the L_2 phase (Fig. 13.6) [104]. The tilt in the L_2 phase is directed toward the nearest neighbor (NN) as opposed to the next-nearest neighbor (NNN) tilt in the L'_2 phase (Fig. 13.4) [101]. The rate of decrease of the chain tilt with increasing surface pressure is greater in the L'_2 phase than in the L_2 phase [101], consistent with the greater increase in surface viscosity with surface pressure. Decreasing the NNN tilt has a greater effect on the surface viscosity than does decreasing an NN tilt.

Within the L'_2 phase, Fig. 13.6 shows that the surface viscosity is strongly enhanced by a decrease in temperature [86, 108]. At a surface pressure of 30 mN/m, η measured at 14 °C is more than an order of magnitude greater than at 18 °C. The same trend occurs in the L_2 phase; however, the difference was at most a factor of 3.

The surface viscosity of the L''_2 and CS phases of palmitic acid (C_{16}) and stearic acid (C_{18}) are shown in Fig. 13.7. In the L''_2 phase for palmitic acid, η appears to be roughly constant; however, this is at the lower limit of accuracy for our MNSV ($Bo \ll 1$). Earlier measurements by Boyd and Harkins [86] showed that for C_{16} the surface viscosity increased exponentially with surface pressure as in the L_2 and L'_2 phases; however, the magnitude of the surface viscosity of C_{16} in the L''_2 phase (0.0003 mN·s/m at 6 mN/m surface pressure [86]) is about 10 times less than for the L_2 phases shown in Figs. 13.5 and 13.6 at the same surface pressure. For stearic acid, the surface pressure does increase roughly exponentially with surface pressure up to the phase transition. The surface viscosity changes discontinuously by more than an order of magnitude at the CS transition surface pressure for both C_{16} and C_{18}, which is consistent with a

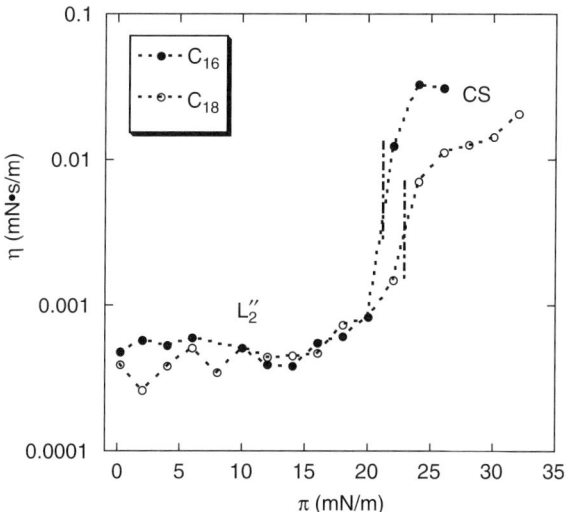

Figure 13.7 Surface shear viscosity measurement across the L_2''–CS phase transition for palmitic (C_{16}) and stearic acid (C_{18}) monolayers on water at 25 °C. A dashed line depicts the phase boundary. η is at the resolution limit for the MNSV in the L_2'' phase of palmitic acid but increases roughly exponentially with surface pressure for stearic acid up to the transition surface pressure. The surface pressure in the L_2'' phase is roughly an order of magnitude less than in the L_2 and L_2' phases (see Figs. 13.5 and 13.6). At the transition surface pressure, the surface viscosity makes an order of magnitude, discontinuous jump, then remains roughly constant in the untitled CS phase, indicative of a first-order phase transition [119].

first-order phase transition (Fig. 13.4) [86]. As with the LS phase, the surface viscosity is relatively independent of surface pressure in the untilted CS phase.

The general structural features of the phase diagram of fatty acids monolayers shown in Fig. 13.4 are widely accepted; however, the order of the tilted to untilted phase transitions is still an unresolved issue. Isotherms, X-ray diffraction data, and Brewster angle imaging do not always agree in determining if the tilted–nontilted transitions L_2''–CS, L_2–LS, and O_v–LS are weakly first order (η discontinuous) or second order (η continuous). The MNSV can measure surface viscosity at constant temperature and surface pressure, which allows us to get very close to the transition to show the nature of phase change. The surface viscosity data presented in Fig. 13.7 shows that, at the L_2''–CS transition, η shows an order of magnitude jump for both palmitic and stearic acids over a range of ~1–2 mN/m in surface pressure. This is in contradiction with the second-order transition suggested in the generic phase diagram (dotted lines in Fig. 13.4). Ghaskadvi et al. [103] show that G'' is discontinuous at the L_2'–S transition, which is also described as second order on the generic phase diagram. For the L_2–LS transitions shown in Figs. 13.5 and 13.6, the surface viscosity is continuous, although the slope of the surface pressure–surface viscosity curve changes discontinuously, suggesting a second-order transition, in agreement with the generic phase diagram (dotted lines in Fig. 13.4). Finally, the surface viscosity is continuous, although the slope changes discontinuously at the L_2-L_2' phase boundary, which is known to be a first-order transition as the lattice symmetry changes from NN to NNN (solid line in Fig. 13.4). Hence it appears that surface viscosity measurements at phase boundaries may contribute additional, perhaps contradictory information on the order of the phase transitions in fatty acid monolayers. It is obvious that the surface viscosity is extremely sensitive to local molecular ordering in monolayers.

13.3 SURFACE VISCOSITY OF PHOSPHOLIPIDS AND LIPID MIXTURES

As discussed previously, lung surfactant must have a large surface viscosity at high surface pressure to maximize the collapse pressure and minimize surfactant flow out of the lungs during exhalation. The surface viscosity must also drop significantly at lower surface pressures to enhance adsorption and respreading of the surfactant on inhalation. How does lung surfactant make the necessary adjustments in the two-dimensional surface viscosity as a function of surface tension? Many of the common mechanisms for enhancing solution viscosity in three dimensions, for example, polymer entanglements, have no analogy in two dimensions.

It could be that the necessary changes in surface viscosity in lung surfactant films arise from changes in monolayer phase behavior that occur with increasing surface pressure as shown in Figs. 13.4–13.7 for fatty acids. Many phospholipids, including DPPC, have less-ordered phases at low surface pressures, with short-range molecular correlations [15–17, 47]. As the surface pressure increases to levels expected in the lungs (>40 mN/m), solid phases, with long-range molecular correlations, form. However, the changes in surface viscosity with surface pressure due to phase transitions occur at too low a surface pressure, are too temperature sensitive, or are too gradual to provide all of the essential requirements of a functional lung surfactant. However, mixing many phospholipids and fatty acids together with lung surfactant specific proteins appears to be key to maintaining the proper relationship between lung surfactant viscosity and lung surfactant function.

The surface viscosities of some typical lipids common to lung surfactants such as DPPC, palmitoyloleoyl phosphatidylglycerol (POPG), palmitic acid (PA), and their mixtures measured at 25 °C [17, 31, 48] are presented in Fig. 13.8. POPG, an unsaturated lipid, is always in the liquid-expanded (LE) phase at physiological conditions and is representative of the unsaturated lipids in lung surfactant [22]. The surface viscosity of POPG remains less than 0.001 mN·s/m over the entire range of surface pressure. In contrast, the surface viscosity of PA increases discontinuously at $\pi = 2$ mN/m, indicative of a first-order phase transition between the L_2'' phase and the CS phase as in Fig. 13.7.

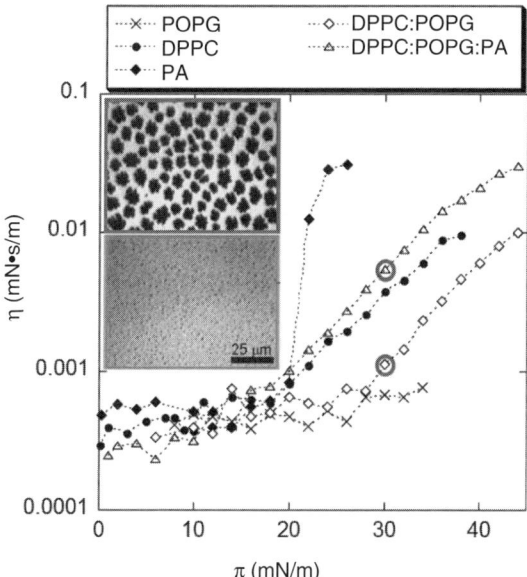

Figure 13.8 Surface viscosity, η_s, as a function of surface pressure, π, for Langmuir monolayers of POPG, PA, DPPC, DPPC:POPG 4:1, and DPPC:POPG:PA 68:22:8 (wt:wt) at 25 °C [17, 119]. The POPG monolayer remains in the fluid state at all π; thus η_s remains low. The PA monolayer undergoes a transition from a tilted to a nontilted phase at 22 mN/m, which triggers a dramatic increase of η_s. DPPC surface viscosity in the LE phase (below $\sim$ 15 mN/m) is similar to POPG. In the LC phase, η_s increases exponentially. Both DPPC:POPG (bottom inset image) and DPPC:POPG:PA (top inset image) films show coexistence between solid phase islands (black in inset images) and a continuous fluid phase (bright in inset images) for surface pressures greater than 10 mN/m. Both images shown are taken at 30 mN/m; PA increases the fraction of solid phase at a given π, which also increases η_s.

DPPC, the dominant lipid in lung surfactant, undergoes a liquid-expanded (LE) to liquid-condensed (LC) first-order transition at 12 mN/m at 25 °C; the coexistence pressure increases with increasing temperature [28–30]. In the LE phase and throughout coexistence, the viscosity is $\sim$0.0005 mN·s/m, similar to the always fluid POPG. In the LC phase, the viscosity increases exponentially with surface pressure [47, 55], increasing by more than an order of magnitude over that in the LE phase by 35 mN/m.

However, what distinguishes lung surfactant films from single-component monolayers is that, over a wide range of surface pressures, ordered and disordered phases coexist [17, 31, 69]. This solid–fluid type of coexistence is only possible for multicomponent mixtures by the Gibbs phase rule and has profound consequences for the surface viscosity. As shown in the lower of the two fluorescence images inset to Fig. 13.8, adding unsaturated POPG to DPPC at 25 °C

(4:1 DPPC:POPG, wt:wt) increases the LE–LC transition pressure, as evidenced by the first appearance of the small, dark solid phase domains in the image at 30 mN/m. This lack of solid phase lowers the surface viscosity by a factor of 5–10 relative to pure DPPC, depending on the surface pressure, although the rate of increase of surface viscosity with surface pressure is similar; the mixture appears to be offset from DPPC by about 10 mN/m in surface pressure.

Adding PA to the DPPC:POPG (68:22:8 DPPC:POPG:PA, wt:wt) mixture at 25 °C decreases the surface pressure of the onset of the LE–LC transition to below that of pure DPPC. At 30 mN/m, the top fluorescence image in the inset shows the coexistence between the dark, solid phase domains and a continuous bright fluid phase. This coexistence between a continuous fluid phase with solid phase islands persists to high surface pressure (Fig. 13.9) and increases the surface viscosity. The added PA leads to a significant increase in both the fraction of solid phase at a given surface pressure and the surface viscosity (Fig. 13.9). PA causes a decrease in the molecular tilt and an increase in the molecular cohesion of DPPC crystals [16, 28–31]. As in the fatty acid phases, a decreased molecular tilt and an increase in the ordering of the film leads to an increase in the surface viscosity; however, the rate of change of surface viscosity with surface pressure is similar to DPPC and the DPPC/POPG mixture.

Depending on the PA fraction, there can be two-phase coexistence over a wide range of surface pressures depending on the composition, surface pressure, and temperature of the film, as shown in the fluorescence images of Fig. 13.9 [16, 28–31]. For a 3:1 wt:wt mixture of DPPC:POPG, solid phase domains are small even at 45 mN/m with no PA present (Fig. 13.9 top row); however, adding 20% by weight PA to the 3:1 DPPC:POPG mixture causes solid phase domains to appear at surface pressures below 5 mN/m (Fig. 13.9 bottom row).

Coexistence between a low surface viscosity LE continuous phase and a solid, high surface viscosity, discontinuous phase as shown in Fig. 13.9 can provide a rather unfamiliar mechanism of altering surface viscosity with changes in surface pressure or composition. Solid phase islands in a continuous fluid monolayer (Fig. 13.9) are the two-dimensional analog to three-dimensional suspensions of hard particles suspended in a solvent. The surface viscosity at phase coexistence (like the bulk viscosity of the suspension) can undergo a dramatic increase from liquid-like to solid-like due to the "jamming" of solid phase at a critical surface area fraction (or volume fraction in the suspension) [31].

The phenomenon is similar to a traffic jam on a freeway or a logjam on a river. Solid phase domains bump into each other constantly and can jam together during the flow, forming transient clusters. If the cluster is large enough to span the channel available for flow, the cluster acts like a dam and retards the flow. As the solid particle fraction increases, the likelihood of a sample-spanning cluster increases. The lifetime of such a cluster is related to the time it takes for the solid particles to diffuse into and out of the cluster, which is the self-diffusivity of the solid particles. This means that in 3D suspensions, the effective viscosity, η_e, scales as the number of particles in contact, divided by

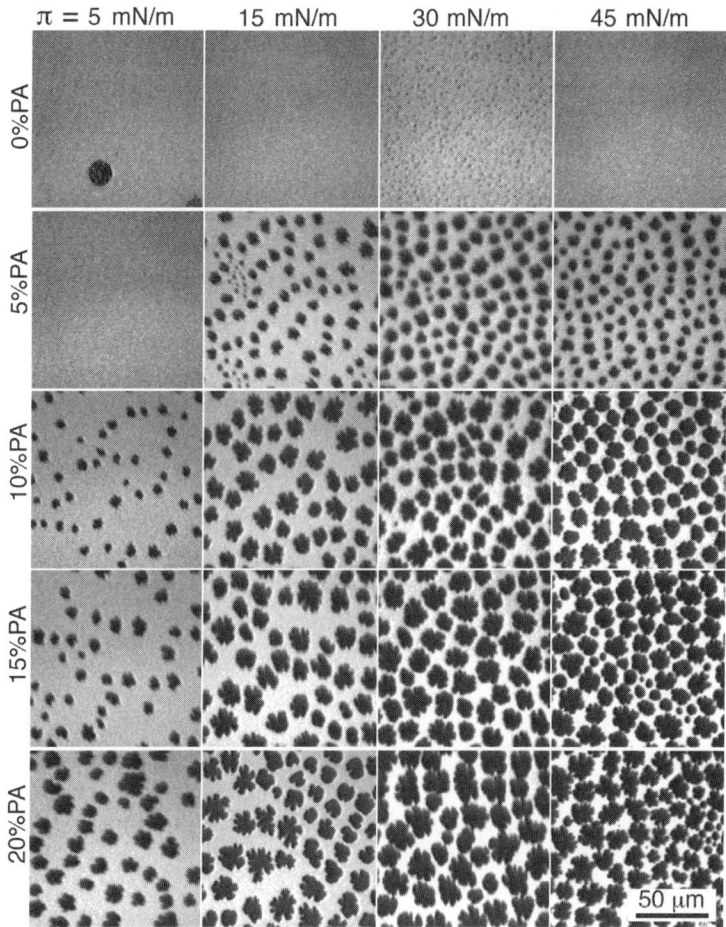

Figure 13.9 Fluorescence optical micrographs of monolayers of mixtures of DPPC/POPG (3:1, wt:wt) with varied amounts of palmitic acid (PA) on a 150-mM NaCl, 5-mM $CaCl_2$, and 0.2-mM $NaHCO_3$ (pH = 6.9) buffer at 25 °C [31]. The PA concentration is labeled on the left of each row, and the top line shows the corresponding surface pressure. Phase coexistence between the bright areas in the images [69] (corresponding to fluid, LE phases) and dark areas (corresponding to better ordered, LC or S phases) occurs over a wide range of surface pressure and composition. Increasing the PA concentration is roughly equivalent to increasing the surface pressure in terms of the fraction of solid phase at a given temperature [28, 29]. Hexadecanol has the same effect on DPPC/POPG mixtures [16, 28].

the short-time self-diffusivity, both of which depend on the solid particle volume fraction, ϕ [109].

For noninteracting hard spheres, as the volume fraction for random close packing, ϕ_c, is approached ($\phi_c = 0.64$ for monodisperse spheres but can be significantly greater for polydisperse spheres or other shapes [110]), the number of particles in contact diverges as $(1 - \phi/\phi_c)^{-1}$. The short-time self-diffusivity vanishes as $(1 - \phi/\phi_c)$ because the particles near contact are held in place by strong hydrodynamic lubrication forces [109]. Thus the reduced viscosity scales as $\eta_e/\eta_{eo} = (1 - \phi/\phi_c)^{-2}$, where η_e is the steady shear viscosity of the dispersion and η_{eo} is the viscosity of the suspending fluid. However, if there are strong repulsive forces between the solid particles on a length scale large compared to the hydrodynamic lubrication forces, the short-time self-diffusivity is independent of the volume fraction. The viscosity still diverges as random close packing is approached, but as $\eta_e/\eta_{eo} = (1 - \phi/\phi_c)^{-1}$ for strongly repulsive particle suspensions [109].

By analogy, monolayers at coexistence can be thought of as two-dimensional "suspensions" of high surface viscosity, solid phase islands in a sea of continuous, low surface viscosity fluid phase. The scaling arguments in 3D suspensions are equally valid in 2D suspensions, with the area fraction of the solid phase replacing the volume fraction of hard spheres [31]. Hence the surface viscosity for noninteracting solid phase islands of area fraction A in a continuous fluid phase should scale as $\eta/\eta_o = (1 - A/A_c)^{-2}$, in which η is the surface viscosity of the monolayer and η_o is the surface viscosity of the continuous, fluid phase. For strongly repulsive solid phase domains, the surface viscosity should scale as $\eta/\eta_o = (1 - A/A_c)^{-1}$. From fluorescence images such as Fig. 13.9, it is simple to calculate the area fraction of solid phase at a given composition and surface pressure. The fluorescent dye partitions preferentially into the liquid-expanded phase, causing it to appear bright and the solid phase dark. Increasing the fraction of PA increases the fraction of solid phase at a given π and temperature [16, 28, 29]. The solid phase islands are primarily composed of an untilted, DPPC/PA cocrystal [28, 30]. POPG is above its critical temperature, so the liquid phase is primarily POPG with any DPPC that does not crystallize [22].

In all of the systems at coexistence, η_o was taken to be the surface viscosity of pure POPG, which from Fig. 13.8 does not change over a wide range of surface pressures. The critical area fraction, A_c, was taken to be the solid phase area fraction at which the surface viscosity was so high that it could no longer be measured with the MNSV (see Section 13.4). A_c depends on the solid domain shapes and polydispersity. Polydisperse circular domains have the highest A_c; more monodisperse, dendritic shaped domains have the lowest A_c. In these experiments, A_c ranged from about 0.5 to about 0.77.

Figure 13.10 shows the surface viscosity of a variety of mixtures of 3:1 wt:wt DPPC:POPG with various weight fraction of palmitic acid or hexadecanol as a function of the solid phase area fraction measured from fluorescence images. PA is added to Survanta, while hexadecanol is one of the three components in Exosurf. All of the measured surface viscosity data can be represented as a single function of A/A_c. The surface viscosity effectively diverges at the critical solid area fraction, in direct analogy with three-dimensional suspensions at the close-packed volume fraction. Over a wide range of surface pressure, temperature, composition, and domain shape, the surface viscosity of monolayers at a

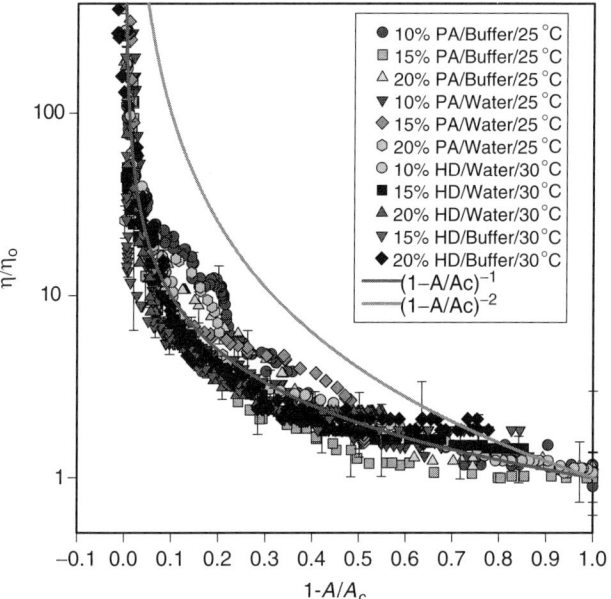

Figure 13.10 Semilog plot of (η/η_o) as a function of $(1 - A/A_c)$ for monolayers of mixtures of DPPC/POPG (3:1, wt:wt) with varied amounts of palmitic acid (PA) or hexadecanol (HD) on a 150-mM NaCl, 5-mM $CaCl_2$, and 0.2-mM $NaHCO_3$ (pH = 6.9) buffer or pure water subphases [31]. Independent of the composition or surface pressure, the surface viscosity normalized to the liquid-expanded phase viscosity at zero surface pressure had a universal scaling with the area fraction, A, of the solid phase islands in the coexistence region. A_c is the critical area fraction defined experimentally to be the solid phase area fraction at which the MNSV could no longer drive the needle across the trough. The solid curves are the theoretical expressions corresponding to an exponent of -1 (dark gray) and -2 (light gray). The excellent fit to the exponent of -1 (no adjustable parameters) shows that there are long-range interactions between the solid domains. The monolayer surface viscosity can change by orders of magnitude as A approaches A_c, even though there is no phase transition or other significant change in the monolayer properties.

solid–liquid coexistence depends only on the area fraction of solid phase present, in direct analogy to three-dimensional dispersions. The effective surface viscosity at coexistence can change by a factor of almost 500 with no significant changes in the monolayer, other than subtle changes in the solid phase area fraction.

The two curves in Fig. 13.10 show the theoretical limiting curves for no repulsion, $\eta/\eta_o = (1 - A/A_c)^{-2}$, and for strong repulsion, $\eta/\eta_o = (1 - A/A_c)^{-1}$, between the solid domains. The exponent of -1 is a surprisingly good fit to the data considering there are no adjustable parameters, which suggests that strong repulsive forces exist between the solid phase domains on a length scale large compared to the hydrodynamic lubrication forces [109]. Both solid and fluid phase domains likely contain charged lipids, PA in the solid domains and POPG in the fluid domains. Even in uncharged monolayers, different densities and orientations of polar molecules in solid and liquid phases lead to a long-range dipole–dipole repulsion that can order solid domains over distances large compared to the domain separation [69]. This universal behavior of lung surfactant lipid mixture viscosity helps explain the trends in Fig. 13.8 as well. DPPC and its mixtures all show the same rate of increase of surface viscosity with surface pressure; the curves are simply offset depending on the surface pressure at which coexistence occurs. From the fluorescence images, the offset in coexistence pressure determines the area fraction of solid phase at a given surface pressure, which in turn, from Fig. 13.10, determines the surface viscosity. Manipulating the surface viscosity of a mixture depends primarily on manipulating the solid phase area fraction by changing lipid composition, temperature, or surface pressure. The solid phase area fraction can also be changed by the action of the surfactant specific proteins SP-B and SP-C.

13.4 EFFECT OF LUNG SURFACTANT PROTEINS ON SURFACE VISCOSITY

For pure lipid films, the surface viscosity within a single-phase domain depends on the local molecular packing—the surface viscosity is high if the ordering is crystalline, less so if the ordering is semicrystalline, and low if there is no molecular order. The surface viscosity is also dependent on the extent and direction of molecular tilt; the surface viscosity is much lower in tilted films than in films with the molecules normal to the interface. In multicomponent monolayers with a solid–fluid phase coexistence, the surface viscosity depends on the fractional distribution of the phases within the film. These relationships between monolayer composition, morphology, and surface viscosity give some interesting clues as to the proper role of surface viscosity in lung surfactant films and also suggest novel roles for both SP-B and SP-C.

On the full compression of the monolayer that results from exhalation, the solid phase fraction of the monolayer could increase past the critical solid phase area fraction, A_c, leading to a rigid, high surface viscosity monolayer that resists further compression or deformation. In addition to maximizing the monolayer collapse pressure (Fig. 13.2), rigid monolayers could ensure that the alveoli retain some fraction of residual air volume and do not collapse on exhalation. The high viscosity might also be important during the recruitment of alveoli in the treatment of premature infants with liquid-filled lungs with replacement surfactants. A mechanism by which the solid phase fraction is adjusted to near A_c should lead to the divergence of the surface viscosity occurring at the appropriate part of the breathing cycle without the large scale changes associated with changing the phase behavior or composition of the entire monolayer.

This suggests a possible function for the lung surfactant specific protein SP-C [111]. As can be seen in the AFM images in Fig. 13.11, SP-C selectively removes

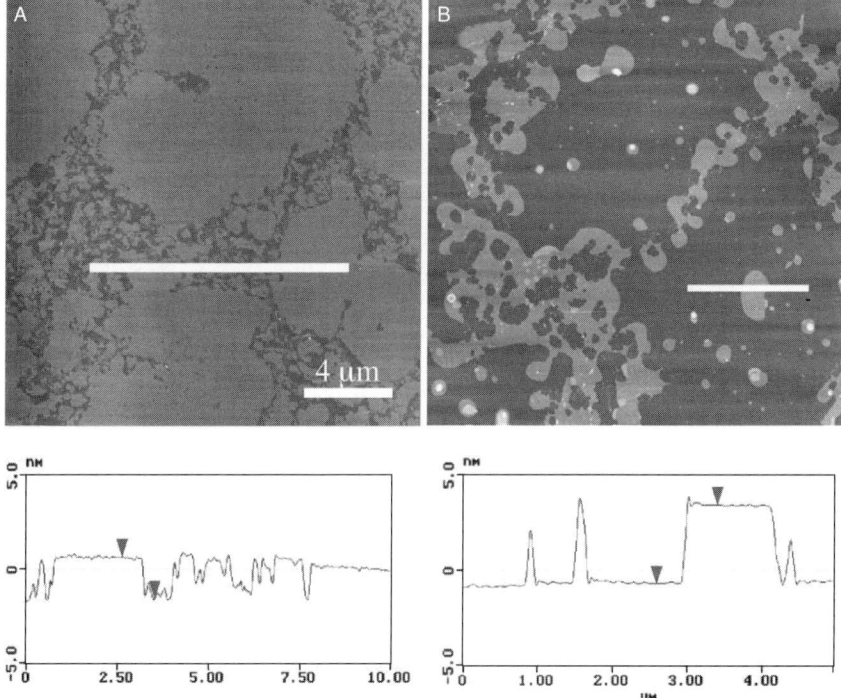

Figure 13.11 AFM images of solvent spread DPPC:POPG:PA + SP-C_{ff} films transferred to mica substrates from a buffered saline subphase [20]. (A) Below a surface pressure of ~40 mN/m, the fluid phase regions are 1 nm lower (see height trace) than the condensed regions. (B) Above ~40 mN/m, the fluid regions have thickened to be about 5 nm higher than the condensed domains. This is consistent with the formation of three-layer thick areas in the fluid phase. The plateau in the isotherms corresponds to the removal of fluid phase lipids from the interface to form these multilayer patches. SP-C and its mutants act to regulate the solid/fluid phase ratio, causing the surface viscosity to diverge [15, 17].

fluid phase lipids from monolayers of DPPC:POPG:PA (68:22:8 by weight) at high surface pressures to form multilayer patches associated with the monolayer [17, 20, 21]. The AFM images show that below a surface pressure of about 40 mN/m (the surface pressure at which unsaturated lipids are usually squeezed-out of the monolayer [22]) the fluid phase regions of the monolayer are about 1 nm lower (darker in image, see height trace below the image) than the solid phase regions. In AFM images, fluid phase regions of monolayers usually appear to be "thinner" than solid phase regions as they are more easily compressed by the AFM tip. The entire image in Fig. 13.11A consists of a single layer at coexistence; the fluid phase is continuous as in the fluorescence images in Figs 13.8 and 13.9.

However, for monolayers transferred just above 40 mN/m, the fluid regions have thickened to be about 5 nm higher than the solid phase domains

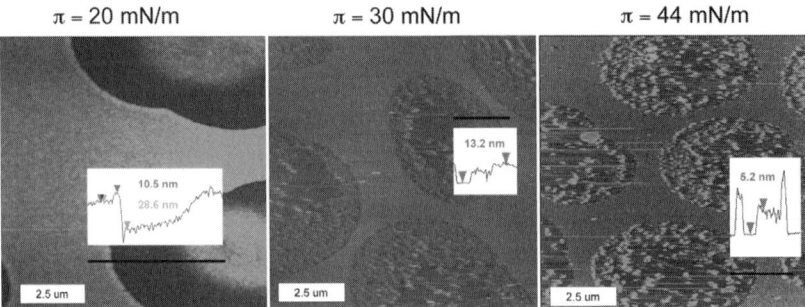

Figure 13.12 AFM images of Survanta monolayers deposited from saline solution onto a Langmuir trough with a buffered subphase transferred onto mica substrates [15]. The brightness in the image increases with the relative height. A cross section of the typical features on each image is shown as the height tracing (inset). The surface pressure at which the film was transferred is given above each image. At 20 mN/m, the film consists of a continuous fluid phase separating circular domains of solid phase. The fluid phase is about 30 nm thicker than the solid phase domains, suggesting the adsorption of multilayers of lipid onto the fluid phase. At 30 mN/m, the morphology is similar, but the height difference has decreased and the solid domains appear less homogeneous and much rougher. At 44 mN/m, the entire film is much rougher and the height difference between the solid and fluid domains is less. Above 44 mN/m, the solid phase domains coalesce and the fluid phase is effectively removed. The sequence of images is similar to that in Fig. 13.11; Survanta contains from 1 to 2 wt% SP-C.

(Fig. 13.11B). This is consistent with the formation of three-layer thick areas in the fluid phase [15, 20–22, 30]. There is a plateau in the isotherms of these films that is consistent with a loss of interfacial area corresponding to the removal of fluid phase lipids from the interface to form these multilayer patches [17, 20, 21]. The clinical surfactants Curosurf (not shown) and Survanta (Fig. 13.12), which both contain native SP-C, also show multilayer patches in the fluid phases [15] separating discrete islands of solid phase. The fluid phase patches also lose connectivity in Figs. 13.11B and 13.12; the solid phase domains appear to have jammed together and have formed larger clusters, as suggested by the model for surface viscosity discussed in Section 13.3.

Figure 13.13 [16] shows that adding 5 wt% of a synthetic version of lung surfactant specific protein SP-C (SP-C_{ff}) (Fig. 13.14) [20, 21] increases the monolayer surface viscosity dramatically from about 0.001 to 0.43 mN·s/m at a surface pressure of 45 mN/m. This is the same surface pressure at which the isotherm shows a plateau and the AFM images show multilayer formation (Fig. 13.11). This monolayer also shows a strong hysteresis—on compression the viscosity stays low up to about 40 mN/m, then increases by about 3 orders of magnitude as the surface pressure is increased slightly. On expansion, the viscosity decreases gradually with decreasing surface pressure. Further increasing the SP-C concentration to 15% does not change the surface viscosity significantly [16]. The lipid

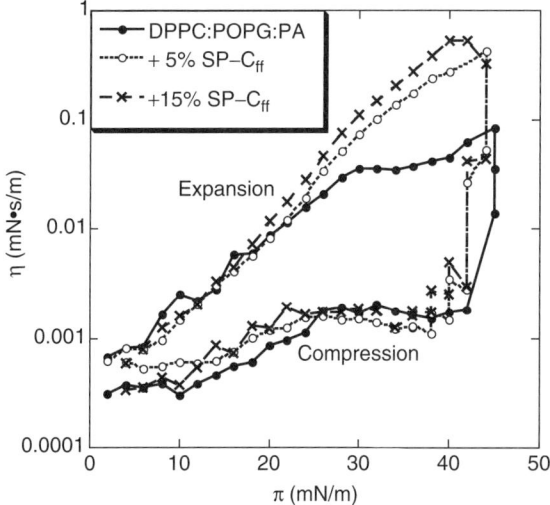

Figure 13.13 Surface shear viscosity, η, as a function of surface pressure, π, for DPPC:POPG:PA (68:22:8 wt%) monolayers with varying weight fractions of SP-C_{ff} [15]. The surface viscosity on compression remains low up to about 40 mN/m, at which it increases by more than 3 orders of magnitude. This is an example of the jamming transition that occurs in monolayers at fluid–solid coexistence (see Fig. 13.19). SP-C_{ff} acts to regulate the fraction of fluid phase by stabilizing bilayers connected to the monolayer at the interface as can be seen in the AFM images in Figs. 13.11 and 13.12 and in the computer simulation in Fig. 13.14. On expansion, the surface viscosity decreases more slowly with decreasing surface pressure, showing a marked hysteresis. Thus 5% SP-C_{ff} increases the surface viscosity by a factor of five over the lipid alone. Increasing the SP-C_{ff} to 15% provides only a marginal increase in the surface viscosity.

film without SP-C has a surface viscosity of ~0.05 mN·s/m, about a factor of 10 below the films containing SP-C. Adding SP-C to pure POPG monolayers or other LE phase monolayers does not increase the surface viscosity; SP-C only seems to have an effect on those films that show a LE–LC coexistence [31].

Figure 13.14 shows how SP-C might be affecting the surface viscosity. Native SP-C is a 4.2-kDa, dipalmitoylated, 35-residue peptide, of which 23 residues are hydrophobic; both SP-C [15, 17, 20, 21] and SP-B [14, 15, 17, 20, 26, 27] are located primarily in the fluid phase domains of the monolayer. While the conformation in the monolayer is not obvious, SP-C likely adopts a transmembrane orientation similar to that of integral membrane proteins in a bilayer. SP-C has an α-helical conformation between residues 9 and 34 that consists primarily of hydrophobic valine residues [111]. In native SP-C, the N-terminal segment includes two palmitoylcysteinyls that can further link to an adjacent monolayer or bilayer. The length of the α-helix is about 3.7 nm, which should fit comfortably along the acyl chains of lipids in a bilayer environment [111, 112].

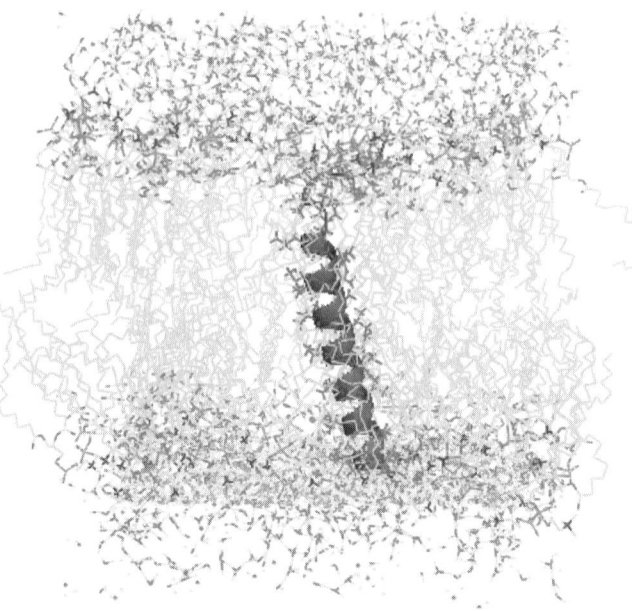

Figure 13.14 Computer graphics models of SP-C$_{ff}$ protein in a palmitoyloleoyl phosphatidylcholine (POPC) (usaturated) lipid bilayer solvated with water (bottom). In SP-C$_{ff}$ (based on work by Byk-Gulden Pharmaceuticals, Konstance, Germany), the palmitoylated cysteines are replaced by phenylalanine residues [17, 20]. The mutant protein performs nearly identically to the native SP-C in animal models and is significantly easier to synthesize [113, 114]. The SP-C$_{ff}$ structure is based on the molecular coordinates of the solution NMR structure of recombinant SP-C [116]. The hydrophobic polyvaline helical sequences are shown as a dark gray ribbon. SP-C has the ability to stitch together adjacent monolayers, thereby increasing the surface viscosity (see Fig. 13.13.)

Molecular dynamics simulations show that the synthetic SP-C$_{ff}$ takes on a trans-bilayer orientation in an unsaturated, fluid POPC lipid bilayer solvated with water (Fig. 13.14). SP-C$_{ff}$ is a synthetic mimic of SP-C based on work by Byk-Gulden Pharmaceutical (Konstance, Germany) in which the palmitoylated cysteines are replaced by phenylalanine residues to simplify the synthesis. This mutant protein performs nearly identically to the native SP-C in animal models [113, 114]. This mutant is similar to dog SP-C, which only has a single palmitoylated cysteine, with the second replaced by phenylalanine [115]. The SP-C$_{ff}$ shown in Fig. 13.14 is based on the molecular coordinates of the solution structure of recombinant SP-C [116]. The hydrophobic (polyvaline) helical sequences are shown as a dark gray ribbon and are located among the light gray alkane segments of the bilayer. As suggested by von Nahmen et al. [21], SP-C has the ability to stitch together opposing sides of a bilayer. Native SP-C, palmitoylated human recombinant SP-C, and synthetic SP-C$_{ff}$ have similar effects in promoting the monolayer to multilayer transition of the fluid fraction of the bilayer at high surface pressures (Figs. 13.11 and 13.12) [15, 17, 20–22].

Increasing the thickness of the fluid phase should increase the surface viscosity, as is observed in Fig. 13.13. In the simplest concept of the surface viscosity, $\eta_b t = \eta_o$, in which η_b is the viscosity of the three-dimensional phase, t is the thickness of the interfacial film, and η_o is the surface viscosity of the fluid phase domains [31]. Hence three-layer films should have roughly three times the surface viscosity of a monolayer, five-layer films should have five times the surface viscosity, and so on. In our model of the viscosity of coexisting films discussed previously, the surface viscosity of the continuous fluid phase is η_o. Increasing η_o by increasing the thickness of the fluid domains should increase the effective surface viscosity of the coexisting films as $\eta = \eta_o (1 - A/A_c)^{-1}$. Hence increasing the surface viscosity of the continuous phase leaves the general shape of Fig. 13.10 intact, and there is the same dramatic increase in the effective surface viscosity as we approach A_c as observed in Fig. 13.13. As the surface pressure reaches ~40 mN/m on compression, the transformation from monolayer to multilayers occurs, and A approaches A_c. The maximum surface viscosity for films with SP-C is increased, relative to the lipid films, by the increase in η_o due to the corresponding increase in the thickness of the continuous, LE phase. In addition to the increase in film thickness, the attachments between the layers shown in Fig. 13.14 likely also increase η_o and therefore the surface viscosity. It is likely that the palmitoylation of native SP-C should increase the viscosity even further as three monolayers could be stitched together by SP-C.

Figure 13.15 shows the effect of various concentrations of dSP-B$_{1-25}$, a peptide mimic of native SP-B, on the surface viscosity of the same DPPC:POPG:PA monolayers. At a given surface pressure [17], Langmuir isotherms show that the average area per molecule increases with the peptide concentration, showing the insertion of the protein within the monolayer. On all the isotherms, there is a smooth kink at low pressure and a shallow plateau at higher pressures, both of which are displaced toward lower pressures as the peptide concentration increases. The plateau goes from 40 mN/m for the pure lipid film to about 32 mN/m with 10% dSP-B$_{1-25}$.

For surface pressures below the plateau, the surface viscosity is low, as is the case for films containing SP-C (Fig. 13.13). At the plateau surface pressure, the surface viscosity increases; however, there is only about a small increase in the maximum η value of the film with 5% dSP-B$_{1-25}$ compared to the lipid-only mixture. This is compared to the factor of 5 with the same amount of SP-C added (Fig. 13.13). However, the surface pressure at which the jump in surface viscosity takes place decreases with dSP-B$_{1-25}$ fraction; the transition appears to be associated with the coexistence plateau in the isotherm discussed previously. SP-B, like SP-C, is primarily located in the fluid phase domains and also removes fluid phase lipids from the monolayer on compression, but not nearly as efficiently as SP-C [14]. As can be seen in the AFM images in Fig. 13.16, dSP-B$_{1-25}$ leads to the formation of isolated protrusions from the fluid phase at high surface

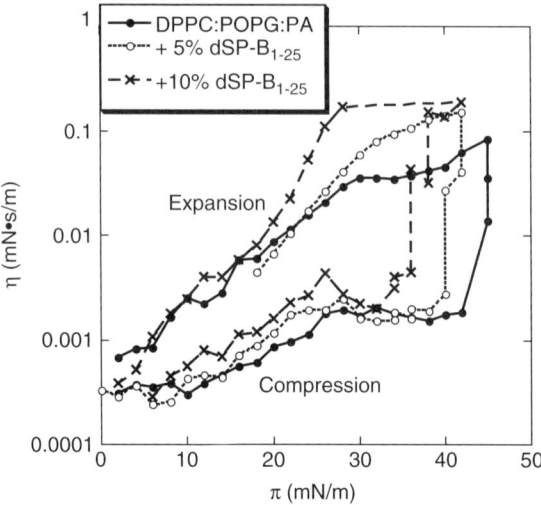

Figure 13.15 Surface viscosity as a function of surface pressure for the dSP-B_{1-25} containing films [15]. Increasing the dSP-B_{1-25} content to 5% only increases the surface viscosity by about 50%, compared to more than a factor of 5 increase caused by an equal amount of SP-C. Increasing the dSP-B_{1-25} fraction increases the surface viscosity, suggesting that more fluid phase lipids are removed from the monolayer, consistent with the images in Fig. 13.16.

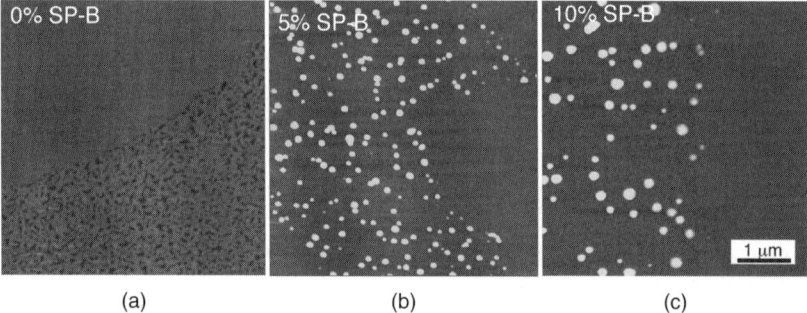

Figure 13.16 AFM images of lipid films transferred to mica substrates at $\pi = 40$ mN/m [14]. (a) shown is 0 wt% dSP-B_{1-25}. With no dSP-B_{1-25}, the solid phase is smooth and slightly higher (about 0.5 nm) than the fluid phase, which is mottled light and dark gray. (b) shown is 5 wt% dSP-B_{1-25}. As is the case for SP-C, SP-B is located in the fluid phase domains; the solid phase domains remain smooth and are free of protrusions. dSP-B_{1-25} leads to the formation of isolated small protrusions from the fluid phase domains, with heights ranging from about 5 to 10 nm. SP-B removes fluid phase lipids from the monolayer on compression, but not nearly as efficiently as SP-C. (c) shown is 10 wt% dSP-B_{1-25}. Increasing the dSP-B_{1-25} to 10 wt% increases the lateral dimensions of the protrusions, but not their density. dSP-B_{1-25} removes material from the fluid phase, as does SP-C, but the location and distribution of the material is quite different, leading to a smaller effect on the surface viscosity.

pressure [14]. With no dSP-B$_{1-25}$ in the films (Fig. 13.16a), the solid phases are smooth and slightly higher (about 0.5 nm) than the fluid phase, which is mottled light and dark gray. With 5 wt% dSP-B$_{1-25}$ in the film (Fig. 13.16b), small white protrusions emerge from the fluid phase domains, with height ranging from about 5 to 10 nm. There are no protrusions in the solid phase, suggesting that SP-B is excluded from the solid phase (as is the dye used in the fluorescence images). Increasing the dSP-B$_{1-25}$ to 10 wt% increases the lateral dimensions of the protrusions, but not their density (Fig. 13.16c). Figure 13.17 shows the computer simulation of the dSP-B$_{1-25}$ protein in an unsaturated, fluid POPC lipid bilayer solvated with water. In contrast to SP-C, SP-B does not extend very far into the bilayer or monolayer and primarily perturbs the polar portion of the monolayer. Hence its effect on the surface viscosity should not be as great, as is observed. SP-B also does not have the ability to link monolayers together as does the trans-membrane SP-C (Fig. 13.14).

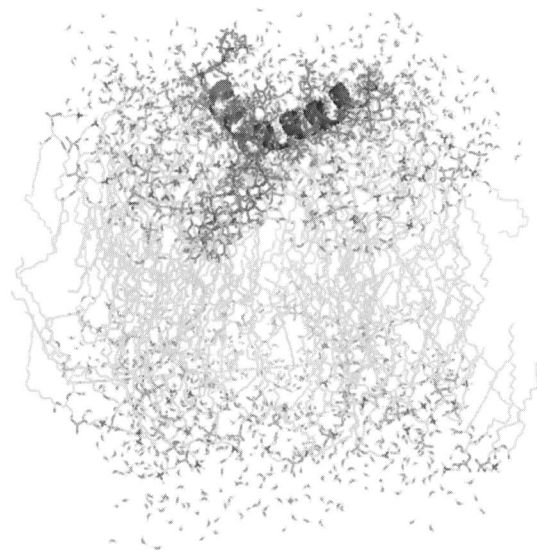

Figure 13.17 Molecular illustration of SP-B$_{1-25}$ dimer in a POPC bilayer. In contrast to SP-C, SP-B does not extend very far into the bilayer or monolayer and primarily perturbs the polar portion of the monolayer. Hence its effect on the surface viscosity should not be as great, as is observed. The molecular graphic image for dSP-B$_{1-25}$ was templated from the structure of monomeric SP-B$_{1-25}$ using the coordinate set of PDB 1DFW and mutating cysteine 11 to alanine. These mutated conformers were then optimized as homodimers using ZDOCK [120]. Both structures were then placed in a simulated solvent box and minimized by steepest decent followed by 1-ns equilibrium molecular dynamics at 300 K using the GROMACS version 3.3 force field to simulate a final approximate molecular conformation for the protein in the POPC bilayer. The final graphic illustrations were prepared using Pymol [121]. The helical sequences are shown as dark gray ribbons.

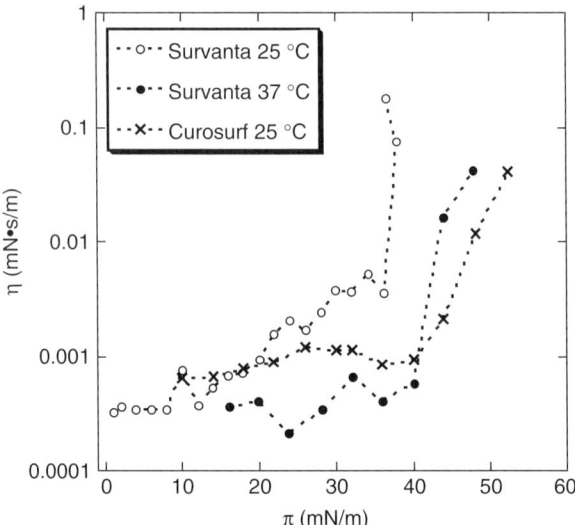

Figure 13.18 Surface viscosity as a function of surface pressure on film compression for Survanta and Curosurf at 25 °C. Both clinical surfactants show a dramatic increase in surface viscosity at about 35–45 mN/m, consistent with the SP-C induced change in the solid phase fraction seen in the lipid mixtures. Increasing the temperature of the Survanta film to 37 °C simply shifts the surface viscosity transition to a higher surface pressure, consistent with the decreased solid phase fraction at higher temperatures.

13.4.1 Surface Viscosity of Clinical Surfactants

Figure 13.18 shows that the clinically used replacement surfactants, Survanta and Curosurf, have very similar surface viscosity as a function of surface pressure as does the lipid/SP-C mixture (Fig. 13.13). Both clinical surfactants show a dramatic increase in surface viscosity at about 35–45 mN/m, consistent with the SP-C induced change in the solid phase fraction seen in the lipid mixtures [15, 17]. As expected, Survanta has a higher viscosity due to the increased DPPC and PA content, which increases the solid phase fraction at a given surface pressure (see Fig. 13.9). Increasing the temperature of the Survanta films just moves the transition to a higher surface pressure, consistent with the temperature dependence of the solid phase area fraction [15, 17]. The Curosurf transition at 37 °C is at too high a surface pressure for our instrument to measure. AFM images show that both Survanta and Curosurf form multilayer patches in the fluid phase domains at high surface pressures (see Fig. 13.12).

Figure 13.19 shows the differences in the surface viscosity between Survanta, Curosurf, Infasurf, and native pig lung surfactant at 25 °C. In comparison to Curosurf and Survanta, Infasurf and native pig surfactant from lavage do not show an increase in surface viscosity. For the surface viscosity to change appreciably, the monolayer must have a critical area fraction, A_c (Figs. 13.8 and 13.9) of

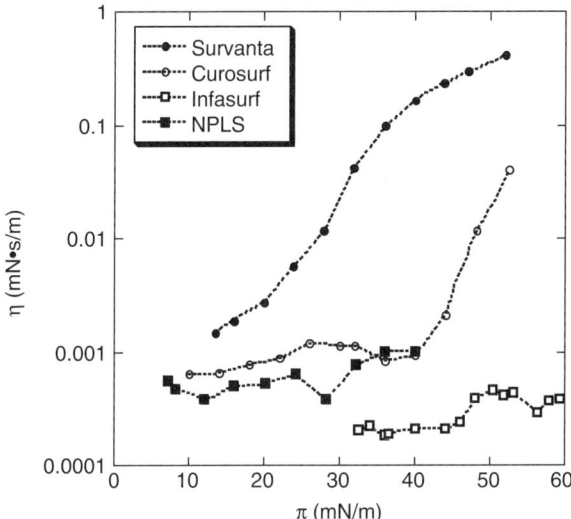

Figure 13.19 Surface viscosity as a function of surface pressure for the clinical surfactants Survanta, Curosurf, Infasurf, and native pig lung surfactant at 25 °C. In comparison to Curosurf and Survanta, Infasurf and native pig surfactant from lavage do not show an increase in surface viscosity. This may be due to their lower saturated lipid content that does not allow for the formation of a critical area fraction of solid phase at the surface pressures tested.

solid phase. For Survanta, Curosurf, and the model lipids plus SP-C, *this critical area* is reached at the plateau surface pressures between 35 and 45 mN/m. This is probably due to the higher DPPC content of Survanta and Curosurf relative to Infasurf [25]. In animals [24, 32], the fraction of DPPC in surfactant changes with age, size, and respiration rate. The surface viscosity increases with solid phase fraction, which in turn depends on the saturated lipid fraction; it may be that this is how nature regulates the surface viscosity to match the needs of the particular animal's lungs or respiration rates. If a viscous film is required at high surface pressure, then the fraction of DPPC (and other saturated lipids such as PA or hexadecanol) must be greater than that present in Infasurf or in the pig lung lavage surfactant.

However, Infasurf and the pig lung surfactant contain more than 5% cholesterol by weight, while there is no cholesterol in Survanta or Curosurf. The effect of cholesterol on surface viscosity is an open question that might shed important light on this controversial component of replacement lung surfactants.

13.4.2 Adjusting the Surface Viscosity of Replacement Surfactants

In general, the surface viscosity of a monolayer increases with an increasing fraction of saturated lipid and ordered phases. Crystalline, untilted packings have the highest surface viscosity; tilted disordered packings have the lowest surface viscosity. In lung surfactants, the fraction of solid phase in the monolayer appears to be regulated by the formation of multilayer patches induced by SP-C and, to a lesser extent, by SP-B. The surface viscosity of Curosurf and Survanta is negligibly low at surface pressures below about 35–40 mN/m, but increases abruptly by two to three orders of magnitude at higher surface pressures. The coupling between monolayers and the multilayer patches induced by SP-C increase the viscosity by an order of magnitude relative to lipid alone.

To see if a low surface viscosity surfactant such as Infasurf could be transformed into one with much higher surface viscosity, hexadecanol was added to the clinical surfactant Infasurf [16]. Infasurf has the smallest fraction of saturated lipids of any of the clinical surfactants and also has the lowest surface viscosity (Fig. 13.19) [16]. Hexadecanol and palmitic acid cocrystallize with DPPC (Fig. 13.9), increasing the fraction of solid phase in the monolayer at a given temperature and surface pressure, as well as decreasing the molecular tilt within the solid phase domains [28, 29]. Figure 13.20 shows that the surface viscosity of Infasurf is relatively constant and low up to the surface pressure at which the molecular tilt (as measured by grazing incidence X-ray diffraction) goes to zero (Fig. 13.21) [16]. The tilt transition in these complex, multicomponent native mixtures also correlates extremely well with the molecular organization of the

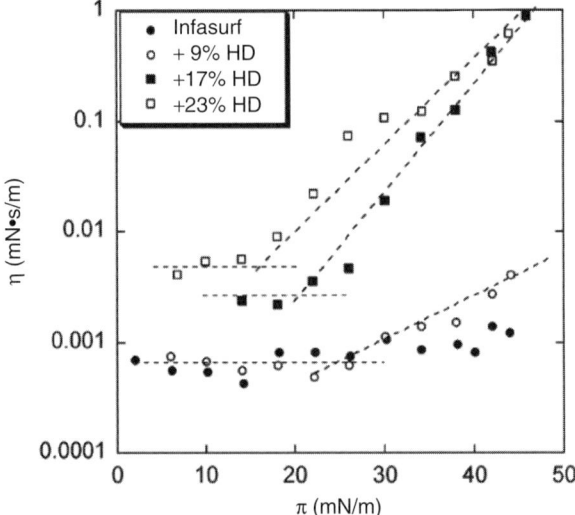

Figure 13.20 Surface shear viscosity as a function of surface pressure, π, for Infasurf films with added HD spread on water at 25 °C [16]. The transition from constant surface viscosity to exponentially increasing (linear on the semilog plot) surface viscosity correlates with the tilt of the molecular lattice going to zero in Fig. 13.21. The overall viscosity increases with the HD concentration in conjunction with the increased crystallinity of the film. Even in complex, multicomponent films, the surface viscosity is dominated by the solid phase fraction, the ordering of the solid domains, and their molecular tilt.

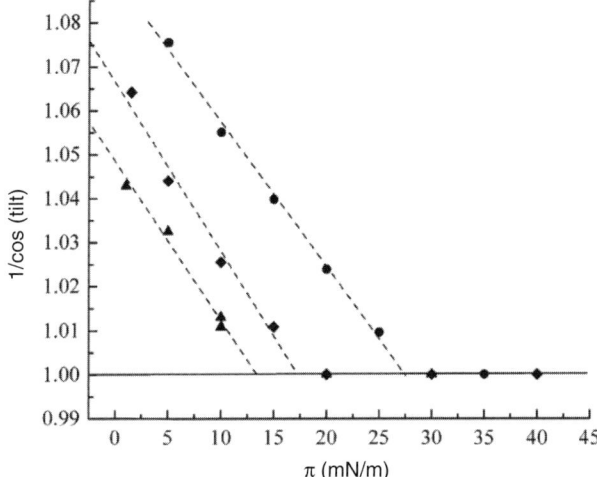

Figure 13.21 Sychrotron X-ray diffraction results for the tilt of the molecular lattice in Infasurf mixed with various amounts of hexadecanol (HD) [16]. The HD complexes with the DPPC in Infasurf create ordered, untilted solid domains. $1/\cos(t)$, in which t is the tilt angle of the molecular lattice, is plotted as a function of the surface pressure, π, of 9% HD (●), 17% HD (♦) and 23% HD (▲) added to Infasurf films at 20 °C. The linear extrapolation toward zero tilt angle $[1/\cos(t) = 1]$ yields the surface pressure π_t of the transition. The tilt transition correlates closely with the onset of the exponential increase in the surface viscosity (see Fig. 13.20.)

film, just as in the simple fatty acid films. This also clearly shows the effects of the added hexadecanol in Exosurf and the added fatty acids and DPPC in Survanta—these additives alter the surface viscosity properties of the films by orders of magnitude [16]. The question remains, however: What is the optimal surface viscosity for a functional lung surfactant film?

13.5 PHYSIOLOGICAL IMPLICATIONS FOR SURFACE VISCOSITY

Relating these dramatic changes in viscosity to their physiological implications is not obvious, as the optimal surface viscosity of a lung surfactant film has not been established. The *in vivo* function of SP-C is also open to question. A further complication is that the surface viscosity has only been measured over a part of the physiologically relevant surface pressure range of 35–70 mN/m due to limitations of the magnetic needle surface viscometer [48]. However, from our previous work and theoretical analyses of the origin of the high surface viscosity [15–17, 47, 48], it is likely that the surface viscosity increases further with surface pressure over the physiological range up to monolayer collapse [15–17, 47, 48]. It is well established that the monolayer structural features that correlate with increased surface viscosity, such as the molecular ordering and cohesion of the solid phase and the area fraction of this solid phase, increase with increasing surface pressure. Hence the resistance to surface tension driven flow should remain high over the entire physiological range of surface pressures. The increased elasticity and viscosity of the monolayers with increasing surface pressure should also help the films resist collapse to higher surface pressures [60].

If the behavior shown by Survanta and Curosurf is optimal, the regulation of surface viscosity is likely to be an important role for SP-C. The physiological effects of surface viscosity might be examined via SP-C deficient animal models. One strain of SP-C deficient mice is viable at birth and grows normally without apparent pulmonary abnormalities [117, 118], but develops pulmonary problems after 1 year of age [117, 118]. The absence of SP-C decreases the hysteresivity, which describes the mechanical coupling between energy dissipative forces and tissue-elastic properties, consistent with a decrease in the surface viscosity. Lower than normal levels of SP-B in SP-C deficient mice led to lung dysfunction [118]. This suggests that the function of SP-C can be carried out to some extent by SP-B in otherwise normal lungs, consistent with our observations that SP-B increases the surface viscosity of monolayers, but not as much as SP-C. In a different strain of SP-C deficient mice, however, there were more severe abnormalities in airway resistance, tissue damping, and hysteresivity, suggesting significant changes in the mechanics of the lung surfactant system. It was suggested that differences in shear forces over time contributed to the disruption of lung structure and function that were observed in the SP-C deficient mice [118]. The control over surfactant viscosity demonstrated by SP-C appears to be important to the mechanics of normal lung function; a lack of SP-C apparently leads to lung damage over time [118].

The ability of SP-C to regulate surface viscosity may be illustrative of a more generic biological control over the mechanical properties of two-dimensional systems including cell membranes [80, 81]. Recent two-photon and fluorescence micrographs show that phase coexistence occurs in lipid bilayers as well as monolayers [80, 81]. This has led to speculation that lipid phase separation is responsible for "raft" formation with its implications for protein localization and function in cell membranes. The shear viscosity of fluid phases in two dimensions is quite low; by comparison the viscosity of condensed phases is very large. However, as shown here, it is not necessary to undergo a complete phase transition in order to switch from low to high viscosity due to the jamming of the solid phase domains. The mechanical properties of bilayer membranes might be controlled by subtle adjustments in the area ratios of a more viscous and a less viscous phase. Large changes in the membrane properties could thus be induced by relatively small changes in the membrane composition or local environment. Proteins similar to the lung surfactant specific SP-B or SP-C, which have the ability to condense or remove unsaturated lipids, may act to tune the solid/liquid phase ratio. The ability of cell membranes to control local membrane phase separation and hence the membrane mechanical properties may be necessary for a variety of cell recognition and binding events [80, 81].

REFERENCES

1. J. A. Clements and M. E. Avery (1998). Lung surfactant and neonatal respiratory distress syndrome. *Am. J. Respir. Crit. Care Med.* **157**: 59–66.
2. R. H. Notter (2000). *Lung Surfactant: Basic Science and Clinical Applications*. Marcel Dekker, New York.
3. J. Goerke (1998). Pulmonary surfactant: functions and molecular composition. *Biochim. Biophys. Acta* **1408**: 79–89.
4. B. A. Holm, G. Enhorning, and R. H. Notter (1988). A biophysical mechanism by which plasma proteins inhibit lung surfactant activity. *Chem. Phys. Lipids* **1988**: 49–55.
5. B. A. Holm, L. Keicher, M. Liu, J. Sokolowski, and G. Enhorning (1991). Inhibition of pulmonary surfactant function by phospholipases. *J. Appl. Physiol.* **71**: 317–321.
6. B. A. Holm, R. H. Notter, and J. N. Finkelstein (1985). Surface property changes from interactions of albumin with natural lung surfactant and extracted lung lipids. *Chem. Phys. Lipids* **38**: 287–298.
7. B. A. Holm, A. R. Venkitaraman, G. Enhorning, and R. H. Notter (1990). Biophysical inhibition of synthetic lung surfactants. *Chem. Phys. Lipids* **52**: 243–250.
8. B. A. Holm, Z. Wang, and R. H. Notter (1999). Multiple mechanisms of lung surfactant inhibition. *Pediatrics Res.* **46**: 85–93.
9. H. W. Taeusch (2000). Treatment of acute (adult) respiratory distress syndrome: the holy grail of surfactant therapy. *Biol. Neonate* **77(Suppl. 1)**: 2–8.
10. H. W. Taeusch, J. Bernadino de la Serna, J. Perez-Gil, C. Alonso, and J. A. Zasadzinski (2005). Inactivaton of pulmonary surfactant due to serum-inhibited adsorption and reversal by hydrophilic polymers: experimental. *Biophys. J.* **89**: 1769–1779.
11. H. E. Warriner, J. Ding, A. J. Waring, and J. A. Zasadzinski (2002). A concentration-dependent mechanism by which serum albumin inactivates replacement lung surfactants. *Biophys. J.* **82**: 835–842.
12. J. A. Zasadzinski, T. F. Alig, C. Alonso, J. Bernadino de la Serna, J. Perez-Gil, and H. W. Taeusch (2005). Inhibition of pulmonary surfactant adsorption by serum and the mechanisms of reversal by hydrophilic polymers: theory. *Biophys. J.* **89**: 1621–1629.
13. Surface tension, γ, is the force per unit area exerted on the interface between two coexisting phases (e.g., air and water). In this work, the surface pressure, π, is defined to be the difference between the γ of the buffered saline subphase at 25 °C (about 72 dyn/cm) and the measured γ for a given surface area. The collapse pressure of a monolayer is defined to be the highest π (lowest γ) attained on continuous compression of the interface. At this point the film has "collapsed." The monolayer typically loses material into the subphase in various forms. This collapse should not be confused with alveolar or lung collapse.
14. J. Ding, I. Doudevski, H. Warriner, A. J. Waring, and J. A. Zasadzinski (2003). Nanostructural changes in lung surfactant monolayers induced by interactions between POPG and surfactant protein B. *Langmuir* **19**: 1539–1550.

15. C. Alonso, T. Alig, H. E. Warriner, J. Yoon, F. Bringezu, and J. A. Zasadzinski (2004). More than a monolayer: relating structure and function to composition. *Biophys. J.* **87:** 4188–4202.

16. C. Alonso, F. Bringezu, G. Brezesinski, A. Waring, and J. A. Zasadzinski (2005). Modifying calf lung surfactant by hexadecanol. *Langmuir* **21:** 1028–1035.

17. C. Alonso, A. J. Waring, and J. Zasadzinski (2005). Keeping lung surfactant where it belongs: protein regulation of surfactant viscosity. *Biophys. J.* **89:** 266–273.

18. S. Schürch, F. H. Y. Green, and H. Bachofen (1998). Formation and structure of surface films: captive bubble surfactometry. *Biochim. Biophys. Acta* **1408:** 180–202.

19. S. Schürch, R. Qanbar, H. Bachofen, and F. Possmayer (1995). The surface-associated surfactant reservoir in the alveolar lining. *Biol. Neonate* **67:** 61–76.

20. J. Q. Ding, D. Y. Takamoto, A. von Nahmen, M. M. Lipp, K. Y. C. Lee, A. J. Waring, and J. A. Zasadzinski (2001). Effects of lung surfactant proteins, SP-B and SP-C, and palmitic acid on monolayer stability. *Biophys. J.* **80:** 2262–2272.

21. A. von Nahmen, M. Schenk, M. Sieber, and M. Amrein (1997). The structure of a model pulmonary surfactant as revealed by scanning force microscopy. *Biophys. J.* **72:** 463–469.

22. D. Y. Takamoto, M. M. Lipp, A. Von Nahmen, K. Y. C. Lee, A. J. Waring, and J. A. Zasadzinski (2001). Interaction of lung surfactant proteins with anionic phospholipids. *Biophys. J.* **81:** 153–169.

23. W. Bernhard, A. Gebert, G. Vieten, A. Rau, J. M. Hohfeld, A. D. Postle, and J. Freihorst (2001). Pulmonary surfactant in birds: coping with surface tension in a tubular lung. *Am. J. Physiol. Regul. Integrative Comp. Physiol.* **281:** R327–R337.

24. W. Bernhard, S. Hoffmann, H. Dombrowsky, G. A. Rau, A. Kamlage, M. Kappler, J. J. Haitsma, J. Freihorst, H. von der Hardt, and C. F. Poets (2001). Phosphatidylcholine molecular species in lung surfactant: composition in relation to respiratory rate and lung development. *Am. J. Respir. Cell Mol. Biol.* **25:** 725–731.

25. W. Bernhard, J. Mottaghian, A. Gebert, G. A. Rau, H. von der Hardt, and C. F. Poets (2000). Commerical versus native surfactants. *Am. J. Crit. Care Med.* **162:** 1524–1533.

26. M. M. Lipp, K. Y. C. Lee, D. Y. Takamoto, J. A. Zasadzinski, and A. J. Waring (1998). Coexistence of buckled and flat monolayers. *Phys. Rev. Lett.* **81:** 1650–1653.

27. M. M. Lipp, K. Y. C. Lee, J. A. Zasadzinski, and A. J. Waring (1996). Phase and morphology changes in lipid monolayers induced by SP-B protein and its amino-terminal peptide. *Science* **273:** 1196–1199.

28. K. Y. C. Lee, A. Gopal, A. Von Nahmen, J. A. Zasadzinski, J. Majewski, G. S. Smith, P. B. Howes, and K. Kjaer (2002). Influence of palmitic acid and hexadecanol on the phase transition temperature and molecular packing of dipalmitoylphosphatidyl-choline monolayers at the air–water interface. *J. Chem. Phys.* **116:** 774–783.

29. F. Bringezu, J. Q. Ding, G. Brezesinski, and J. A. Zasadzinski (2001). Changes in model lung surfactant monolayers induced by palmitic acid. *Langmuir* **17:** 4641–4648.

30. F. Bringezu, J. Q. Ding, G. Brezesinski, A. J. Waring, and J. A. Zasadzinski (2002). Influence of pulmonary surfactant protein B on model lung surfactant monolayers. *Langmuir* **18:** 2319–2325.

31. J. Q. Ding, H. E. Warriner, and J. A. Zasadzinski (2002). Viscosity of two-dimensional suspensions. *Phys. Rev. Lett.* **88:** 168102.

32. S. Orgeig and C. B. Daniels (2001). The roles of cholesterol in pulmonary surfactant: insights from comparative and evolutionary studies. *Comp. Biochem. Phys. A* **129:** 75–89.

33. J. Bernadino de la Serna, J. Perez-Gil, A. C. Simonsen, and L. A. Bagatolli (2004). Cholesterol rules: direct observation of the coexistence of two fluid phases in native pulmonary surfactant membranes at physiological temperatures. *J. Biol. Chem.* **279:** 40715–40722.

34. U. Pison, R. Herold, and S. Schürch (1996). The pulmonary surfactant system: biological functions, components, physicochemical properties and alterations during lung disease. *Coll. Surf.* **114:** 165–184.

35. F. R. Poulain and J. A. Clements (1995). Pulmonary surfactant therapy. *West. J. Med.* **162:** 43–50.

36. B. T. Bloom, J. Kattwinkel, R. T. Hall, P. M. Delmore, E. A. Egan, J. R. Trout, M. H. Malloy, D. R. Brown, I. R. Holzman, C. H. Coghill, W. A. Carlo, A. K. Pramanik, M. A. McCaffree, P. L. Toubas, S. Laudert, L. L. Gratny, K. B. Weatherstone, J. H. Seguin, L. D. Willett, G. R. Gutcher, D. H. Mueller, and W. H. Topper (1997). Comparison of Infasurf (calf lung surfactant extract) to Survanta (Beractant) in the treatment and prevention of respiratory distress syndrome. *Pediatrics* **100:** 31–38.

37. R. Ramanathan, M. Rasmussen, D. Gerstmann, N. Finer, K. Sekar, and G. Nas (2004). A randomized, multicenter masked comparison of poractant alfa (Curosurf) versus beractant (Survanta) in the treatment of respiratory distress syndrome in preterm infants. *Am. J. Perinatol.* **21:** 109–119.

38. F. R. Moya, J. Gadzinowski, E. Bancalari, V. Salinas, B. Kopelman, A. Bancalari, M. K. Kornacka, T. A. Merritt, R. Segal, C. J. Schaber, H. Tsai, J. Massaro, R. d'Agostino, and I. S. C. S. Group (2005). A multicenter, randomized, masked, comparison trial of lucinactant, colfosceril palmitate, and beractant for the prevention of respiratory distress syndrome among very preterm infants. *Pediatrics* **115:** 1018–1029.

39. S. K. Sinha, T. Lacaze-Masmonteil, A. Valls i Soler, T. E. Wiswell, J. Gadzinowski, J. Hadju, G. Bernstein, M. Sanchez-Luna, R. Segal, C. J. Schaber, J. Massaro, and R. d'Agostino (2005). A multicenter, randomized controlled trial of lucinactant versus poractant alfa among very premature infants at high risk for respiratory distress syndrome. *Pediatrics* **115:** 1030–1038.

40. B. Guyer, D. M. Strobino, S. J. Ventura, and G. K. Singh (1995). Annual summary of vital statistics—1994. *Pediatrics* **96:** 1029–1039.

41. G. K. Singh and S. Yu (1995). Infant mortality in the US: trends, differentials and projections 1950–2010. *Am. J. Pub. Health* **85:** 957–964.

42. G. K. Suresh and R. F. Soll (2005). Overview of surfactant replacement trials. *J. Perinatol.* **25:** S40–S44.

43. J. A. Clements (1956). Dependence of pressure–volume characteristics of lungs on intrinsic surface-active material. *Am. J. Physiol.* **187:** 592–592.

44. J. A. Clements (1957). Surface tension of lung extracts. *Proc. Soc. Exp. Biol. Med.* **95:** 170–172.
45. J. A. Clements (1961). Pulmonary edema and permeability of alveolar membranes. *Arch. Environ. Health* **2:** 280.
46. J. A. Clements, E. S. Brown, and R. P. Johnson (1957). Alveolar surface tension. *Fed. Proc.* **16:** 23–23.
47. C. Alonso and J. Zasadzinski (2004). Linear dependence of surface drag on surface viscosity. *Phys. Rev. E* **69:** 021602.
48. J. Q. Ding, H. E. Warriner, J. A. Zasadzinski, and D. K. Schwartz (2002). Magnetic needle viscometer for Langmuir monolayers. *Langmuir* **18:** 2800–2806.
49. S. Schürch, H. Bachofen, J. Goerke, and F. Green (1992). Surface properties of rat pulmonary surfactant studied with the captive bubble method: adsorption, hysteresis, stability. *Biochim. Biophys. Acta* **1103:** 127–136.
50. S. Schürch, H. Bachofen, J. Goerke, and F. Possmayer (1989). A captive bubble method reproduces the *in situ* behavior of lung surfactant monolayers. *J. Appl. Physiol.* **67:** 2389–2396.
51. S. B. Hall, M. S. Bermel, Y. T. Ko, H. J. Palmer, G. Enhorning, and R. H. Notter (1993). Approximations in the measurement of surface tension on the oscillating bubble surfactactometer. *J. Appl. Physiol.* **75:** 468–477.
52. G. Enhorning (1977). Pulsating bubble technique for evaluating pulmonary surfactant. *J. Appl. Physiol.* **43:** 198–203.
53. C. F. Brooks, G. G. Fuller, C. W. Frank, and C. R. Robertson (1999). An interfacial stress rheometer to study rheological transitions in monolayers at the air–water interface. *Langmuir* **15:** 2450–2459.
54. A. D. Bangham, C. J. Morley, and M. C. Phillips (1979). The physical properties of an effective lung surfactant. *Biochim. Biophys. Acta* **573:** 552–556.
55. M. Sacchetti, H. Yu, and G. Zografi (1993). In-plane steady shear viscosity of monolayers at the air/water interface and its dependence on free area. *Langmuir* **9:** 2168–2171.
56. P. C. Hiemenz (1986). *Principles of Colloid and Surface Chemistry*, 2nd ed. Marcel Dekker, New York.
57. A. Krishnan, J. Sturgeon, C. A. Siedlicki, and E. A. Vogler (2003). Scaled interfacial activity of proteins at the liquid–vapor interface. *J. Biomed. Mater. Res.* **68A:** 544–557.
58. C. Ybert, W. Lu, G. Moller, and C. M. Knobler (2002). Kinetics of phase transitions in monolayers: collapse. *J. Phys. Condensed Matter* **14:** 4753–4762.
59. M. Longo, A. Bisagno, J. Zasadzinski, R. Bruni, and A. Waring (1993). A function of lung surfactant protein SP-B. *Science* **261:** 453–456.
60. W. Lu, C. M. Knobler, R. F. Bruinsma, M. Twardos, and M. Dennin (2002). Folding Langmuir monolayers. *Phys. Rev. Lett.* **89:** 146107.
61. D. Vollhardt (1993). Nucleation and growth in supersaturated monolayers. *Adv. Colloid Interface Sci.* **47:** 1–23.
62. H. Diamant, T. A. Witten, A. Gopal, and K. Y. C. Lee (2000). Unstable topography of biphasic surfactant monolayers. *Europhys. Lett.* **52:** 171–177.
63. A. Gopal and K. Y. C. Lee (2001). Morphology and collapse transitions in binary phospholipid monolayers. *J. Phys. Chem. B* **105:** 10348–10354.

64. R. D. Smith and J. C. Berg (1980). Collapse of surfactant monolayers at the air–water interface. *J. Colloid Interface Sci.* **74:** 273–286.
65. A. Gopal, V. A. Belyi, H. Diamant, T. A. Witten, and K. Y. C. Lee (2006). Microscopic folds and macroscopic jerks in compressed lipid monolayers. *J. Phys. Chem. B* **110:** 10220–10223.
66. P. Tchoreloff, A. Gulik, B. Denizot, J. E. Proust, and F. Puisieux (1991). A structural study of interfacial phospholipid and lung surfactant layers by transmission electron microscopy after Blodgett sampling: influence of surface pressure and temperature. *Chem. Phys. Lipids* **59:** 151–165.
67. D. Y. Takamoto, E. Aydil, J. A. Zasadzinski, A. Ivanova, D. K. Schwartz, T. Yang, and P. Cremer (2001). Stable ordering in Langmuir–Blodgett films. *Science* **293:** 1292–1295.
68. V. M. Kaganer, H. Mohwald, and P. Dutta (1999). Structure and phase transitions in Langmuir monolayers. *Rev. Mod. Phys.* **71:** 779–819.
69. H. McConnell (1991). Structures and transitions in lipid monolayers at the air–water interface. *Annu. Rev. Phys. Chem.* **42:** 171–195.
70. C. M. Knobler and R. C. Desai (1992). Phase transitions in monolayers. *Annu. Rev. Phys. Chem.* **43:** 207–236.
71. S. Schürch, J. Goerke, and J. A. Clements (1976). Direct determination of surface tension in the lung. *Proc. Natl. Acad. Sci. USA* **73:** 4698–4702.
72. S. Schürch, J. Goerke, and J. A. Clements (1978). Direct determination of volume and time-dependence of alveolar surface tension in excised lungs. *Proc. Natl. Acad. Sci. USA* **75:** 3417–3421.
73. V. Im Hof, P. Gehr, V. Gerber, M. M. Lee, and S. Schurch (1997). *In vivo* determination of surface tension in the horse trachea and *in vitro* model studies. *Respir. Physiol.* **109:** 81–93.
74. M. Geiser, S. Schürch, and P. Gehr (2003). Influence of surface chemistry and topography of particles on their immersion into the lung's surface-lining layer. *J. Appl. Physiol.* **94:** 1793–1801.
75. H. A. Stone (1995). Fluid motion of monomolecular films in a channel flow geometry. *Phys. Fluids* **7:** 2931–2937.
76. J. L. Bull, L. K. Nelson, J. T. J. Walsh, M. R. Glucksberg, S. Schurch, and J. B. Grotberg (1999). Surfactant-spreading and surface-compression disturbance on a thin viscous film. *J. Biomed. Eng.* **121:** 89–98.
77. F. F. Espinosa and R. D. Kamm (1999). Bolus dispersal through the lungs in surfactant replacement therapy. *J. Appl. Physiol.* **86:** 391–410.
78. F. F. Espinosa, A. H. Shapiro, J. J. Fredberg, and R. D. Kamm (1993). Spreading of exogenous surfactant in an airway. *J. Appl. Physiol.* **75:** 2028–2039.
79. E. R. Weibel (1963). *Morphometry of the Human Lung*. Academic Press, New York.
80. T. Baumgart, S. T. Hess, and W. W. Webb (2003). Imaging coexisting fluid domains in biomembrane models coupling curvature and line tension. *Nature* **425:** 821–824.
81. S. L. Keller (2003). Miscibility transitions and lateral compressibility in liquid phases of lipid monolayers. *Langmuir* **19:** 1451–1456.
82. J. A. Zasadzinski, R. Viswanathan, L. Madsen, J. Garnaes, and D. K. Schwartz (1994). Langmuir–Blodgett films. *Science* **263:** 1726–1733.

83. I. Langmuir (1936). Two-dimensional gases, liquids and solids. *Science* **84:** 379–383.
84. W. D. Harkins and R. J. Meyers (1937). Viscosity of monolayer films. *Nature* **140:** 465–467.
85. R. J. Myers and W. D. Harkins (1937). The viscosity (or fluidity) of liquid or plastic monomolecular films. *J. Chem. Phys.* **5:** 601–603.
86. E. Boyd and W. D. Harkins (1939). Molecular interactions in monolayers: viscosity of two-dimensional liquids and plastic solids. *JACS* **61:** 1188–1195.
87. L. Fourt and W. D. Harkins (1938). Surface viscosity of long-chain alcohol monolayers. *J. Phys. Chem.* **42:** 897–910.
88. G. C. Nutting and W. D. Harkins (1939). Surface viscosity. *JACS* **61:** 1180.
89. M. Joly (1956). Non-Newtonian surface viscosity. *J. Colloid Science* **11:** 519–531.
90. N. L. Jarvis (1965). Surface viscosity of monomolecular films of long-chain aliphatic amides, amines, alcohols and carboxylic acids. *J. Phys. Chem.* **69:** 1789–1797.
91. A. K. Doolittle (1951). Studies of Newtonian flow. 2. The dependence of the viscosity of liquids on free-space. *J. Appl. Phys.* **22:** 1471–1475.
92. A. K. Doolittle and D. B. Doolittle (1957). Studies in Newtonian flow. 5. Further verification on the free-space viscosity equation. *J. Appl. Phys.* **28:** 901–905.
93. L. E. Copeland, W. D. Harkins, and G. E. Boyd (1942). A superliquid in two dimensions and a first order change in a condensed monolayer. *J. Chem. Phys.* **10:** 357.
94. I. Langmuir (1933). Oil lenses on water and the nature of monomolecular expanded films. *J. Chem. Phys.* **1:** 756–776.
95. I. R. Peterson, V. Brzezinski, R. M. Kenn, and R. Steitz (1992). Equivalent states of amphiphilic lamellae. *Langmuir* **8:** 2995–3002.
96. S. Stallberg-Stenhagen and E. Stenhagen (1945). Surface viscosity of monomolecular films. *Nature* **156:** 239–241.
97. A. M. Bibo, C. M. Knobler, and I. R. Peterson (1991). A monolayer phase miscibility comparison of long-chain fatty acids and their ethyl esters. *J. Phys. Chem.* **93:** 5591–5599.
98. A. I. Kitaigorodskii (1961). *Organic Chemical Crystallography*. Consultant Bureau, New York.
99. J. Garnaes, D. K. Schwartz, R. Viswanathan, and J. Zasadzinski (1992). Domain boundaries and buckling superstructures in Langmuir–Blodgett films. *Nature* **357:** 54–57.
100. D. K. Schwartz, J. Garnaes, R. Viswanathan, and J. Zasadzinski (1992). Surface order and stability of Langmuir–Blodgett films. *Science* **257:** 508–511.
101. V. M. Kaganer, I. R. Peterson, R. M. Kenn, M. C. Shih, M. Durbin, and P. Dutta (1995). Tilted phases of fatty acid monolayers. *J. Chem. Phys.* **102:** 9412–9422.
102. W. D. Harkins and J. G. Kirkwood (1938). Fatty alcohols and surface viscosity. *J. Chem. Phys.* **6:** 53.
103. R. S. Ghaskadvi and M. Dennin (2000). Alternate measurement of the viscosity peak in heneicosanoic acid monolayers. *Langmuir* **16:** 10553–10555.
104. R. S. Ghaskadvi, J. B. Ketterson, and P. Dutta (1997). Nonlinear shear response and anomalous pressure dependence of viscosity in a Langmuir monolayer. *Langmuir* **13:** 5137–5140.

105. I. Denicolo, J. Doucet, and A. F. Craievich (1983). X-ray study of the rotator phase of paraffins (III): even-numbered paraffins $C_{18}H_{38}$, $C_{20}H_{42}$, $C_{22}H_{46}$, $C_{24}H_{50}$ and $C_{26}H_{54}$. *J. Chem. Phys.* **78:** 1465–1469.

106. J. Doucet, I. Denicolo, and A. F. Craievich (1981). X-ray study of the "rotator" phase of the odd-numbered paraffins $C_{17}H_{36}$, $C_{19}H_{40}$, $C_{21}H_{44}$. *J. Chem. Phys.* **75:** 1523–1529.

107. A. F. Craievich, I. Denicolo, and J. Doucet (1984). Molecular motion and conformational defects in odd-numbered paraffins. *Phys. Rev. B* **30:** 4782–4787.

108. W. J. Moore, Jr. and H. Eyring (1938). Theory of the viscosity of unimolecular films. *J. Chem. Phys.* **6:** 391–394.

109. J. F. Brady (1993). The rheological behavior of concentrated colloidal dispersions. *J. Chem. Phys.* **99:** 567–581.

110. A. Donev, I. Cisse, D. Sachs, E. A. Variano, F. H. Stillinger, R. Connelly, S. Torquato, and P. M. Chaikin (2004). Improving the density of jammed disordered packings using ellipsoids. *Science* **303:** 990–993.

111. J. Johansson (1998). Structure and properties of surfactant protein C. *Biochim. Biophys. Acta* **1408:** 161–172.

112. A. Gericke, C. R. Flach, and R. Mendelsohn (1997). Structure and orientation of lung surfactant SP-C and L-α-dipalmitoylphosphatidylcholine in aqueous monolayers. *Biophys. J.* **73:** 492–499.

113. A. J. Davis, A. H. Jobe, D. Hafner, and M. Ikegami (1998). Lung function in premature lambs and rabbits treated with a recombinant SP-C surfactant. *Am. J. Respir. Crit. Care Med.* **157:** 553–559.

114. M. Ikegami and A. H. Jobe (1998). Surfactant protein-C in ventilated premature lamb lung. *Pediatr. Res.* **44:** 860–864.

115. J. J. Batenburg and H. P. Haagsman (1998). The lipids of pulmonary surfactant: dynamics and interactions with proteins. *Prog. Lipid Res.* **37:** 235–276.

116. V. Kairys, M. K. Gilson, and B. Luy (2004). Structural model for an AxxxG-mediated dimer of surfactant-associated protein C. *Eur. J. Biochem.* **271:** 2086–2092.

117. S. W. Glasser, M. S. Burhans, T. R. Korfhagen, C. L. Na, P. D. Sly, G. F. Ross, M. Ikegami, and J. A. Whitsett (2001). Altered stability of pulmonary surfactant in SP-C-deficient mice. *Proc. Natl. Acad. Sci. USA* **98:** 6366–6371.

118. S. W. Glasser, E. A. Detmer, M. Ikegami, C. L. Na, M. T. Stahlman, and J. A. Whitsett (2003). Pneumonitis and emphysema in SP-C gene targeted mice. *J. Biol. Chem.* **278:** 14291–14298.

119. C. Alonso and J. A. Zasadzinski (2006). A brief review of the relationships between monolayer viscosity, phase behavior, surface pressure and temperature using a simple monolayer viscometer. *J. Phys. Chem. B* **110:** 22185–22191.

120. R. Chen, L. Li, and Z. Weng (2003). ZDOCK: an initial stage protein docking algorithm. *Proteins: Structure, Function and Genetics* **52:** 80–87.

121. W. L. DeLano (2002). *The PyMOL User's Manual*. Delano Scientific, San Carlos, CA.

CHAPTER 14

A Cursory Glance at the Phyiscochemical Properties of Oppositely Charged Surfactants in Solution and at the Air–Water Interface

AMIYA KUMAR PANDA

Department of Chemistry, University of North Bengal, Darjeeling, West Bengal, India

KAUSHIK NAG

Memorial University, St. Jhon's, Newfoundland and Labrador, Canada

CONTENTS

14.1	Introduction	385
14.2	Catanionic Surfactants	386
14.3	Molecular Structural Features of Various Surfactants	387
14.4	Surfactant Membranous Polymorphism	389
14.5	Phase Diagrams and Mixing Behavior	393
14.6	Domains and Rafts in Surfactant Monolayers	396
14.7	Application of Catanionics and Coacervates	408
Acknowledgment		410
References		410

14.1 INTRODUCTION

Amphiphiles have the properties to aggregate into various forms of nanostructures in aqueous media and at solid–solid, solid–liquid, liquid–liquid, and liquid–gas interfaces. Such aggregates include micelles, reverse micelles, microemulsion,

Structure and Dynamics of Membranous Interfaces, edited by Kaushik Nag
Copyright © 2008 John Wiley & Sons, Inc.

vesicles, and monomolecular films [1–3]. In most practical applications/systems, a mixture of amphiphiles always performs better than a single-component system [4–8]. Biological systems also follow the same rule, by judicious choice of composition (e.g., lipid mixtures in cell membranes, which control the life processes). Also in the lung, there is a thin lining of a mixture of phospholipid molecules, which helps in reducing the surface tension at the lung–air interface [9–14]. Under this context, systems containing a mixture of cationic and anionic (oppositely charged) single-tailed (specially) surfactants have been found to be important for their multifaceted application potentials and hence have become the subject of extensive research. Since the last century, several researchers have been involved in this particular field of research.

14.2 CATANIONIC SURFACTANTS

Cationic–anionic surfactant mixtures are commonly known as catanionics. Catanionics usually get precipitated in aqueous media, especially when they are in an equimolar ratio. These precipitates are also known as coacervates. These coacervates find several applications and could also mimic model membranes like phospholipids. Catanionics have two hydrocarbon chains (when two single-tailed oppositely charged surfactants are used) and headgroups are electrostatically bound. When oppositely charged surfactants are mixed together, due to strong synergistic interactions, precipitation occurs. This precipitation is maximum when they are present in an equimolar ratio. The ion pairs could mimic phospholipids: the only difference is ion pair formation through headgroups. Catanionics can aggregate into several forms like phospholipids when two oppositely charged surfactants are not in equimolar ratio. They can form mixed micelles with strong synergistic interactions [15, 16]. They can also form vesicles. Formation of vesicles from totally synthetic amphiphiles was first reported by Kunitake and Okahata in 1977 [17]. In that experiment, didodecyldimethylammonium bromide (DDAB) was dispersed into water by sonication. Formation of such vesicles was confirmed by electron microscopy and other techniques. Later on, Evans and co-workers [18–24] also examined vesicular properties of the same double-tailed cationic surfactant. Three basic criteria for a vesicle formation, namely, (1) spontaneous formation upon dispersion of dry surfactant powder into water, (2) no aggregation with time, and (3) reversibility, were not fully met for the double-tailed cationic surfactants. The problem with using a single component of double-tailed surfactant was the strong repulsive forces between the headgroups, which later on could easily be overcome using surfactant ion pairs or catanionics. This sort of argument has been extensively studied by Kaler and co-workers [25–29]. In their studies, they could simply overcome the problems by using anionic surfactants as the opposite ions to the DDAB. In their work, they claimed that a many-fold improvement in spontaneous vesicle formation could be ensured using two single-tailed oppositely charged surfactants.

Since the catanionic surfactant mixtures behave entirely differently than their single components, extensive work has been done on catanionics during the last two decades. Micellization, dispersion, and liquid crystal behavior of different catanionics have also been studied by several workers [30–43]. In some review works and monographs, Khan and co-workers [44–50] described in detail different aspects. Hoffmann and co-workers have extensively studied calorimetric and electron microscopic measurements on different catanionics [51–58], as have others [16, 59–62]. On the other hand, differential scanning calorimetry and optical texture changes were also carried out [63–68]. In all the aforementioned experiments, researchers tried to shed light on the properties and functionalities of catanionics. All the studies were done to find an acceptable substitute for naturally occurring phospholipids. Naturally occurring phospholipids suffer from serious limitations of biodegradability. Moreover, their tolerance under varied physicochemical conditions is poor; for example, liposomes are highly pH and temperature sensitive. So to have a good substitute with a better functionality, a judicious choice of catanionics is essential. As already mentioned, there are several methods for characterizing the solid/solution/monolayer properties of catanionics. In this chapter, we confine ourselves to the following studies:

1. FTIR and/or XRD measurements to check the formation of catanionic coacervates.
2. Phase diagrams (ternary) for detecting different zones (e.g., mixed micelle, vesicles, precipitates, or liquid crystals).
3. Polarization and/or fluorescence microscopic studies to visualize the appearance of different textures/phases.
4. Surface pressure–area measurements to focus on the monomolecular films of coacervates. Subsequently, epifluorescence microscopic measurements are discussed to check the structural changes that takes place at the air–water interface.

Before going into the details of different studies to characterize catanionics, a brief discussion regarding the structural features that play a key role in the formation of different aggregated species would be helpful.

14.3 MOLECULAR STRUCTURAL FEATURES OF VARIOUS SURFACTANTS

A double-tailed surfactant, which can aggregate into several forms, could be any one of the following type:

1. A double-tailed cationic, for example, didodecyldimethylammonium bromide (DDDAB) or ditetradecyldimethylammonium bromide (DTDAB).
2. A double-tailed zwitterionic surfactant (e.g., a phospholipid, dipalmitoylphosphatidylcholine (DPPC)).

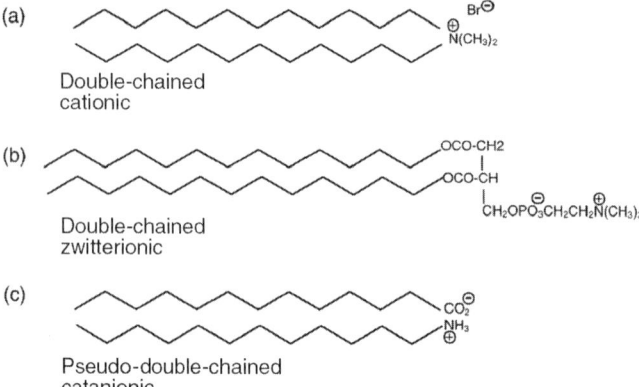

Figure 14.1 The chemical structure of some swelling surfactants: (a) didodecyldimethyl ammonium bromide, (b) dipalmitoyl phosphatidylcholine (lecithin), and (c) dodecylammonium dodecylcarboxylate.

3. A pseudo-double-tailed catanionic surfactant (e.g., dodecylammoniumdodecyl carboxylate). Individual components are depicted schematically in Fig. 14.1.

As already mentioned, first the two categories of double-tailed surfactants suffer from some practical limitations (e.g., lack of three basic criteria for vesicles), and the surfactants in the second category are very susceptible to biodegradation. Pseudo-double-tailed catanionics could easily overcome those shortcomings.

Formation of aggregates of different shapes are basically controlled by the mean curvature for a sequence of surfactant aggregates. Different possible geometries are shown in Fig. 14.2.

Needless to mention, a spherical micelle or a reverse micelle is formed when the length of the surfactant is comparable to the dimension of the micelle. On the other hand, a cylindrical aggregate is generated when the surfactant chain

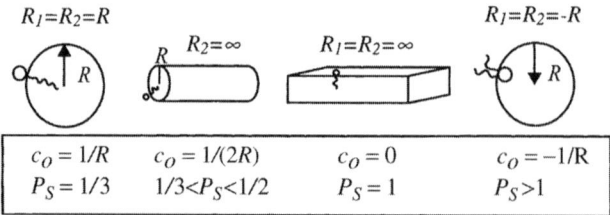

Figure 14.2 Mean curvature for a sequence of surfactant aggregates, represented by geometric models: from left to right, spherical micelles, cylindrical micelles, planar bilayers, and inverted structures.

length is relatively smaller. Other possibilities, as presented in Fig. 14.2, are self-explanatory.

Attractive and repulsive interactions occurring at the interfacial region of amphiphilic structures were classically described by Israelachvili [69]. This idea was further exemplified by Clint [70]. He summarized as follows: if a_0 is the headgroup area, v is the alkyl chain volume, and l_c is the maximum chain length to which the alkyl chain can extend, then the following possibilities could occur:

1. $v/a_0 l_c < \frac{1}{3}$ spherical micelle
2. $\frac{1}{3} < (v/a_0 l_c) < 1/2$ nonspherical micelle
3. $1/2 < (v/a_0 l_c) < 1$ vesicles or bilayers
4. $1 < (v/a_0 l_c)$ reverse micelle

Here, we concentrate on the third category of aggregation.

14.4 SURFACTANT MEMBRANOUS POLYMORPHISM

Yu et al. [71] diagrammatically described the formation of a catanionic surfactant ion pair. The particular size and shape of self-assembled aggregates were the product of a fine balance between different free-energy contributions. However, a simple and common conceptual way of relating the geometric shape of an amphiphile to the preferred type of aggregation is governed by the packing parameter, P_c. Formation of a catanionic vesicle can be achieved if the effective interfacial headgroup area of each partner of an ion pair amphiphile (herein catanionic) is substantially smaller than that for the individual surfactant, due to electrostatic attraction between oppositely charged headgroups and a reduction in hydrations. That is why, unlike their micelle-forming parent surfactants, catanionics prefer to assemble into bilayer vesicles when the chain lengths of cationic and anionic surfactants are different, than into the structure of an equimolar mixture of catanionics. Long-chain containing surfactants get tilted into the bilayer. This is schematically shown in Fig. 14.3.

Because of their close resemblance to naturally occurring phospholipids, the catanionics easily (spontaneously in most cases, with few exceptions) form lamellar phases. Subsequently, because there is no charge repulsion between the headgroups (as the ion pairs are formed through electrostatic interaction), they can also form bilayers. One of the advantages in using such bilayers for catanionics is that the exact distance between the headgroups is known, as shown in Fig. 14.4 [72].

In some cases (although very rarely), catanionics aggregate into flat disks. This phenomenon usually occurs in a salt-free environment. When a trace amount of excess cationic surfactant is added, it reduces the equilibrium size of the aggregates, and part of the excess cationic surfactants form the edge of the aggregate; thus a nanodisk is formed. Therefore one could alter the shape of

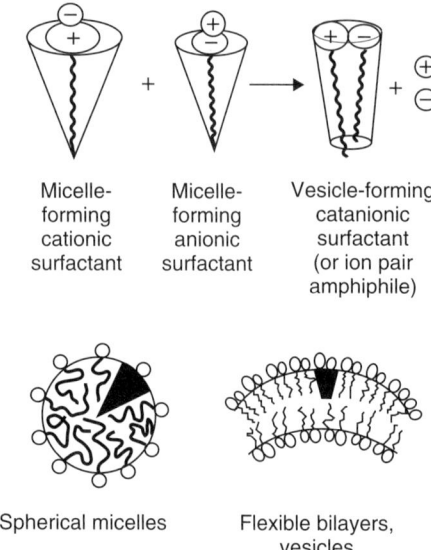

Figure 14.3 Formation of vesicle-forming catanionic surfactant from micelle-forming cationic and anionic surfactants. (From Ref. 71 with permission.)

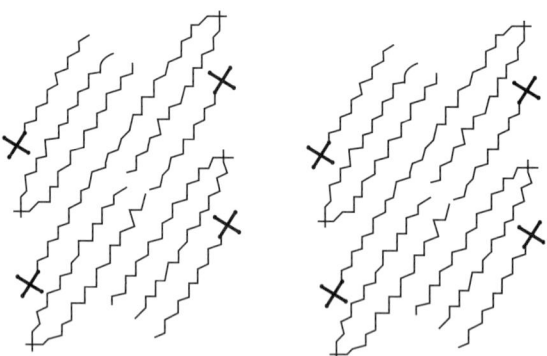

Figure 14.4 A possible structure of the wet equimolar SDS + DODAB complex. (Adapted from Ref. 72.)

the catanionic aggregate by judicious choice of cationic anionic surfactant mole ratio. A schematic model of such a nanodisk is given in Fig. 14.5 [73].

It has also been reported that when catanionics are dilute enough, they can even self-assemble into a new type of bilayer organization—a hollow aggregate of regular icosahedral shape formed at certain compositions in the absence of salts. Such an aggregate is shown schematically in Fig. 14.6 [74].

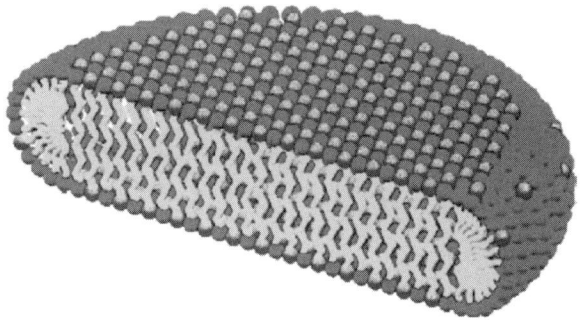

Figure 14.5 Schematic model of nanodisks formed by catanionic surfactants; the excess cationic surfactant molecules construct the edges of the disks. (From Ref. 73, with permission.)

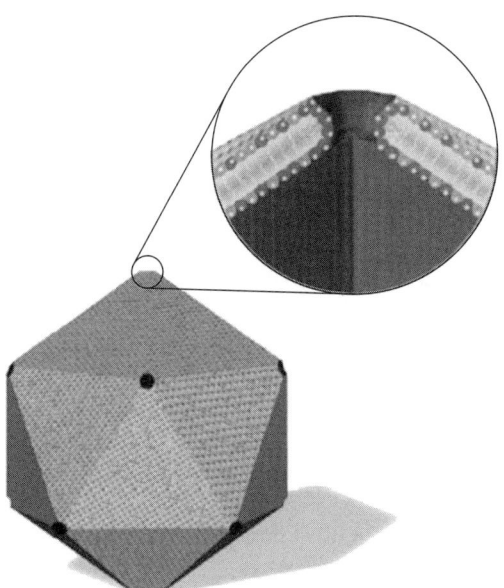

Figure 14.6 Schematic model of icosahedra in dilute salt-free catanionic surfactant mixtures. (From Ref. 74, with permission.)

After obtaining a preliminary idea about the shape and size of different catanionic aggregates, one needs to know how to characterize the catanionics. The first and foremost characterization process is to study the phase diagram of the catanionics in aqueous media. Several workers have performed such studies but, they are considered fragmentary. Usually phase diagrams (also known as Gibbs phase diagrams) are constructed on a triangle. To start, let us consider the ternary

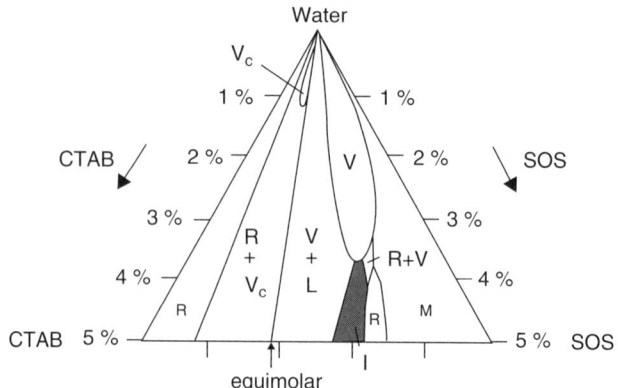

Figure 14.7 Microstructures formed at room temperature for the ternary system of CTAB/SOS/H$_2$O. R, rod-like micelles; M, spherical micelles; VC, cationic-rich unilamellar vesicles; VA, anionic-rich unilamellar vesicles; L, planar bilayers; and I, an unidentified multiaggregate region. On the equimolar line, a 1:1 crystalline precipitate forms. (Adapted from Ref. 28, with permission.)

phase diagram of the CTAB/SOS/H$_2$O system. Strictly speaking, such a system is a pseudoternary one because in reality such a system is composed of five components—CTAB, SOS, CTA$^+$, SOS$^-$, and NaBr. Such a system requires a phase pyramid to accurately express all of the phase behavior [65]. To avoid such a complication, a two-dimensional plot in a triangle is preferred, as shown in Fig. 14.7.

Several microstructures are formed, as evidenced from Fig. 14.7. They include rod-like micelles (R), spherical micelles (M), cationic-rich unilamellar vesicles (VC), anionic-rich unilamellar vesicles (VA), planar bilayers (L) and also some unidentified regions (I). This region is probably a multiaggregated region. Also, a coacervation (formation of precipitate) phenomenon occurs when surfactants are mixed in an equimolar ratio [28]. How can we detect these different regions? Usually after surfactants are mixed in different ratios, they are kept for a substantial period of time (up to 6–8 weeks). Then different solutions, kept in test tubes, are checked visually. This long equilibration time is necessary for the samples containing vesicles to distinguish between single-phase vesicle regions and two-phase vesicle/lamellar regions. This is because small amounts of lamellar structure development take relatively longer compared to other phases. The color, scattering capacity, transparency, and optical birefringerence of different solutions are characteristics of different phases. Vesicle samples, which are isotropic, appear bluish in color. Spherical and globular micelles appear colorless, but their scattering capabilities are higher than water. Rod-like micelles are viscous and viscoelastic. Vesicular phases also appear as birefringerent clouds. Similar findings were reported by O'Connor et al. [67] some other representative diagrams are shown in Fig. 14.8.

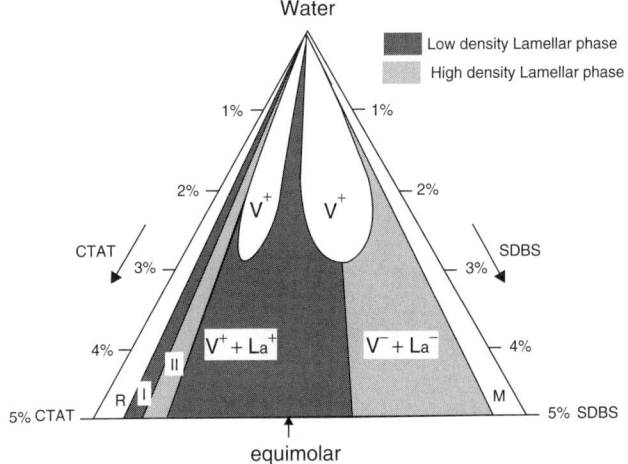

Figure 14.8 Ternary phase diagram for CTAT/SDBS/H$_2$O at 25 °C. One-phase regions are unshaded and are CTAT-rich vesicles (V'), CTAT-rich rod-like micelles (R), SDBS-rich vesicles (V$^-$), and SDBS-rich micelles (M). Two-phase regions are shaded and are V$^+$ and CTAT-rich lamellar phase (La$^+$)V$^-$ and SDBS-rich lamellar phase (La$^-$)R, and hexagonal liquid crystalline (zone I), and an isotropic liquid and a precipitate along the equimolar line. There is also an unresolved multiphase region called zone 11. Compositions are on a weight percent basis. (Adapted from Ref. 6.)

14.5 PHASE DIAGRAMS AND MIXING BEHAVIOR

These sorts of pseudoternary phase diagrams have been further simplified by Nan et al. [66], who studied the ternary phase diagram of the CTAB/SOS/H$_2$O system under varied salt concentrations and temperatures. They have observed altogether seven different regions, as shown in Fig. 14.9. These include two aqueous two-phase regions (Regions 1 and 4), one rich with anionic and vice versa. Then there were two liquid crystalline phases, enriched with either cation or anion surfactant species (Regions 2 and 3). On the two extreme ends there appear mixed micelles, which have an excess of either surfactant. (Regions 5 and 6). In between the liquid crystalline phases, there also appears a bluish region, which, by virtue of its scattering property, could be defined as a vesicular region (Region 7). As already mentioned, in the absence of any free salts, a disk-like vesicular aggregate could be generated [73] or even hollow icosahedral structures could be obtained when a very dilute solution of catanionics is prepared. Therefore, it could be concluded that added salts have a drastic effect on the vesicular structure, which in turn could affect the pseudoternary phase diagram. In the same work of Nan et al. [66], they studied the effect of KCl on the phase diagram of the CTAB/SDS/H$_2$O system. The added salt could reduce the solubility of catanionics; hence an increase in salt concentration results in the increment of the two-phase region. Besides the phase diagrams

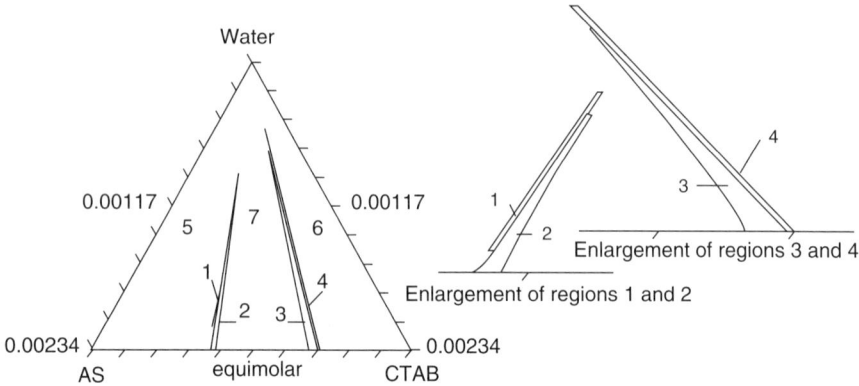

Figure 14.9 Pseudoternary phase diagram of CTAB/AS/water systems at 313.15 K. (1) ATPS-A; (2) liquid crystal; (3) liquid crystal; (4) ATPS-C; (5 and 6) transparent homogeneous solution; (7) lightly bluish or turbid liquid, or precipitate coexisting with liquid. (Adapted from Ref. 66.)

mentioned, some additional self-explanatory phase diagrams are also presented in Fig. 14.10.

After the construction of the phase diagram, one could examine the solubility limit as well as the usability limit of the oppositely charged surfactant ion pairs in solution. The next task is to isolate the coacervates (if possible) from the solution and to characterize them. This task is very simple and straightforward. Consider the coacervate formation between CTAB and SDS. When mixed in equimolar concentration, a 1:1 stoichiometric surfactant ion pair will be precipitated. Now the coacervate could easily be extracted with a chloroform:methanol mixture (3:1, v/v) [77]. The organic extract is then dried with nitrogen. The powdery (fluffy) material can then be stored under dry conditions for at least 1 year.

Once the material has been synthesized, one can characterize them through spectroscopic measurements. FTIR and XRD are the conventional tools for such characterization. A representative XRD pattern is shown in Fig. 14.11. In Fig. 14.11, diffractograms for decylammonium-decylsulfate are shown for various temperatures. It is worthwhile to mention that the coacervate also shows liquid-crystal-like behavior. At lower temperature, specific crystallinity was retained as revealed by the appearance of sharp spectral lines. On further heating, a transition from solid to liquid-crystal-like structure was observed, as evidenced Fig. 14.11c, 14.11d. The reversible nature of the coacervate in responding to the XRD was observed when the substances were cooled down, as shown in Fig. 14.11e, 14.11f.

Upon progressive addition of one of the components, the XRD patterns could also change significantly. This sort of observation were noted by Filipović-Vinceković et al. [37], where the effect of progressive addition of one component to the other has been shown in a self-explanatory manner

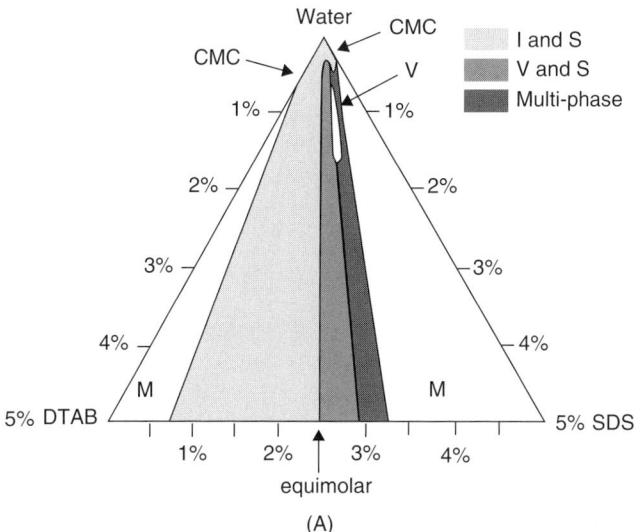

Figure 14.10 (A) Water-rich corner of the pseudoternary phase diagram for DTAB/SDS/water at 25 °C. One-phase regions are unshaded and contain micelles (M) or vesicles (V). Two-phase regions are shaded and consist of either clear liquid (I) and precipitate (S) or vesicles and precipitate. There is an unresolved region (shaded) in which samples appear bluish and viscoelastic at low surfactant concentrations, and turbid and birefringent at higher concentrations. Some samples apparently slowly separate with time. Compositions are on a weight percent basis. (b) Model prediction of the micelle/precipitate phase boundary for the system DTAB/SDS/water at 25 °C. Compositions are on a weight percent basis, and only the dilute (<1 wt% surfactant) region is shown. The nomenclature is the same as for part (A). (Parts (A) and (B) adapted from Ref. 27.) (C) Ternary phase diagram for SDS/DTAB/water at 25 °C. The unshaded regions contain isotropic single phases (micelles, mixed micelles, and vesicles); the shaded regions consist of two or more phases. Enthalpy of binding and enthalpy of formation of mixed surfactant aggregates are measured by choosing surfactant concentrations in the syringe and titration cell along the lower edge (1 wt% total surfactant) of this phase diagram. (Adapted from Ref. 61.) (D) Phase diagram of the ternary system SDBS/DDAB/H_2O at 30 °C is represented in the highly diluted concentration region with a maximum concentration of surfactant amounting to 2 wt%; CMC, CVC, and coacervates are denoted by arrows; I = isotropic solution, M = SDBS-rich micelles, V^+ = one-phase DDAB rich vesicles, $S + V^+$ = two-phase region (i.e., the mixture of a solid phase and DDAB-rich vesicles). Compositions of the mixture are given on the mole fraction basis, x(DDAB), and on the weight fraction basis, ω_{SDBS} and ω_{DDAB}. (Adapted from Ref. 78.) (E) Phase diagrams for mixtures of DDAB with different anionic surfactants, at 25 °C: (a) sodium dodecyl sulfate (SDS); (b) sodium bis(2-ethyl-hexyl)sulfosuccinate (AOT); and (c) sodium taurodeoxycholate (STDC). Phase notations: D, DI, and DII—lamellar phases; I, I1, and I2—cubic phases; L1—isotropic solution; F1—normal hexagonal phase; F2—reverse hexagonal phase; L3—sponge phase; GI and GII—crystal phases; disp—dispersion; and hyd cryst—hydrated crystals. (Adapted from Ref. 68.)

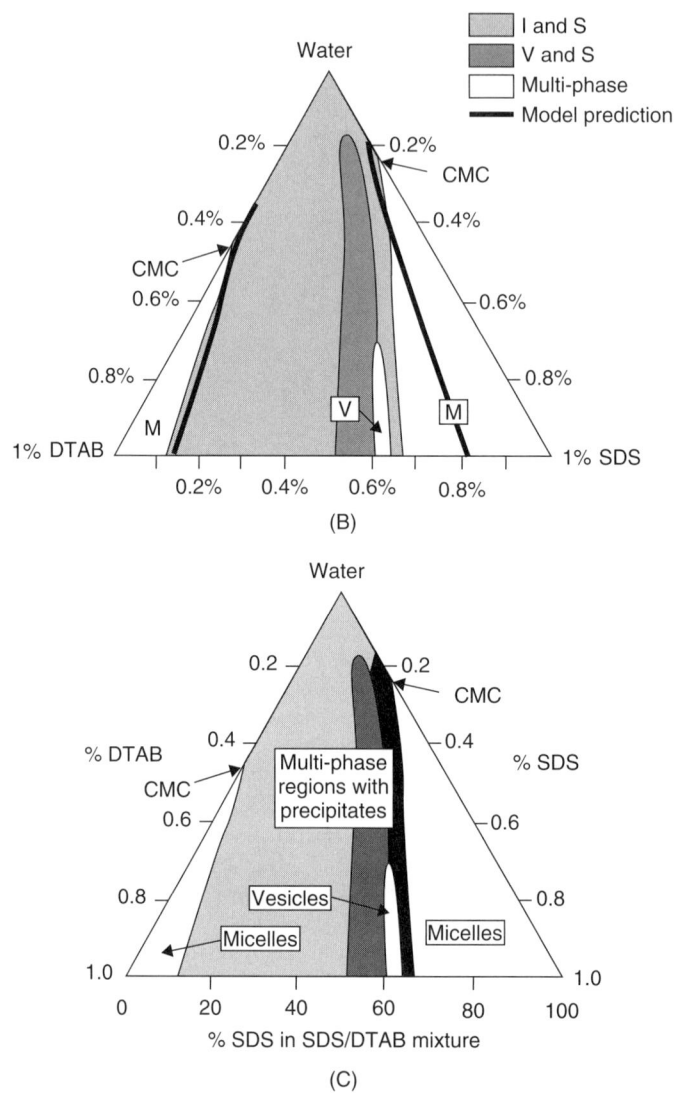

Figure 14.10 (*Continued*)

in Fig. 14.12. Several reports from the same research groups were published [15, 30–38].

14.6 DOMAINS AND RAFTS IN SURFACTANT MONOLAYERS

As already mentioned, most catanionics undergo phase transition through the formation of liquid crystals. This is further supported by results from differential

DOMAINS AND RAFTS IN SURFACTANT MONOLAYERS   397

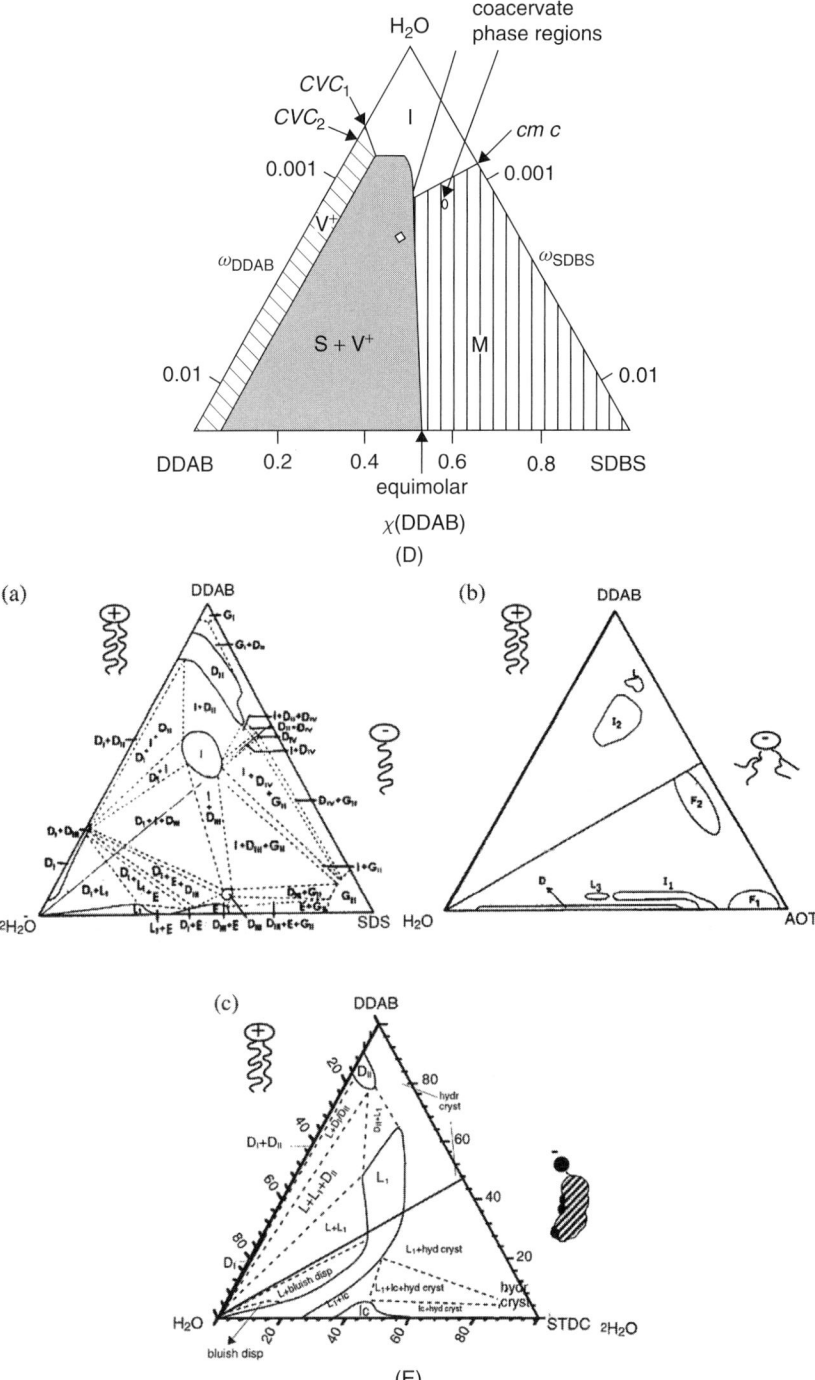

Figure 14.10 (*Continued*)

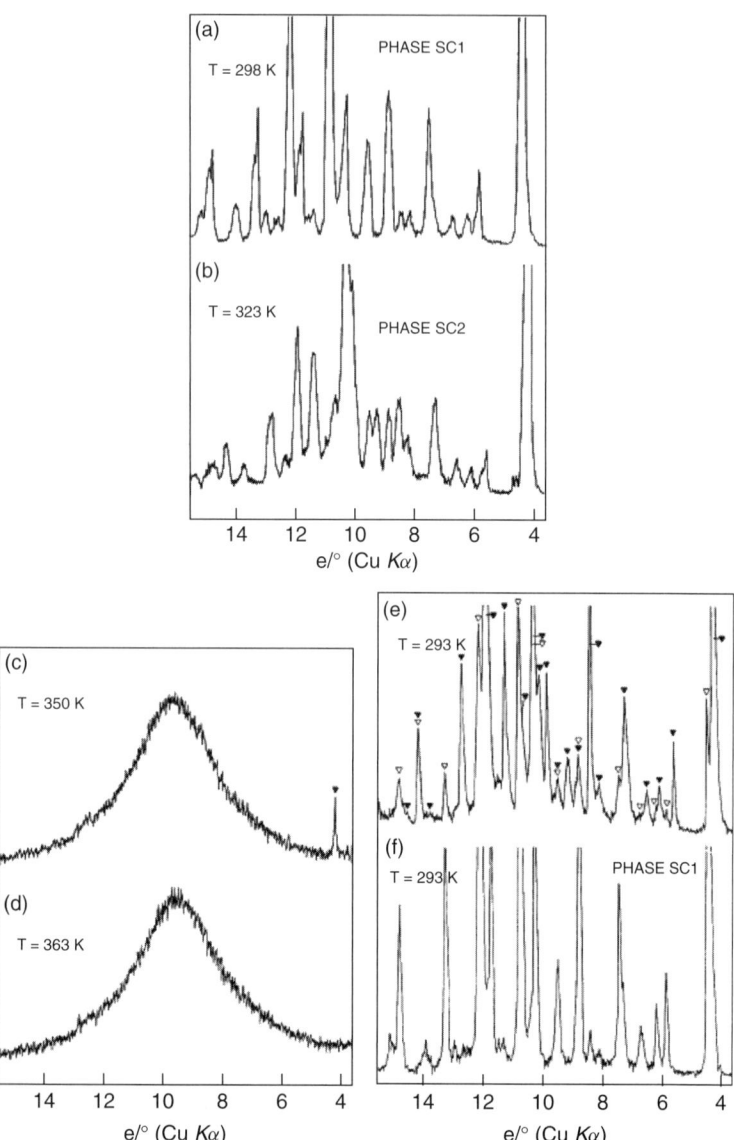

Figure 14.11 Characteristic parts of XRD patterns of DeADeS, taken at different temperatures (T) during the heating (a–d) and cooling (e, f) cycles. Phases SC1 and SC2 are denoted. (Adapted from Ref. 63.)

scanning calorimetric measurements. One should expect changes in the thermograms at two points. The first one corresponds to the transition from the solid to liquid crystalline phase, whereas the second break is known as the clearing temperature, whereby an isotropic liquid is formed. A representative differential scanning calorimetric thermogram is shown in Fig. 14.13.

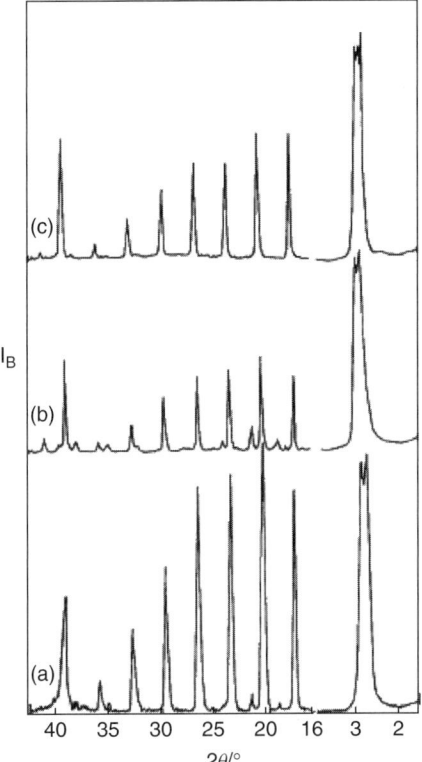

Figure 14.12 Diffractograms of DeACl/SDeS dry samples: (a) c(DeACl)/mol dm^{-3} Å 0.01, c(SDeS)/mol dm^{-3} Å 0.002; (b) c(DeACl)/mol dm^{-3} Å, c(SDeS)/mol dm^{-3} Å 0.01; (c) c(DeACl)/mol dm^{-3} Å 0.01, c(SDeS)/mol dm^{-3} Å 0.02. (Adapted from Ref. 37.)

It is evident from Figure 14.13, that there is a chain length dependence on the thermogram. With the increase in chain length of the cationic surfactant, the transition temperature increases. This is not uncommon, one could expect a rise in the transition temperature as the molecular weight of the complex increases. Similar observations have been observed also by Hoffmann et al. [58] (Fig. 14.14).

Similar work is in progress in our laboratory, where we are now studying the coacervates comprising cationic surfactants and inorganic anions like dichromate, permanganate, vanadate, and tungstate. From the unpublished data, we could conclude that the dichromate containing coacervates showed a huge weight loss, probably due to thermal decomposition at ~350 °C. Further work is still ongoing before a final conclusion can be reached.

Most of the catanionics are composed of two oppositely charged single-tailed surfactants. Hence it is quite expected that the surfactant ion pairs will have behavior similar to the phospholipid molecules. Attempts were made to check whether the coacervates could form an insoluble monolayer at the

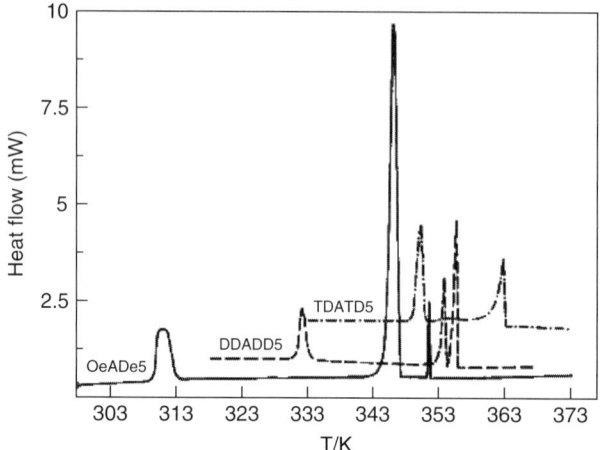

Figure 14.13 Typical DSC heating curves of decylammonium decylsulfate (DeADeS), dodecylammonium dodecylsulfate (DDADDS), and tetradecylammonium tetradecylsulfate (TDATDS). Scans are for 6.57 mg of DeADeS, 9.36 mg of DDADDS, and 5.64 mg of TDATDS [63].

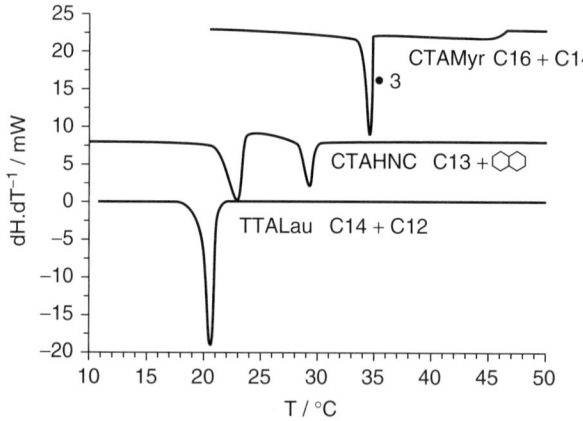

Figure 14.14 DSC curves for TTALau, CTAHNC, and CTAMyr for a heating rate of 0.2 K/min. (Adapted from Ref. 58.)

air–water interface. Panda et al. [75] studied the surface pressure (π)–area (A) behavior of different catanionics like alkyltrimethylammonium–dodecylsulfate, alkyltrimethylammonium–deoxycholate, and alkyltrimethylammonium–dioctyl-sulfosuccinate of the solvent spread films at the air–water interface. Like the phospholipids, lift-off areas were found to be dependent on the chain length of the components. Catanionics with longer hydrocarbon tails have a larger limiting area (lift-off area). The lift-off area was detected from the point after which the surface pressure started rising. The film compressibility, maximum attainable surface pressure of the films, and so on were determined from the surface pressure–area measurements, as shown in Fig. 14.15.

On the other hand, the nature of the headgroup played a crucial role in controlling the film behavior of the surfactants. A catanionic with a larger headgroup and a shorter hydrocarbon tail should show a higher lift-off area, whereas its tolerance limit to high surface pressure could be low. This was reflected through

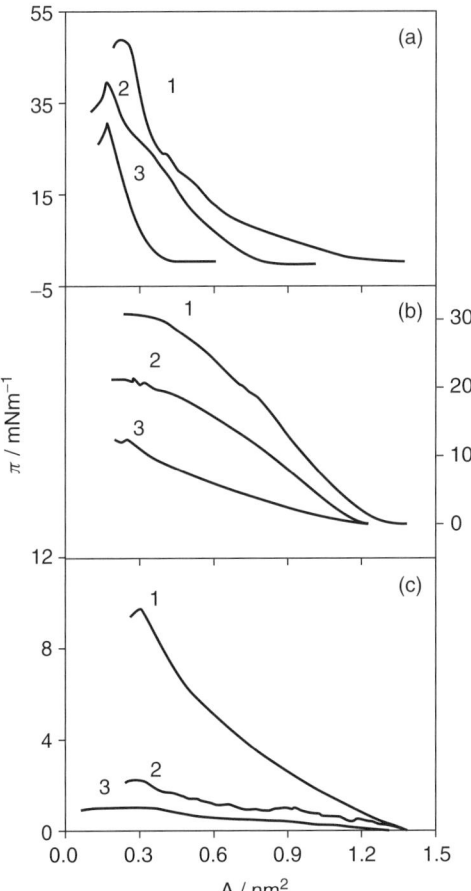

Figure 14.15 Surface pressure–area behavior of the monolayers of (a) N-alkyltrimethylammonium bromide–dodecylsulfate, (b) deoxycholate, and (c) bis[2-ethylhexylsulfosuccinate] complexes at the air–water interface at 298 K. Cationic surfactants: 1, CTA^+ ion; 2, TTA^+ ion; 3, DTA^+ ion. (Adapted from Ref. 75.)

the π–A measurement of N-alkyltrimethylammonium–dioctylsulfosuccinate ion pair. Results are graphically represented through Figure 14.16.

In a similar way, formation of coacervates at the interface was studied through π–A measurements using different combinations of catanioncs by Viseu et al. [79], as shown in Fig. 14.17.

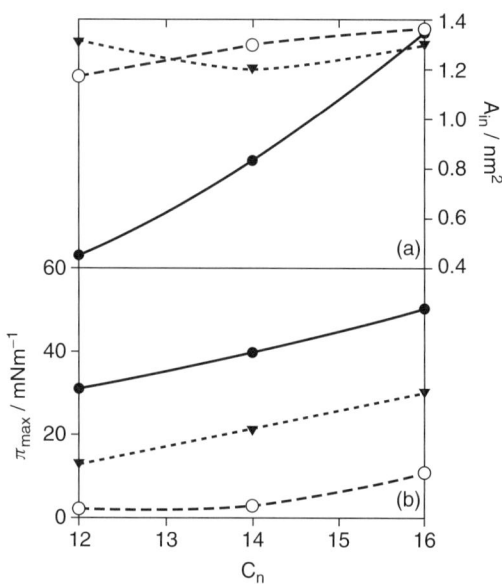

Figure 14.16 (a) A_{min} versus C_n and (b) π_{max} vs. C_n profiles for the (●) ATA$^+$–DS$^-$, (▼) ATA$^+$–DC$^-$, and (○) ATA$^+$–DC$^-$ complexes at 298 K [75].

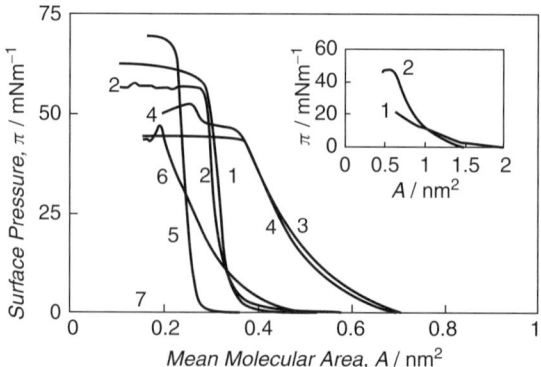

Figure 14.17 π–A isotherms at the air–water interface and 298 K for systems (**1**) DODAB-SHS (**2**) DODAB-SDS, (**3**) DDABSHS, (**4**) DDAB-SDS, (**5**) HTAC-SHS, (**6**) TTAC-STS, and (baseline, **7**) DTAC-SDS. Inset: π–A isotherms for (**8**) pure DODAB and (**9**) DDAB. (Adapted from Ref. 79.)

In another experiment [4], surface pressure–area behavior was studied for a double-tailed cationic surfactant didodecyldimethylammonium bromide using varying concentrations of anionic surfactant SDS in the subphase, as shown in Fig. 14.18.

From the analysis of the isotherms, it is clear that, like phospholipids or pulmonary surfactants, these coacervates (or catanionics) undergo two-dimensional phase transition. Change in compressibility supported this further. Therefore it is quite expected that the catanionics would have structural similarity at the air–water interface. This was evidenced from the epifluorescence measurements

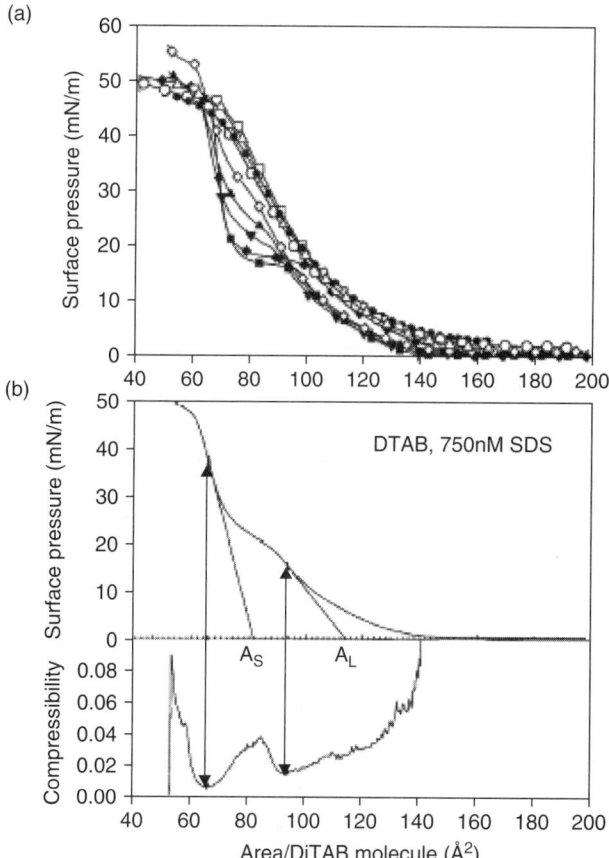

Figure 14.18 (a) π(surface pressure)–A isotherms of DiTAB monolayers compressed over subphases containing 0–50,000 nM SDS at 25 °C: (●) pure water, (□) 50 nM SDS, (△) 400 nM SDS, (○) 500 nM SDS, (◇) 625 nM SDS, (▲) 750 nM SDS, (▼) 1250 nM SDS, (◆) 5000 nM SDS, (■) 50,000 nM SDS. (b) ∂–A isotherm of DiTAB over 750 nM SDS(top), illustrating the determination of the limiting areas A_S and A_L. The intercepts of the tangents to the compression isotherm at the compressibility minima (bottom) were defined as the limiting areas A_S and A_L. (Adapted from Ref. 4.)

of solvent spread films of the catanionics. When doped with the fluorescent probe NBD-PC [1-palmitoyl-2-{12-[(7-nitrobenz-2-oxa-1,3-diazo-4-yl)amino] dodecanoyl}-sn-glycero-3-phosphocholine], a solvent spread film was generated; the bulky fluorescent probe cannot partition itself into organized solid-like regions. Therefore the organized zones should appear black when observed under the epifluorescence microscope [77]. Some images are shown in Fig. 14.19.

Growth of two-dimensional crystals at the air–water interface could be visualized from microscopic studies. Here also the onset of formation of crystals/organized regions (also known as domains) is dependent on the surface pressure and chain length of the catanioncs. A long chain containing surfactant could undergo such a process at a relatively low surface pressure. Here, unlike the phospholipids or lung surfactants, the domains are not spherical. In a mixture

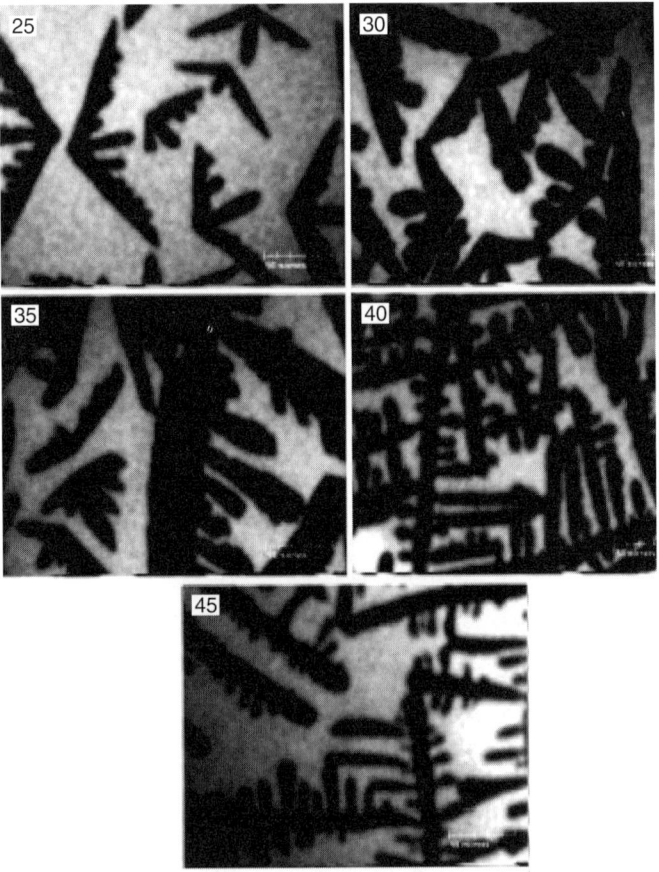

Figure 14.19 Fluorescence behavior of monolayers of NBD-PC doped CTA^+–DS. Ion-pair complex at the air–water interface at different π values (indicated in the top left corner) at 298 K. (Adapted from Ref. 75.)

of phospholipids (e.g., pulmonary surfactant), cholesterol controls the line tension across the domains. Line tension lowering along the domain boundary leads to a circular shape (which is the usual/common structure). Absence of such line tension lowering agents like cholesterol in catanionics leads to the formation of flower petal or flying kite-like structures.

Catanionics are also highly organized in solution phases, not just at the air–water interface. In the presence of trace amounts of impurities, or sometimes salts, and/or a small excess of the second component, the solubility of catanionics could drastically be increased in water. Alcohols could also increase the solubility significantly [71]. The solubility is drastically enhanced due to the formation of vesicles in solution. Like phospholipids, formation of vesicles by catanionics could be visualized through electron microscopic measurements. Usually a drop of the vesicular solution is placed on the grid of a transmission electron microscope (TEM) and the stained grid is used for imaging. Beautiful bilayer vesicular structure could easily be seen using the TEM. Shape, size, and structure of the vesicles were found to be dependent on the salinity, surfactant concentration, and composition. In fact, TEM measurement is believed to be the most useful tool to characterize catanionics. Some self-explanatory TEM images are shown in Fig. 14.20–14.24.

As already mentioned, the presence of trace amounts of impurities could greatly control the shape, size, and structure of the aggregates. In the preparation of catanionics, usually the smaller anion and cation of the cationic and anionic surfactants, respectively, remain in solution. They then act as impurities. Sometimes, for practical applications, additives like salts, alcohols, or glycerols are essential for formulations. The effect of glycerol on the microstructure of the aggregates is shown in Fig. 14.25.

It is common practice to study the pseudoternary phase diagram for a substantial time period. This is essential because sometimes the interactions between

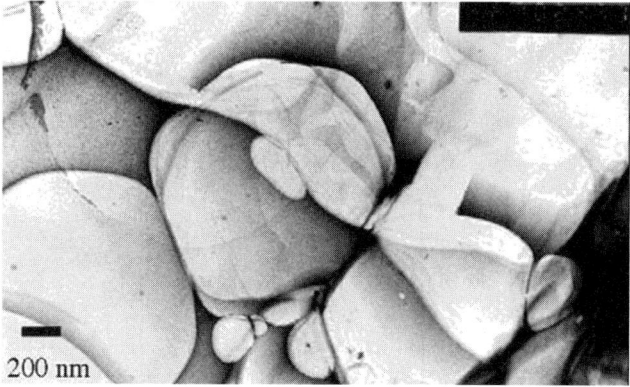

Figure 14.20 TEM micrograph of the bottom phase of ATPS-C in the CTAB/AS/water system ($xG = 0.00180$, MRCTAB/AS $= 2.177$) by negative stained method at 318.15 K. (Adapted from Ref. 66.)

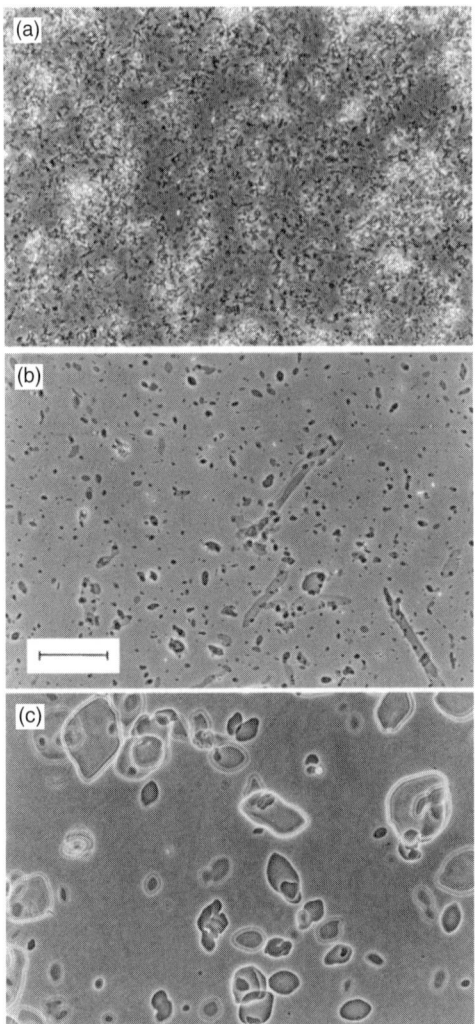

Figure 14.21 Microphotographs of (a) crystalline hexadecyltrimethylammonium dodecylsulfate (CTADS) formed from an equimolar mixture [c(CTAB): c(NaDS), 0.025 mol dm^{-3}]; (b) vesicles from a mixture with c(CTAB), 0.015 mol dm^{-3} and c(NaDS), 0.025 mol dm^{-3}, and (c) vesicles from a mixture c(CTAB), 0.025 mol dm^{-3} and c(NaDS), 0.015 mol dm^{-3}. The temperature is 303 K and the aging time is 1 day. The bar represents 100 nm. (Adapted from Ref. 15.)

two oppositely charged surfactant ions are very slow. It sometimes requires a few months' time to achieve equilibrium. Moreover, the concentration of the components plays a key role in controlling the shape, size, and structure of the vesicles. A lower catanionic concentration would lead to the formation of bigger vesicles, with lower polydispersity, and vice versa, as shown in Fig. 14.26.

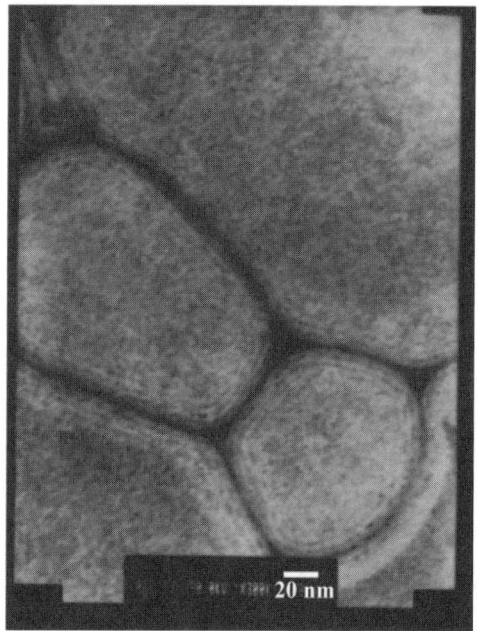

Figure 14.22 TEM image in the 1:1 STS-OTMAB/5% ethanol solution observed by the negative staining technique (concentration 10 mM). (Adapted from Ref. 71.)

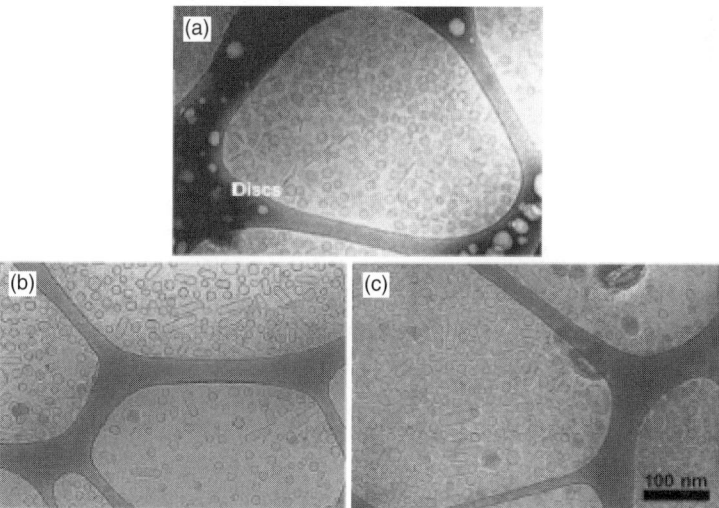

Figure 14.23 Three cryo-TEM images of 2 wt% total surfactant (CTAB and FC7) mixtures at different rCTAB/FC7 ratios. (a) At rCTAB/FC7 = 25:75; (b) at rCTAB/FC7 = 20:80; and (c) at rCTAB/FC7 = 15:85. (From Ref. 80, with permission.)

408   A CURSORY GLANCE AT THE PHYISCOCHEMICAL PROPERTIES

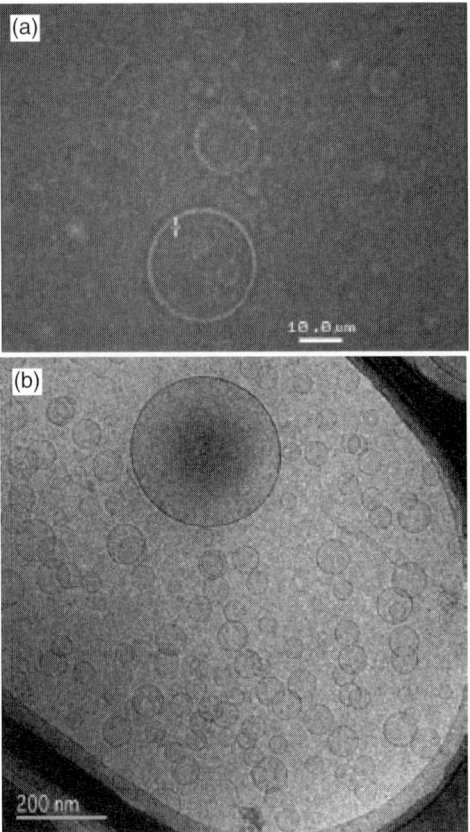

Figure 14.24 Light microscope and cryo-TEM images showing vesicles of CTAB-rich solutions in the aqueous CTAB-SOS system: (a) [S^+] 3.77 mM and (b) [S^+] 2.27 mM. ([CTAB]/[SOS] 1.27.) (Adapted from Ref. 76.)

14.7 APPLICATION OF CATANIONICS AND COACERVATES

Coacervates have an added advantage over conventional phospholipids in terms of their stability, biodegradability, and salt/acid tolerance. Thus such systems could easily be used as a substitute for phospholipids [77]. They could also be used as model membranes and drug carriers. Because the oppositely charged surfactant ion pairs are formed through electrostatic attraction, the catanionics could substitute for sphingolipids. The difference and added advantage of catanionics is that in them the distance between two opposite charges is not fixed. For phospholipids, the rotation of individual hydrocarbon tails is possible; and hydration can become a problem, but these occurrences are almost absent in catanionics. Large swelling of membrane bilayers could be overcome using catanionics as no electrostatic headgroup repulsion is present. Moreover, they have a better efficiency

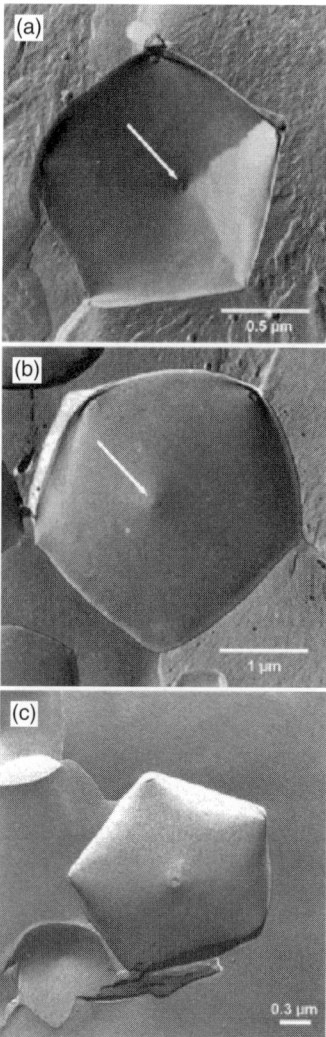

Figure 14.25 Typical freeze-fracture electron microscopy images of catanionic icosahedral aggregates taken in the presence of (a,b) 30% and (c) 70% glycerol. The arrows point out an open and a closed faceted vertex on images (a) and (b) respectively. (Adapted from Ref. 60.)

for the formation of microemulsion and could, in turn, be used in tertiary oil recovery. Catanionics could also be used as templates for nanoparticle synthesis at the air–water interface. Such a method has the advantage that a very thin film of nanoparticles could be prepared using solvent spread/adsorbed catanionics at the air–water interface. Usually such catanionics can hold the nanoparticles at the interface [77].

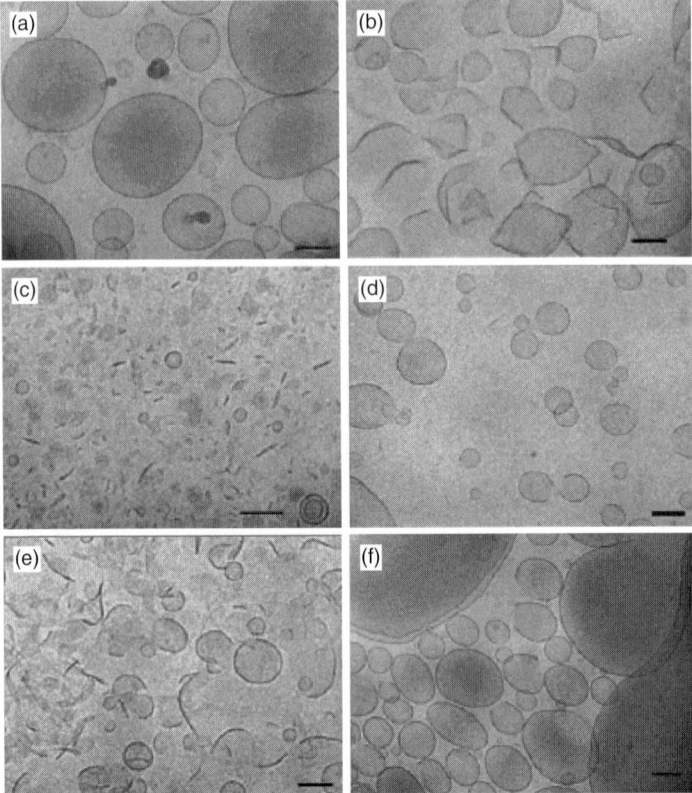

Figure 14.26 Cryo-TEM micrographs of samples aged for about 60 days at different concentrations and compositions. CTAB/SOS (a) 30:70, 1.6 wt%; (b) 30:70, 1.3 wt%; (c) 30:70, 3.0 wt%; (d) 20:80, 2.0 wt%; (e) 33:67, 2.0 wt%; (f) 37:63, 2.0 wt%. Bar length 200 nm. (Adapted from Ref. 76.)

ACKNOWLEDGMENT

AKP thanks the Department of Science and Technology, Government of India, for a research grant. The authors are thankful to Elsevier Science, Springer Verlag, American Chemical Society, Science Magazine, Nature, and the National Academy of Science USA for kindly permitting them to use the figures and images in the text.

REFERENCES

1. H. Maeda (2005). A thermodynamic analysis of charged mixed micelles in water. *J. Phys. Chem. B* **109**: 15933–15940.

2. D. F. Evans and H. Wennerström (1994). *The Colloidal Domain*. VCH, Weinheim.
3. C. Tanford (1980). *The Hydrophobic Effect: Formation of Micelles and Biological Membranes*, 2nd ed. Wiley, Hoboken, NJ.
4. A. Berman, M. Cohen, and O. Regev (2002). Catanionic vesicle–PEG–lipid system: Langmuir film and phase diagram study. *Langmuir* **18**: 5681–5686.
5. K. L. Herrington, E. W. Kaler, D. M. Miller, J. A. Zasadzinski, and S. Chiruvolu (1993). Phase behavior of aqueous mixtures of dodecyltrimethylammonium bromide (DTAB) and sodium dodecyl sulfate (SDS). *J. Phys. Chem.* **97**: 13792–13802.
6. E. W. Kaler, K. L. Herrington, M. Kamalakara, and J. A. Zasadzinski (1992). Phase behavior and structures of mixtures of anionic and cationic surfactants. *J. Phys. Chem.* **96**: 6698–6707.
7. A. Khan and E. Marques (1997). In *Catanionic Surfactants* (I. D. Robb, Ed.). Blackie Academic and Professional, London, p. 37.
8. S. V. Pingali, T. Takiue, G. Luo, A. M. Tikhonov, N. Ikeda, M. Aratono, and M. L. Schlossman (2005). X-ray reflectivity and interfacial tension study of the structure and phase behavior of the interface between water and mixed surfactant solutions of $CH_3(CH_2)_{19}OH$ and $CF_3(CF_2)_7(CH_2)_2OH$ in hexane *J. Phys. Chem. B* **109**: 1210–1225.
9. K. Nag, R. R. Harbottle, A. K. Panda (2000). Molecular architecture of a self assembled bio-interface: lung surfactant. *J. Surf. Sci. Technol.* **16**: 156–171.
10. K. Nag, A. K. Panda, B. A. Hu, D. V. Heyd, R. R. Harbottle, M. Schoel, N. O. Petersen, and L. A. Bagatolli (2002). Biophysical studies of nano-structured interfaces as models of lung surfactant membranes. *Recent Res. Dev. Biophys.* **1**: 53.
11. A. K. Panda, K. Nag, R. R. Harbottle, F. Possmayer, and N. O. Petersen (2008). *J. Colloid Interface Sci.* **311**: 551–555
12. A. K. Panda, K. Nag, R. R. Harbottle, F. Possmayer, and N. O. Petersen (2004). Thermodynamic studies on mixed molecular Langmuir films. Part 2. Mutual mixing of DPPC and bovine lung surfactant extract with long-chain fatty acids. *Colloids Surf. A: Physicochem. Eng. Aspects* **247**: 9–17.
13. A. K. Panda, K. Nag, R. R. Harbottle, K. Rodriguez-Capote, R. A. W. Velduizen, F. Possmayer, and N. O. Petersen (2004). Effect of acute lung injury on structure and function of pulmonary surfactant films. *Am. J. Respr. Biol.* **30**: 641–650.
14. C. J. Lang, A. D. Postle, S. Orgeig, F. Possmayer, W. Bernhard, A. K. Panda, K. D. Jürgens, W. K. Milsom, K. Nag, and C. B. Daniels (2005). Dipalmitoylphosphatidylcholine is not the major surfactant phospholipid species in all mammals. *Am. J. Physiol. Regul. Integrative Comp. Physiol.* **289**: 1426–1439.
15. V. Tomašić, I. Štefanić, and N. Filipović-Vinceković (1999). Adsorption, association and precipitation in hexadecyltrimethylammonium bromide/sodium dodecyl sulfate mixtures. *Colloid Polym. Sci.* **277**: 153–163.
16. R. Talhout and J. B. F. N. Engberts (1997). Self-assembly in mixtures of sodium alkyl sulfates and alkyltrimethylammonium bromides: aggregation behavior and catalytic properties. *Langmuir* **13**: 5001–5006.
17. T. Kunitake and Y. Okahata (1977). A totally synthetic bilayer membrane. *J. Am. Chem. Soc.* **99**: 3860–3861.
18. Y. Talmon, D. F. Evans, and B. W. Ninham (1983). Spontaneous vesicles formed from hydroxide surfactants: evidence from electron microscopy. *Science* **221**: 1047–1048.

19. B. W. Ninham, D. F. Evans, and G. J. Wei (1983). The curious world of hydroxide surfactants. Spontaneous vesicles and anomalous micelles. *J. Phys. Chem.* **87**: 5020–5025.
20. J. E. Brady, D. F. Evans, R. Kachar, and B. W. Ninham (1984). Spontaneous vesicles. *J. Am. Chem. Soc.* **106**: 4279–4280.
21. D. F. Evans and B. W. Ninham (1986). Molecular forces in the sew-organization of amphiphiles. *J. Phys. Chem.* **90**: 226–234.
22. J. E. Brady, D. F. Evans, G. G. Warr, F. Grieaer, and B. W. Ninham (1986). Counterion specificity as the determinant of surfactant aggregation, *J. Phys. Chem.* **90**: 1853–1859.
23. D. D. Miller, J. R. Bellare, T. Kaneko, and D. F. Evans (1988). Control of aggregate structure with mixed counterions in an ionic double-chained surfactant. *Langmuir* **4**: 1363–1367.
24. D. D. Miller, L. J. Magid, and D. F. Evans (1990). Fluorescence quenching in double-chained surfactants. 2. Experimental results. *J. Phys. Chem.* **94**: 5921–5930.
25. E. W. Kaler, K. L. Herrington, A. K. Murthy, and J. A. N. Zasadzinski (1992). Phase behavior and structures of mixtures of anionic and cationic surfactants. *J. Phys. Chem.* **96**: 6698–6707.
26. E. W. Kaler, K. L. Herrington, D. D. Miller, and J. A. N. Zasadzinski (1992). In *Phase Behavior of Anionic and Cationic Surfactants Along a Dilution Path: Structure and Dynamics of Strongly Interacting Colloids and Supramolecular Aggregates in Solution* (S. H. Chen, J. S. Huang, and P. Tartaglia, Eds.). Kluwer Academic Publishers, Dordrecht, The Netherlands, p. 571.
27. K. L. Herrington, E. W. Kaler, D. D. Miller, J. A. N. Zasadzinski, and S. Chiruvolu (1993). Phase behavior of aqueous mixtures of dodecyltrimethylammonium bromide (DTAB) and sodium dodecyl sulfate (SDS). *J. Phys. Chem.* **97**: 13792–13802.
28. M. T. Yatcilla, K. L. Herrington, L. L. Brasher, E. W. Kaler, S. Chiruvolu, and J. A. N. Zasadzinski (1996). Phase behavior of aqueous mixtures of cetyltrimethylammonium bromide (CTAB) and sodium octyl sulfate (SOS). *J. Phys. Chem.* **100**: 5874–5879.
29. O. Soderman, K. L. Herrington, E. W. Kaler, and D. D. Miller (1997). Transition from micelles to vesicles in aqueous mixtures of anionic and cationic surfactants. *Langmuir* **13**: 5531–5538.
30. N. Filipović-Vinceković and D. Škrtić (1988). Counterion binding on mixed anionic/cationic micelles. *Colloid Polym. Sci.* **266**: 954–957.
31. M. Vincekovic, D. Jurasin, V. Tomasic, M. Bujan, and N. Filipovic-Vincekovic (2006). Interactions in aqueous mixtures of alkylammonium chlorides and sodium cholate. *J. Disp. Sci. Technol.* **27**: 1099–1111.
32. N. Filipović-Vinceković, D. Škrtić, and V. Tomašić (1991). Interactions in dodecylammonium/dodecyl sulfate systems. *Ber. Bunsenges. Phys. Chem.* **95**: 1646.
33. D. Škrtić, V. Babic-Ivancic, M. Bujan, L. Tusek-Bozic, and N. Filipovic-Vincekovic (1993). On the formation of catanionic surfactant precipitate. *Phys. Chem. Chem. Phys.* **97**: 121–127.
34. N. Filipović-Vinceković, M. Bujan, Đ. Dragčević, and N. Nekić (1995). Phase behavior in mixture of cationic and anionic surfactants in aqueous solutions. *Colloid Polym. Sci.* **273**: 182–187.

35. Đ. Dragčević, M. Bujan, Ž. Grahek, and N. Filipović-Vinceković (1995). Adsorption at the air/water interface in dodecylammonium chloride/sodium dodecyl sulfate mixtures. *Colloid Polym. Sci.* **273**: 967–973.
36. M. Bujan, N. Vdović, and N. Filipović-Vinceković (1996). Phase transitions in cationic and anionic surfactant mixtures. *Colloids Surf. A: Physicochem. Eng. Aspects* **118**: 121–126.
37. N. Filipović-Vinceković, M. Bujan, I. Šmit, L. Tušek-Božić, and I. Štefanić (1998). Phase transitions from catanionic salt to mixed cationic/anionic vesicles. *J. Colloid Interface Sci.* **201**: 59–70.
38. V. Tomašić, I. Štefanić, and N. Filipović-Vinceković (1999). Adsorption, association and precipitation in hexadecyltrimethylammonium bromide/sodium dodecyl sulfate mixtures. *Colloid Polym. Sci.* **277**: 153–163.
39. R. A. Salkar, D. Mukesh, C. D. Samant, and C. Manohar (1998). Mechanism of micelle to vesicle transition in cationic–anionic surfactant mixtures. *Langmuir* **14**: 3778–3782.
40. M. S. Vethamuthu, M. Almgren, W. Brown, and E. Mukhtar (1995). Aggregate structure, gelling, and coacervation within the L_1 phase of the quasi-ternary system alkyltrimethylammonium bromide–sodium desoxycholate–water. *J. Colloid Interface Sci.* **174**: 461–469.
41. M. S. Vethamuthu, M. Almgren, B. Bergensthähl, and E. Mukhtar (1996). The hexagonal phase and cylindrical micelles in the system alkyltrimethylammonium bromide–sodium desoxycholate–water as studied by X-ray diffraction and fluorescence quenching. *J. Colloid Interface Sci.* **178**: 538–548.
42. J. Gustaffsson, T. Nylander, M. Almgren, and H. Ljudbberg-Wahren (1999). Phase behavior and aggregate structure in aqueous mixtures of sodium cholate and glycerol monooleate. *J. Colloid Interface Sci.* **211**: 326–335.
43. M. Dubois, V. Lizunov, A. Meister, T. Gulik-Krzywicki, J. M. Verbavatz, E. Perez, J. Zimmerberg, and T. Zemb (2004). Shape control through molecular segregation in giant surfactant aggregates. *Proc. Natl. Acad. Sci.* **101**: 15082–15087.
44. E. Marques, A. Khan, M. D. Gracia Miguel, and B. Lindman (1993). Self-assembly in mixtures of a cationic and an anionic surfactant: the sodium dodecyl sulfate-didodecyldimethylammonium bromide–water system. *J. Phys. Chem.* **97**: 4729–4736.
45. E. F. Marques, O. Regev, H. Edlund, and A. Khan (2000). Micelles, dispersions, and liquid crystals in the catanionic mixture bile salt double-chained surfactant: the bile salt-rich area. *Langmuir* **16**: 8255–8262.
46. E. F. Marques, H. Edlund, C. La Mesa, and A. Khan (2000). Liquid crystals and phase equilibria in binary bile salt–water systems. *Langmuir* **16**: 5178–5186.
47. P. Jokela, B. Jonsson, and A. Khan (1987). Phase equilibria of catanionic surfactant–water systems. *J. Phys. Chem.* **91**: 3291–3298.
48. O. Regev and A. Khan (1994). Cryo-TEM and NMR studies of solution microstructures of double-tailed surfactant systems: didodecyldimethylammonium hydroxide, acetate and sulphate. *J. Phys. Chem.* **98**: 6619–6625.
49. A. Khan and E. Marques (1997). Catanionic surfactants. In *Specialists Surfactants* (I. D. Robb, Ed.). Blackie, Academic and Professional, an imprint of Chapman & Hall, London, p. 37.

50. A. Khan and E. F. Marques (2000). Synergism and polymorphism in mixed surfactant systems. *Curr. Opin. Colloid Interface Sci.* **4**: 402–410.
51. H. Hoffmann, M. Bergmeier, M. Gradzielski, and C. Thunig (1998). Preparation of three morphologically different states of a lamellar phase. *Prog. Colloid Polym. Sci.* **109**: 13–20.
52. M. Bergmeier, H. Hoffmann, and C. Thunig (1997). Preparation and properties of ionically charged lamellar phases that are produced without shearing. *J. Phys. Chem. B* **101**: 5767–5771.
53. J. Hao, H. Hoffmann, and K. Horbaschek (2000). A vesicle phase that is prepared by shear from a novel kinetically produced stacked L_α-phase. *J. Phys. Chem. B* **104**: 10144–10153.
54. K. Horbaschek, H. Hoffmann, and J. Hao (2000). Classic L_α-phases as opposed to vesicle phases in cationic–anionic surfactant mixtures. *J. Phys. Chem. B* **104**: 2781–2784.
55. J. Hao, H. Hoffmann, and K. Horbaschek (2001). A novel cationic/anionic surfactant system from a zwitterionic alkyldimethylamine oxide and dihydroperfluorooctanoic acid. *Langmuir* **17**: 4151–4160.
56. J. Hao, Z. Yuan, W. Liu, and H. Hoffmann (2004). *In situ* vesicle formation by a kinetic reaction in aqueous mixtures of single-tailed catanionic surfactants. *J. Phys. Chem. B* **108**: 5105–5112.
57. K. Horbaschek, H. Hoffmann, and C. Thunig (1998). Formation and properties of lamellar phases in systems of cationic surfactants and hydroxynaphthoate. *J. Colloid Interface Sci.* **206**: 439–456.
58. C. Vautrin, M. Dubois, Th. Zemb, St. Schmölzer, H. Hoffmann, and M. Gradzielski (2003). Chain melting in swollen catanionic bilayers. *Colloids Surf. A: Physicochem. Eng. Aspects* **217**: 165–170.
59. K. Tsuchiya, H. Nakanishi, H. Sakai, and M. Abe (2004). Temperature-dependent vesicle formation of aqueous solutions of mixed cationic and anionic surfactants. *Langmuir* **20**: 2117–2122.
60. K. Glinel, M. Dubois, J.-M. Verbavatz, G. B. Sukhorukov, and T. Zemb (2004). Determination of pore size of catanionic icosahedral aggregates. *Langmuir* **20**: 8546–8551.
61. R. J. Meagher, T. A. Hatton, and A. Bose (1998). Enthalpy measurements in aqueous SDS/DTAB solutions using isothermal titration microcalorimetry. *Langmuir* **14**: 4081–4087.
62. J. Cocquyt, P. Saveyn, M. Declercq, H. Demeyere, and P. Van der Meeren (2007). Interaction kinetics of anionic surfactants with cationic vesicles. *Colloids Surf. A: Physicochem. Eng. Aspects* **298**: 22–26.
63. N. Filipović-Vinceković, I. Pucić, S. Popović, V. Tomašić, and Đ. Težak (1997). Solid-phase transitions of catanionic surfactants. *J. Colloid Interface Sci.* **188**: 396–403.
64. V. Tomašić, S. Popović, and N. Filipović-Vinceković (1999). Solid state transitions of asymmetric catanionic surfactants. *J. Colloid Interface Sci.* **215**: 280–289.
65. L. L. Brasher, K. L. Herrington, and E. W. Kaler (1996). Electrostatic effects on the phase behavior of aqueous cetyltrimethylammonium bromide and sodium octyl sulfate mixtures with added sodium bromide. *Langmuir* **11**: 4267–4277.

66. Y. Nan, H. Liu, and Y. Hu (2005). Aqueous two-phase systems of cetyltrimethylammonium bromide and sodium dodecyl sulfonate mixtures without and with potassium chloride added. *Colloids Surf. A: Physicochem. Eng. Aspects* **269**: 101–111.
67. A. J. O'Connor, T. A. Hatton, and A. Bose (1997). Dynamics of micelle–vesicle transitions in aqueous anionic/cationic surfactant mixtures. *Langmuir* **13**: 6931–6940.
68. E. F. Marques, O. Regev, A. Khan, and B. Lindman (2003). Self-organization of double-chained and pseudodouble-chained surfactants: counterion and geometry effects. *Adv. Colloid Interface Sci.* **100–102**: 83–104.
69. J. Israelachvili (1987). Physical principles of surfactant self-association into micelles, vesicles, and microemulsion droplets, in *Surfactants in Solution*, Vol. 4 (K. L. Mittal and P. Bothorel, Eds.). Plenum Press, New York, p. 3.
70. J. H. Clint (1987). In *Surfactant Aggregation.* Chapman and Hall, New York, p. 85.
71. W.-Y. Yu, Y.-M. Yang, and C.-H. Chang (2005). Cosolvent effects on the spontaneous formation of vesicles from 1:1 anionic and cationic surfactant mixtures. *Langmuir* **21**: 6185–6193.
72. J. Cocquyt, U. Olsson, G. Olofsson, and P. Van der Meeren (2004). Temperature quenched DODAB dispersions: fluid and solid state coexistence and complex formation with oppositely charged surfactant. *Langmuir* **20**: 3906–3912.
73. Th. Zemb, M. Dubois, B. Demé, and Th. Gulik-Krzywicki (1999). Self-assembly of flat nanodiscs in salt-free catanionic surfactant solutions. *Science* **283**: 816–819.
74. M. Dubois, B. Demé, Th. Gulik-Krzywicki, J. C. Dediu, C. Vautrin, S. Désert, E. Perez, and Th. Zemb (2001). Self-assembly of regular hollow icosahedra in salt-free catanionic solutions. *Nature* **411**: 672–675.
75. A. K. Panda, F. Possmayer, N. O. Petersen, K. Nag, and S. P. Moulik (2005). Physico-chemical studies on mixed oppositely charged surfactants: their uses in the preparation of surfactant ion selective membrane and monolayer behavior at the air–water interface. *Colloids Surf. A: Physicochem. Eng. Aspects* **264**: 106–113.
76. M. Almgren and S. Rangelov (2004). Spontaneously formed nonequilibrium vesicles of cetyltrimethylammonium bromide and sodium octyl sulfate in aqueous dispersions. *Langmuir*, **20**: 6611–6618.
77. B. Prélot and T. Zemb (2005). Calcium phosphate precipitation in catanionic templates. *Mater. Sci. Eng.* C **25**: 553–559.
78. S. Šegota, S. Heimer, and D. Težak (2006). *Colloid Surf. A Physicochem. Eng. Aspects* **274**: 91.
79. M. I. Viseu, A. M. Gonçalves da Silva, and S. M. B. Costa (2001). Reorganization and desorption of catanionic monolayers. Kinetics of $\pi-t$ and $A-t$ relaxation. *Langmuir* **17**: 1529–1537.
80. H. T. Jung, S. Y. Lee, E. W. Kaler, B. Coldren, and J. A. Zasadzinski (2002). Gaussian curvature and the equilibrium among bilayer cylinders, spheres, and discs. *Proc. Natl. Acad. Sci. USA* **99**: 15318–15322.

CHAPTER 15

Phase Transitions, Cholesterol and Raft Structures in Films and Bilayers of a Natural Membranous System

KAUSHIK NAG, MAURICIA FRITZEN-GARCIA, RAVI DEVRAJ, ASHLEY HILLIER, and DOYLE ROSE

Departments of Biochemistry and Physics & Physical Oceanography, Memorial University, St. John's, Newfoundland and Labrador, Canada

CONTENTS

15.1	Introduction: Membrane Models	417
15.2	Structures in Model Membranous Systems	418
15.3	Imaging Two-Dimensional Phase Transitions in Flatland	420
15.4	Lung Surfactant as Ubiquitous Membranes?	422
15.5	Functional Implication of Supramolecular Domains	424
15.6	Structural Rafts in LS Bilayers	426
15.7	Thermotropic Transitions and Lipid Dynamics in Bilayers	428
15.8	Structural Transitions in Domains at the Molecular Lattice Dimensional Level	430
15.9	Cholesterol in Domain Organizations	432
Acknowledgments		436
References		436

15.1 INTRODUCTION: MEMBRANE MODELS

The cell membrane structure–function models are currently once more being revised due to some breakthroughs in the last two decades. From the initial

---

Structure and Dynamics of Membranous Interfaces, edited by Kaushik Nag
Copyright © 2008 John Wiley & Sons, Inc.

development of the previous model where cellular cytoplasm was encased by a lipid–protein "bilayer" as developed by Gorter and Grendall [1] by studying Langmuir monomolecular films of red blood cells, to the later development by Singer and Nicholson [2] of the fluid-mosaic model, the last decade has seen a complete revision of these based on raft structures discovered in cell membranes [3]. The liquid crystalline properties of membranes, surfactants, and other colloidal systems have been defined theoretically as soft materials by DeGennes [4], allowing for understanding of the organization of these systems as types of condensed matter. These systems show properties that cannot be understood from just the fluid anisotropy of crystalline systems, since ordered structure–function units or "lipid rafts" have been discovered in membranes by Simons and Ikonen [5]. These discoveries have led to understanding that, although membranes are fluid mosaics and act as barriers for exchange of materials, specific membrane components can phase segregate into momentarily structured units for function. The fluid-mosaic model has thus been revised and currently a paradoxical dynamic, "yet" structured model has been proposed [6].

15.2 STRUCTURES IN MODEL MEMBRANOUS SYSTEMS

From previous studies performed by Langmuir on various surfactants and organic and inorganic amphipathic molecules in films in the 1930s, the phase segregation and structural organization of model membranes could only be suggested from indirect inference from monolayer isothermal and bilayer enthalpy phase transition data. Attempts to understand the organization of amphipathic molecules at the water interface using some fascinating techniques, such as spreading of talcum powder on surfaces of monolayers under various stages of packing density, to measurement of optical diffraction pattern of light from membrane surfaces (or light blockage by black lipid membranes), had led to limited information about membrane structure. The breakthrough in directly imaging domains came about perhaps accidentally in some of the early studies performed by Peters and Beck [7] and Von Tscharner and McConnell [8] in the early 1980s [9]. In studying the simple diffusion of fluorescent lipid probes to measure the lateral motions of lipids in Langmuir films under compression (Fig. 15.1b), a glimpse of membrane domains were obtained by these groups [7–9]. The monolayer films of a phospholipid were observed to have micron scale supramolecular organization of gel-like lipids in condensed domains (Fig. 15.1a,c) existing in a sea of fluid anisotropic phase, close to the intermediate compressed packing densities (Fig. 15.1b, left panel), which suggested a phase transition occurring in such films from the isothermal breaks in surface pressure–area profiles [9, 10]. The gel-like domains formed due to regions of the film undergoing a phase separation (or segregation), where the ordered phospholipid acyl chains organized-out into condensed domains from a more fluid disordered phase with an increase in lipid lateral packing density. Over the last two decades probably a few hundred papers have been published by various groups on domain imaging using the fluorescence

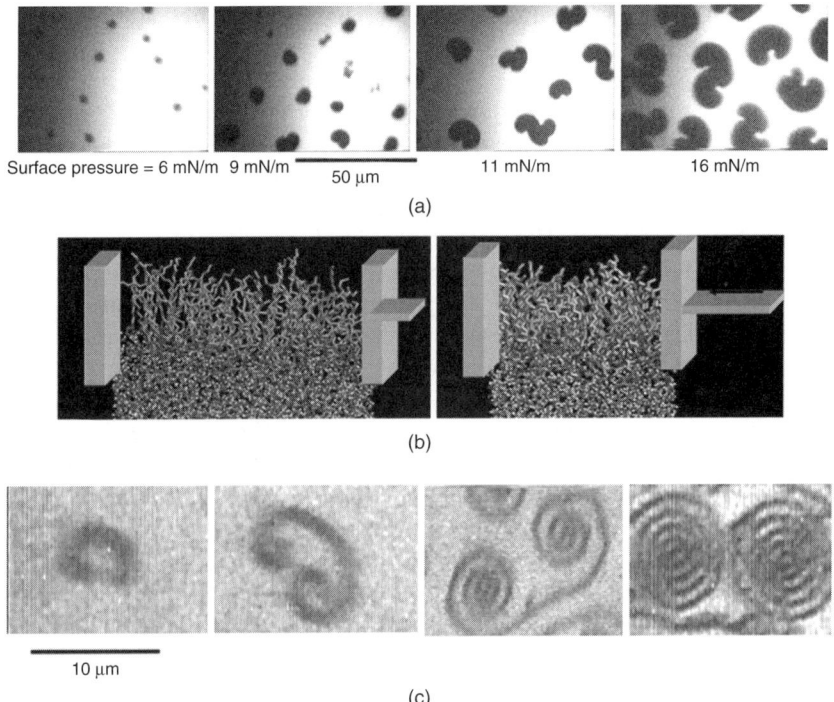

Figure 15.1 Fluorescence microscopy images of (a) DPPC and (c) DPPC + 2% cholesterol monomolecular films undergoing lateral phase transition due to as increase in packing density (b). The fluorescent probe partitioned in the fluid-disordered regions (bright areas in (a) and (c)) of the film due to the lower packing density, allowing for imaging of the gel domain (black in (a)). These images were obtained during quasistatic film compression directly from the air–water interface [12]. Panel (b) shows the molecular dynamics simulation of the lipid molecules at 60 Å^2/molecule (left) and 50 Å^2/molecule (right) correlating to a change of 4–9 mN/m surface pressure [46]. Cholesterol reduces the line tension of the DPPC gel–fluid domain boundary by accumulating in such boundaries, which is manifested by the growth of the domains in length or in one dimension (c) [9].

microscopy of Langmuir films of phospholipids, membranes, lung surfactant, organic and inorganic amphipathic molecules, drugs, proteins, and liquid crystals (reviewed in Refs. 9–11), including some of our studies [12–24]. The gel domains of a phospholipid (16:0/16:0-PC) monomolecular film of 1,2-dipalmitoyl phosphatidycholine (DPPC) imaged using fluorescence microscopy directly from an air–water interface are shown in Fig. 15.1a,c [12, 13]. With addition of a second component or cholesterol (Fig. 15.1c) to DPPC, the organization of the gel-domain structures was observed to shift from ellipsoids to spirals (shape transition), suggesting that the complexity of lipid mixing in membrane domains is not as simple as those previously understood [20]. The role of cholesterol in such domain structures has also become an intense area of research due to the discovery of lipid rafts.

Despite these breakthroughs in observing structures in monomolecular films or model membranes, the suggestion of using planar two-dimensional systems as "good" models for the three-dimensional cell "bilayer" organization was initially received with much skepticism [9, 10, 25]. One of the major reasons was that a specific lateral packing density or "surface pressure" of a bilayer membrane could not be unequivocally demonstrated to be those of a planar monolayer at intermediate density and was only theoretically predicted to be ~30 mN/m [25]. This is due to the fact that the bilayers are confined in three dimensions by polar solvents, spherical in nature, as well as there being hydrophobic interactions at the acyl chain core region, which are not found in lipid planar films at the air–water interface [25]. Thus the two opposed hemilayers in the cell bilayers may exhibit packing arrangements and structures that are completely different from those of the more nonconfined monolayer films at the air–water interface, as well as the cell membranes being "asymmetric" in composition. Also, the curvature and complexity in composition of a cell membrane would make the monolayer structural models overly simplistic and can only be compared with caution. The second breakthrough in this controversy came from some seminal studies by Bagtolli, Gratton, and co-workers [26, 27], where, using similar fluorescent probe principles as for the monolayer, giant unilamellar vesicles (GUVs) or freestanding bilayers were directly surface imaged. The bilayer surfaces were visually observed also to contain gel–fluid domain organization observed during thermal annealing, suggesting that the monolayer (or hemilayer) studies were not all in vain [9–25]. Recently, a number of native cell membrane extracts and model lipid–protein bilayers, as well as those of diverse membranous systems such as lung surfactants, have been observed to have supramolecular domain organizations in GUV bilayers [26, 27]. In one study we also showed that in LS the gel domain organization was coupled across each hemilayer (monolayer) of the bilayer. Due to the advent of the models of lipidrafts as well as direct imaging of such raft structures in cellular membranes, these systems have once more become an area of intensive research. This is due to the fact that rafts have been determined to be the functional units for transmembrane protein functions as well as being linked to membrane-related disease and dysfunction [3, 28].

15.3 IMAGING TWO-DIMENSIONAL PHASE TRANSITIONS IN FLATLAND

Historically, some of the observations made on monolayers possibly dates back to about three thousand years ago, when certain visual observations were made while using Chinese ink (an oil–water–pigment emulsion) to write text, which required one to dip the tip of the ink pen in clean water to remove debris. The light reflected from the surface of this water showed rainbow colored patterns, suggesting that there was some sort of diffraction of the white light from the interfaces. Later, the seminal studies of Irving Langmuir, Pockels, and Blodgett clearly showed that the films of some organic and fatty substances formed monolayers, which suggested specific molecular organization of the materials at the air–water interface [29]. To observe this organization, some studies were attempted where Brownian motion of insoluble particulates (talcum), floating paper strips, and circular strings were observed to change either motion or shape, as the interface of the films were compressed. Since monomolecular films are two-dimensional systems [9–11, 29], these processes were considered to be "phenomena occurring in flatland" (Alice's observations in Wonderland!) [30].

Lipids and a host of other amphipathic molecules undergo a phase separation or transition when compressed laterally at the air–water interface (as shown in Fig. 15.1b). One of the most studied phospholipids that is not present in eukaryotic cell membranes, DPPC (16:0/16:0-PC), has a chain melting temperature (41 °C) well above ambient laboratory temperatures used to study bilayer membranes and is very thermodynamically suitable for studying synthetic lipid–protein bilayers using scanning calorimetry [16]. The hydrocarbon chains of DPPC alter in the perpendicular axial tilt as well as decrease in motion at an air–water interface upon an increase in packing density of the films upon compression [11]; similar processes are assumed in the thermal chain freezing process in bilayers. The surface pressure–molecular area isotherms show a reasonable number of well-identified breaks or nonlinear surface pressure changes, suggesting discontinuity in phase change and some sort of supramolecular organizations of these frozen lipids in the films. By incorporating a small amount of fluorescent probes in such films, Von Tscarner and McConnell [8], Peters and Beck [7], and Möhwald [10] were able to show that the lipids organized in gel-like or condensed domains coexisting as islands in a sea of fluid-expanded phase. The domains appear dark due to their tighter packing excluding the fluorescent probe from the condensed regions after initial nucleation; such domains grow in size with increasing lipid packing density, eventually merging into a single phase upon high packing states [12, 13]. Some of these studies have revealed complex processes that occur not only at a membranous interface but also apply to the electronic properties of liquid crystalline materials, the thermodynamics of nerve system propagation, the hydrodynamic flows at an interface, and novel pathways in explaining Einstein's fluid electrohydrodynamics [4, 9–11, 30].

The phospholipid DPPC is not found in any significant amount in any eukaryotic cell membrane [31, 32]. Most biological membrane phospholipids have a high degree of unsaturation in their fatty acyl chains; therefore, due to their melting temperature being below 0 °C, they do not display any transitions in films studied at ambient temperatures of 15–50 °C [31]. Also, monolayer films can only be studied above the freezing point of water to maintain a consistent air–water interface. The monounsaturated phospholipid 1-palmitoyl-2-oleyl-phsophatidylcholine (16:/18:1-PC), which constitutes a significant portion of the outer red blood cell plasma membrane, exists as a continuous fluid phase in film when laterally packed to high pressure [20]. Thus the "real" membrane phospholipids are difficult to study using monolayer films and limited information is available regarding the structural organization in cell membranes from such monolayer studies [9–11, 31]. However, DPPC is present in significant amounts in an extracellular lipid–protein secretion found in the lungs, termed "lung surfactant." We have studied this membranous "system" extensively not only as a model of a surface active agent of the lung but also as a model of some cellular membranes [12–24, 31]. This is due to the fact that surfactant also contains equivalent amounts of cell membrane components such as a fluid phospholipid (POPC) and significant amounts of cholesterol [12–24, 32].

15.4 LUNG SURFACTANT AS UBIQUITOUS MEMBRANES?

Lung surfactant (LS) is a lipid–protein secretory material that stabilizes the alveoli by reducing the surface tension of the lung at the air–water interface to values near 1–2 dynes/cm [33–35]. Most mammalian LS contains a significant proportion (about 30–35%) of 16:0/16:0-PC (DPPC) and 16:0/18:1-PC (POPC) as well as ~10% cholesterol [32]. Thus LS is a highly ubiquitous membranous system in eukaryotes due to this unusual lipid composition, is found in all air-breathing vertebrates and has also been evolutionarily conserved over a quarter of a million years. The duck-billed platypus has similar LS composition as humans and most other mammals [32, 36]. However, some recent studies have shown that in a few species of mammals unusual phospholipid compositions are detected and DPPC may not be the major surfactant component [37]. In fact, the physical properties of the LS are closer to some cellular membrane components. Also, some of the LS lipid associated surfactant proteins are structurally very similar to some cellular transmembrane proteins, and their main function may be to enhance the "surface" activity of LS lipids. Somehow peanut butter (!) seems to be able to do similar things to LS lipids as these proteins, and it has been a paradox of surfactant researchers over the last five decades in trying to explain why such complicated lipid–protein systems are required to perform a simple "surface activity" change in the lung [38, 39]. The saturated acyl chain component of LS (DPPC) can sufficiently reduce the surface tension of an air–water interface to values near 1–2 mN/m, and the process does not require other lipids and proteins [39].

Recently, studies have suggested that some surfactant proteins (SPs), SP-B and SP-C, may function as channels (SP-C) or antimicrobial porins and possibly take part in lipid signaling (SP-D), such as exhibited by some cellular transmembrane proteins [40]. The LS complexes in the lung also have structural features seen in various lipid polymorphic forms such as multilamellar and unilamellar vesicles, tubular hexagonal type-II phases (tubular myelin), and some planar multilayered films and bilayers. Despite these structures being discovered about four decades ago using electron microscopy, clinical "surfactologists" have ignored comparison of the system with cell membranes, as LS is considered to be the holy grail—an "extracellular surface tension reducing agent" [38]. In the vast literature that pertains to models of cellular membranes, the physical chemistry and lipid polymorphic behavior have never been compared to the LS system [40]. This has led to a lack of understanding of this system from the viewpoint of lipid–protein interactions and the polymorphic behavior of membrane components, as well as the function of LS.

Recently, this narrow view of the LS system as just a surface tension reducer has led to more paradoxical findings and some completely contradictory ideas [39] about LS lipid and protein functions. Recent studies have pointed out that the classical "laboratory assigned" roles attributed to some of the major LS components, such as DPPC, require reevaluation [40, 41]. Certain species of mammals that do not have DPPC as the major component of LS, however, can reduce surface tension of an air–water interface to near zero values [37]. Fluid phospholipids (e.g., POPC) of LS, when rapidly compressed in captive bubbles, can also reduce surface tension to very low values [41]. Also, LS films with cholesterol have shown lipid raft-like structures in bilayers and films, upon which the surfactant proteins may reside (to perform some yet unknown functions) [42]. Some new researchers in diverse fields are interested in studying LS as a membranous system, rather than a surface active agent (see other chapters in this book).

Systematic studies of surfactant components, with single or multiple component mixtures of lipids and surfactant proteins, have shown similarities between lipid–protein arrangements and membrane rafts [42]. In the case of LS, the cholesterol-rich areas form fluid-ordered domains with DPPC and other lipids [42]. In other studies, we have observed raft-like domain structures in a native LS system extracted from bovine lungs (bovine lipid extract surfactant or BLES), which contains all components of surfactant except the water-soluble glycoprotein (SP-A and SP-D) and cholesterol. The mass spectral composition of the material shows that this system consists mainly of about 30% DPPC (16:0/16:0-PC), 28% POPC (16:0/18:1 PC), and ~20% of other phospholipids such as phosphatidylglycerol (PG, about 8%) and small amounts (1–4%) of phosphatidylethanolamine (PE), phosphatidylinositol, and sphingomyelin among others [32, 43]. The hydrophobic proteins SP-B and SP-C are also present in BLES (~2–3 wt% of the lipids). BLES films as well as those of other surfactants extracted from bovine, porcine, and rat lungs show phase transition and specific DPPC-rich domain structures as shown in Figs. 15.2 and 15.3 [38, 43]. The supramolecular domains are formed due to organization of the gel lipids into isotropic phases, as those previously observed in some cell membrane systems. The gel domains in BLES films could also be chemically mapped and were found to contain mainly DPPC [43].

The atomic force microscopy (AFM) image in Fig. 15.2 shows that the gel domains have a higher (0.9-nm) height profile than their surrounding fluid phase (inset of Fig. 15.2); also, some internal organization inside the domains could be observed (Fig. 15.3). The domains are formed by phase segregation of DPPC into condensed domains, when the BLES films are packed above a surface pressure of 15 mN/m. Since AFM can be used to image structures at nanoscales, the image shown in Fig. 15.3 shows that the terminal ends of the hydrocarbon chains of the

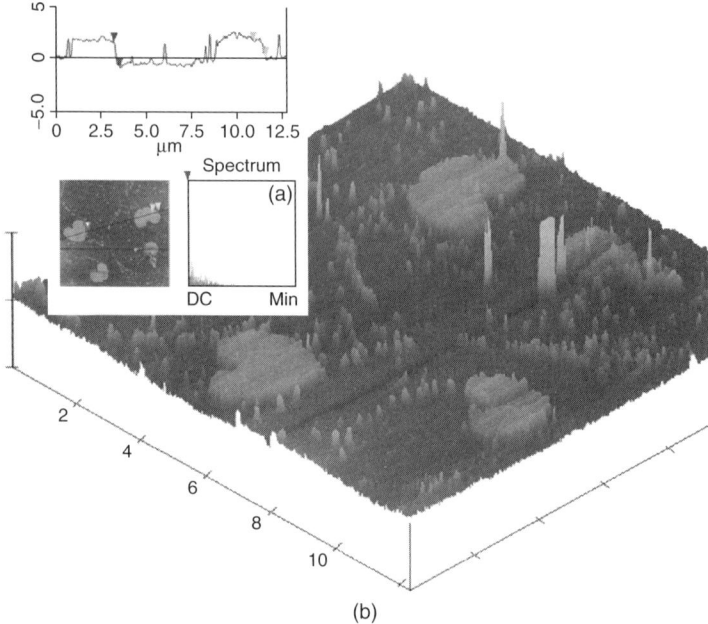

Figure 15.2 (a) Atomic force microscopy (AFM) images and section analysis and (b) three-dimensional topography of domains observed in bovine lung surfactant extract (BLES) films [68]. The films were deposited on mica by the Langmuir–Blodgett method and imaged in air using the contact mode of AFM [68]. The section analysis in (a) was performed at the gel–fluid domain boundary (marked by arrows) and suggests that the gel domains are about 1 nm higher than the surrounding fluid phase [44]. These domains are rich in the gel-phase lipid DPPC [43]. Other microscopic nanoscale structures are also observed in the fluid phase of BLES films (b), which are aggregates of other lipids or molecules of a single protein (SP-B or SP-C) present in BLES [44, 68].

phospholipid DPPC inside the gel domains have further nanoscale organization [38, 44, 45]. The possible acyl chain conformation of the phospholipids in the gel and fluid phase in a BLES film is shown in the snapshot (inset of Fig. 15.3), using molecular dynamics simulations of such layers [46].

15.5 FUNCTIONAL IMPLICATION OF SUPRAMOLECULAR DOMAINS

The discovery of lipid rafts in cell membranes has boosted research efforts in further comprehending the functional implication of such lipid organization in diverse models of cell membranous systems [47–49]. These membrane rafts are known to contain mostly saturated chain sphingomyelin, cholesterol, and one or more membrane fluid phospholipids (e.g., POPC) and acylated GPI-anchored proteins [3]. Obviously, this process of raft formation is dependent on the phase

Figure 15.3 (a) High resolution AFM images of a DPPC-rich domain in a BLES film, imaged at (b) atomic resolution, where the terminal methyl groups of the fatty acyl chains are observed, suggesting nanoscale organization of this molecule in the gel phase [68–70]. The arrow in (b) shows the 0.9-nm height difference between the gel (tabloid shaped domain) and the fluid phase, which is due to the more perpendicular tilt of the fatty acyl chains of DPPC in gel domains. This tilt is shown in the molecular dynamics (MD) simulation snapshot of a gel–fluid domain boundary in (c) [46, 69]. The MD simulations are performed by lateral confinement of the molecules into two different areas (per molecule) of packing density or by allowing a larger number of molecules to equilibrate at an intermediate packing density [46, 69]. Previously, others have been able to observe specific gel domain formations in DPPC using MD over longer equilibration time directly, although the dynamic nanostructuring occurs over extremely short time scales (femtoseconds) [69].

mixing of specific lipids and proteins, as well as on the lateral packing density fluctuations, which are dynamic processes [3, 6]. Various studies have shown that in model membranes of lipid raft mixtures, cholesterol–phospholipid interactions lead to the fluid–fluid phase immiscibility in a gel–sphingolipid and fluid–liquid crystalline environment [44, 47–50]. Cholesterol acts mainly as a rigidifier of the fluid lipids and as a fluidizer of the gel lipids inside the rafts, leading to such dynamic structures being considered as domains of "liquid-ordered" (L_o) phases [3, 47, 50].

Recently, these same properties have been observed in natural porcine LS systems, where additional cholesterol was found to induce a raft-like phase, and the proteins were found to reside in such phases [42]. However, in the case of most LS with natural amounts of cholesterol (7–10%), the L_o phase is probably not of primary importance, since the functional implications of the domains (Figs. 15.2 and 15.3) that are rich in DPPC can allow LS films to reach low surface tensions at the lung air–water interface [44, 45]. The phospholipid DPPC when packed to the molecular limit (van derWaal radii) in films (40 Å^2/molecule) allows for almost "solid-like" films at the air–water interface [39]. However, others have suggested that the DPPC solid phase can actually change the air–water interface to rather a solid–water interface, and thus the term surface tension may become redundant at high packing densities of the films at the lung air–water interface [39]. Despite these specific controversies of a solid, gel, or L_o phase in LS function, there is recent evidence to support the belief that this supramolecular structural arrangement of lipid–protein domains in LS is abolished in some cases of lung diseases [51], as well as by an excess of cholesterol, which increases in LS in some of these diseases [44, 45, 51, 52]. This may allow for comparison of specialized cell membrane rafts and gel and L_o domains in the LS system induced by cholesterol. There are also reports in the LS literature suggesting that some surfactant proteins may have similar functions as cell membrane proteins. The surfactant glycoprotein SP-A may help membrane receptor binding and immune functions; SP-C is acylated and may play a role in transmembrane anchoring of other proteins and lipid layers; while SP-B allows for membrane fusogenic processes since this is a saposin with antimicrobial properties. Some of these processes may occur via the domains formed in LS layers. Recently, SP-B/C has been found to reside in these liquid-ordered domain regions [42], and other studies have shown specific association of some of these proteins with DPPC-rich gel domains [12–24]. Thus raft-like structures in LS may play some critical role in the functioning of various surfactant proteins, and some of these functions may well be "non-surface-active" [40].

15.6 STRUCTURAL RAFTS IN LS BILAYERS

There is only a limited amount of literature available on LS models in bilayers and thus comparison with various models previously developed for cellular systems is difficult. Recently, the focus on bilayer studies of LS has been prompted by discovery of direct imaging of freestanding bilayers or GUVs [42, 53, 54]. In a previous study we showed that, during cooling of BLES bilayers, the DPPC-rich gel-like domain phase segregates out of the fluid phase and the gel domains can be observed by imaging the GUV surface [53, 54]. Figure 15.4 shows the effect of cooling of a BLES freestanding bilayer surface; the bright regions of the surface of the GUV are the fluid phase and the dark-spotted areas are the gel domains (Fig. 15.4b). In a previous study we also observed (by using two-photon confocal fluorescence imaging of the BLES bilayers) that the DPPC-rich gel domains

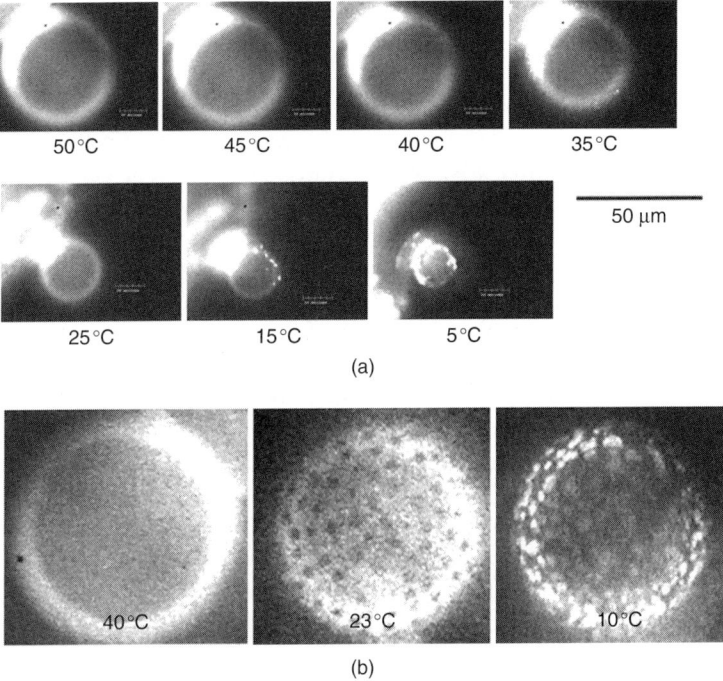

Figure 15.4 Phase transition in freestanding bilayers of BLES observed by cooling a giant unilamellar vesicle (GUV) from 45 °C to 5 °C (a), as noted from shrinkage of the vesicle's diameter. The dark regions in (b) are the DPPC-rich gel-condensed domain and the bright areas are the fluorescent probe containing the fluid phase. The gel domains are coupled across each hemilayer of the bilayer in (b) and at low temperatures possibly a third subgel or liquid ordered phase is formed as shown by a complete rearrangement of the fluorescent probe [42, 53, 70]. The GUV normally ruptures at low temperatures or extremely high packing densities and some fluid-probe containing lipids can dissociate from these bilayer surfaces as other smaller vesicular forms [54].

were coupled across each hemilayer of the bilayers [53, 54]. The vesicles shrink in size upon cooling, suggesting a cooperative transition between the fluid and gel phases [53, 55, 56]. The diameter or size of the GUV decreases since the gel phase formation reduces the area (molecular volume) occupied by the lipids in the bilayer [55]; a similar property is induced in the films by lateral compression [55, 56]. The structures have also been observed in multilamellar vesicles of porcine LS and may suggest that lamellar bodies found in the lungs may actually have specific presecretory lipid–protein organization, which is related to functional lipid sorting, as found in some cell membrane systems [42]. Others have shown that these thermotropic transitions and lipid demixing may be critical for functioning of LS as well as in other cellular membrane systems [47–49, 55–57].

15.7 THERMOTROPIC TRANSITIONS AND LIPID DYNAMICS IN BILAYERS

The structural reorganization of lipids in LS bilayers reveals features where the phase transition of the whole complex lipid–protein system can be measured as a disorder–order transition, as those measured previously in gel–fluid models of cell membranes [55–57]. To study such phase transitions, one has to take into consideration the specific lipid dynamics, such as translational or rotational motions of the raft lipids as well as the critical transitions and thermodynamics of such systems [56]. It was suggested about three decades ago that these structures and transitions in LS are "critical" for respiration [56], as well as in other cell membrane systems such as bacteria [57]. Some studies have established that the accurate cholesterol composition and thermodynamics of lipid–lipid interactions may be "critical" to the functioning of many diverse membranous systems [9, 47–50, 54, 55, 58]. Others have established structural phase segregation of lipids in naturally extracted cell membrane systems such as those of red blood cells, neural membranes, and bacteria [47–49, 57, 58]. Due to these facts, it is appropriate to correlate details of structure–function properties of rafts in diverse membranous bilayer systems and their phase transitional processes.

Currently, structural studies of various membrane systems can be used to understand lipid–protein supramolecular organization and associations only at the macroscopic level, since the equilibrium properties of such domain structures are not clear. Also, rafts formed in cell membranes (and probably in LS) are dynamic structures due to various lipid motions, from the bond vibrations observed at the nano- to femtosecond time scale to the translational motions observed at the microsecond time scale. We have applied deuterium nuclear magnetic resonance (^{2}H-NMR) to understand the dynamics of the ordering of surfactant lipids in gel–fluid regions of BLES bilayers [54, 59]. Using multilamellar dispersions of BLES doped with a small amount of chain predeuterated DPPC (DPPC-d_{62}), the NMR profiles suggested a broad liquid crystalline to gel transition in such bilayers upon cooling (Fig. 15.5), in the same temperature range as those in GUV bilayers [53]. The sharpness of the peaks and the higher width of the terminal methyl splitting at high temperature suggest that the gel lipids in BLES had a liquid crystalline motional environment at high temperature and these motions freeze to slower speeds in the gel-like environment upon cooling. This process is mainly observed as a reduction of the sharpness of the methyl and methylene frequency peaks upon entering the gel phase and can be defined more clearly in the order parameter profiles (Fig. [5]b). The profile shows that with decreasing temperature an increase of the order parameters or of DPPC chains occurs [59]. The increase of ordering upon cooling showed a more abrupt change between 20 and 30 °C. The peak or midpoint of the transition is located at $\sim$27 °C, as is also shown by the differential scanning calorimetry profile (inset in Fig. 15.5) of BLES bilayers [53, 59]. This study suggested that the structural features of domains observed in the films and GUVs, can also be monitored at the dynamic level and can possibly be applied to other natural membrane systems, although

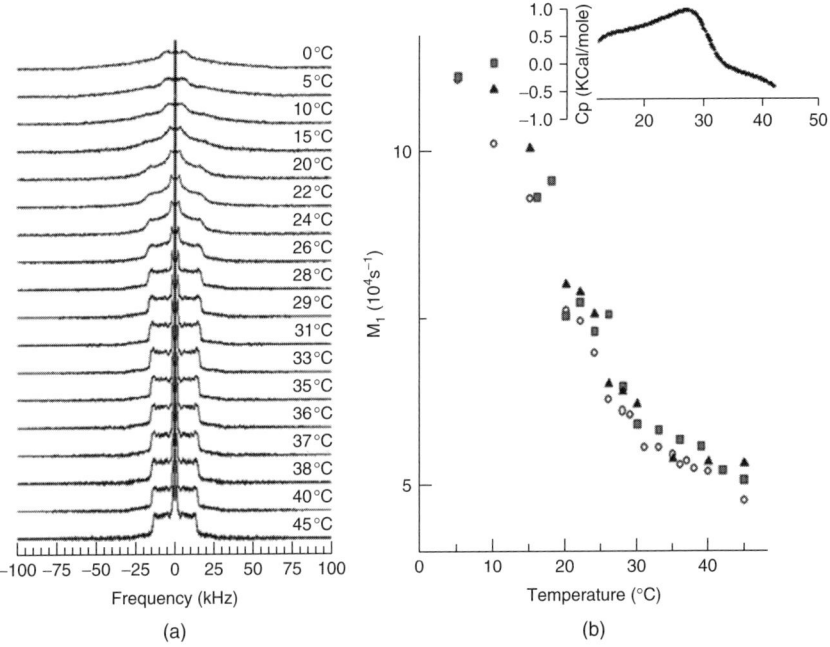

Figure 15.5 (a) Deuterium ^{2}H-NMR frequency shift and (b) order parameter profiles of BLES bilayers in MLV dispersions plotted as a function of temperature, suggesting the fast and slow motions of an acyl predeuterated DPPC (DPPC-d$_{62}$) incorporated in the membrane [59]. These profiles suggest that the microscale lateral and rotational motions of the DPPC lipids in the gel and fluid phase environments are induced by temperature change. The differential scanning calorimetry profile of the thermal (enthalpy change) phase transition is shown in the inset of (b). The methylene and methyl peaks of DPPC acyl chains in (a) are more prominent and separated in the fluid anisotropic phase; they become more diffuse upon entering the gel phase at lower temperature, suggesting slowing of the lateral and rotational motions of the phospholipid in the gel phase [54, 59]. The midpoint of this disorder–order (fluid–gel) transition observed by NMR (b) is similar to that observed in the DSC enthalpy profile (inset). The order parameter profiles of BLES (open circles) are very similar to those previously obtained from a DPPC:POPC bilayer; addition of cholesterol (square symbol) slightly shifts the order parameter values at lower temperatures [54].

the average rotational and translational motions of the deuterated DPPC suggest ordering changes of its fatty acyl chains. Addition of cholesterol, calcium, and a soluble protein to BLES was found to alter the NMR observed motional ordering profiles, suggesting that the dynamics of the system can be altered because of the additions [59].

The ordering profiles of membrane lipids in bilayers at an even faster (10^{12} s) bond vibrational time scale have recently been studied using Raman spectroscopy, and some structural information on raft-like domains has been deciphered

[60, 61]. Studying membranes using Raman spectroscopy allows the various bond vibrations of phospholipids to be independently monitored without the use of probes. Although Raman signals are quite weak for structural studies, Raman microscopy of model membranes of skin cells allowed domains or rafts to be visualized [61]. By heating the BLES bilayer, dispersions of the vibrations of the C—H symmetric and asymmetric modes could be observed to shift in frequency (wavenumber) throughout the 20–30 °C range of the fluid to gel phase transition (Fig. 15.6) [60]. The Raman shift of the wavenumbers from 2847 to 2850 cm^{-1} (hydrocarbon methylene) plotted as a function of temperature shows similar order parameter profiles for BLES as those seen by ^{2}H-NMR (Fig. 15.5). However, here the average motions were obtained from vibration of all bonds and not specifically from gel or the fluid lipids in BLES. These studies suggest that the structural organization of specific lipids in the supramolecular domain in bilayers can be monitored at the dynamic level and may allow future studies of other natural bilayer membrane systems as well as the dynamic properties of lipid–protein associations in raft domains.

15.8 STRUCTURAL TRANSITIONS IN DOMAINS AT THE MOLECULAR LATTICE DIMENSIONAL LEVEL

Although supramolecular organization of domains can be inferred indirectly from the dynamic aspects of the phase transition as well as from imaging such materials at the micrometer scale in bilayers, the crystalline nature of these membranes allows studies at the molecular or atomic scales using X-ray diffraction [62–64]. Small angle X-ray (SAX) scattering from the BLES bilayers suggests that an increase of bilayer width (thickness) could be observed during cooling of the BLES dispersions, as shown in Fig. 15.7. The bilayer width increases monotonously as the system enters the gel phase with a small inflection between 25 and 30 °C suggesting that the DPPC gel domains grow in size probably coupled across each hemilayer, which can lead to an increase of bilayer thickness [64] . The thickening of the bilayers is mainly due to the saturated chains of DPPC decreasing in the chain motion and increasing in order upon reaching higher packing densities, as well as possibly acyl chains having more vertical conformations to the plane of the bilayers, as observed in the monolayer system using AFM (Figs. 15.2 and 15.3). Addition of a small amount of perturbant such as soluble protein (or cholesterol) to the BLES bilayers leads also to changes (thinning) of the bilayers, suggesting that the structural organization of lipid proteins can be observed directly at the molecular level [64]. We have also shown that using the wide angle X-ray scattering (WAX) profiles from a BLES bilayer surface between 25 and 30 °C, the 4.5-Å crystalline hexagonal lattice of DPPC molecules in the gel phase was dissolved at higher temperature in the fluid phase [65]. Cholesterol was shown to dramatically change these lattice parameters in another porcine LS system [62, 63]. The SAX profile changes in BLES bilayers

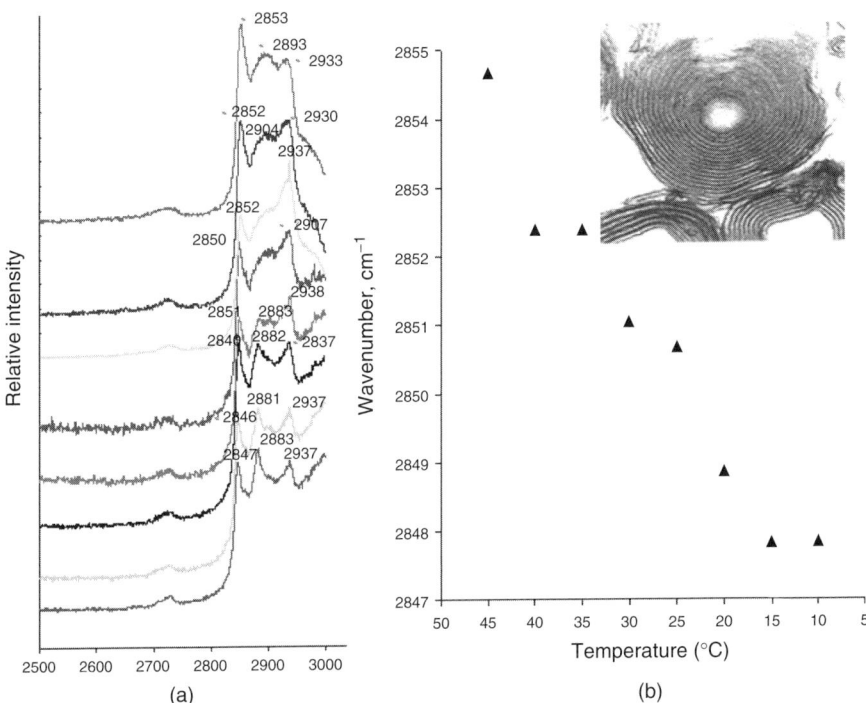

Figure 15.6 (a) Raman spectral (frequency shift) profiles of the symmetric and asymmetric CH_2 and CH_3 bond vibrations of phospholipids in BLES bilayers and (b) order parameter profiles of the average methylene vibrations plotted as a function of temperature. This method allows for measurement of vibrational dynamics of lipid acyl chain bonds at very small time scales (10^{-12} s). The typical structure of MLVs used in this study are shown in the transmission electron micrograph in the inset of (b), where each 45-nm thick bilayer can be observed via lipid staining. The acyl chain C—H vibrational frequencies shift at 2847 cm^{-1} to higher wavenumbers (or to the right in (a)) with increasing fluidity (b) and show changes of hydrocarbon chains from an all-*trans* to a more *trans-gauche* conformation when gel domains melt into the fluid phase [60, 69, 71].

induced by a soluble protein perturbant was suggested to be due to an interaction of the perturbant directly with surfactant proteins (SP-B and SP-C), thus allowing for studies of protein–protein interactions in a structured lipid environment by X-ray diffraction of natural membranous systems [64]. In a previous study, it was observed that LS from diseased lungs had a twofold increase of cholesterol as well as an increase of some serum soluble proteins that leaked into the lungs; the raft-like structural organization in these films was completely abolished [51]. Other studies have shown that organization of lipid rafts in cellular membranes may also be altered in certain diseases [3, 28, 65].

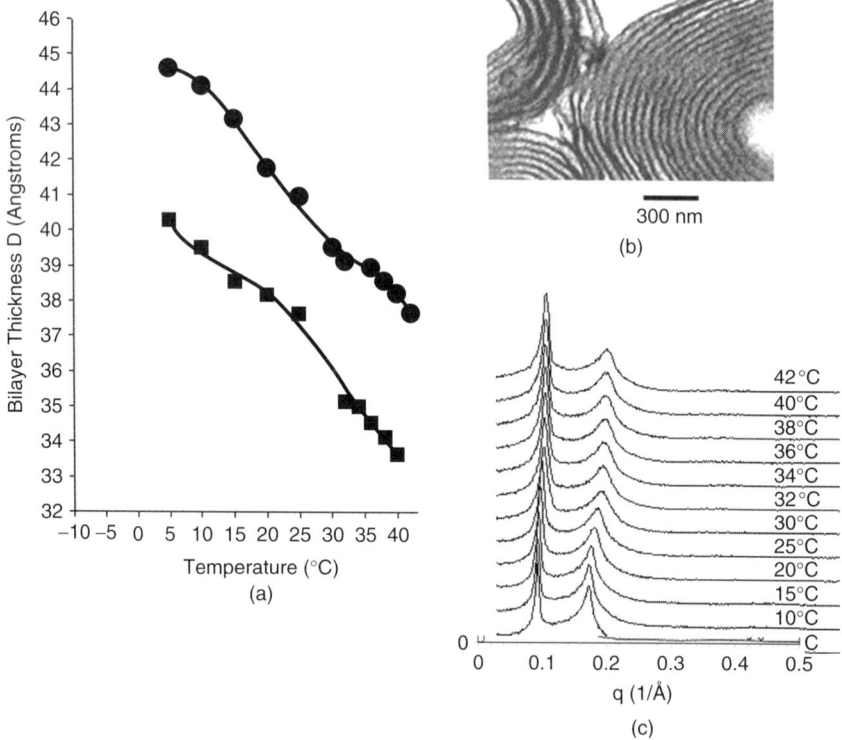

Figure 15.7 (a) X-ray scattering profiles of BLES bilayer thickness change of (b) multilamellar layers and (c) SAX scattering as a function of temperature. The bilayers become thicker with increase of packing in the gel phase (b) and follow similar ordering patterns as those seen in Figs. 15.6. This suggests dynamic molecular rearrangements at the nanoscale occurring during the fluid to gel transition process either in the monolayer [11] or bilayers [62–64]. Addition of a perturbant such as soluble protein or cholesterol shifts the profiles, further thinning the bilayers and suggesting an increase of fluidity in the system and some form of structural disruption of the internal gel domain organization in the individual surfactant bilayers [62–64].

15.9 CHOLESTEROL IN DOMAIN ORGANIZATIONS

Cholesterol is known to play a critical role in the organization of lipids in bilayers, as well as in the structure–function properties of lung surfactant [42, 52, 62–64]. In cellular membrane systems, cholesterol content varies from 5% to 23% of the phospholipids, and some studies have suggested that a specific concentration of this lipid may be "critical" to the functioning of some of the membrane proteins and raft formation [3, 5, 6, 47, 57, 65]. The cholesterol content of plasma membranes [56, 57] and LS is altered in membranes in certain disease situations and in the homeoviscous adaptation of certain mammals [36, 37]. However, in normal functioning of both systems, cholesterol content is highly regulated. As shown in Fig. 15.1, only small amounts (2%) of cholesterol dramatically altered the structural features of the condensed-gel DPPC domains. The domains are altered to a more spiral "stripe" phase, and this process is considered to be due to specific localization of cholesterol in gel–fluid domain or phase boundaries [9]. Small amounts of cholesterol reduce the one-dimensional line tension of the gel–fluid domains. However, excess cholesterol may completely abolish this boundary and thereby cause changes in the phase segregation of lipids in the specific gel–fluid phases. These processes may be reversed in systems containing high amounts of fluid or unsaturated lipids. In a previous study, we showed that in an equimolar mixture of gel–fluid phospholipids (DPPC:POPC, 50:50), cholesterol completely abolished the formation of DPPC domains; however, this also induced formation of other domains that did not have characteristics of either the gel or the fluid phase and thus these domains are termed liquid-ordered phase [20]. The effect of cholesterol on gel–fluid domains has also been observed in films of LS from multiple species [41, 44, 51, 52, 64], and the role of cholesterol in LS function requires further research.

In LS, cholesterol not only reduces the gel–fluid domain or phase boundaries in films but may further induce structures that are units of multiple bilayers, as shown in Fig. 15.8a. In normal surfactant in the lung interface, it is presumed that LS films have to attain a certain level of saturation with DPPC upon compression to reach "very low" surface tension ($\sim$1 mN/m) or attain high surface pressure or packing density (70 mN/m). Films of pure DPPC can be compressed to reach such low surface tension; however, films of other fluid lipids in LS such as POPC may collapse at higher surface tensions ($\sim$30 mN/m) and cannot reach such high packing density. Since most mammalian LS contains almost equivalent amounts of gel and fluid phospholipids, it is not clear to date how the fluid lipids sustain such high surface pressures in films, and some studies have suggested that the fluid lipids may be "squeezed-out" of the plane of the films [33–35, 44, 45]. The squeeze-out process of specific materials from films (or portions or domains in the film) can result from a buckling instability that occurs in the films. In the case of buckling, monolayers of fluid lipids can collapse into bilayer liposomal structures in the hypophase of the films [66], or they may stack into bilayer structures at the air–water interface in air, a direction perpendicular to the plane of the film's subphase [44]. Whatever the case, cholesterol may induce these non-monolayer or bilayer structures and such interactions have been suggested to be due to specific interaction of LS lipids and the hydrophobic SP-B/C [44, 52, 66]. As shown in Fig. 15.8a, the stacks formed (and imaged using AFM deposits of such film on mica) are generally about 45 Å in height, suggesting that they are mainly planar bilayers, and multiple units of these bilayers protrude out of the film into air at high packing density [44]. In other studies it was suggested that these stacks also contain acylated surfactant protein SP-C, where the acyl chains may anchor the stacks (G-protein coupled receptors?) to the initial monolayer at the air–water interface [52]. Some studies have also indicated that a cholesterol-rich environment in LS is extremely important for the functioning of SP-B/C, and

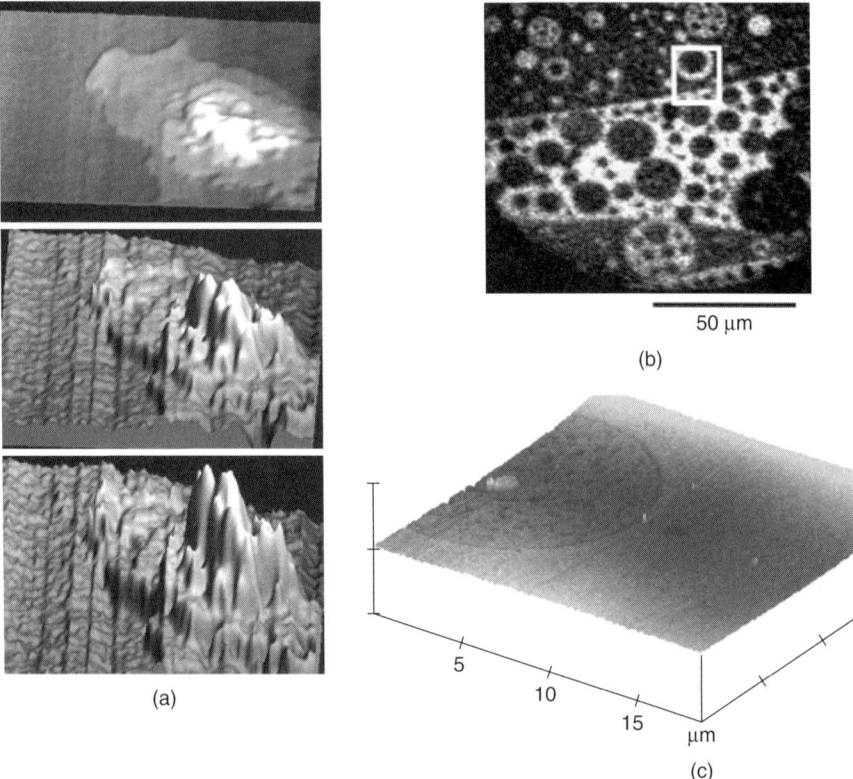

Figure 15.8 Cholesterol induced multilayer domain formation and cholesterol crystallization in BLES films imaged using (a) AFM; and liquid-ordered phase (lipid raft) formations in films of (b, c) POPC: sphingomyelin [9, 47–49, 71]. The stacking in (a) occurs due to squeeze-out of the fluid lipids from the almost completely gel phase environment, as well as probably needle-shaped solid-phase cholesterol crystal formations [67] in the center of such stacks [bright region in (a)]. In the case of the membrane–raft lipid system (b, c), cholesterol induces a liquid-ordered phase [dark regions in the fluorescence in (b)] in a sea of more fluid-disordered phase with different height profile than the gel or fluid phase seen in BLES films (as in Fig 15.3) in the AFM image (c) [44].

these proteins reside in the cholesterol-rich phase [41, 42]. Although it is not completely clear to date what the exact role of cholesterol or its excess in LS is, most studies indicate that the surfactant protein prefers the cholesterol-rich phase, similar to cell membrane proteins in lipid rafts.

Lipid rafts are formed in cellular membranes by a combination of fluid lipids such as POPC (16:0/18:1-PC), saturated chain gel lipid sphingomyelin (18:0/18:0-Sph), and cholesterol. It is also known that cholesterol probably rigidifies the fluid lipid environment, and fluidizes the gel lipid environment in the rafts, reaching an equilibrium status of liquid-ordered (L_o) phase coexisting with a fluid disordered phase. The fluorescent (Fig. 15.8b) and AFM (Fig. 15.8c) images of a film with POPC: Sph: cholesterol (40:40:20) are shown in Fig. 15.8. The L_o phases are observed as the black or probe excluded regions in a sea of fluid disordered lipids where the fluorescent probe resides, due to a looser packing of this phase. The AFM image in Fig. 15.8c also shows that the L_o phase has a different height profile than those observed for either the gel–fluid phases (Fig. 15.2) or DPPC: cholesterol films [44].

It is yet unclear if excess cholesterol in a membranous system LS or plasma membrane would lead to nonfunctioning protein or possibly the formation of other complex phases, and we must determine how this can be related to the molecular basis of some membrane-related diseases. In some studies we have shown that cholesterol in a gel–fluid environment not only alters the structural phase boundaries or reorganizes the lipids but also alters the phase transition process [20, 44]. Cholesterol cannot withstand high packing density in films due to its rigid ring structure and may self-aggregate out of the monolayer films of BLES, as shown in Fig. 15.8a. The center of the stacked regions contains peculiar needle-shaped structures, which may be cholesterol aggregates [44]. Previously, others have shown that these needle-shaped structures found in films of gel phospholipids are actually solid crystals of cholesterol [67]. Cholesterol may crystallize out (be squeezed out) of phospholipid films upon increased packing density [44, 67, 68]. This process was suggested as a possible mechanism by which formation of atherogenic plaques in epithelial cell membranes may occur [20, 67]. In arterial circulation, an excess cholesterol in lipoprotein particles may be taken up my macrophages, and after deesterification the excess cholesterol may be deposited or accumulate in the plasma membranes of epithelial cells lining the arterial walls [20]. Although the process is far more complicated than proposed here, the critical role of cholesterol and its concentration in membrane structures can be monitored from model systems such as LS [67–72] or from other cell membrane models [47–49]. It is perhaps too speculative to suggest at this point that, in the lung surfactant system, these processes observed with cholesterol in normal or excess amounts may hint at a certain molecular mechanism involved in disease similar to that which may occur in plasma membrane-related disease [72]. Whatever the case, future studies could be directed toward detailed understanding of the processes of cholesterol-induced phase segregation of lipids and proteins in diverse membranous systems, using the overly simplistic models suggested here.

In summary, this chapter defines some model membrane studies done on a natural extracellular membranous system of the lungs and compared to the general bilayer membrane system of the cell. The structures formed as domains or rafts of specific lipid–lipid–protein associations may suggest specific molecular mechanisms involved in the function of diverse membrane systems as well as the specific role of cholesterol in these systems. Biophysical techniques utilized to study such model membranes allowed us to study molecular lipid–protein organization involved in lung surfactant function and these methods may also allow for models of rafts in the cell membranes. Future studies using natural extracts of various membranous systems in monolayer films followed up with their bilayer phase properties may give a further glimpse into the molecular nanoscale architecture of these soft material systems.

ACKNOWLEDGMENTS

This study was supported by a Canadian Institute for Health Research (CIHR) New Investigator Award and operating grants from CIHR, National Scientific and Engineering Research Council of Canada (NSERC), and Janeway Children's Health Foundation to K.N. The authors would also like to acknowledge the major instrumental infrastructure award to K.N. from the Canada Foundation for Innovation (CFI) for the purchase of the Raman spectral microscope and the scanning probe microscope. We acknowledge the following colleagues for their advice and help regarding some of the technical aspects of instrumentation and data interpretation: Drs. Michael Morrow, Kevin M.W. Keough, Valerie Booth, and Joseph Banoub from Memorial University, as well as Drs. Amiyo K. Panda, Christopher Yip, Nils O. Petersen, Marcus Larsson, Mathias Amrein, Fred Possmayer, Robert Harbottle, and Kyle T. Vanderlick. The authors thank Dr. Dave Bjarnson of BLES Biochemical (London, Ontario) for the generous supply of BLES. D.R., A.H., and R.D acknowledge financial support from the School of Graduate Studies, Memorial University.

REFERENCES

1. E. Gorter and F. Grendall (1925). On biomolecular layers of lipid on the chromocyte of blood. *J. Exp. Med.* **41**: 439–443.
2. S. J. Singer and G. L. Nicholson (1972). The fluid mosaic model of the structure of cell membranes. *Science* **175**: 720–731.
3. E. London (2002). Insights into lipid raft structure and formation from experiments in model membranes. *Curr. Opin. Struct. Biol.* **12**: 480–486.
4. P. G. DeGennes (1992). Soft matter (Nobel Lecture). *Angew. Chem.* **31**: 842–845.
5. K. Simons and E. Ikonen (1997). Functional rafts in cell membranes. *Nature* **387**: 569–572.
6. G. Vereb, J. Szollosi, J. Matko, P. Nagy, T. Farakas, L. Vigh, L. Matyus, T. A. Waldmann, and S. Damjanovich (2003). Dynamic, yet structured: the cell membrane three decades after the Singer–Nicholson model. *Proc. Natl. Acad. Sci. USA* **100**: 8053–8058.
7. R. Peters and K. Beck (1983). Translation diffusion in phospholipid monolayers measured by fluorescence microphotolysis. *Proc. Natl. Acad. Sci. USA* **80**: 7183–7187.
8. V. Von Tscharner and H. M. McConnell (1981). An alternative view of phospholipid phase behaviour at the air–water interface. *Biophys. J.* **36**: 409–419.
9. H. M. McConnell (1991). Structures and transition in lipid monolayers at the air–water interface. *Annu. Rev. Phys. Chem.* **42**: 171–195.
10. H. Möhwald (1990). Phospholipid and phospholipid–protein monolayers at the air/water interface. *Annu. Rev. Phys. Chem.* **41**: 441–476.

11. V. M. Kaganer, H. Mohwald, and P. Dutta (1999). Structure and phase transition in Langmuir monolayers. *Rev. Mod. Phys.* **71**: 779–819.
12. K. Nag, C. Boland, N. H. Rich, and K. M. W. Keough (1990). Design and construction of an epifluorescence microscopic surface balance for the study of lipid phase transition. *Rev. Sci. Instrum.* **61**: 3425–3430.
13. K. Nag, C. Boland, N. H. Rich, and K. M. W. Keough (1991). Epifluorescence microscopic observation of monolayer of dipalmitoylphosphatidylcholine: dependence of domain size on compression rates. *Biochim. Biophys. Acta* **1068**: 157–160.
14. J. Perez-Gil, K. Nag, S. Taneva, and K. M. W. Keough (1992). Pulmonary surfactant protein SP-C causes packing rearrangements of dipalmitoylphosphatidylcholine in spread monolayers. *Biophys. J.* **63**: 197–204.
15. K. Nag, N. H. Rich, and K. M. W. Keough (1994). Interaction between dipalmitoyl phosphatidylglycerol and phosphatidylcholine and calcium. *Thin Solid Films* **244**: 841–844.
16. M. T. Montero, J. Hernandez-Borrel, K. Nag, and K. M. W. Keough (1994). Fluoroquinone distribution in a phospholipid environment studied by fluorimetry. *Anal. Chim. Acta* **290**: 58–64.
17. K. Nag, J. Perez-Gil, A. Cruz, and K. M. W. Keough (1996). Epifluorescence microscopic studies on interaction of fluorescently labelled pulmonary surfactant protein-C (SP-C) with phospholipids and calcium in spread monolayers. *Biophys. J.* **71**: 246–256.
18. K. Nag, J. Perez-Gil, A. Cruz, N. H. Rich, and K. M. W. Keough (1996). Spontaneous interfacial monolayers formation during adsorption of lipid–protein vesicles. *Biophys. J.* **71**: 1356–1363.
19. K. Nag, K. M. W. Keough, M. T. Montero, J. Trias, M. Pons, and J. Hernandez-Borrel (1996). Segregation of a quinoline antibiotic in dipalmitoylphosphatidylcholine bilayers and monolayers. *J. Lipos. Res.* **6**: 713–736.
20. L. A. Worthman, K. Nag, P. J. Davis, and K. M. W. Keough (1997). Cholesterol in condensed and fluid phosphocholine monolayers studied by epifluorescence microscopy. *Biophys J.* **72**: 2569–2580.
21. K. Nag, S. Taneva, J. Perez-Gil, A. Cruz, and K. M. W. Keough (1997). Combinations of fluorescently labeled pulmonary surfactant proteins SP-B and SP-C in phospholipid films. *Biophys. J.* **72**: 2638–2650.
22. M. L. F. Ruano, K. Nag, C. Casals, J. Perez-Gil, and K. M. W. Keough (1998). Differential partitioning of pulmonary surfactant protein SP-A in liquid-expanded and liquid-condensed regions of spread monolayers of dipalmitoylphosphatidylcholine and dipalmitolyphosphatidylcholine/dipalmitoylphosphatidylglycerol. *Biophys J.* **74**: 1101–1109.
23. K. Nag, J. Perez-Gil, M. L. F. Ruano, S. G. Taneva, N. H. Rich, J. Stewart, and K. M. W. Keough (1998). Phase transitions in films of lung surfactant at the air–water interface. *Biophys. J.* **74**: 2983–2995.
24. K. Nag, R. R. Harbottle, A. K. Panda, S. A. Hearn, and N. O. Petersen (2004). Physico-chemical mapping of phase heterogeneity in biomembrane films. *Microscopy Anal.* **18**: 13–15.
25. M. Edidin (2003). The state of lipid rafts: from model membranes to cells. *Annu. Rev. Phys. Chem.* **32**: 257–283.

26. C. Dietrich, L. A. Bagatolli, Z. N. Volovyk, N. L. Thompson, M. Deri, K. Jacobson, and E. Gratton (2001). Lipid rafts reconstituted in model membranes. *Biophys. J.* **80**: 1417–1428.
27. L. A. Bagatolli and E. Gratton (2001). Direct observation of lipid domains in free-standing bilayers using two-photon excitation fluorescence microscopy. *J. Fluor.* **11**: 141–160.
28. E. E. Bennaroch (2007). Lipid rafts, protein scaffolds, and neurologic disease. *Neurology* **69**: 1635–1639.
29. I. Langmuir (1936). Two dimensional gasses, liquids and solids. *Science* **84**: 379–383.
30. C. M. Knobler (1990). Seeing phenomena in flatland: studies of monolayer by fluorescence microscopy. *Science* **249**: 870–874.
31. K. Nag and K. M. W. Keough (1993). Epifluorescence microscopic studies of monolayers containing mixtures of dioleoyl and dipalmitoyl phosphatidylcholines. *Biophys. J.* **65**: 1019–1026.
32. R. A. W. Veldhuizen, K. Nag, S. Orgeig, and F. Possmayer (1998). The role of lipids in pulmonary surfactant. *Biochim. Biophys. Acta* **1408**: 90–108.
33. R. J. King and J. A. Clements (1972). Surface active material from dog lungs. II. Composition and physiological correlation. *Am. J. Physiol.* **223**: 715–726.
34. S. Schurch, J. Goerke, and J. A. Clements (1976). Direct determination of surface tension in the lungs. *Proc. Natl. Acad. Sci. USA* **73**: 4698–4702.
35. K. M. W. Keough, E. Farrel, M. Cox, G. Harrel, and H. W. Taeusch (1985). Physical, chemical, and physiological characteristics of isolates of pulmonary surfactant from adult rabbits. *Can. J. Physiol. Pharmacol.* **63**: 1043–1051.
36. C. B. Daniels and S. Orgeig (2001). The comparative biology of pulmonary surfactant: past, present and future. *Comp. Biochem. Physiol.* **129**: 9–36.
37. C. Lang, A. D. Postle, S. Orgeig, F. Possmayer, W. Bernhard, A. K. Panda, K. D. Jurgens, W. K. Milsom, K. Nag, and C. B. Daniels (2005). Dipalmitoylphosphatidylcholine is not the major surfactant phospholipid species in all mammals. *Am. J. Physiol.* **289**: 1426–1439.
38. K. Nag, S. Vidyasankar, A. K. Panda, and R. R. Harbottle (2005). Chain dancing, super-cool surfactant and heavy breathing: membranes, rafts & phase transitions, in *Lung Surfactant Function and Disorder, Lung Biology in Health and Disease*, Vol. 201 (K. Nag, Ed.). CRC Press, Boca Raton, FL, Chap. 6, pp. 145–152.
39. A. D. Bangham (1987). Lung surfactant: how it does and does not work. *Lung* **165**: 17–25.
40. K. Nag (Ed.) (2005). *Lung Surfactant Function and Disorder, Lung Biology in Health and Disease*, Vol. 201. CRC Press, Boca Raton, FL.
41. B. Piknova, V. Schramm, and S. B. Hall (2002). Pulmonary surfactant: phase behavior and function. *Curr. Opin. Struct. Biol.* **12**: 487–494.
42. J. B. Serna, J. Perez-Gil, A. Smith, A. C. Simonsen, and L. A. Bagtolli (2004). Cholesterol rules: direct observation of the co-existence of two fluid phases in native pulmonary surfactant membranes at physiological temperatures. *J. Biol. Chem.* **279**: 40715–40722.
43. R. Harbottle, K. Nag, N. S. McIntyre, F. Possmayer, and N. O. Petersen (2003). Molecular organization revealed by time of flight secondary ion mass spectrometry of a clinically used extracted pulmonary surfactant. *Langmuir* **19**: 3698–3704.

44. K. Nag, M. Fritzen-Garcia, R. Devraj, and A. K. Panda (2007). Interfacial organization of gel phospholipid and cholesterol in bovine lung surfactant. *Langmuir* **23**: 4421–4431.
45. A. K. Panda, K. Nag, R. R. Harbottle, F. Possmayer, and N. O. Petersen (2007). Thermodynamic studies of bovine lung surfactant extract mixing with cholesterol and its palmitate derivative. *J. Colloid Interface Sci.* **311**: 551–555.
46. D. Rose (2007). Numerical simulation of lung surfactant phospholipid monolayers. MSC Thesis, Memorial University, St. John's, Newfoundland.
47. S. L. Keller (2003). A closer look at the canonical "raft mixture" in model membrane studies. *Biophy. J.* **84**: 725–726.
48. S. L. Keller, W. L. Pitcher, W. H. Huestis, and H. M. McConnell (1998). Red blood cell membrane lipids form immiscible liquids. *Phys. Rev. Lett.* **81**: 5019–5022.
49. S. L. Veatch and S. Keller (2003). Separation of liquid phases in giant unilamellar vesicles of ternary mixtures of phospholipid and cholesterol. *Biophys. J.* **85**: 3074–3083.
50. H. M. McConnell and M. Vrljic (2003). Liquid–liquid immiscibility in membranes. *Annu. Rev. Biophys. Biomol. Struct.* **32**: 469–492.
51. A. K. Panda, K. Nag, R. R. Harbottle, K. Rodriguez-Capote, R. A. W. Veldhuizen, F. Possmayer, and N. O. Petersen (2004). Effect of acute lung injury on structure and function of pulmonary surfactant films. *Am. J. Respir. Cell Mol. Biol.* **30**: 641–650.
52. L. Gunasekara, S. Schurch, W. M. Schoel, K. Nag, Zleonenko, M. Haufs, and M. Amrein (2005). Pulmonary surfactant function is abolished by an elevated level of cholesterol. *Biochim. Biophys. Acta* **1737**: 27–35.
53. K. Nag, J. S. Pao, R. R. Harbottle, F. Possmayer, N. O. Petersen, and L. A. Bagatolli (2002). Segregation of saturated chain lipids in pulmonary surfactant films and bilayers. *Biophys. J.* **82**: 2041–2041.
54. K. Nag, M. Morrow, and K. M. W. Keough (2004). Biophysical studies of lung surfactant. *Phys. Canada* **60**: 141–149.
55. H. Ebel, P. Grabitz, and T. Heimbeurg (2001). Enthalpy and volume change in lipid membranes. I. The proportionality of heat and volume change in the lipid melting transition and its implication for the elastic constants. *J. Phys. Chem. B.* **105**: 7353–7360.
56. H. Trauble, H. Eibl, and R. Sawada (1974). Respiration—a critical phenomenon? *Naturwissenschaften* **61**: 344–354.
57. P. Overath and H. Trauble (1973). Phase transitions in cells, membranes and lipids of *Escherichia coli*. Detection by fluorescnt probes, light scattering and dialatometry. *Biochemistry* **12**: 2625–2634.
58. R. Wallace (1996). Microcomputational evolution of the neural membranes. *Nanobiology* **4**: 25–37.
59. K. Nag, K. M. W. Keough, and M. R. Morrow (2006). Probing perturbation of bovine lung surfactant extracts by albumin using DSC and ^{2}H-NMR. *Biophys. J.* **90**: 3632–3642.
60. K. Nag, A. K. Panda, B. H. Au, D. Heyd, R. R. Harbottle, M. Schoel, N. O. Petersen, and L. A. Bagtolli (2002). Biophysical studies of nano-structured interfaces as models of lung surfactant membranes. *Rec. Res. Dev. Biophys.* **1**: 53–70.
61. A. Percot and M. Lafleur (2001). Direct observation of domains in model stratum corneum lipid mixtures by Raman microspectroscopy. *Biophys. J.* **81**: 2144–2153.

62. M. Larsson, K. Larsson, T. Nylander, and P. Wollmer (2003). The bilayer melting transition in lung surfactant bilayers: role of cholesterol. *Eur. Biophys. J.* **31**: 633–636.
63. M. Larsson, O. Terasaki, and K. A. Larsson (2003). A solid state transition in the tetragonal lipid bilayer structure at the lung air–water interface. *Solid State Sci.* **5**: 109–114.
64. M. Larsson, T. Nylander, K. M. W. Keough, and K. Nag (2007). An X-ray diffraction study of alterations of bovine lung surfactant bilayer structures induced by albumin. *Chem. Phys. Lipids* **144**: 137–145.
65. J. A. G. Briggs, T. Wilk, and S. A. Fuller (2003). Do lipid rafts mediate virus assembly and pseudotyping? *J. Gen. Virol.* **84**: 757–768.
66. J. A. Zasadzinski, J. Ding, H. E. Warriner, F. Bringezu, and A. J. Waring (2001). The physics and physiology of lung surfactants. *Curr. Opin. Colloid Interface Sci.* **6**: 506–513.
67. S. Lafont, H. Rapaport, G. J. Sömjen, A. Renault, P. B. Howes, K. Kjaer, J. Als-Nielsen, L. Leiserowitz, and M. Lahav (1998). Monitoring the nucleation of crystalline films of cholesterol on water and in the presence of phospholipid. *J. Phys. Chem. B.* **102**: 761–765.
68. K. Nag, R. R. Harbottle, and A. K. Panda (2003). Atomic force microscopy of interfacial films of pulmonary surfactant, in *Methods in Molecular Biology: Atomic Force Microscopy, Biomedical Methods and Applications* (P. C. Braga and D. Ricci, Eds.). Humana Press, Totowa, NJ, Vol. 242, pp. 231–343.
69. Y. N. Kaznessis, S. Kim, and R. G. Larson (2002). Simulations of zwitterionic and anionic phospholipid monolayers. *Biophys. J.* **82**: 1731–1742.
70. K. Nag, A. K. Panda, R. Devraj, and M. Fritzen-Garcia (2007). Nano-structuring and molecular domain organizations in lipid–protein membranous interfaces, in *Modern Research and Educational Topics in Microscopy* (A. Mendez-Vilas and J. Diaz, Eds). Formatex, Badazoz, Spain, pp 483–490.
71. K. Nag, H. Au, D. Heyd, and K. M. W. Keough (2003). Raman spectral imaging and mapping of phase heterogeneity in pulmonary surfactant dispersions. *Biophys J.* **84**(2): 197a.
72. K. Nag, A. Hillier, K. Parsons, and M. Fritzen-Garcia (2007). Interactions of serum with lung surfactant extract in the bronchiolar and alveolar airway model. *Respir. Physiol. Neurobiol.* **157**: 411–424.

INDEX

Abiotic stress 229
Action potential 324
Acylated protein 169, 177, 368
Acylated-recoverin 168
Adiabatic compressibility 320, 330
Aeromonas salmonicida 306
Air-alveolar interface 138
Air-liquid interface 342
Air-water interface 387
Airways 349
Albumin 138, 151
Aligned bilayers 60, 61, 121, 126
Amide I & II bands 142, 151, 173–177
Amino acid fragments 33, 34
Amino acid osmolytes 229
Amino diol 209
Anesthetic 335
Antimicrobial peptides 276
Aquaporin x
Aqueous channels 272
Arrhenius activation PLA 9, 11
Atomic force microscopy (AFM) 6–8, 364–366, 370, 424–426, 429
   bovine surfactant films 423–426
   DPPC films 425
   domains in monolayer 364–366
   gel domains 423
   height profiles 424
   solid supported bilayer 7
Atomistic simulation 272
Attenuated total reflection-Fourier transformed infrared spectroscopy (ATR-FTIR) 181
Azimuthal orientations 193, 203, 217
Azimuthal projections 212

β-Barrel protein 272
Bacteriorhodopsin 270
Balanced spring model 46
Bending elasticity 48, 49
Bending modulus 53, 63, 67
Bicelles 118, 120
Bilayered micelles 117, 118
Bilayer deformation 47
Birefringent clouds 392
Bodipy-PC 8, 15
Boltzmann equation 251
Bovine lipid extract surfactant (BLES) 91, 423–426
Bragg equation 85
Bragg law 115, 116
Bragg peaks 55, 56
Brewster angle microscopy (BAM) 147, 160, 183, 193–195, 197
Buckling, monolayer 346, 347, 433
Buckminsterfullerene 22
Bulk phase dispersion 139
Bunched mode spectra 23
Burst aligned mode-SIMS 23

Calcium binding loops 278
Calcium-bound states 179
Calcium coordination 256
Calcium-palmitate complex 171–173, 176
Capacitive current 322
Capillary electrophoresis 295
Cardiolipin 291
Catanionics 386, 399, 408
Catanionic surfactants 386
Chain melting transition 9, 332, 333
Chalk River laboratories 111
Chaotropic ions 244

Structure and Dynamics of Membranous Interfaces, edited by Kaushik Nag
Copyright © 2008 John Wiley & Sons, Inc.

Chemical imaging, monolayer 20–34
Chiral discrimination 192
Chirality, molecular 196
Chiral lipids 192
Cholesterol 8, 38, 51, 92, 128, 195, 268, 419, 433
  chirality 202
  lipid rafts 425
  line tension 419
  rich domains 38
  rigidifier 425
  spiral domains in films 419
  solid crystals 435
  topography in films 38
Cholesterol-deuterium labeled 128
Clausius-Clapeyron equation 233
Clearing temperature 398
Coacervates 387
Coarse grained model 206
Coherent scattering-X ray 115
Cold neutrons 110
Cold unfolding of proteins 319
Collapsed film 346
Collapse nuclei 346
Collapse phase 346–348
Collapse plateau 143
Collapse structures 348
Collision cascade 24
Collision induced dissociation analysis 298, 303–305
Compatibility paradigm 229
Compressibility-area isotherms 403
Compression-expansion isotherms 347
Computer simulations 235, 268, 421
Condensed to solid transition 351–354
Copartitioning 236
Critical micellar concentrations 395–397
Cryoprotectant 234, 246
Cryo-transmission electron microscopy (Cryo-TEM) 84, 97
Crystalline anisotropy, films 197
C terminal α-helix- 174
Cubic bilayers 89, 102
Cubic phase 88, 395
Curosurf 342
Curvature 62, 99
Curvature strain 277
Cylindrical membranes 328

Cylindrical micelles 388
Cytosolic PLA 278

Deprotonated molecules 36–40, 305
Deoxycholate 400
Desorbed particles 24
Deuterium isotope 108
Didodecyldimethylammonium bromide (DDAB) surfactant 386
Dielectric constant 215
Differential scanning calorimetry 72, 329, 400, 428, 429
DiIC18 fluorescence 6, 15
Dimethylsulphoxide (DMSO) 234, 239
Dimyrisitoyl-phosphatidylethanolamine (DMPE) 199
Dipalmitoylphosphatidylcholine (DPPC) 31, 84, 138, 145, 168, 211, 331, 342, 421
  chiral discrimination 192
  dipolar interactions 214–218
  energy minimized conformation 218
  fragment ions 31
  headgroup derivative 31
  kidney shaped domains 145
  protonated ion (DPPC+H) 30
  RS-isomers 197–199
Dipalmitoylphosphatidylethanolamine (DPPE) 196, 197, 274
Dipalmitoylphosphatidylglycerol (DPPG) 31–33
DPP(Me)E 199
Dipolar interactions 214–218
Dipole 192, 201, 211, 269
Dipole density 211
Dipole moment 212
Distorted hexagonal lattice 354
Docking mechanism 278
Docosohexanoic acid (DHA) 274, 275
Domain shapes
  chirality 196
  elongation 201, 210, 220
  Fractal like 197
  triskelions 196, 198, 213
DTDAB surfactant 387, 388
Dynamic monolayer expansion 345
Dynamic surface tension 139
Dynamic yet structured model xii, 4, 418

E. coli 272, 287, 305, 329
Egg lecithin 53, 71, 151
Electron density distribution 127–129
Electric double layer model 249
Electro-spray ionization mass spectrometry (ESI-MS) 22–24, 289, 296
Embedded peptides 278
Entropic pulse 336
Enzyme activity 165
Enzyme kinetics 11
Enzymology of PLA 9
Equilibrium spreading pressure 343
Equilibrium surface tension 139
Evaporative light scattering detection (ELSD) 289
ELSD- Liquid chromatography, lipids 289

Fast atom bombardment mass spectrometry (FAB-MS) 296, 297
Fatty acylated proteins 166
Fibrinogen 138, 151
Fluctuation analysis 51
Fluid flexible sheet, bilayer 3
Fluid-fluid phase immiscibility 425
Fluid ordered phase 423
Fluorescence anisotropy 327
Fluorescence microscopy 6, 7, 347, 418, 419, 420
  bovine lipid extract surfactant 418–420
  collapse structures in films 347, 361
  DPPC/cholesterol films 418–420
  free standing bilayers 7
  giant vesicles 7
  ionic detergent surfactant films 404
  POPC-ceramide films 7
  solid supported layer 6
Flow behavior, monolayer 350
Fourier transform infrared spectroscopy (FTIR) 141, 153, 170
Fractal like structures 196, 197
Fragment ions 27
Free standing bilayers 426
Freeze concentration 246
Freeze fracture electron micrographs 409
Frozen lipids 421
Fructose 243
Fullerene ion 22

Gallium ion 22, 27
Gas chromatography (GC) of lipids 289
Gel domains in neural firing 334
Gel-fluid phase coexistence 9, 121, 329, 419, 421
Generalized polarization (GP) values 13, 254
Giant uni-lamellar vesicles (GUV) 7–14, 51, 420, 421, 427
  bovine surfactant 420, 427
  DMPE-DMPC 14
  DPPC-POPC-cholesterol 8
  pulmonary surfactant 8, 420
Gibbs phase diagrams 392–395
Gibbs phase rule 359
Gibbs-surface excess 243
Glassy sugar matrix 247
GlcN II ring structure 306
Globulins 157–159
Glycerol 237, 405, 409
Glycolipids 291
Glycoprotein 245
Golden Gate stage 181
Gold coated glass 29
Gouy-Chapman model 249–252
G-protein coupled receptor (GPCR) 166, 273, 433
Grotthuss-like mechanism 272
Grazing incidence angle X-ray diffraction (GIXD) 193–204, 207, 214

Headgroup distances 58
Helmholtz model 249
Hemilayer 420, 427
Herringbone motif 351
Heneicosanoic acid (C_{21}) 352–355
Hexadecanol 362, 363, 364
Hexagonal lattice 86, 117
Hexagonal packing 352
Hexagonal phase 418, 435
Hexagonal symmetry 355
Heyter-Penfold structures 119
High performance liquid chromatography (HPLC) 292–294
HPLC-ESI-MS 300
Hodgkin-Huxley model 318, 321–324
Hofmeister salts 243

Hydrocarbon-air interface 348
Hydrocarbon ion 31
Hydrogen bonding in shape formation 209
Hydrolysis products, PLA 14
Hysteresis curves- monolayer isotherms 144

Icosahedral shapes 390, 391, 409
Infrared spectroscopy (IRS) 141–144, 153–156, 168, 174, 212
Infrared vibrational bands 142, 148, 153–156, 170, 430
Interface expansion 155
Ion channel 321
Ion fluxes 228
Ion gun 21

Kirchoff circles 323
Kosmotropes 243–246
K-shell electrons 113

Lamellar repeat distance 65
Lamellar to hexagonal transition 244
Langmuir-Blodgett films 6, 28, 36
Langmuir films xi, xii, 25–38, 214–218, 347, 350, 418, 421
Langmuir film balance 140
Langmuir trough 344
Lateral organization, space & time 4
Lateral packing density 425
LAURODAN 13, 15, 254
Laurodan, general polarization 13, 15
Lecithin retinol acyl transferases (LRAT) 167, 174
Left handedness, lipid 208
Light microscopy 408
Line tension 210, 419
Lipid A 287, 288
Lipid demixing by calcium 37
Lipidomics 4
Lipid-protein system 33, 34, 426
Lipid rafts viii, ix, 3, 8, 37, 350, 396, 418, 423
Lipoproteins viii, 29, 301, 435
Liquid condensed to solid transition 347–349
Liquid crystalline phase 86, 87, 94, 101, 393

Liquid expanded (LE) to liquid condensed (LC) transitions 38, 95, 213, 351, 359, 419, 421
Liquid metal ion gun 21
Liquid ordered (L_o) phase 7, 8, 37, 94, 425, 426
Low density lipoprotein 301
Lung surfactants vii, 8, 96, 349, 357, 422
  airway distribution 349
  bilayer swelling 97
  bovine lipid extract 423
  cholesterol 97
  curvature 99
  evolutionary conservation 432
  hexadecanol 342
  hydrophobic proteins 8, 366
  imaging in GUV 8, 427
  palmitic acid 342, 357
  surfactant proteins 346, 401–403, 422
  tubular structures 96–101, 422
Lyso-phosphatidic acid 288
Lyso-phosphatidylcholine (Lyso-PC) 169, 176, 193, 302

Magainins 276
Magnetic moment 109, 110
Magnetic needle surface viscometer 343
Maja nerve bundle 324
Mass by charge ratios (m/z) 21–33, 297, 299, 300
Mass spectra 21–33, 297–300
Mass resolution, lipid ions 22
Matrix assisted laser desorption-ionization mass spectrometry (MALDI) 298
Matrix effect 39
Melting transitions 9, 329, 330, 332, 333
Membrane asymmetry 420
Membranes capacitor 326
Membrane-glycerol interactions 237–239
Membrane mimetic molecules 192
Membrane proteins 4, 8, 167
Mesophase 203, 205
Mirror symmetry 203
Mixed Langmuir monolayers 139–143
Modal projections 396
Molecular adsorption 345
Molecular dynamics simulation (MD) 9, 41, 268–273, 425

Molecular dynamics -snapshots 276, 424, 425
Molecular graphics image 371
Monoacylglycerol 210
Monolayer 14, 25–38, 99, 214–218, 347, 350, 418, 421
   AFM 6–8, 364–366
   coacervates 387
   hydrolysis by PLA 14, 169
   imaging using ToF-SIMS 35–39
   lipid-protein systems 37–40
   two-dimensional lattice 193
   two-phase coexistence 194
Monolayer associated reservoir 99
Monolayer hydrolysis 169, 176
Monosaccharide 242
Multi-lamellar vesicles (MLV) 5, 55, 118, 431
Multibilayer stacks 119, 121, 126
Multilayer patches 365
Myelinated nerve 321

Naja naja venom 171, 172
Nanodisks 391
Nanoparticles 409
Nanoscale defect structure 12
Nanoscale imaging 423
Nanoscale organization 425
Nanoscale phase separation 350
Nanotechnology 20
Nerve membrane 318
Nerve pulse propagation 320
Next nearest neighbor (NNN) 203, 352, 356
Neutron scattering 108
NIDDM patient platelet membrane 289
Nodal surfaces 88, 89, 99
Nuclear spins 109

OmpF porin 272
Optical anisotropy 141
Optical microscopy 407, 408

Palmitoylated cysteines 368
Palmitoylation 167
Palmitoyl aspartic acid 206
Palmitoyl-oleyl-phosphatidylcholine (POPC) 49, 59, 202, 234, 271, 371, 421, 422, 434
Peptide mimics 369

Peripheral proteins 137
Phase diagrams 194, 352, 387, 391–394
Phosphatidic acid (PA) 201, 275, 286, 297, 301
Phosphatidylcholines (PC) 7–10, 15, 30, 31, 53, 118, 122, 142–144, 150, 170, 177, 181, 230, 234, 286, 303, 419–430
Phosphatidylethanolamines (PE) 38, 54, 58, 60, 63, 196, 199, 211, 218, 244, 274, 287
Phosphatidylglycerols (PG) 31, 32, 35, 254, 256, 275, 301, 347, 358, 359, 362, 423
Phosphatidylinositols (PI) 286, 287, 291–294, 423
Phosphatidylserines (PS) 181, 254, 287, 288, 294, 423
Phosphocholine fragment 39
Phospholipid asymmetry 286
Phospholipid derivation, chart 296
Phospholipid-protein interaction 138
Phospholipases 166, 167
Phospholipase A_2 (PLA$_2$) 4–6, 10–15, 169
   binding model 15
   enzymology of PLA 5
   hydrolysis of mixed bilayers 10, 14, 15
   hydrolysis product 14
   interfacial activation 4
   lag burst behavior 10
   lag phase 11
   solid phase 14
Phospholipase, secretory (sPLA) 4–6
Plasma protein 138
Planar model membranes 7
Planar tripod 205
Platelet phospholipids 290
Polarization microscopy 387
Polyhydroxy osmolytes 234
Polyunsaturated fatty acids (PUFA) 128, 129, 274, 291
Polyvaline helical segment 368
Porcine lung surfactant extracts 10, 93, 95
Potential energy profile 206
Potential function 270
Potassium channel 275, 321–323
Preferential binding parameter 232
Primary ions 21, 24
Protein denaturation model 319

Protein-solute system 235
Protonated ion, lipid 30
Proton conduction pathway 271
Proton pumps x, 270
Pulsating area condition 159
Pyronine B fluorescence 328

Quantum mechanical approach 271

Raman spectroscopy xii, 430–432
Recoverin 177
Replacement surfactant 342
Respiratory distress syndrome 342
Reverse micelles 385, 389
Ripple phase 12, 13, 123, 124
Rhodopsin 166, 273, 274
Rod-like micelles 392
Rod photoreceptor 181
Rotator phase 352

Scattering length 109
Sciatic nerve 325
Secondary ion emission yields 26
Secondary ion mass spectrometry (SIMS) 20–38
Second order transition 358
Seminal plasma 293
Serum albumin 138
Serum lipids 297
Shape transition 419
Small angle neutron scattering (SANS) 108, 116
Small angle X-ray (SAX) scattering 48, 54, 55, 431–433
Sodium channels 323
Sodium dodecyl sulphate (SDS) 390, 394
Sound velocity in membranes 331
Solid-like films 426
Solid phase domains 14, 360
Solid phase islands 360
Solitary waves 332
Soliton 332, 336
Soliton propagation in nerve 336
Spallation neutrons 110–112
SP-B_{1-25} peptide 369–371
Spermatozoa 293
Spherical micelles 390, 392
Sphingomyelin 37, 268, 287, 291–294, 423, 434

Sponge phase 395
Squid axon 321, 326
Stearic acid 357
Stern-Grahame model 249, 252
Sterols 268
Subatomic particles 110–112
Sucrose 244
Superliquid films 351
Supramolecular organization ix, xii, 421, 424
Surface active material 144
Surface pressure-area isotherms 140, 345, 346, 401–403
Surface shear viscosity 353, 357
Surface tension 139
Surface tension gradients 349
Surfactant protein- A (SP-A) 99–101, 343, 426
Surfactant protein–B (SP-B) 8, 32, 84, 342, 347, 364, 422, 426
Surfactant protein-C (SP-C) 8, 84, 364, 422, 426
Surfactant protein- D (SP-D) 343
Synchrotron radiation 113
Synchrotron X-ray diffraction 375
Swelling 64, 68, 97

Tandem mass spectrometry 298, 302
Taurine 229
Thermodynamic, laws 318–320
Thin layer chromatography 289–291
Tilted phase 351
Tilt transition 374
Time of flight-secondary ion mass spectrometry (ToF-SIMS) 20–23, 38
Total ion chromatography 301
Transfer thermodynamics 241
Trans-gauche conformation 47, 64
Transient clusters 360
Transiently fluctuating dipoles 269
Transmembrane orientation 367, 368
Transmission electron microscopy (TEM) 84, 97, 101, 405–407
Triglycerides viii
Trimethylaminoxide (TMAO) 228
Triple chain phosphocholine 195
Tubular myelin 96–101, 422
Turnit method 150, 156
Two dimensional lattice 193

Two dimensional phase transition 351
Two dimensional suspensions 362
Two phase coexistence 194
Two-photon fluorescence microscopy, 14
Tyloxapol 343

Ubiquitous membrane 422
Undulation 69
Unilamellar vesicles (ULV) 120, 127, 238, 331
Unit cell 85
Urea 228

Vacuum permittivity 215
Vandate ions 399
Vesicle fluctuation analysis 51–53
Viscosity 327

Water adsorption band 170
Water replacement 247
Water-rich corners 395

Wide angle X-ray (WAX) diffraction 54, 62, 87, 113, 375, 431–433
Wilhelmy plate 344

Xenopus laevis 276
X-ray diffraction 83–92, 127–129, 193, 202, 204, 207, 394, 431–433
  2D-pattern 125
  grazing incidence angle 193–204
  inorganic surfactants 394, 398, 399
  lung surfactant 431
  powder pattern 123
  small angle 48, 54
  synchrotron 374–376
  wide angle 54, 62, 431

Young-Laplace equation 139

Zwitterionic surfactants 387
Zwitterions 278, 387